Jim Donoughe

Industrial Electronics: Design and Application

Merrill's International Series in Electrical and Electronics Technology

SAMUEL L. OPPENHEIMER, *Consulting Editor*

BEATY	*Basic ET Laboratory—An Audio-Tutorial Program*
BOYLESTAD	*Introductory Circuit Analysis, Second Edition*
DAVIS	*Industrial Electronics: Design and Application*
DIXON	*Network Analysis*
FABER	*Introduction to Electronic Amplifiers*
HARTKEMEIER	*Fortran Programming of Electronic Computers*
HERRICK	*Introduction to Electronic Communications*
HERRICK	*Mathematics for Electronics*
LENERT	*Semiconductor Physics, Devices, and Circuits*
OPPENHEIMER	*Semiconductor Logic and Switching Circuits, Second Edition*
PRISE	*Electronic Circuit Packaging*
SCHWARTZ	*Survey of Electronics*
TOCCI	*Fundamentals of Electronic Devices*
TOCCI	*Fundamentals of Pulse and Digital Circuits*
TOCCI	*Introduction to Electric Circuit Analysis*
TURNER	*Digital Computer Analysis*
TURNER	*Fundamental Electronics*

Industrial Electronics: Design and Application

Charles A. Davis

Western Michigan University

CHARLES E. MERRILL PUBLISHING COMPANY
A Bell & Howell Company
Columbus, Ohio

Published by
Charles E. Merrill Publishing Co.
A Bell & Howell Company
Columbus, Ohio 43216

International Standard Book Number: 0-675-09010-5

Library of Congress Catalog Card Number: 72-92570

6 79
Printed in the United States of America

To CLEMETINE
Lisa
Karen
Glen

Preface

This text presents a design-oriented approach to the understanding of electronic devices, circuits, and systems used in industry. All of the important solid state devices used in industry are presented in design situations with appropriate applications.

The material in this text was developed and class-tested in a course in Industrial Electronics taught by the author at Western Michigan University. The philosophy is that one understands electronics much more thoroughly if the design approach is used in teaching. Both analog and digital electronics are covered, including schematics of commercial equipment.

The text assumes the reader understands basic algebra and has some familiarity with the graphical analysis of semiconductor circuits using ac and dc load lines.

Chapter 1 reviews the terminal characteristics of solid state devices. Chapter 2 is a detailed presentation of the design of semiconductor circuits. It can be skipped if the reader feels he already has a strong background in the subject. Chapter 3 discusses some control relays used in industry. It has been the author's experience that the most confusing component in industrial electronic circuits is the capacitor. Chapter 4 covers capacative circuits thoroughly. The whole family of PNPN switching devices is presented in Chap. 5. The parameter that designers are trying to control most often is time. Chapter 6 discusses timing and time delay circuits. The important subject of phase shift control is handled in detail in Chap. 7. For the reader with no background in digital electronics, Chap. 8 presents the concepts basic to digital circuits. Chapter 9 covers the design and application of digital sequence controls. The relay ladder diagram is used as a starting point and the text progresses to the modern programmable controller. Chapter 10 presents a review of motors and their characteristics. This material is not usually included in electronic texts but the author has found it to be a weakness of many electronics people. The different types of electronic controls for motors are included in Chap. 10. A commercial welder is analyzed in Chap. 11. Some of the welder design techniques applicable to other electronic systems are emphasized. Chapter 12 covers the detailed design of dc power supplies with both analog and digital control. For the reader with no understanding of semiconductor physics, Appendix A is an introduction to that subject.

Exercises at the end of each chapter allow the reader to apply the text material to a problem.

The author wishes to thank the following companies for providing material and reading portions of the manuscript: Bodine Electric Company; Delco Electronics Division, General Motors Corporation; *Electronic Engineering;* Fairchild Corp.; General Electric Company; ITT Semiconductors; Kepco Inc.; Lambda Electronics Corp.; Loyola Industries, Inc.; McGraw-Hill Book Company; Modicon Corporation; Motorola Corporation; Radio Corporation of America; Reliance Electric Company; Robotron Corporation; Signetics Corporation; Superior Electric Company; and Technical Publishing House. A special thanks to Mrs. Ruth Barrett, who typed most of the manuscript.

March, 1973 *Charles A. Davis*

Contents

Industrial Electronics: Design and Application

1

Semiconductor Devices

1–1 Introduction

Our study of modern industrial electronics will begin with a discussion of the semiconductor devices about which most circuits are designed. You might recognize some of these devices from courses you have had or equipment you have come in contact with. We will learn to think about these devices in terms of terminal voltages and currents just as though we had measured them in the lab. You will find this simple approach quite adequate for the design and applications to be discussed in later chapters.

The list of semiconductor devices on the market today would probably fill this page and the next. We will not attempt to cover such a multitude of items. Instead, we will concentrate on a number of devices whose characteristics are representative of the types you are likely to be faced with in industrial electronics. (A condensed listing of the most commonly used devices appears in Appendix D.)

1–2 Diodes

The simplest of semiconductor devices is the diode. It is a two-terminal device. The terminals are called the *anode* and *cathode*. Diodes are made for different applications.

Let's look at the different types of diodes and their characteristics.

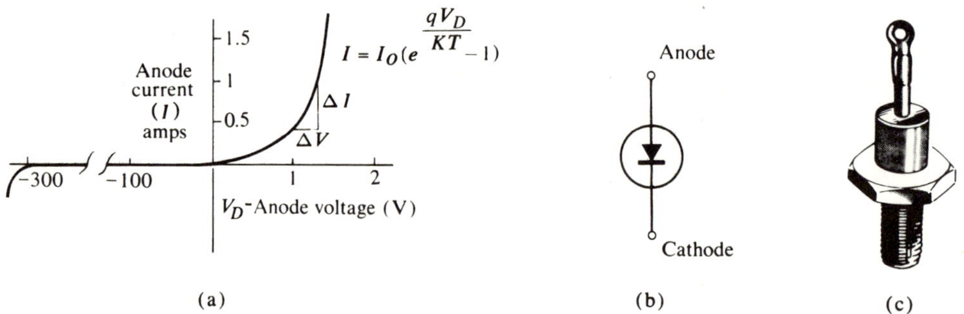

Fig. 1–1 (a) Rectifier diode characteristics, (b) circuit symbol, (c) typical industrial diode.

Rectifier diodes—Typical characteristics of a rectifier diode are shown in Fig. 1–1. When the anode voltage is more positive than the cathode voltage, current flows from anode to cathode. Less than one volt is usually required. Negative voltages of fifty volts or more may be applied at the anode with respect to the cathode without drawing measurable current.

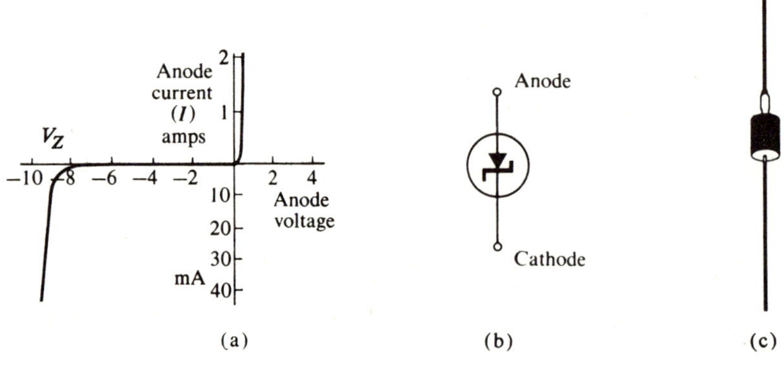

Fig. 1–2 (a) Zener diode characteristics, (b) circuit symbol, (c) typical zener diode.

Zener diodes—The zener diode is similar to the rectifier diode when the anode voltage is more positive than the cathode voltage. When the anode voltage is made more negative than the cathode voltage, no current flows until a preset value of voltage is reached. At the value of the preset voltage, the anode current rises rapidly but the voltage remains fairly constant (Fig. 1–2). The preset voltage or "zener voltage" is determined by the design of the diode and the materials used.

Thyrector diodes—The thyrector diode has characteristics similar to the zener diode when the anode is negative with respect to the cathode. It has the same characteristics when the anode is more positive than the cathode. In other words, the thyrector diode acts like two zener diodes in series with the cathodes tied together. No current flows through the diode until a preset value of voltage is reached regardless of the polarity of the voltage (Fig. 1–3).

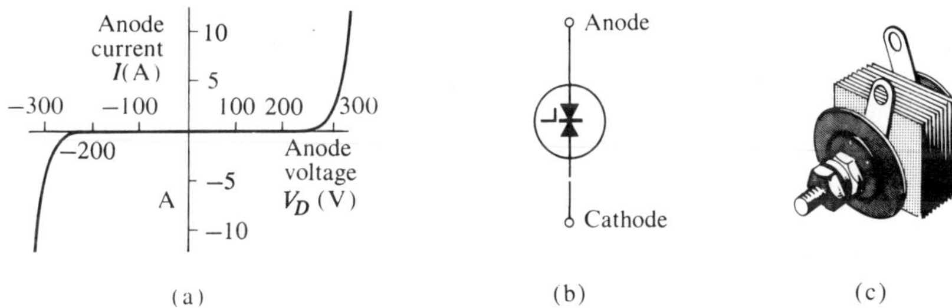

Fig. 1–3 (a) Thyrector diode characteristics, (b) circuit symbol, (c) typical thyrector.

DIACS—The DIAC has characteristics similar to the thyrector. The noticeable difference is that when the preset voltage is reached, the DIAC current increases but the voltage drops to about ten volts. DIAC characteristics are shown in Fig. 1–4.

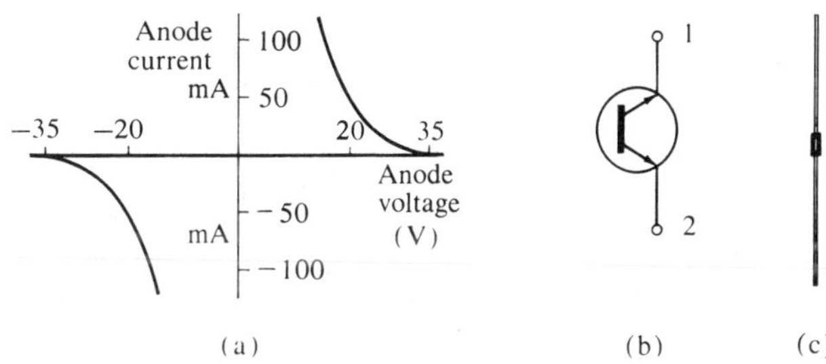

Fig. 1–4 (a) DIAC characteristics, (b) circuit symbol, (c) commercial DIAC.

Shockley diodes—The Shockley diode is a 4-layer PNPN device. It has two stable states, a high-resistance "off" state, and a low-resistance "on" state. These are labeled regions I and III respectively in Fig. 1–5(a). The unusual property of this diode is the negative resistance region between the "on" and "off" states, region II in Fig. 1–5(a). This negative resistance property makes the Shockley diode very useful in oscillator circuits. The diode switches "on" when positive anode voltage exceeds the predesigned value and remains "on" as long as the anode current is above the *holding current* level.

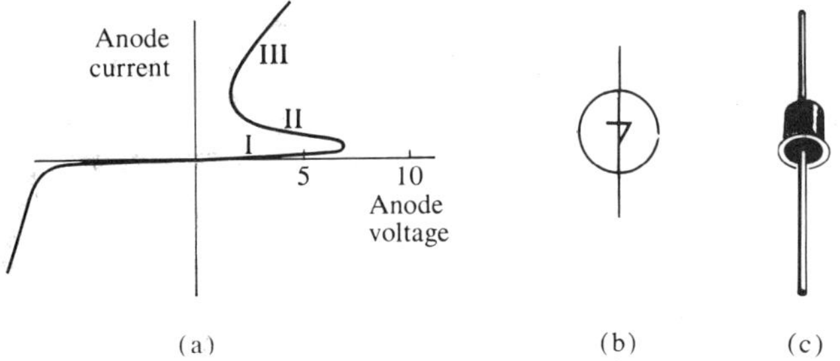

Fig. 1–5 (a) Shockley diode characteristics, (b) circuit symbol, (c) typical Shockley diode.

1–3 Bipolar Junction Transistors

The most common three-terminal semiconductor device is the *bipolar junction transistor* (Fig. 1–6). The most useful approach is to describe the device by its terminal voltages and currents.

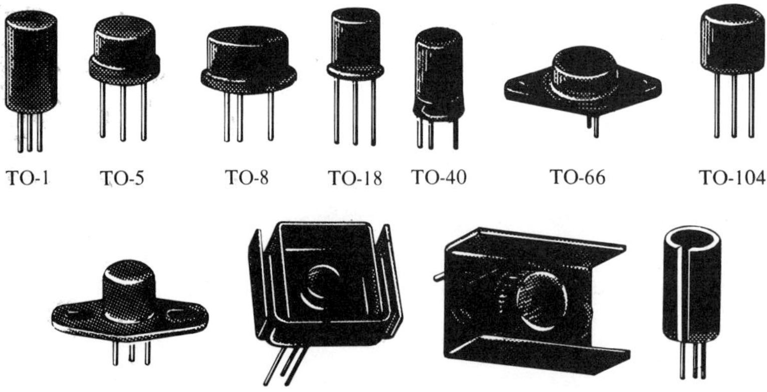

Courtesy of Newark Electronics Corp.

Fig. 1–6 Commercial bipolar transistors.

The two types of bipolar transistors are **PNP** and **NPN**. The transistor type is determined by the semiconductor materials in the construction of the device. Figures 1–7 and 1–8 show the semiconductor materials used in the collector, base, and emitter regions of the two types of transistors. We will use base current (I_B), collector current (I_C), base-to-emitter voltage (V_{BE}), and collector-to-emitter voltage (V_{CE}), to describe the terminal characteristics of the bipolar transistor. The circuit of Fig. 1–9 might be used to make laboratory measurements of I_B, I_C, V_{BE}, and V_{CE}. Two graphs are required to plot the measured values. Figure 1–10 represents the two *x-y* plots that are required in most design and application situations in industry. Since the base and emitter terminals are usually the inputs to the transistor, the I_B vs V_{BE} characteristics are called input characteristics. The collector and emitter terminals are usually the output of transistors; therefore, plots of I_C vs V_{CE} are called output characteristics. Figures 1–11 and 1–12 are output and input characteristics of a commercial transistor as presented in the

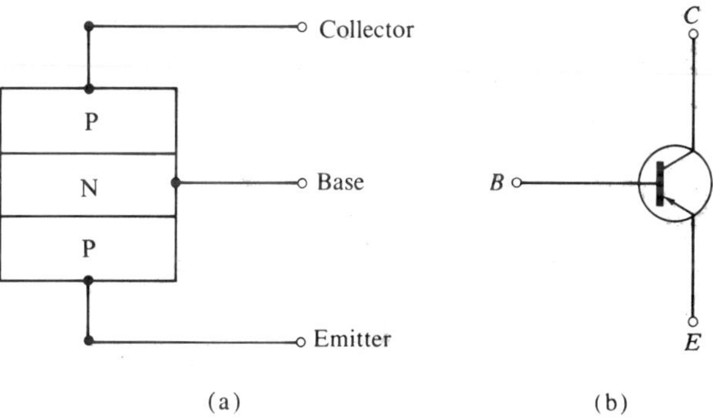

Fig. 1–7 (a) PNP transistor junction schematic, (b) circuit symbol.

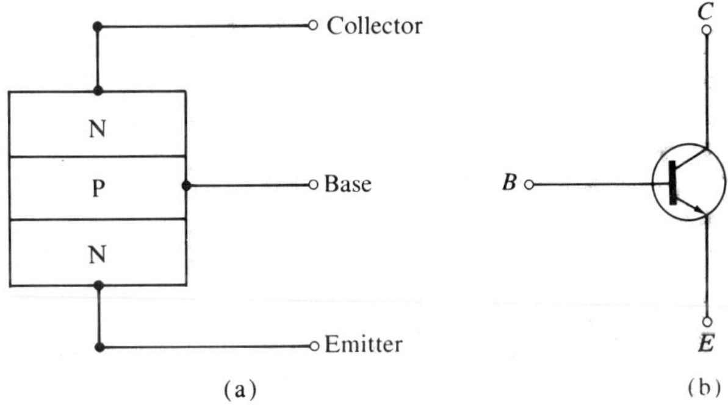

Fig. 1–8 (a) NPN transistor junction schematic, (b) circuit symbol.

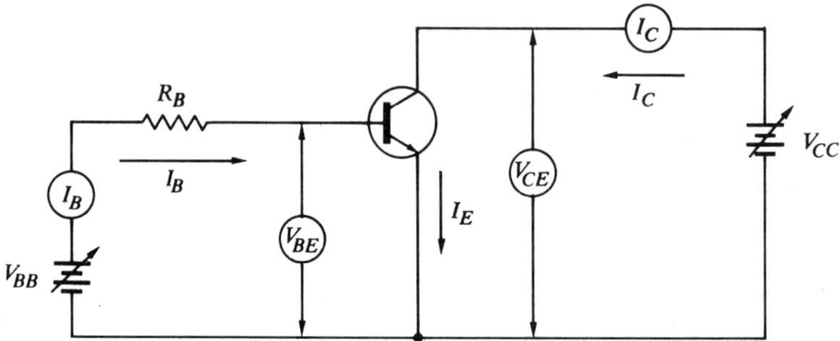

Fig. 1–9 Circuit for the measurement of bipolar transistor terminal characteristics.

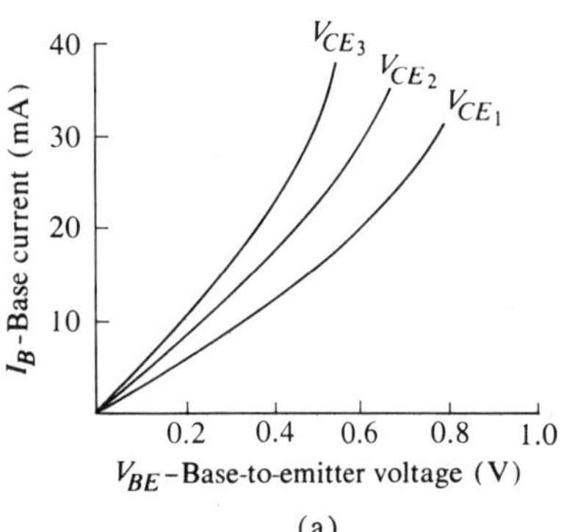

(a)

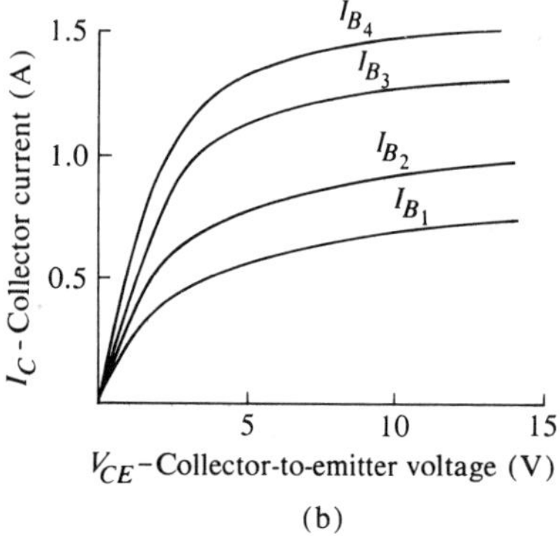

(b)

Fig. 1–10 Bipolar junction transistor terminal characteristics: (a) common emitter input characteristics, (b) common emitter output characteristics.

manufacturer's manual. The other graphical representation of transistor terminal characteristics that we find useful in later chapters is output current vs input current, I_C vs I_B. This plot is called the *forward current transfer characteristic curve.* It is from this characteristic curve that we can determine the all important current gain of the transistor. The forward current transfer characteristic curve for a commercial transistor is shown in Fig. 1–13. We will spend more time discussing transistor characteristics in the next chapter. The thing that should be planted firmly in our minds at this point is that for design and application in electronic circuits, the transistor is best described by its terminal characteristics.

1–4 Field-Effect Transistors

The *field-effect transistor* is a three-terminal semiconductor device similar to the bipolar transistor we discussed in the previous section. There are some major differences in the terminal characteristics of the two devices. As a matter of fact, there are two types of field-effect transistors with different terminal characteristics. We shall discuss the two types of field-effect transistors (FETs) separately and identify those terminal characteristics that contain most of the information necessary for design and analysis in industrial electronic applications.

Junction Field-Effect Transistors (JFETs)

The JFET is basically a bar of doped semiconductor material with a junction formed near the center across the surface of the bar (Fig. 1–14). The bar is called the *channel* and may be of either P-type or N-type material. The de-

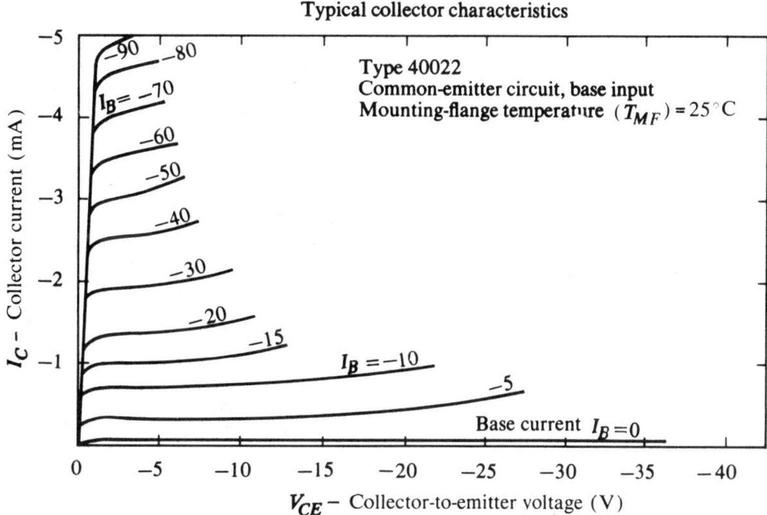

Courtesy of Radio Corporation of America
Solid State Division

Fig. 1–11 PNP transistor output characteristics.

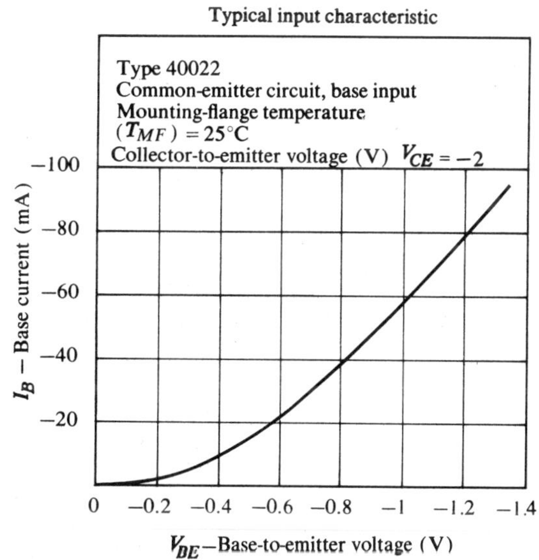

Courtesy of Radio Corporation of America
Solid State Division

Fig. 1–12 PNP transistor input characteristics.

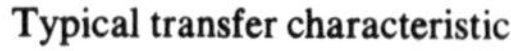

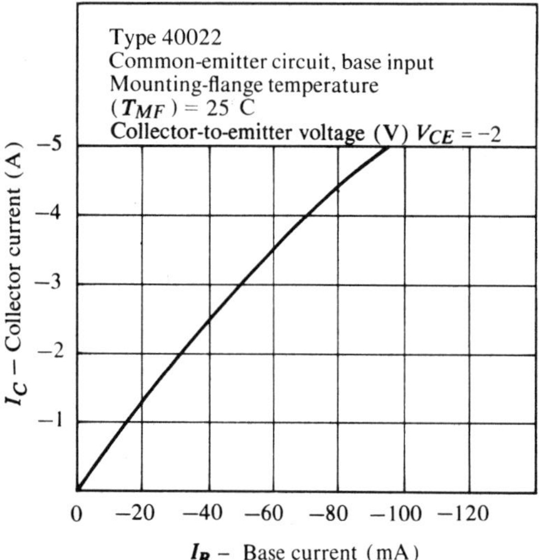

Courtesy of Radio Corporation of America
Solid State Division

Fig. 1–13 PNP transistor transfer characteristics.

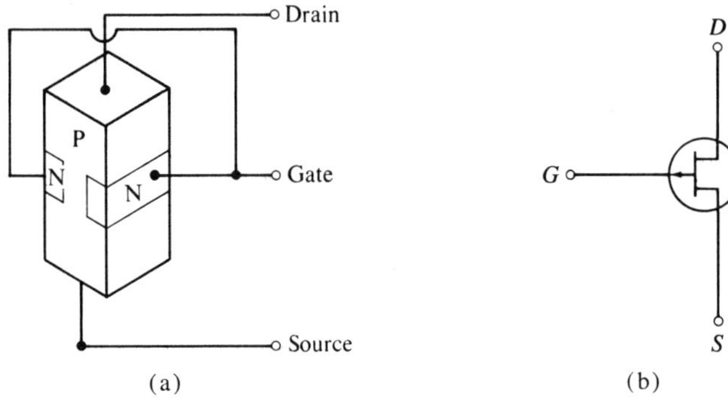

Fig. 1–14 (a) Junction field-effect transistor structure, (b) circuit symbol.

vices are called P-channel or N-channel respectively. The polarity of the junction material is opposite that of the channel material. Leads attached to either end of the channel are called the *drain* and *source* terminals. The junction terminal is called the *gate*. The usual circuit configuration has the gate and source terminals at the input and the drain and source terminals at the output. Figure 1–15 shows the dc circuit for measuring the characteristics of an N-channel JFET. If the gate-to-source voltage is set at zero, the drain

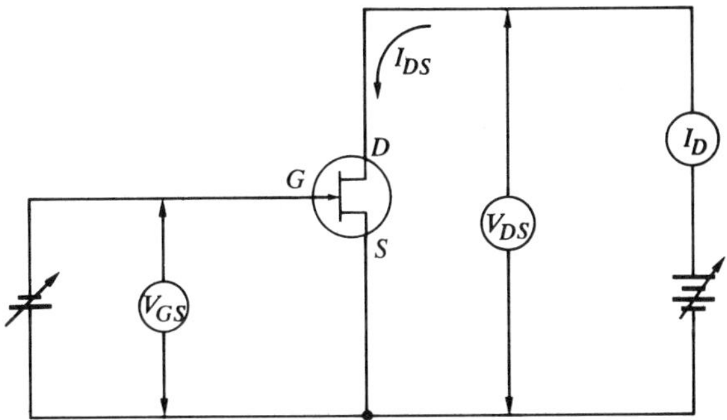

Fig. 1–15 Circuit for measurement of JFET terminal characteristics.

current vs drain-to-source voltage characteristic curve decreases in slope to almost zero with increasing current [Fig. 1–16(a)]. By varying the gate-to-source voltage, a family of I_D vs V_{DS} characteristic curves is generated [Fig. 1–16(b)]. These characteristics are similar to the output characteristics of the junction transistor of Fig. 1–10. The major difference between the JFET and the normal junction transistor is at the input terminals. The normal junction transistor input parameter is "input current," usually base current. There is essentially no input current in the JFET circuit.

The other graphical data usually required for design of JFET circuits is input voltage vs output current or V_{gs} vs I_d. These are called the transfer characteristics. Figure 1–17 is typical of N-channel JFET transfer characteristics. Characteristics of a commercial JFET are shown in Fig. 1–18.

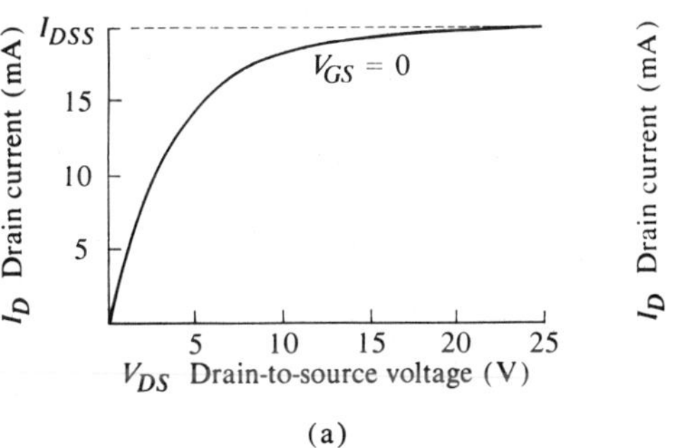

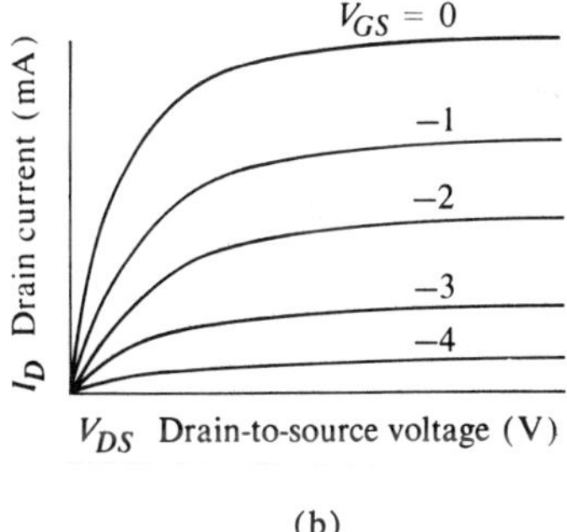

Fig. 1–16 (a) JFET output characteristic curve with gate shorted to source, (b) JFET output characteristics.

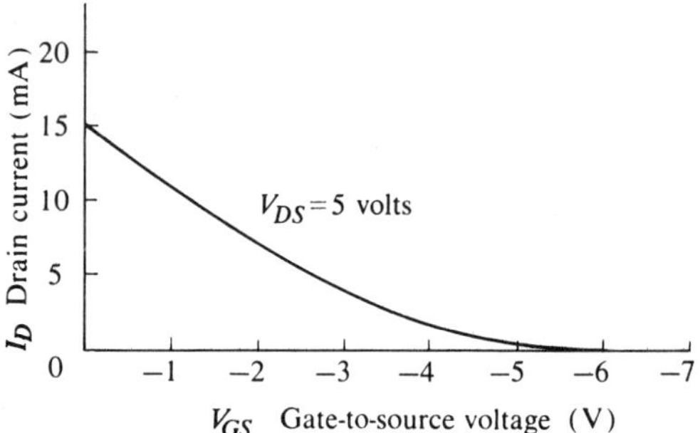

Fig. 1–17 JFET transfer characteristics.

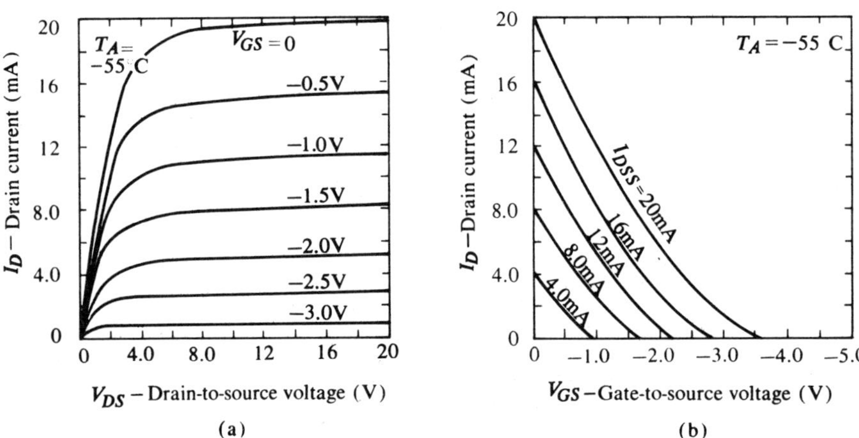

Courtesy of Texas Instruments

Fig. 1–18 (a) Commercial JFET output characteristics, (b) commercial JFET transfer characteristics.

Surface Field-Effect Transistors

The surface field-effect transistor has characteristics very similar to the JFET even though the two devices are very different in structure. Surface FETs are commonly called IGFETs or MOSFETs. These names are descriptive of the *M*etal-*O*xide-*S*emiconductor construction and *I*nsulated *G*ate configuration. The drain and source are formed on the surface of a lightly doped semiconductor; an oxide layer is formed on the surface for insulation from the metallic plate that acts as the gate. Figure 1–19 is a cross section of a typical surface FET. MOSFETs and IGFETs are available with N-channel or P-channel construction. The symbols are shown in Fig. 1-20. Output and transfer characteristics of a typical P-channel MOSFET are in Fig. 1-21. Notice that the gate-to-source voltage may be either positive or negative.

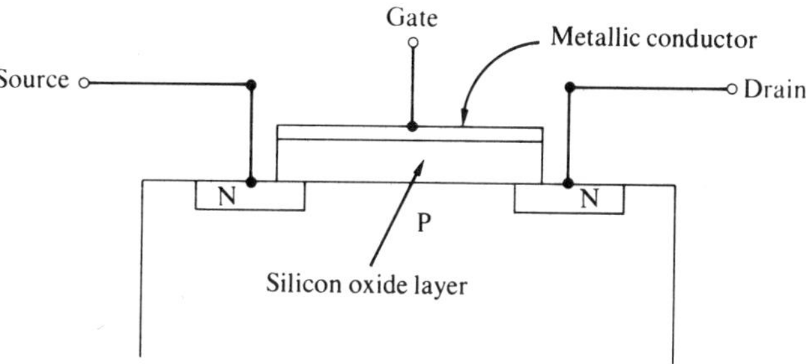

Fig. 1-19 Cross section of a surface FET or MOSFET.

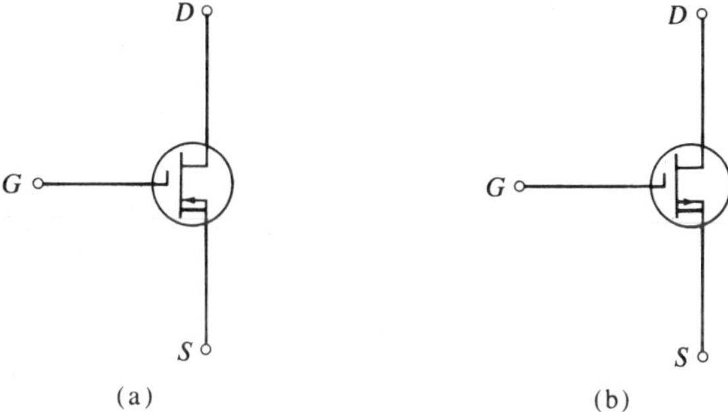

Fig. 1-20 Circuit symbols for the IGFET: (a) P-channel, (b) N-channel.

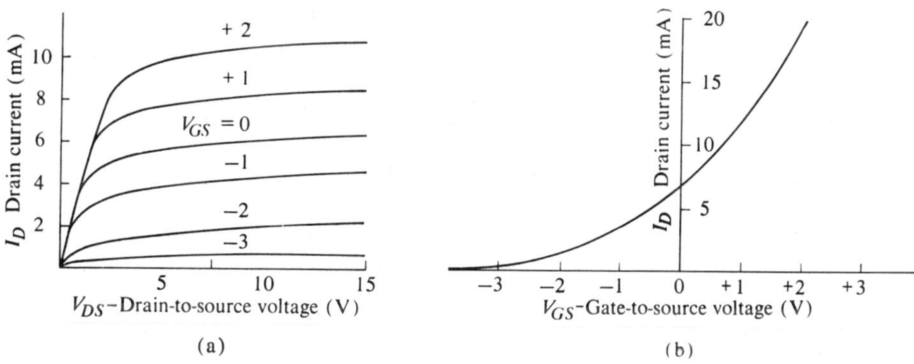

Fig. 1-21 MOSFET characteristics: (a) output characteristics, (b) transfer characteristics.

1–5 Silicon Controlled Rectifiers (SCRs) and TRIACs

SCRs and TRIACs belong to a family of semiconductor devices called *thyristors*. These devices are used in switching applications where current flow is either allowed or prevented; there is no partial or amplitude control by the device. Thyristors are more efficient and economical than transistors when low-frequency, high-power switching is required.

SCRs

The SCR is a three-terminal switching device made of four layers of alternate P- and N-type semiconductor materials (Fig. 1–22). The terminal connections are anode, cathode, and gate. The anode connection is to the outer

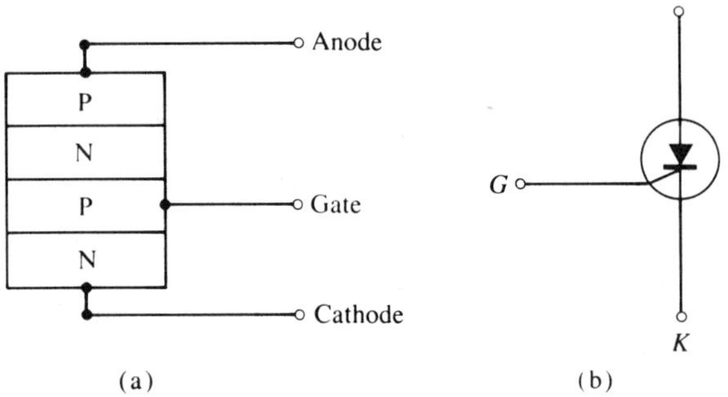

Fig. 1–22 (a) SCR junction construction, (b) circuit symbol.

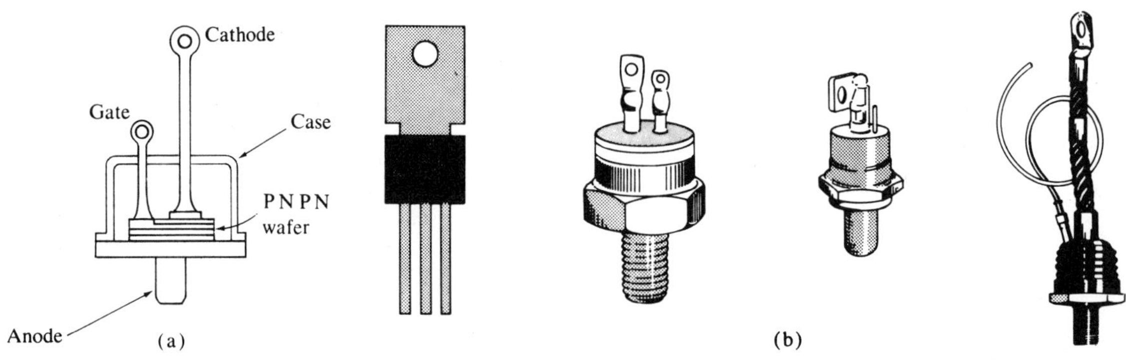

Courtesy of General Electric Company

Fig. 1–23 (a) Cross section of typical SCR construction, (b) commercial SCRs.

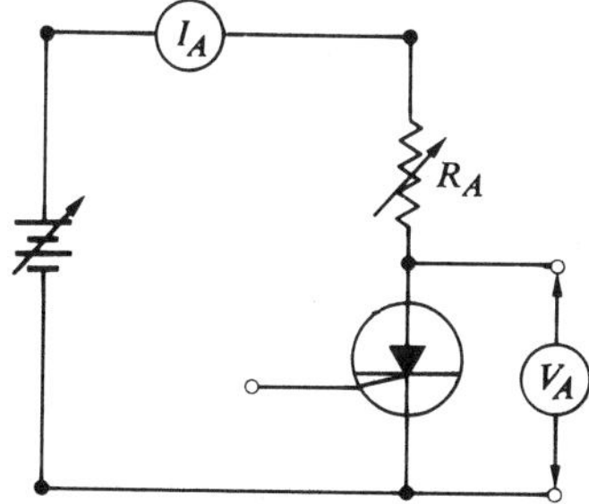

Fig. 1–24 Test circuit for measuring SCR forward-blocking voltage.

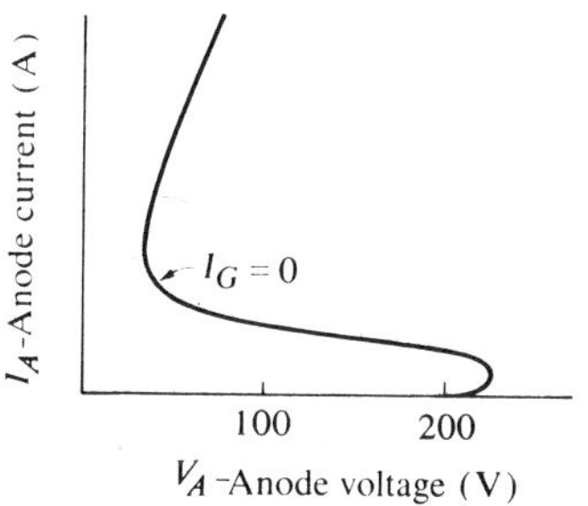

Fig. 1–25 SCR anode voltage vs anode current with gate circuit open.

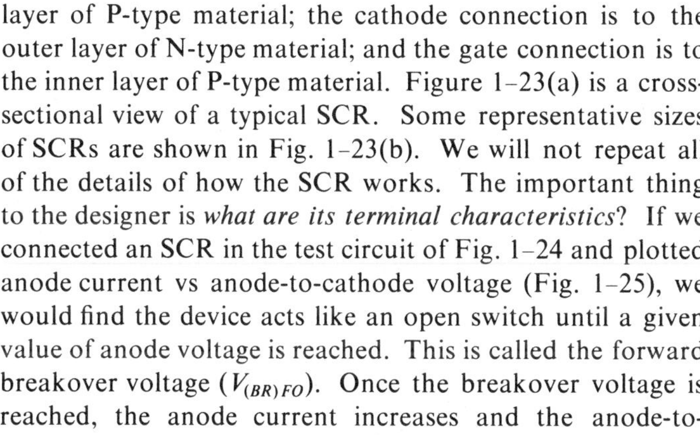

layer of P-type material; the cathode connection is to the outer layer of N-type material; and the gate connection is to the inner layer of P-type material. Figure 1–23(a) is a cross-sectional view of a typical SCR. Some representative sizes of SCRs are shown in Fig. 1–23(b). We will not repeat all of the details of how the SCR works. The important thing to the designer is *what are its terminal characteristics*? If we connected an SCR in the test circuit of Fig. 1–24 and plotted anode current vs anode-to-cathode voltage (Fig. 1–25), we would find the device acts like an open switch until a given value of anode voltage is reached. This is called the forward breakover voltage ($V_{(BR)FO}$). Once the breakover voltage is reached, the anode current increases and the anode-to-

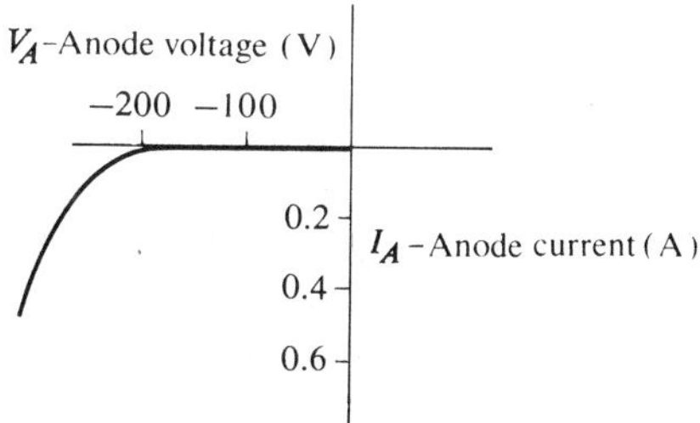

Fig. 1–26 SCR reverse voltage characteristics.

cathode voltage decreases. This represents a negative resistance similar to the Shockley diode that we discussed in Sec. 1–2. After the anode current reaches a given level, the SCR acts like a regular rectifier diode. Reversing the polarity of the power supply in Fig. 1–24, we find that the SCR acts exactly like a reverse-biased rectifier diode (Fig. 1–26). The anode and cathode are the output terminals of

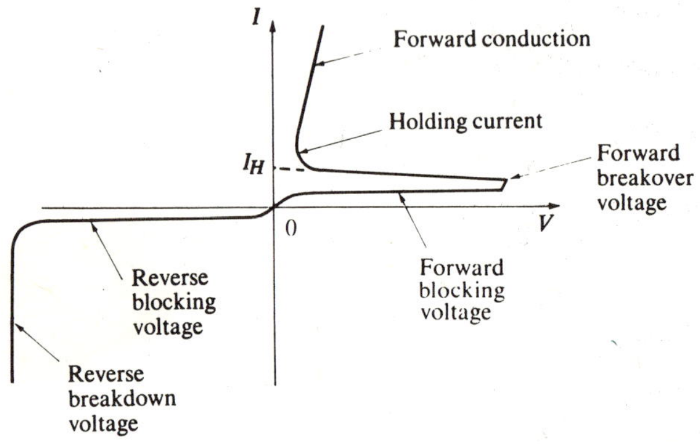

Courtesy of Fairchild Semiconductor

Fig. 1–27 SCR output characteristic curve.

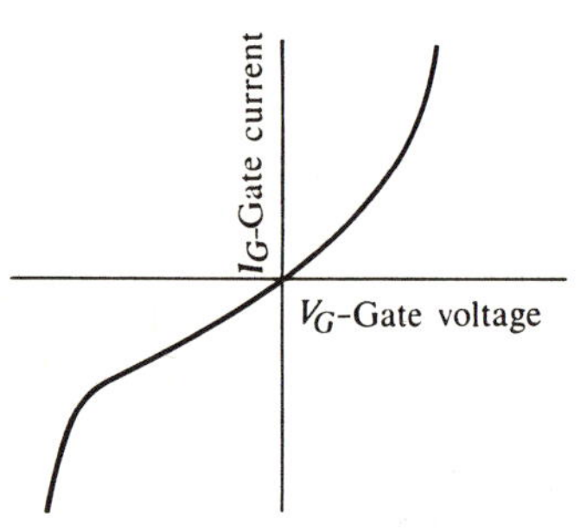

Fig. 1–28 Gate-cathode characteristic curve with anode disconnected.

the SCR. Figure 1–27 is a complete anode characteristic curve. Let's now look at the gate and cathode, or input terminal characteristics. We saw in Fig. 1–22 that the gate-to-cathode represents a P-N diode-type junction. Typical gate-to-cathode characteristics are very similar to diode characteristics (Fig. 1–28). Application of a positive, or forward, bias to the gate of the SCR changes the output characteristics. The positive gate current flow reduces the forward breakover voltage. The result is that a family of output characteristic curves may be plotted with gate current being the control variable. A typical family of characteristic curves is shown in Fig. 1–29. If the SCR goes into the "on," or conducting, state, the gate no longer controls the current. The SCR must be forced into the "off" state by reducing the anode current to nearly zero or by reversing the polarity of the anode-to-cathode voltage.

There are variations of the basic SCR that we should note here.

LASCR The light-activated SCR operates the same as the basic SCR except it can be forced to switch into forward conduction by light shining on the gate-to-cathode junction.

CSCR The complementary SCR has the gate terminal connected at the inner negative layer so that negative gate voltages may be used for switching.

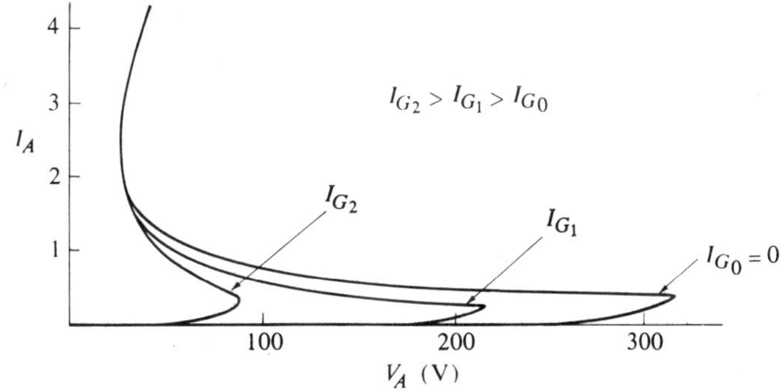

Fig. 1–29 SCR output characteristics with variable gate current.

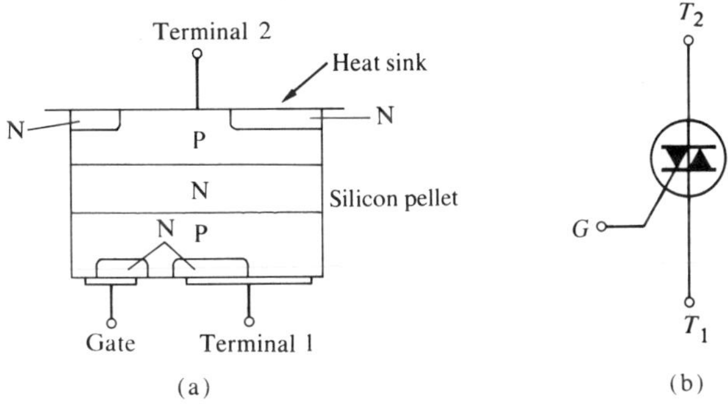

Courtesy of General Electric Semiconductor Dept., Syracuse, N.Y.

Fig. 1–30 The TRIAC: (a) pellet structure, (b) circuit symbol.

TRIACs

The TRIAC is an extension of the SCR to allow gate control when the anode is positive or negative with respect to the cathode. The structure of the TRIAC is much more complex than other devices we have looked at so far. A number of precisely arranged P-N junctions are required to achieve the bidirectional control. Figure 1–30 is a cross section of the TRIAC developed by General Electric engineers. Notice that the terminal connections are Terminal 2, Gate and Terminal 1. Equally acceptable is Anode 2, Gate and Anode 1. Terminal 1 is usually the reference point for voltage measurements. The case of the TRIAC is internally connected to Anode 2 (Fig. 1–31). The TRIAC acts like two SCRs in inverse parallel, back-to-back connection.

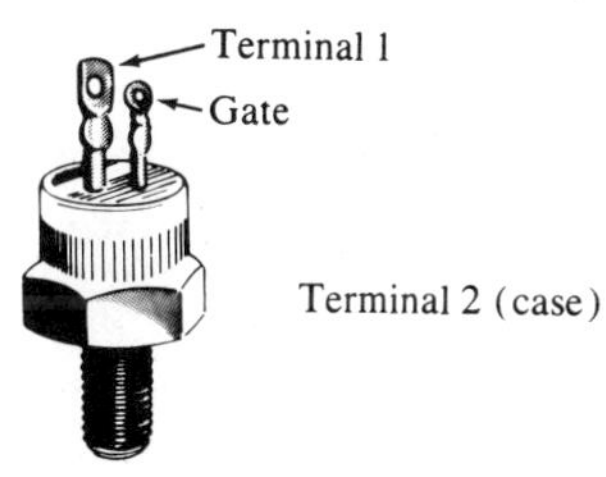

Fig. 1–31 Commercial TRIAC.

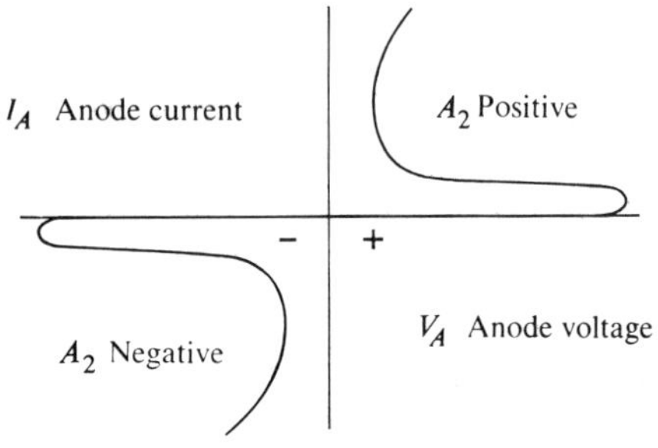

Fig. 1–32 TRIAC output characteristics with gate circuit open.

The output characteristics of Fig. 1–32 are typical for a TRIAC with the gate circuit open. TRIAC gate characteristics are very similar to SCR gate characteristics. TRIACs have rapidly replaced SCRs in low-power control circuits where greater range is required.

1–6 PNPN Silicon Switches

We have considered the SCR and TRIAC as special PNPN switches because they are the devices about which many industrial electronic circuits are designed. There are several other switching devices based on PNPN semiconductor structure. Figure 1–33 is a display of various PNPN

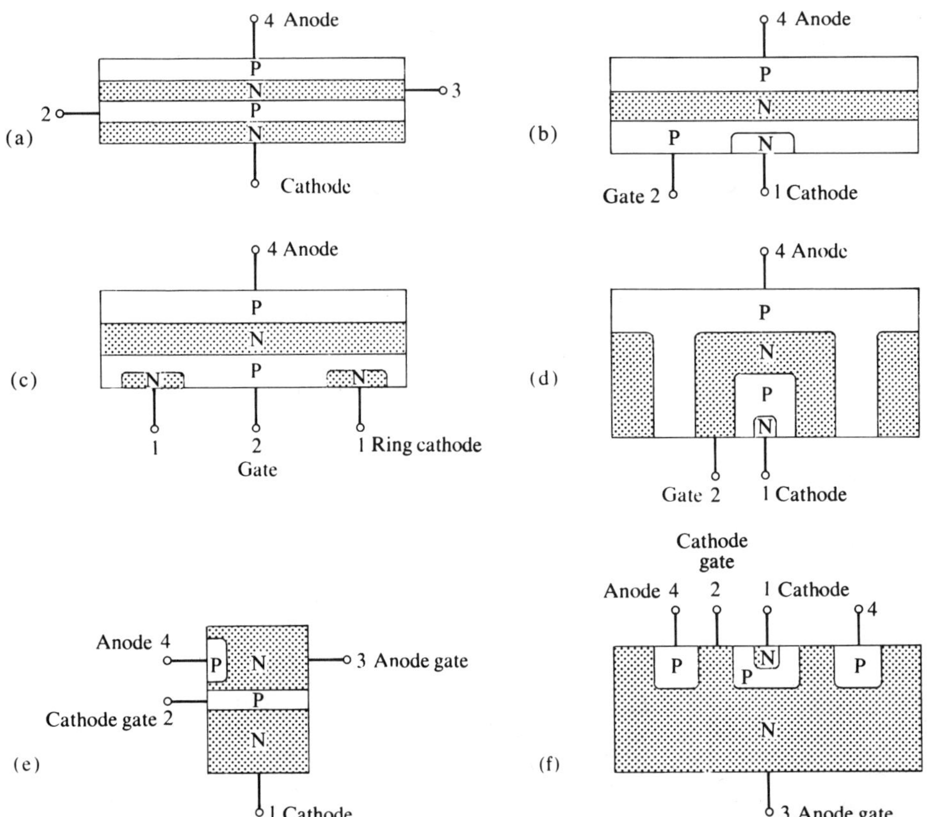

Courtesy of General Electric Semiconductor Dept., Syracuse, N.Y.

Fig. 1–33 Sectional views of PNPN geometries.

geometries. The terminal characteristics of these devices are quite similar. We will list some of the most common devices and distinguishing features of their characteristics.

Silicon Controlled Switch (SCS)

The SCS is a four-terminal device. External connections are made to the anode, cathode, and both gate sections of the PNPN structure. The shape of the output characteristic curve is similar to that of the SCR output characteristic curve; however, the SCS can be switched "on" or "off" (Fig. 1–34).

Silicon Unilateral Switch (SUS)

The SUS is a three-terminal device quite similar to the SCR except for a built-in zener diode junction at the anode gate. Typical output characteristics are shown in Fig. 1–35.

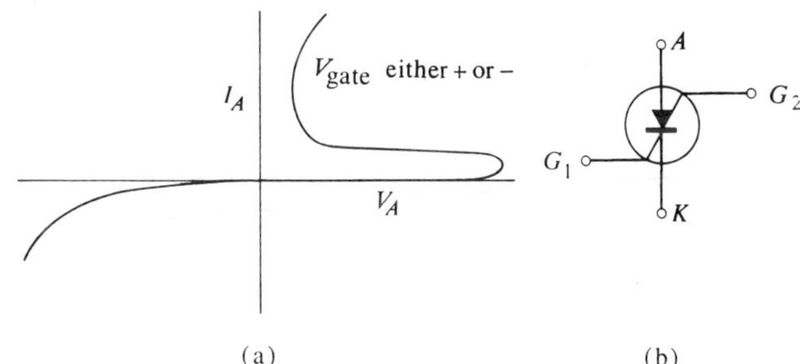

Fig. 1–34 (a) SCS output characteristics, (b) circuit symbol.

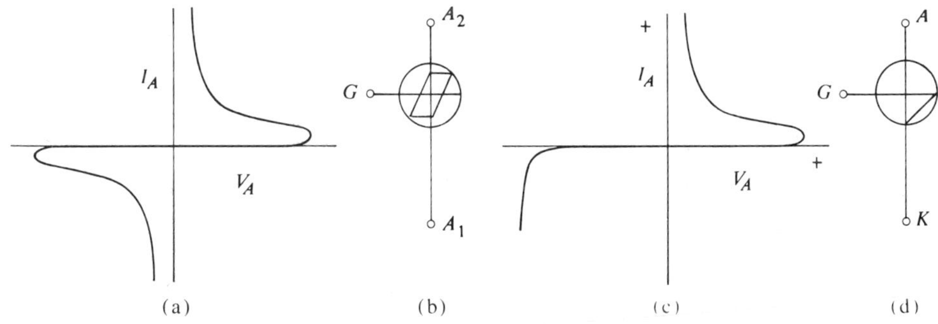

Fig. 1–35 (a) SBS output characteristics, (b) circuit symbol, (c) SUS output characteristics, (d) circuit symbol.

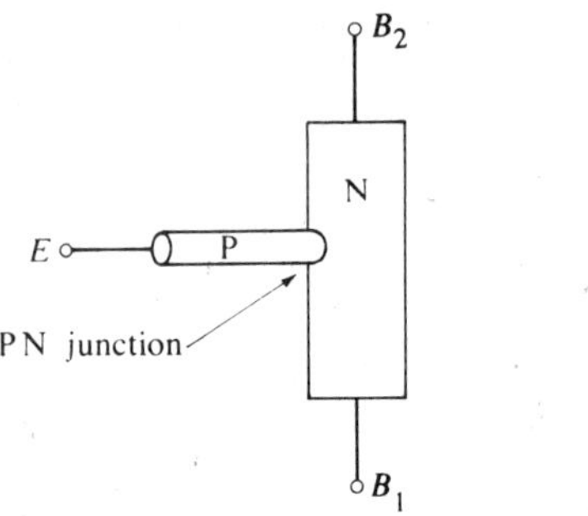

Fig. 1–36 Unijunction transistor construction.

Silicon Bilateral Switch (SBS)

The SBS acts like two SUS devices connected in an inverse parallel configuration. The SBS can be switched with either polarity of applied voltage. The terminal characteristics are shown in Fig. 1–35(a).

There will be occasional industrial applications where these PNPN switching devices are chosen. We will point these out as we go along.

1–7 Unijunction Transistors (UJT)

Another semiconductor device whose characteristics we should be familiar with is the UJT. This device is particularly important in industrial electronics because it is used so often in triggering circuits. The UJT is a three-terminal device. Its structure consists of a silicon bar with a single P-N junction formed at a point approximately two-thirds of its length. Metallic contacts are formed at each end of the bar and base terminals are connected. The third terminal at the junction is called the *emitter*. Figure 1–36 is a cross-sectional view of a typical UJT. The lower base terminal (B_1) is the reference; a positive voltage is applied at the upper base (B_2). The emitter terminal is the input connection to the UJT. When a positive voltage is applied at the input, very little current flows into the emitter junction until a specific level of voltage is reached. When the required emitter voltage level is reached, the emitter current increases rapidly as the emitter voltage decreases (Fig. 1–37). This action represents the negative resistance that we no-

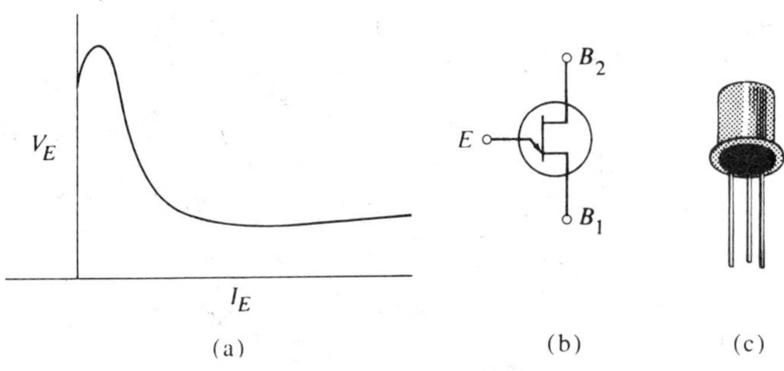

Fig. 1–37 (a) UJT emitter characteristics, (b) circuit symbol, (c) typical UJT.

ticed in SCRs and other devices. As the emitter current continues to increase, a point is reached where the emitter voltage increases slightly. It is in this higher emitter current region that operation of the device is stable. By varying the voltage at the upper base terminal, a family of I_E vs V_E characteristics may be generated.

The emitter characteristics of a commercial UJT are shown in Fig. 1–38. The current that flows from the emitter to the lower base will be used as the output of the UJT.

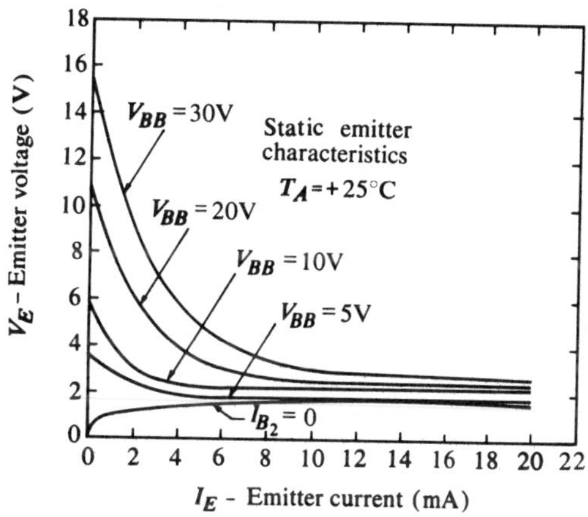

Courtesy of General Electric Semiconductor Dept., Syracuse, N.Y.

Fig. 1–38 Commercial UJT emitter characteristics.

1–8 Summary

In this chapter, we have discussed diodes, junction (or bipolar) transistors, field-effect transistors, SCRs, TRIACs, PNPN switches, and the UJT. You are likely to use each of these devices at one time or another in the design of industrial electronic components or systems. The most useful approach is to describe the action of these devices by terminal voltages and currents, or terminal characteristics. We will use graphical plots of terminal characteristics and manufacturers' specifications in the design and application of semiconductor circuits for industrial electronics.

EXERCISES

1. Sketch the *V-I* characteristics for a rectifier diode.
2. How are thyrector diode characteristics different from rectifier diode characteristics?
3. Draw the symbol for a PNP bipolar transistor.
4. Draw the symbol for an NPN bipolar transistor.
5. Explain the difference between the output characteristics of the NPN and PNP bipolar transistor.
6. What is the major difference between the output characteristics of the MOSFET and the JFET?
7. Compare the input impedance of the MOSFET to that of the bipolar transistor.
8. What are the advantages of using thyristors in high-power switching circuits instead of bipolar transistors?
9. What are the differences between the *V-I* characteristics of the SCR, SCS, SUS, and SBS?
10. What is the usual output parameter of the UJT?
11. A given SCR was tested in the lab and the following data was recorded:

I_G	$V_{(BR)F}$	I_H
0	400V	0.5 mA
0.01 mA	350V	0.5 mA
0.05 mA	300V	0.5 mA
0.1 mA	200V	0.5 mA
0.2 mA	100V	0.5 mA

Sketch the *V-I* anode characteristics.

12. Why is the TRIAC referred to as two "back-to-back parallel SCRs"?
13. What do we mean by the terminal characteristics of a semiconductor device?

2

Semiconductor Circuit Design

2–1 Introduction

It has been my experience that most industrial electronic designs can be successfully completed without getting deeply involved in the physics of the devices. The combination of terminal characteristics, device specifications, and simple graphical analysis is usually sufficient. We have discussed the terminal characteristics of semiconductor devices in the previous chapter. In this chapter, we will include device specifications and apply some simple graphical analysis to achieve circuit designs.

2–2 Rectifier Diode Circuits

The diode is the simplest semiconductor device. It is the logical device to use as an example for establishing design practices, not because of the importance of diode circuits but because the device characteristics are so simple that design principles are clearly illustrated. These same principles will carry over to more complicated devices.

Consider the circuit of Fig. 2–1. A rectifier diode controls the current flow into a resistive load, R_L, due to a dc voltage source, E, and an ac voltage source, $V(t)$. [The notation $V(t)$ means that the voltage varies as a function of time.] The first step in the design of this circuit is to determine what performance is required. This seems like a trivial statement but the success of any design will be determined by how well the expected performance is achieved. Learn what performance is expected before you start. Let's assume that the circuit of Fig. 2–1 must be designed for an operating point at 0.4 V with a peak sinusoidal current output of 5 mA. You will recall that the operating point refers to the dc or quiescent voltage drop across the diode. The next step then is to establish the required quiescent conditions, or the Q point. The 33-ohm load line is drawn on the diode characteristic curve (which is obtained from the manufacturer or from laboratory measurement). Figure 2–2 shows the load line and characteristics. The required battery voltage, E, is 0.73 V. We are now ready to look at the ac operation of the circuit. The ac source voltage, $V(t)$, is superimposed (combined linearly) with the dc source voltage. The operating point moves back and forth about the Q point at the same frequency as $V(t)$. The peak input voltage required for the 5-mA output can be read from Fig. 2–3 as 0.25V. The ac source can be expressed mathematically as $V(t) = 0.25 \sin \omega t$. This is a good point to

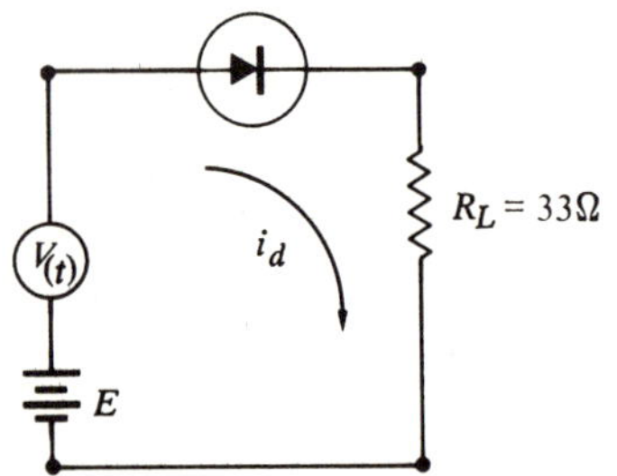

Fig. 2–1 Diode circuit.

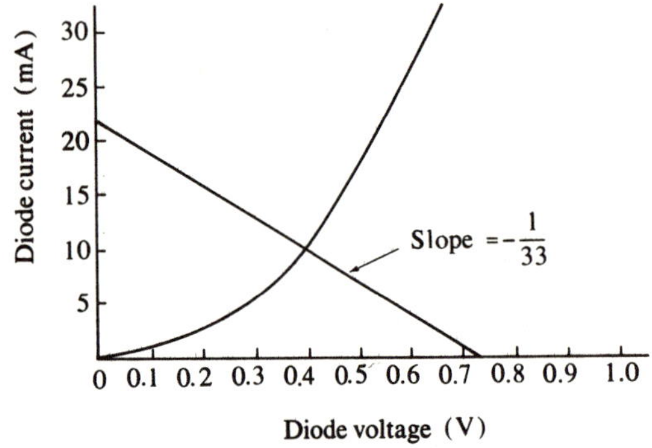

Fig. 2–2 DC load line for diode circuit.

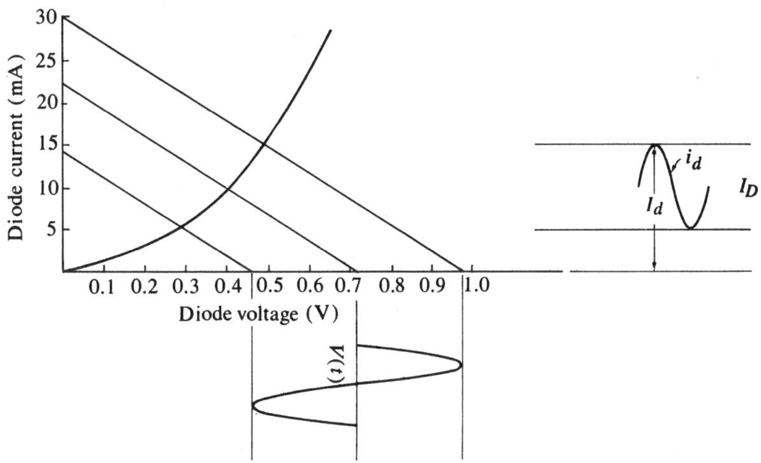

Fig. 2–3 AC analysis of circuit of Fig. 2–1.

agree on our symbology for discussing time-varying waveforms. The dc or quiescent values will be in capital letters with capital subscript, i.e., I_D. The time-varying values will be in lowercase script with lowercase subscript, i.e. i_d. The sum of dc and time-varying values will be in capital letters with lowercase script, i.e., I_d. The current in the circuit of Fig. 2–1 can be expressed as $I_d = I_D + i_d$. The load line of Fig. 2–2 can be expressed by the equation

$$I_d = -\frac{V_d}{R_L} + \frac{E}{R_L} \tag{2.1}$$

at the Q point

$$I_{DQ} = -\frac{V_{DQ}}{R_L} + \frac{E}{R_L} \tag{2.2}$$

We have gone as far as we need go with the paperwork design of this diode circuit. We should now proceed to the lab and construct the circuit model and test it for the predicted output current based on the ideal voltage sources.

The steps that we have discussed are very basic to the design of any electronic circuit. Let's enumerate them for emphasis:

1. Determine the performance requirements.
2. Establish the dc operating conditions.
3. Establish and evaluate the ac operating conditions.
4. Build a model of the circuit and make adjustments for the required performance.

You will find these four elements in every complete industrial electronic design.

The diode circuit of Fig. 2–4(a) is often used as a half-wave rectifier to convert ac voltage to dc voltage. The diode only allows current to flow through the load resistor when the secondary voltage of transformer $1T$ is positive at the anode of diode $1D$. The load sees half-wave pulses of current as shown in Fig. 2–4(b). The current pulses are sometimes smoothed by a reactive filter circuit. We will look at filtering techniques later in Chapter 12.

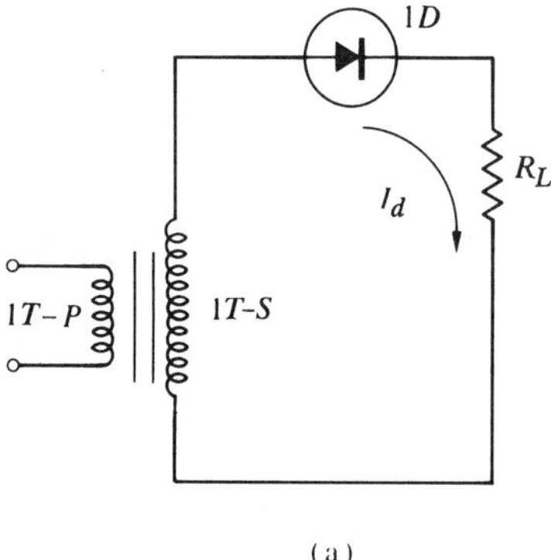

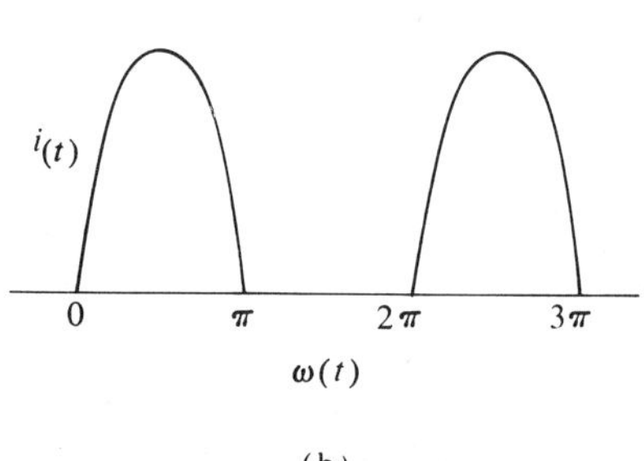

Fig. 2–4 (a) Half-wave rectifier circuit, (b) load current waveform.

Let's consider some of the commonly used parameters associated with the current and voltage seen by a resistive load in the half-wave rectifier circuit.

Average Current—The average current can be calculated from the equation

$$I_{av} = \frac{1}{2\pi}\int_0^{\pi} I_{max} \sin \omega t \, d\omega t = \frac{I_{max}}{\pi} \quad \textbf{(2.3)}$$

since $1/\pi = 0.318$,

$$I_{av} = 0.318 I_{max} \quad \textbf{(2.4)}$$

The average current can be easily measured with a dc ammeter placed in series with the load.

Average Voltage—The average voltage can be determined from the equation

$$V_{av} = I_{av}R = 0.318 I_{max} R \quad \textbf{(2.5)}$$

since $I_{max}R = V_{max}$ (assuming $V_D = 0$ during conduction).

$$V_{av} = 0.318 \mathrm{V}_{max} \quad \textbf{(2.6)}$$

RMS Current—The root mean square current, sometimes called the "effective current," can be calculated from the equation

$$I_{rms} = \sqrt{\frac{1}{2\pi}\int_0^{\pi} I_{max}^2 \sin^2 \omega t \, d\omega t} = \frac{I_{max}}{2} \qquad \textbf{(2.7)}$$

or

$$I_{rms} = 0.5\, I_{max} \qquad \textbf{(2.8)}$$

The rms current is used to determine the average power dissipated in the load. Since most ac meters are calibrated to indicate the rms value of a full sine wave, a more expensive "true rms" ammeter would be required to measure I_{rms} in the half-wave rectifier circuit. A "peak reading" ac ammeter could be used to measure I_{max}. I_{rms} could then be determined from Eq. (2.8).

RMS Voltage—The root mean square, or effective, voltage can be calculated from the equation

$$V_{rms} = I_{rms}R = 0.5 I_{max}R \qquad \textbf{(2.9)}$$

since $V_{max} = I_{max}R$

$$V_{rms} = 0.5\, V_{max} \qquad \textbf{(2.10)}$$

The rms voltage can be determined by measuring V_{max} and then using Eq. (2.10).

Dissipated Power—The average power dissipated in the load can be determined by the equation

$$P_{av} = V_{rms} \times I_{rms} \qquad \textbf{(2.11)}$$

as measured at the load or calculated from Eqs. (2.8) and (2.10).

Peak Inverse Voltage—Since the diode must block current flow through the load when the secondary voltage of transformer $1T$ is negative at the anode of diode $1D$, the peak inverse voltage rating of the diode must be at least as large as the maximum value of the transformer secondary voltage.

$$\text{PIV} > V_{max} \qquad \textbf{(2.12)}$$

Don't be confused by the terminology. Average power *is* the product of rms voltage and rms current! Average power *is not* the product of average voltage and average current!

Let's look at an example of the design of a half-wave rectifier circuit.

Example 2–1. In the circuit of Fig. 2–4, the transformer secondary voltage is 170 sin ωt, $R_L = 40$ ohms, and $V_D = 0$ during conduction. Specify the diode average current rating, the diode peak current rating, and PIV rating.

Solution: The maximum voltage can be determined from 170 sin ωt. $V_{max} = 170$ V. The diode average current is the same as the average load current.

From Eq. (2.4)

$$I_{av} = 0.318 I_{max}$$

$$I_{max} = V_{max}/R = 170/40 = 4.25 \text{ A}$$

$$I_{av} = (0.318)(4.25) = 1.35 \text{ A}$$

The diode peak current is the same as the maximum load current, 4.25 A. The diode PIV $= V_{max}$ or 170 V. The specified diode should have the following ratings:

$$I_{av} = 1.35 \text{ A}$$

$$I_{max} = 4.25 \text{ A}$$

$$\text{PIV} = 170 \text{ V}$$

A more practical specification would be:

$$I_{av} = 1.5 \text{ A}$$

$$I_{max} = 4.5 \text{ A}$$

$$\text{PIV} = 200 \text{ V}$$

The rectifier circuit of Fig. 2–4(a) can be modified to provide full-wave rectified current flow through the load resistor. Figure 2–5 shows the center-tapped transformer full-wave rectifier circuit. The specifications for design of such a rectifier might be:

$$V_{ac} = 600 \sin \omega t$$

$$I_{max} = 200 \text{ A}$$

$$T_A = 25°\text{C}$$

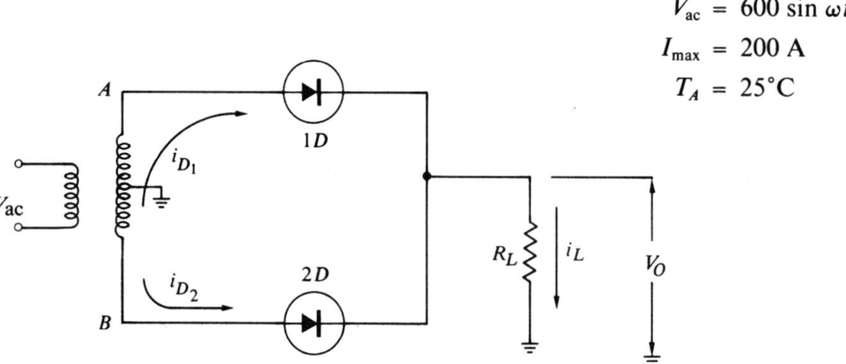

Fig. 2–5 Diode full-wave rectifier.

where T_A is the temperature of the environment in which the rectifier must operate, or *ambient temperature*. The designer must choose diodes $1D$ and $2D$ and specify the heat sinking to cool the diodes.

Figure 2–6 shows the input and output voltages of the rectifier.

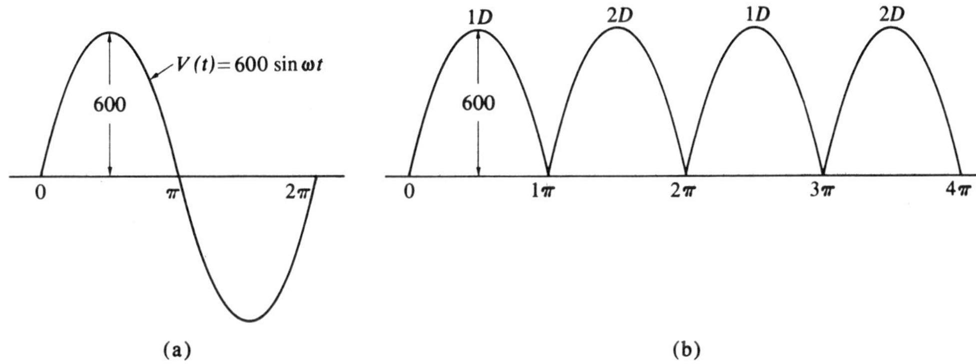

Fig. 2–6 (a) Rectifier input voltage, (b) rectifier output voltage.

When point A of Fig. 2–5 is positive with respect to ground, diode $1D$ is forward-biased and diode $2D$ is reverse-biased. The load current will only flow through $1D$ as indicated in Fig. 2–6. When point B is positive, $2D$ is forward-biased and $1D$ is reverse-biased. Load current only flows through $2D$. The diodes continue to alternately conduct depending on the polarity of the applied voltage. We can look at the performance of one of the diodes for purposes of design, since the other diode will perform exactly the same. On the diode characteristic curve of Fig. 2–7, the

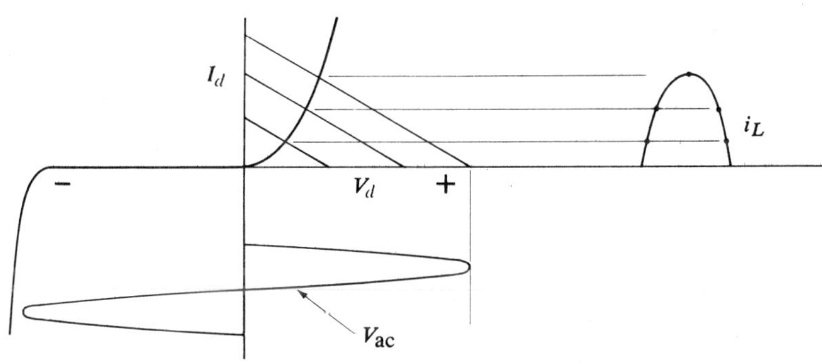

Fig. 2–7 Graphical analysis of diode full-wave rectifier.

load line is drawn for one of the diodes. Two design conditions are obvious from this simple load-line analysis:

1. The diode peak reverse voltage must be greater than the peak ac voltage.
2. The diode must be able to conduct a 200-A peak, half-wave rectified current.

The Delco Radio DRS-250 series silicon rectifiers are hard soldered, double diffused devices intended for high current, high voltage applications. The diffusion results in low leakage, and the hard solder permits more extreme thermal cycling. All units are hermetically sealed in a cold welded, nickel-plated copper package. A ceramic insulator is used between cap and mounting stud (white ceramic for positive base units, pink ceramic to identify negative base units). To specify negative, add "R" after the part number. These rugged units are mounted in a modified DO-9 case with 2 inch extra braid length.

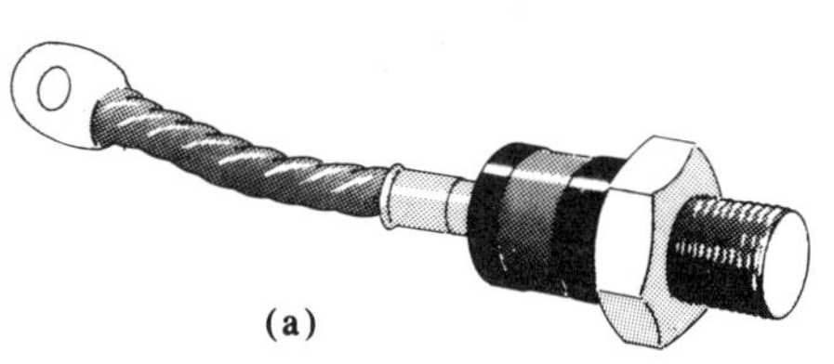

(a)

SYMBOL	PARAMETER		DRS-250	251	252	253	254	UNITS
V_{RM}	Peak Reverse Voltage (A.C.)		800	1000	1200	1400	1600	Volts
V_{tr}	Transient Reverse Voltage (A.C.) $\leq$ 5 ms		1050	1300	1600	2000	2400	Volts
$V_{F(av)}$	Average Forward Voltage Drop (A.C.) at 250 Amps avg. at $T_C = 110°C$	typ.	....	0.48	...			Volts
		max.	....	0.55	...			Volts
I_O	Average Forward Current (A.C.) at $T_C = 110°C$		....	250	...			Amps.
i_{surge}	½ Cycle Surge Current, Peak at $T_C = 110°C$		....	4500	...			Amps.
I_{RM}	Peak Reverse Current (A.C.) at rated PRV at $T_j = 25°C$	max.	....	2	...			mA
	at rated PRV at $T_j = 175°C$	max.	....	10	...			mA

(b)

T_s	Storage Temperature (T_6 to T_7)	−50 to +200	°C
T_j	Junction Operating Temperature (T_1 to T_5)	−40 to +175	°C
Θ_{jc}	Thermal Resistance—junction to case	0.22	°C/Watt

(c)

Courtesy of Delco Electronics Division, General Motors Corporation

Fig. 2–8 Semiconductors: (a) dimensions and connections, (b) electrical ratings and characteristics, (c) thermal ratings and characteristics,

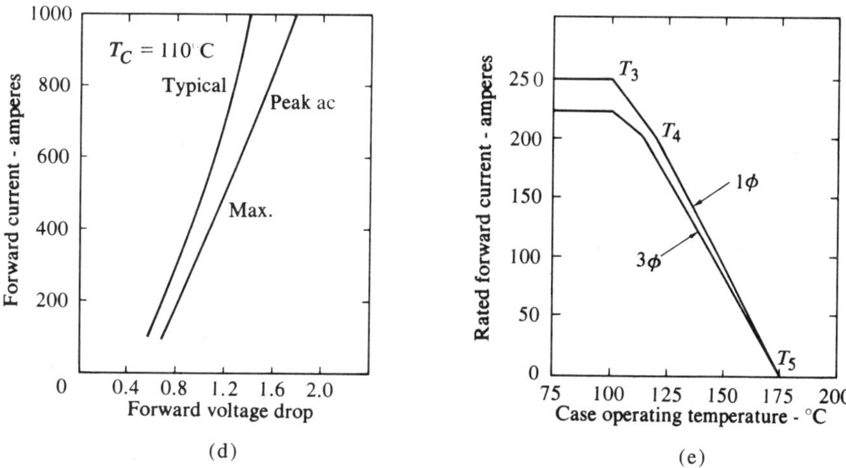

Courtesy of Delco Electronics Division, General Motors Corporation

Fig. 2–8 Semiconductors (continued): (d) forward characteristics, and (e) current derating curve.

A diode that meets both of those conditions is DELCO DRS-250. The engineering data for this diode are shown in Fig. 2–8. When the maximum current of 200 A is flowing, according to Fig. 2–8(a), the voltage drop across the diode is 0.8 V. The product of V_F and I_F is $200 \times 0.8 = 160$ W. We must now be concerned with how this power will flow away from the diode junction to prevent overheating or possible burnout. Figure 2–9 indicates how the heat caused by diode junction power dissipation will flow from the junction to air by way of a heat sink. As the heat flows, there will be temperature drops along its path analogous to voltage drops

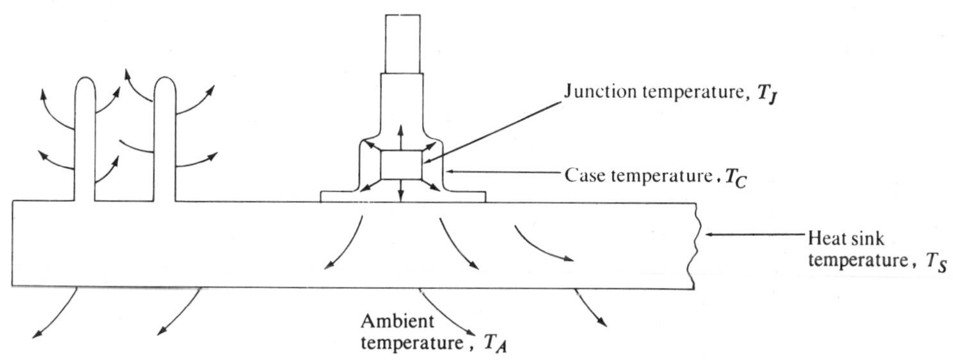

Fig. 2–9 Heat flow in diode and sink.

along the path of current flow. The circuit analogies are shown in Fig. 2–10. The equations are:

$$V = IR \tag{2.13}$$

$$T_J - T_A = P \quad \Theta_T = P(\Theta_{JC} + \Theta_{CS} + \Theta_{SA}) \tag{2.14}$$

where P = power dissipated, and Θ_T = thermal resistance.

The three components of thermal resistance, Θ_{JC}, Θ_{CS}, and Θ_{SA} represent the resistance-to-heat flow from junction to case, case to sink, and sink to air. The maximum allowable junction temperature and Θ_{JC} are usually included in the manufacturer's data sheet [Fig. 2–8(c)]. The power dissipated is determined by the electrical performance requirements of the circuit and ambient temperature is determined by the environment in which the circuit will operate.

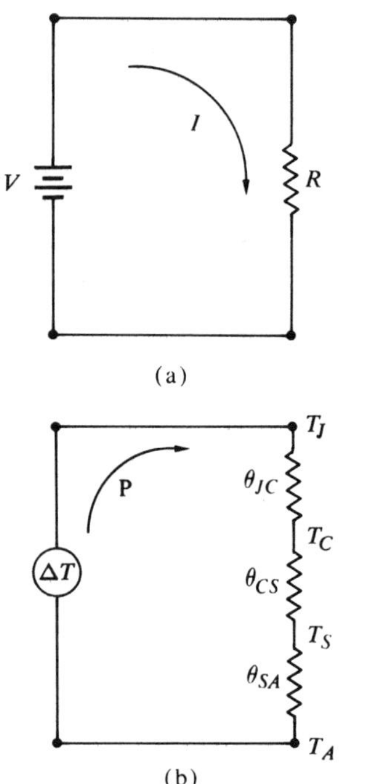

Fig. 2–10 (a) Electrical and (b) thermal series circuits.

Then Eq. (2.14) can be written as

$$\Theta_{CS} + \Theta_{SA} = \frac{T_J - T_A}{P} - \Theta_{JC} \tag{2.15}$$

where all parameters on the right-hand side of the equation are fixed. Our job as designers is to control Θ_{CS} and Θ_{SA} such that the diode junction temperature does not exceed the rated maximum value. The case-to-sink thermal resistance depends on the physical mating of the two surfaces. The surfaces should be flat, free of any ridges, burrs, or high spots. No paint, plating, or anodization should be allowed between the diode case and sink. Adequate pressure must be applied at the surfaces to maintain minimum thermal resistance. Typical values of 0.2 to 0.7°C/W can be obtained. Electrical insulating washers will increase Θ_{CS} by a factor of five or more. Heat sinks are generally black aluminum structures designed for maximum cooling effect with extruded fins and specially treated surfaces. Figure 2–11 shows some commercial heat sink extrusions. They may be specified by thermal resistance or surface area. Aluminum plate may also be used as heat sink material. A close approximation to the thermal resistance of the square aluminum of Fig. 2–12 is $\Theta_{SA} = 32.6A - 0.472$ where A is the surface area in square inches. The data of Fig. 2–12 were obtained from laboratory measurements of Θ_{SA} vs surface area for aluminum plate and various heat-sink configurations.

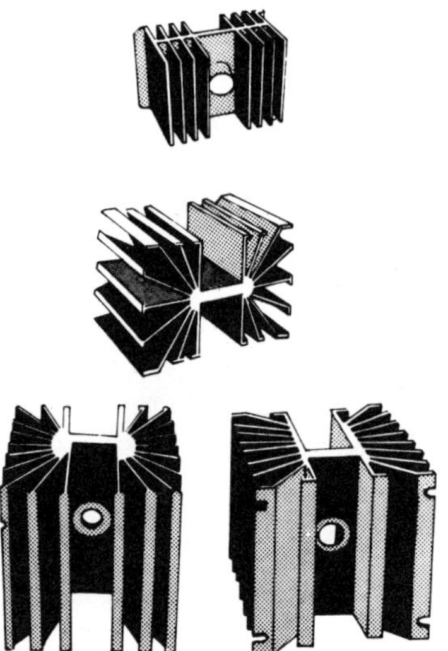

Fig. 2–11 Commercial heat sinks.

Manufacturers sometimes provide additional data in the form of derating curves such as Fig. 2–8(e). As long as the

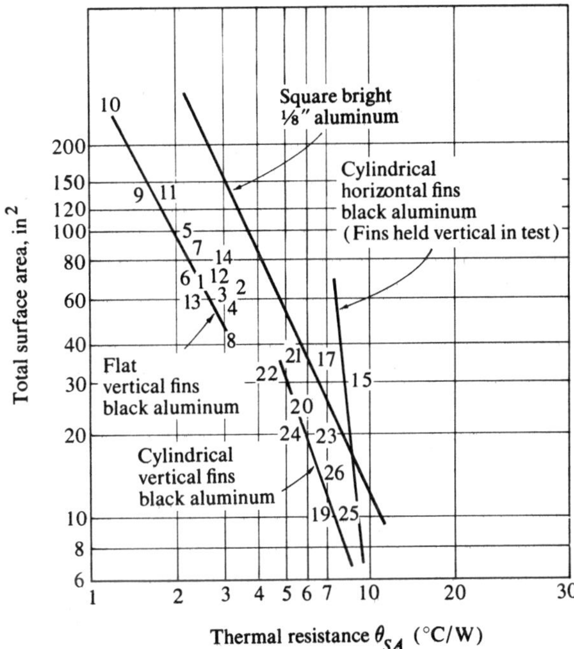

Courtesy of Motorola Inc.

Fig. 2–12 Thermal resistance of aluminum sinks.

diode case temperature is below 100°C, maximum current flow is allowed through the diode.

As the case temperature increases beyond 100°C, the current flow through the diode must be reduced to prevent damage to the P-N junction.

Let's now return to the rectifier of Fig. 2–5 and the design specifications from Eq. (2.15).

$$\Theta_{CS} + \Theta_{SA} = \frac{175 - 25}{P_J} - 0.22 \tag{2.16}$$

for the half-wave rectified diode current of Fig. 2–7.

$$P_J = I_F V_F \times 1.1 = 200 \times 0.48 \times 1.1 = 53 \text{ W} \tag{2.17}$$

$$\Theta_{CS} + \Theta_{SA} = \frac{175 - 25}{53} - 0.22 = 2.42\text{°C/W} \tag{2.18}$$

If we assume a nominal value of 0.3°C/W for Θ_{CS}, then Θ_{SA} must be 2.12°C/W. Since there are two diodes, both can be mounted on the same sink if the thermal resistance is reduced by a factor 2 or the sink surface area is double and the diodes are judiciously placed to avoid any hot spots.

From Fig. 2–12, a flat heat sink with vertical fins or black aluminum and approximately 250 sq in. would be sufficient.

All semiconductor devices used at high currents and voltages must be designed for junction temperature control. The same design consideration we have just applied to the diodes applies equally well to transistors, SCRs, and TRIACs.

2–3 Zener Diode Circuits

The zener diode is often used as a voltage regulator in dc circuits with varying loads and sources. A typical circuit is shown in Fig. 2–13.

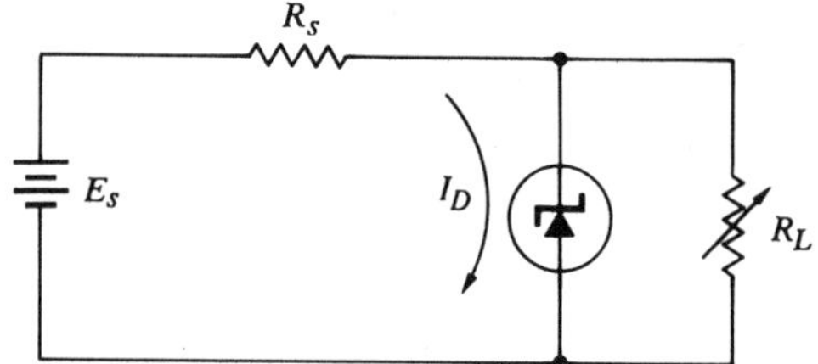

Fig. 2–13 Zener voltage regulator.

Let's stop for a minute and discuss the definition of *loads*. Most engineers in industry will refer to the amount of current drawn from the source as the load. There are occasions when we refer to the actual value of the load resistance. It is equally acceptable to speak of a 10-ohm load or a 2-amp load so long as we understand each other. Remember, maximum load current is drawn when load resistance is minimum!

In the circuit of Fig. 2–13, the source current divides between the load resistor and the zener diode, or

$$I_S = I_Z + I_L \tag{2.19}$$

also

$$V_Z = E_S - R_S(I_Z + I_L) \tag{2.20}$$

If $E_s = 70$ V and I_L varies between 100 and 200 mA, at $V_Z = 45 \pm 0.5$ V, design of the voltage regulator requires that we specify R_S and the zener diode. The characteristic curve for the zener diode is shown in Fig. 2–14.

The total current drawn from the battery under maximum load will be

$$I_S = 200 \text{ mA} + I_{Z(\text{min})} \tag{2.21}$$

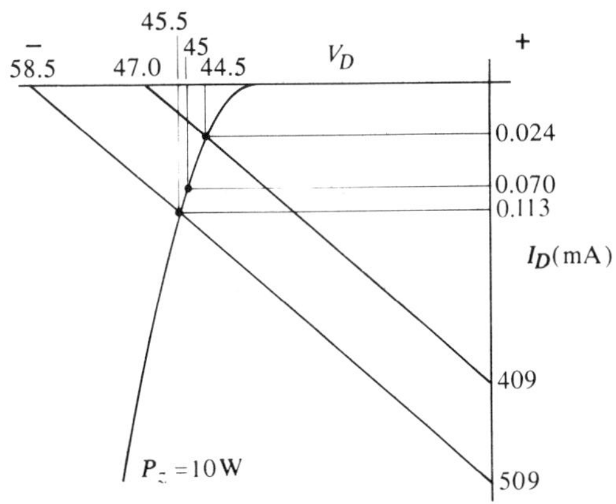

Fig. 2–14 Graphical analysis of zener voltage regulator.

It is good practice to make the minimum diode current at least 10% of the maximum load current, so

$$I_S = 200 \text{ mA} + 20 \text{ mA} = 220 \text{ mA}. \qquad (2.22)$$

The voltage dropped across R_S will be

$$V_{R_S} = E_S - V_Z \qquad (2.23)$$

The diode voltage, V_Z, will be almost constant due to the zener action. Then

$$V_{R_S} = E_S - V_Z = 70 - 45 = 25 \text{ V} \qquad (2.24)$$

and

$$R_S = \frac{V_{RS}}{I_S} = \frac{25}{0.220} = 115 \ \Omega$$

The load line must be drawn at maximum and minimum load currents.

From Eq. (2.20), the equations are

$$V_Z = 70 - 115(I_Z + 0.100) \qquad (2.25)$$

and

$$V_Z = 70 - 115(I_Z + 0.200) \qquad (2.26)$$

Equations (2.25) and (2.26) can be simplified to

$$V_Z = 58.5 - 115 I_Z \qquad (2.27)$$

$$V_Z = 47.0 - 115 I_Z \qquad (2.28)$$

The load lines are shown on the characteristic curve in Fig. 2–14.

We must still guarantee that the load voltage is regulated between 44.5 and 45.5 V. This can be done by calculating the change in diode current for the one-volt variation from Eqs. (2.27) and (2.28).

$$45.5 = 58.5 - 115I_Z \tag{2.29}$$

$$44.5 = 47.0 - 115I_Z \tag{2.30}$$

then

$$I_Z = \frac{58.5 - 45.5}{115} = 0.113\text{A} \tag{2.31}$$

at 45.5 V, and

$$I_Z = \frac{47.0 - 44.5}{115} = 0.024\text{A} \tag{2.32}$$

at 44.5 V.

$$\Delta I_Z = 0.113 - 0.024 = 89 \text{ mA} \tag{2.33}$$

$$R_Z = \frac{\Delta V_Z}{\Delta I_Z} = \frac{1}{89} \times 10^3 = 11\ \Omega \tag{2.34}$$

The zener diode should be able to handle the entire source current in case the load resistor is disconnected.

$$P_Z = 0.22 \times 45 = 9.9 \text{ W} \tag{2.35}$$

One other specification that is often overlooked in zener regulators is the wattage rating of R_S.

$$P = \frac{(25)^2}{115} = 5.4 \text{ W} \tag{2.36}$$

The design specifications for the regulator are

$$R_S = 115\ \Omega,\ 6\text{ W}$$
$$V_Z = 45 \text{ V at } I_Z = 0.070\text{A}$$
$$R_Z \leq 11\ \Omega$$
$$P_Z = 10 \text{ W}$$

A final comment about this type of voltage regulation is that the efficiency is quite low. The input power is

$$P_{\text{in}} = I_S \times E_S = 70(0.220) = 15.4 \text{ W} \tag{2.37}$$

The maximum power delivered to the load is

$$P_{L_{\max}} = V_Z I_L = (45)(0.200) = 9.0 \text{ W} \tag{2.38}$$

$$\text{Efficiency} = \eta = \frac{9}{15.4} \times 100 = 58.5\%.$$

2–4 Transistor Switches

The transistor finds many applications in industrial electronics as a switching device. Earlier applications found the transistor replacing the vacuum tube. Transistors began replacing mechanical relays in "static controls." Transistors are at the heart of the solid-state gating circuits being used in digital controls and computer-oriented programmable electronic controls. We will look at the basic design of switching circuits using bipolar transistors.

The basic bipolar junction PNP transistor switch is shown in Fig. 2–15. When the input voltage is at $+V$, the base-to-emitter junction is reverse-biased; no current flows in the load circuit so the transistor acts like an open switch.

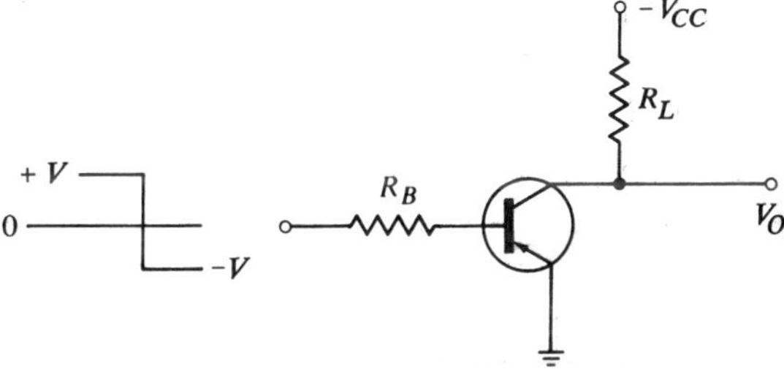

Fig. 2–15 Transistor switch.

When the input voltage is at $-V$, the base-to-emitter junction is forward-biased; the load current is then determined by the load line and base current flow; the transistor acts like a closed switch. The load-line analysis is shown in Fig. 2–16. The load line is drawn on the output characteristics with the slope $-1/R_L$. The operating point is at A

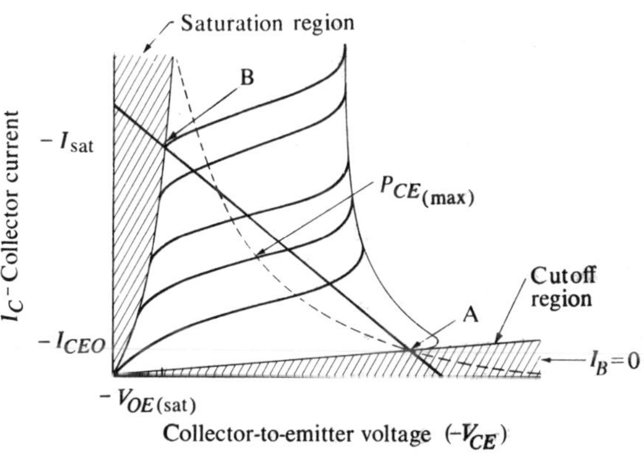

Fig. 2–16 Switching transistor characteristics and load line.

when the input voltage is $+V$. The load current is limited to the small leakage current $-I_{CEO}$. The operating point is at B when the input voltage is $-V$. The voltage drop across the collector to emitter is limited to the saturation voltage, $V_{CE(\text{sat})}$. The saturation voltage is approximately -0.2 V for germanium transistors and -0.4 V for silicon transistors. The shaded region at the left is called the saturation region. The shaded region at the bottom is called the cutoff region. It is in these regions that transistor switches operate. Notice that the load line lies partially in the forbidden region where $P_{CE} > P_{CE(\max)}$. This is safe for switching applications since the operating point never comes to rest on that portion of the load line. Let's look at the output of the transistor switch as a function of time. The output will not follow exactly the input voltage due to inherent delay properties of the transistor. Figure 2–17 is a comparison of V_{in} and V_{CE}. These delay times are identified in Fig. 2–17. Remember this figure; we will refer to it in the future.

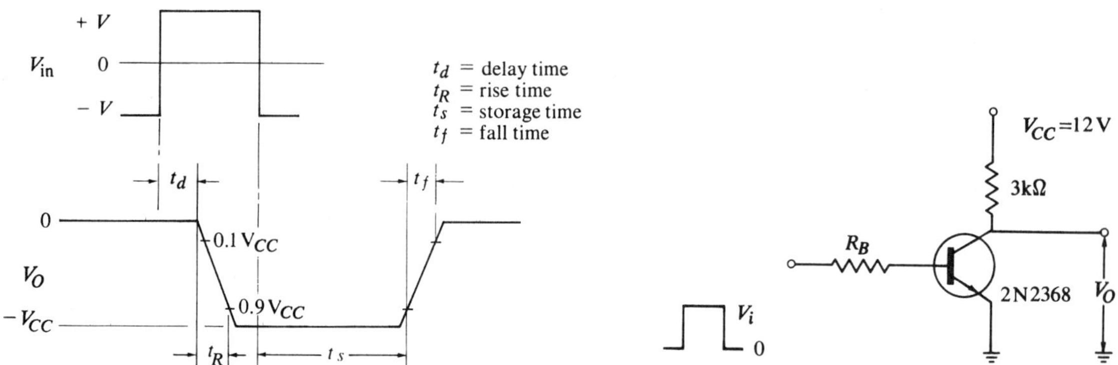

Fig. 2–17 Pulse response of transistor switch.

Fig. 2–18 Transistor switch.

Figure 2–18 is the circuit for the basic NPN transistor switch. Let's look at the design based on the following specifications:

$$V_i = 2 \text{ V peak}$$
$$V_o = \text{less than 200 mV in ``on'' state}$$
$$T_A = 25°\text{C}$$

The output characteristics for the transistor are shown in Fig. 2–19(a). The load line is drawn on the output characteristics and the operating points A and B are located.

In the "on" state, the transistor is in saturation at operating point *B*. $I_C = 3.9$ mA, $I_B = 0.09$ mA, from Fig. 2–19(b). $V_{CE(\text{sat})} = 0.16$ V. From Fig. 2–19(c), $V_{BE(\text{sat})} = 0.68$ V.

Writing the input loop equation

$$V_i = R_B I_B + V_{BE} \tag{2.39}$$

or

$$R_B = \frac{V_i - V_{BE}}{I_B} \tag{2.40}$$

$$R_B = \frac{2 - 0.68}{0.09 \times 10^{-3}} = \frac{1.32}{0.09} \times 10^3 = 14.6\ \text{k}\Omega \tag{2.41}$$

Using the closest standard 10% resistor, $R_B = 15\ \text{k}\Omega$.

The final step in the design is still to build the circuit and check its performance in the lab.

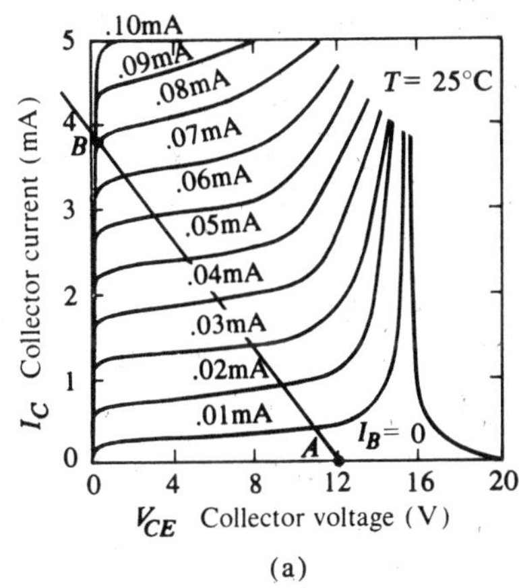

(a)

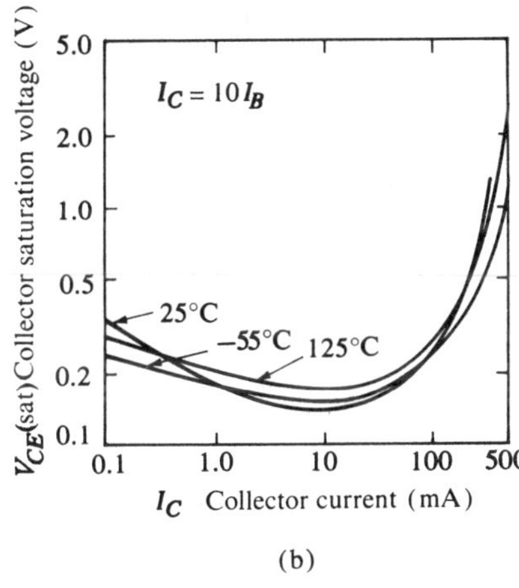

(b)

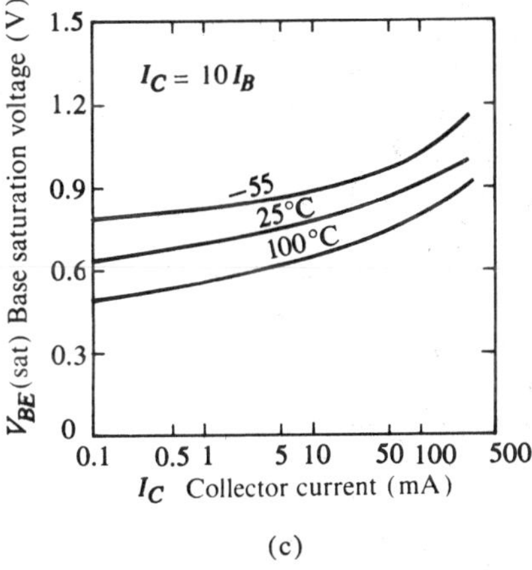

(c)

Courtesy of Fairchild Semiconductor

Fig. 2–19 2N2368 transistor characteristics.

The switching circuit of Fig. 2–20 is often used. Notice the addition of the resistor R_1 and a second supply voltage, V_{BB}. Two of the advantages of this circuit are:

(a) The input voltage does not have to go below zero.

(b) The transistor is reverse-biased when $V_i = 0$ so that leakage current is minimized.

The transistor characteristics are modified as shown in Fig. 2–21. The normal operating points are at *A* and *B* for load R_L. One of the disadvantages of this circuit can be seen by recognizing that if V_{BB} and V_i are both at ground potential then the operating point moves to *C* and the

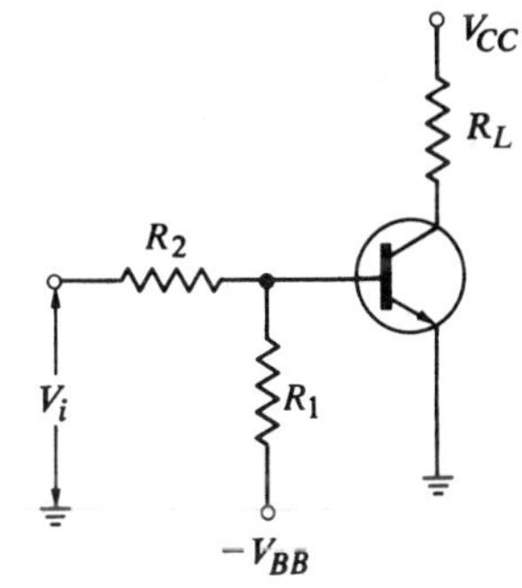

Fig. 2–20 Transistor switch with base bias supply.

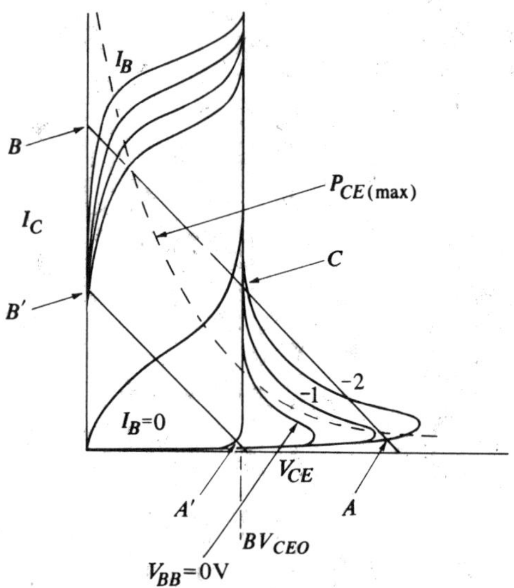

Fig. 2–21 Graphical analysis of transistor switch of Fig. 2–20.

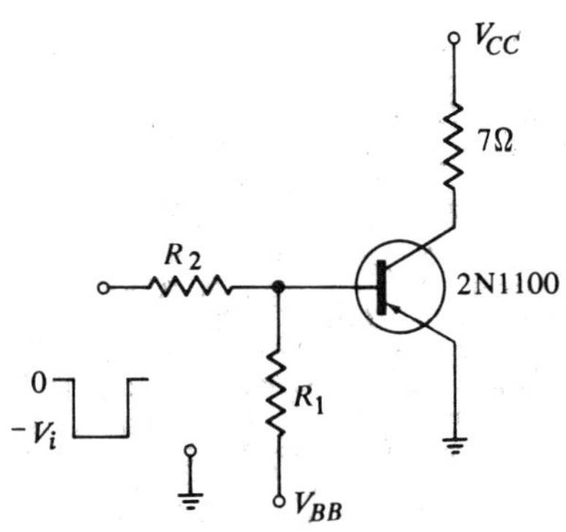

Fig. 2–22 Power transistor switch.

transistor will be destroyed by thermal runaway. To avoid this condition, V_{CC} should be kept below BV_{CEO} so that the operating points will be at A' and B'.

Let's look at the design of a PNP power switching circuit of Fig. 2–22 according to the following specifications:

$$T_A = 25°\text{C}$$
$$V_{BB} = 5 \text{ V}$$
$$R_2 = 50 \ \Omega$$
$$I_{L(\text{min})} = -10.0 \text{ A in "on" state}$$
$$I_{\text{in(max)}} = -10 \text{ A}$$

The transistor characteristics are shown in Fig. 2–23.

Since $BV_{CEO} \approx -70$ V, we can safely let $V_{CC} = -70$ V. The load line is drawn in Fig. 2–23(a). The "on" state operating point is at $I_C = -10.0$ A, $I_B = -0.5$ A. From Fig. 2–23(b), $V_{BE(\text{sat})} = -0.97$ V.

Let's first look at the biasing circuit when $V_i = 0$ and the base-to-emitter voltage is positive. From Fig. 2–23(b), we see that $V_{EB} = +1$ V is sufficient. The circuit is redrawn in Fig. 2–24.

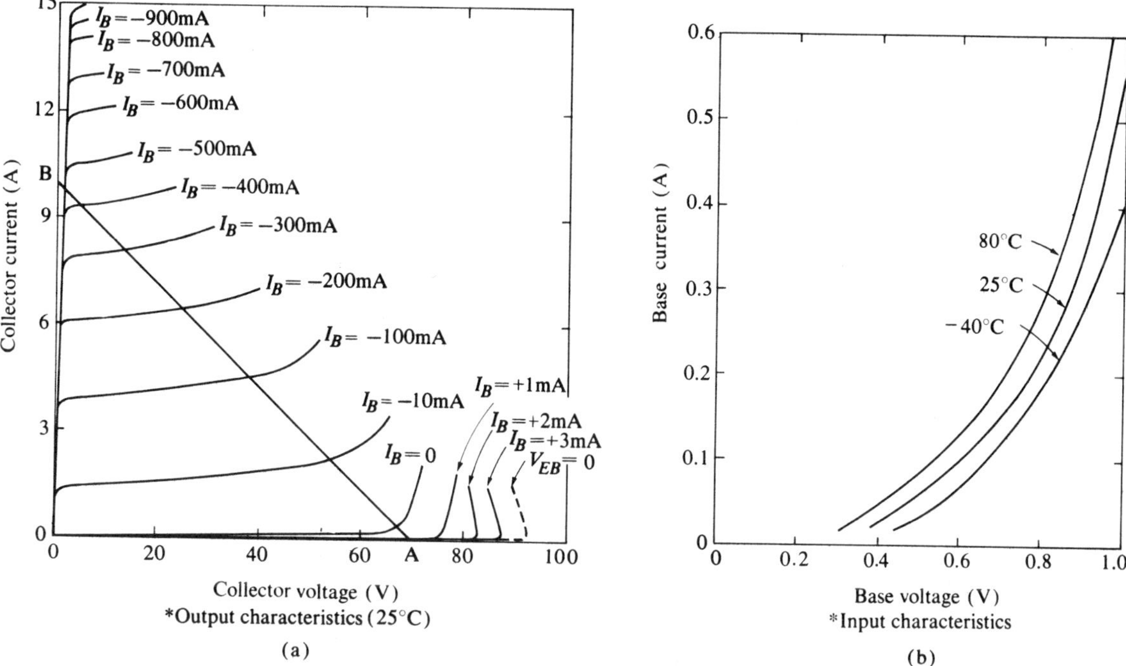

Courtesy of Delco Electronics Division, General Motors Corporation

Fig. 2–23 Transistor characteristics for 2N1100: (a) output, (b) transfer.

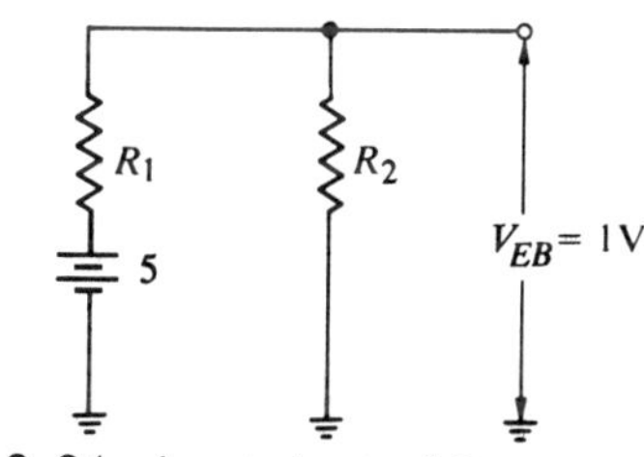

Fig. 2–24 Input circuit of Fig. 2–22, $V_i = 0$.

Using the voltage divider rule

$$\frac{R_1}{R_2} = \frac{5 - 1}{1} = 4 \quad \text{or} \quad R_1 = 4R_2 \tag{2.42}$$

$$R_1 = 200\ \Omega$$

The "on" state circuit is shown in Fig. 2–25. The circuit equations are:

$$I_2R_2 - V_i = -0.97 \tag{2.43}$$

$$-I_1R_1 + 5 = -0.97 \tag{2.44}$$

$$I_2 = I_1 + I_B \tag{2.45}$$

From Eq. (2.44),

$$I_1 = \frac{5 + 0.97}{R_1} = \frac{5.97}{200} \tag{2.46}$$

$$I_1 = 0.03\ \text{A}$$

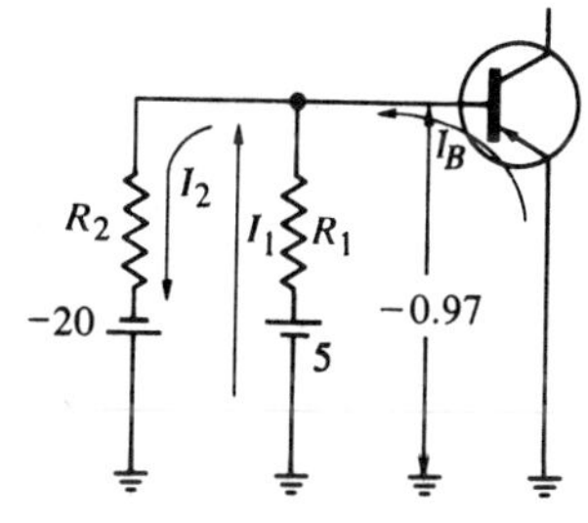

Fig. 2–25 Input circuit of Fig. 2–22 with $V_i = -20$ V.

From Eq. (2.45),

$$I_2 = I_1 + I_B = 0.0201 + 0.500 \tag{2.47}$$
$$I_2 = 0.5201 \text{ A}$$

From Eq. (2.43),

$$V_i = I_2R_2 + 0.97 = 50(0.5201) + 0.97$$
$$V_i = 26.98 \text{ V}$$

The design is completed. It should be noted that the power dissipated in R_2 is $(0.5201)^2 50 = 13.5$ W. A small power resistor should be used.

The application of transistor switches to digital circuits and industrial logic systems has introduced another problem that the designer must consider: multiple input and output switches. The circuit of Fig. 2–26 is an example of a transistor switch, Q_2, with N possible inputs and L possible loads. Q_2 must be "on" if any of the input transistors are "off." Q_2 must also be capable of providing the current to keep all L load transistors in the "on" state. Let's separate the design into two simple cases and look at them separately.

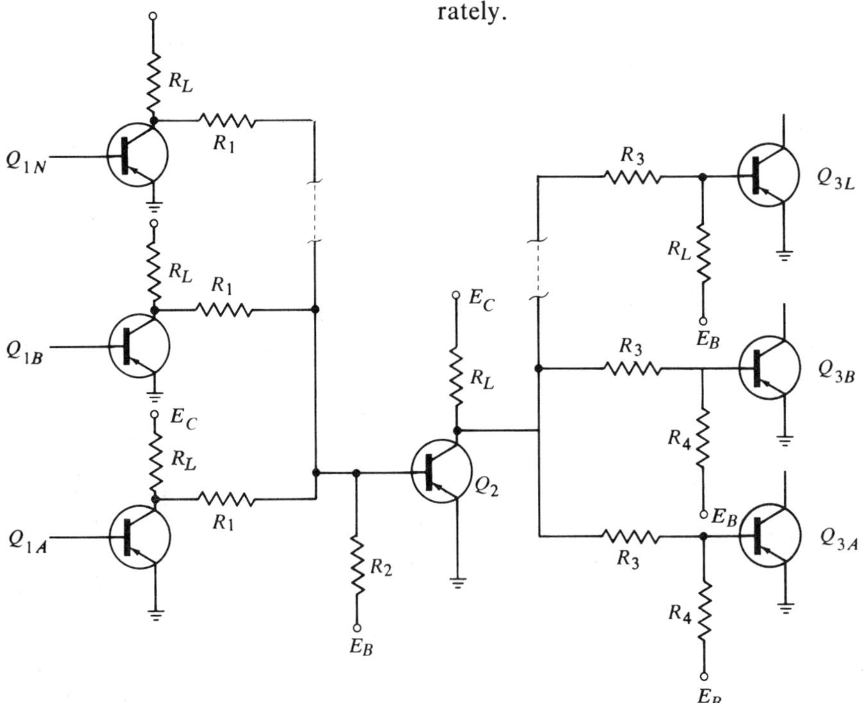

Fig. 2–26 Multiple input-output transistor switch.

We will deal first with the multiple input switch with a single load as shown in Fig. 2–27. Assume that all transistors are the 2N404 whose characteristics are shown in Fig. 2–28. Let $E_c = -20$ V, $E_b = 6$ V, $R_C = 2$ kΩ. The "on" and "off" conditions at 25°C are:

ON	OFF
$V_{CE(sat)} = -125$ mV (max)	$V_{CE} = -19.8$ V
$I_B = -0.18$ mA (max)	$I_B = I_{CBo} = -10\ \mu$A (min)
$I_B = -0.16$ mA (min)	$I_C = -10^{-4}$ A (min)
$I_C = -10$ mA (min)	$V_{BE} = +1$ (max)
$V_{BE} = -210$ mV (min)	
$V_{BE} = -340$ mV (max)	

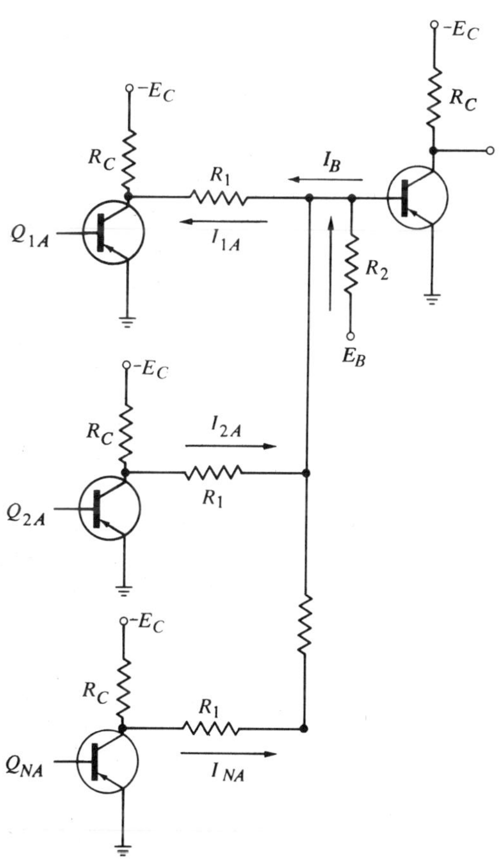

Fig. 2–27 Multiple input transistor switch.

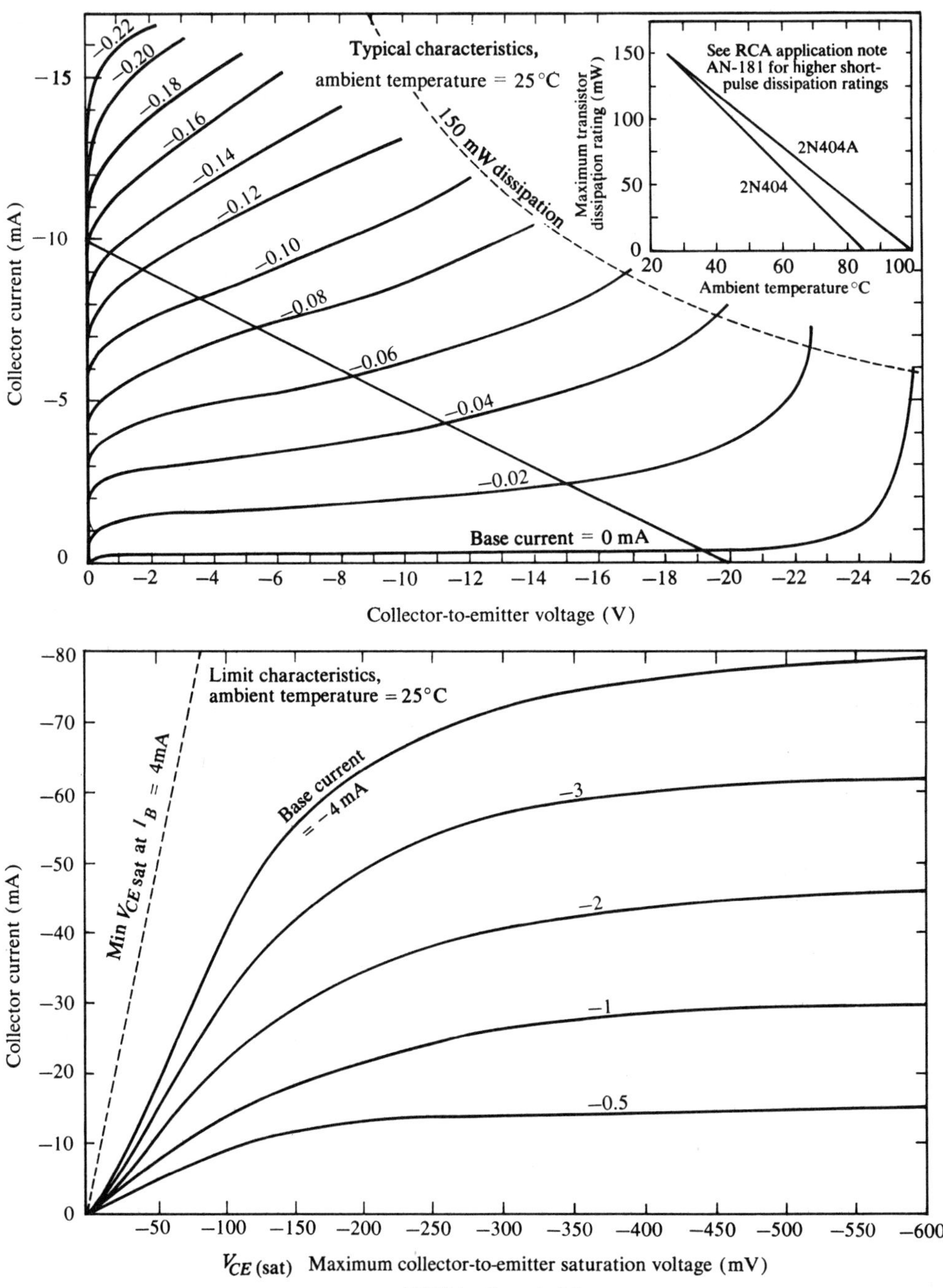

Courtesy of Technical Publishing House

Fig. 2–28 Characteristics of the 2N404 transistor.

The conditions are taken from the load line and characteristics of Fig. 2–28. Let's consider the circuit when Q_{1A} is "off" and all other input transistors are "on." Then worst case conditions are

$$I_{1A} = (N - 1) I_{2A} + I_B + I_2 \tag{2.48}$$

or

$$\frac{V_{CE(\min)} - V_{BE(\max)}}{R_1} = (N - 1)\left[\frac{V_{BE\max} - V_{CE\,\text{sat}}}{R_1}\right] + \left[\frac{E_B - V_{BE\max}}{R_2}\right] + I_B \tag{2.49}$$

Putting in the specified numbers

$$\frac{19.8 - 0.340}{R_1} = (N - 1)\left[\frac{0.340 - 0.125}{R_1}\right] + \left[\frac{6 - 0.340}{R_2}\right] + I_B \tag{2.50}$$

The condition required to maintain V_{BE} at cutoff is:

$$\frac{R_2(E_B - V_{CE\,\text{sat}})}{R_2 + R_1/N} = V_{BE(\text{off})} = 1 \tag{2.51}$$

Using Eqs. (2.50) and (2.51) at $I_B = 0.18$ mA and $I_B = 0.16$ mA, the calculated values for R_1 and R_2 for $N = 1$, 2, 3, and 4 are:

$I_B = 0.16$ mA

N	R_1	R_2
1	114 kΩ	555 kΩ
2	112.5 kΩ	545 kΩ
3	111.3 kΩ	542 kΩ
4	110 kΩ	540 kΩ

$I_B = 0.18$ mA

N	R_1	R_2
1	100 kΩ	487.5 kΩ
2	98.8 kΩ	482 kΩ
3	97.6 kΩ	476 kΩ
4	96.4 kΩ	470 kΩ

These data are plotted in Fig. 2–29. The shaded region represents values of R_1 and R_2 that could be used according to the stated specifications.

The second part of the problem is then to consider the transistor switch driving multiple loads such as in Fig. 2–30. Transistor Q_2 must be able to supply base current to drive all the loads at the same time without allowing its own collector-to-emitter voltage to rise above a maximum value. Let's assume the following specifications when all Q_3 transistors are "on" and Q_2 is "off."

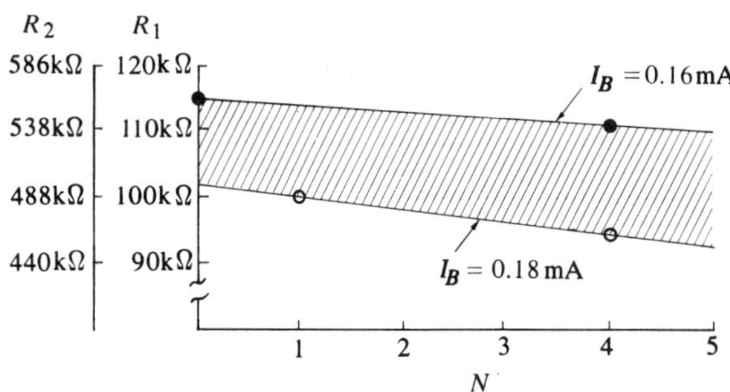

Fig. 2–29 Bias resistor variation for circuit of Fig. 2–27.

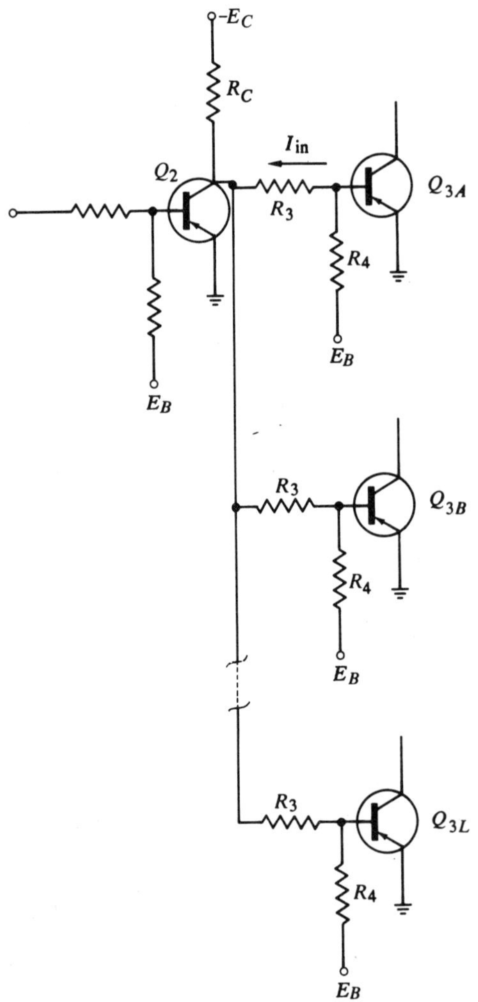

Fig. 2–30 Multiple output transistor switch.

$$V_{CC} = -20\text{ V}, \quad E_B = 6\text{ V}$$
$$I_{\text{in (min)}} = -0.2\text{ mA}, \quad V_{BE\,(\text{max})} = -0.340\text{ V}$$
$$I_{CBO} = -10^{-5}\text{ A}$$

Our design problem is to specify R_c and R_3 for $L = 1$, 2, 3, 4. From Fig. 2–30, worst case conditions are:

$$V_{CC} - R_c(LI_{\text{in}} + I_{CBO}) - R_3 I_{\text{in}} - V_{BE\,\text{max}} = 0 \quad \mathbf{(2.52)}$$

or

$$20 = R(0.2 \times 10^{-3} + 10^{-5}) \quad \mathbf{(2.53)}$$
$$+0.2 \times 10^{-3} R_3 - 0.340 \quad \mathbf{(2.53)}$$

Values of L and R_3 for chosen values of R_C are shown in the following table:

$R_C = 10\text{ k}\Omega$		$R_C = 5\text{ k}\Omega$		$R_C = 2\text{ k}\Omega$	
L	R_3	L	R_3	L	R_3
1	88.3 kΩ	1	93.3 kΩ	1	96.3 kΩ
2	78.3 kΩ	2	88.3 kΩ	2	94.3 kΩ
3	68.3 kΩ	3	83.3 kΩ	3	92.3 kΩ
4	58.3 kΩ	4	78.3 kΩ	4	90.3 kΩ

These data are plotted in Fig. 2–31.

Transistor switching circuit design becomes complicated rapidly as higher order combinations of switches are required. We have looked at the simplest cases but the same techniques are used at more sophisticated levels.

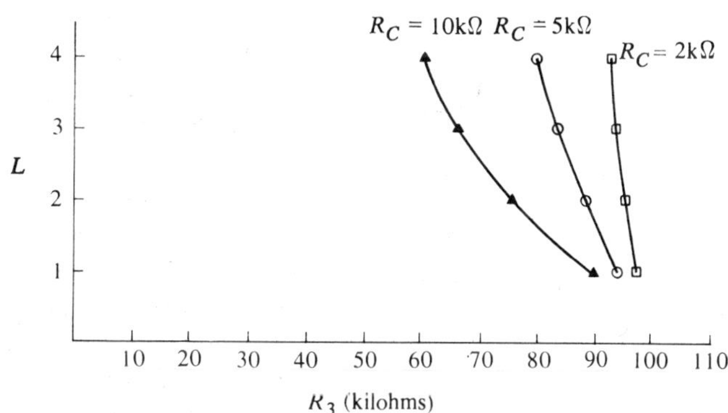

Fig. 2–31 Bias resistor variation for circuit of Fig. 2–30.

2–5 Transistor Amplifiers

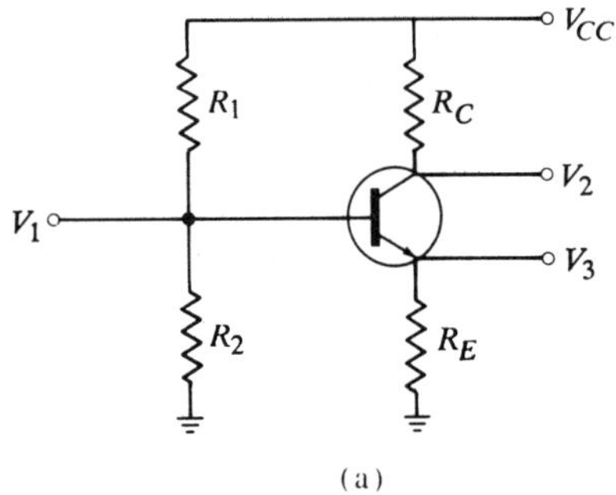

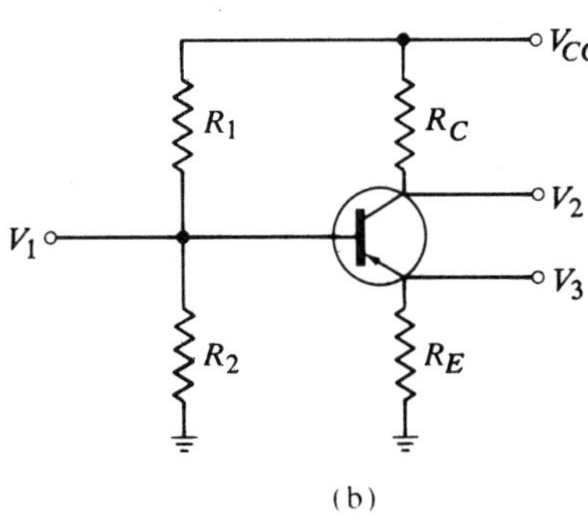

Fig. 2–32 Common emitter amplifier (a) NPN, (b) PNP.

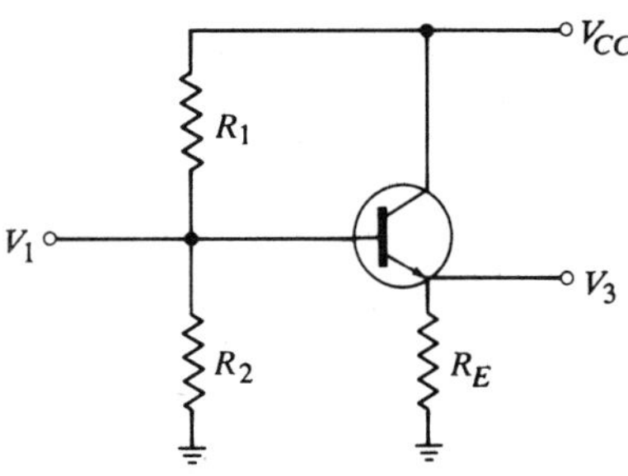

Fig. 2–33 NPN common collector amplifier.

The transistor switching circuits of Sec. 2–4 were always operated in the "saturation region" or "cutoff region" of Fig. 2–16. There are many applications where the transistor is operated in the "active region" of the output characteristics. In this region, the collector current varies in proportion to the base current. If the signal level is very small, the collector current will be some constant, m, times the base current. Mathematically

$$i_c = mi_b \tag{2.54}$$

That is what we mean by "linear" amplification. As the signal level increases, the nonlinear characteristics of the transistor will introduce distortion in the output so that

$$i_c = mi_b + \Delta i_c' \tag{2.55}$$

where $\Delta i_c'$ represents deviation from true linear amplification. We are often in a position where $\Delta i_c'$ is not significantly large so we "live with it" in order to get the required signal amplitude. This is usually called "large signal amplification."

Most of the amplifier design techniques we will discuss are accurate for linear operation but will be extended to large signal class-A operation realizing that we must live with the $\Delta i_c'$ contribution. The approximation is usually adequate for industrial electronic applications.

The basic transistor amplifier is shown in Fig. 2–32. The NPN and PNP configurations are the same except for the polarity of the bias supply. This is the "common emitter" configuration. The input is at V_1 and the output is taken at the V_2 terminal. By letting R_C go to zero and taking the output from V_3, the emitter follower or common collector configuration is developed as in Fig. 2–33. By laying the amplifier on its back, using V_3 as the input terminal and V_2 as the output terminal, the common base configuration of Fig. 2–34 is developed. As you see, there is only one transistor amplifier. By manipulating the terminal connections and varying the resistor values, we can develop different configurations to take advantage of gain and impedance characteristics of the amplifier. We will use the common emitter configuration 90% of the time because it offers the highest power gain, has input and output impedances that are more nearly the same, and manufacturers' terminal characteristics are more readily available. We will concentrate on the design of the CE amplifier.

You will recall that in Sec. 2–2 we outlined four steps to be followed in designing electronic circuits:

1. determine the performance requirements
2. establish the dc operating conditions
3. establish and evaluate ac operating conditions
4. build a model and make adjustments for the required performance

Let's look at each of the steps and apply them to the transistor amplifier.

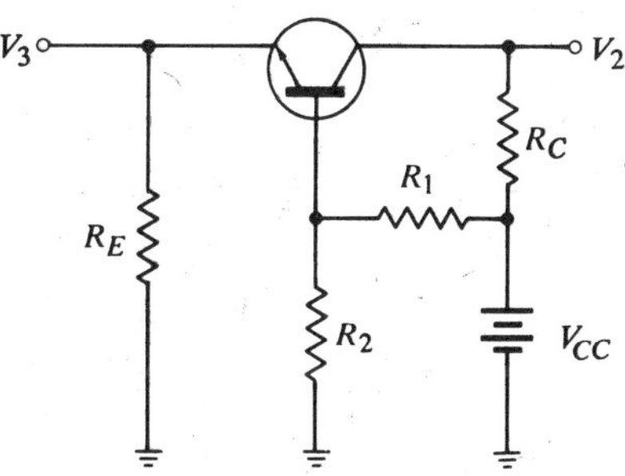

Fig. 2–34 NPN common base amplifier.

The "performance requirements" of a transistor amplifier include:

1. peak load voltage
2. peak load current
3. voltage gain
4. current gain
5. power gain
6. input impedance
7. output impedance

Peak load voltage is determined by V_{CC}, $V_{CE(\text{sat})}$ R_L, and R_E.

$$V_{L(\text{peak})} = \frac{R_L}{R_E + R_L} V_{o(\text{peak})} \tag{2.57}$$

where

$$V_{o(\text{peak})} = \frac{V_{CC} - V_{CE(\text{sat})}}{2} \tag{2.58}$$

or

$$V_{L(\text{peak})} = \left(\frac{R_L}{R_E + R_L}\right)\left(\frac{V_{CC} - V_{CE(\text{sat})}}{2}\right) \tag{2.59}$$

The designer can change the four variables to achieve required peak load voltage.

Example 2–2. Using the 2N2614 transistor, R_L = 1 kilohm, and R_E = 270 ohms; specify V_{CC} to allow 8 V peak across R_L.

Let $V_{CE(\text{sat})}$ – 0.5 V to bc surc wc stay away from saturation distortion.

Rearranging Eq. 2.59,

$$V_{CC} = 2\,\frac{(R_E + R_L)}{R_L}\,V_{\text{peak}} + V_{CE(\text{sat})}$$

$$= 2\,\frac{(1000 + 270)}{1000}\,(8) + 0.5 \qquad \textbf{(2.60)}$$

$$= 20.8\text{ V}$$

Peak load current for the direct coupled amplifier of Fig. 2–32 is determined by R_E, R_L, and peak collector-to-emitter voltage, or

$$i_{L(\text{peak})} = \frac{V_{CC} - V_{CE(\text{sat})}}{2(R_E + R_L)} \qquad \textbf{(2.61)}$$

Example 2–3. Specify the value of R_E for a 6-mA peak load current if $V_{CC} = 20$ V, $R_L = 1$ kΩ, and $V_{CE(\text{sat})} = 0.5$ V. Rearranging Eq. (2.61)

$$\mathrm{R_E} = \frac{V_{CC} - V_{CE(\text{sat})}}{2\,i_{\text{peak}}} - R_L$$

$$= \frac{20 - 0.5}{2 \times 6 \times 10^{-3}} - 10^3$$

$$= \frac{19.5}{12} \times 10^3 - 10^3$$

$$= 625\,\Omega \qquad \textbf{(2.62)}$$

Voltage gain can be determined by drawing the ac circuit of the amplifier and applying a voltage source as in Fig. 2–35. From Fig. 2–35(b), $i_b = \dfrac{V_{\text{in}}}{R_{\text{in}}}$ where R_{in} is the base-to-ground impedance of the transistor.

$$R_{\text{in}} = (\beta + 1)\,R_E + R_{BE} \qquad \textbf{(2.63)}$$

where

$$R_{BE} = \frac{25}{I_b\,(\text{mA})}$$

also

$$V_o = i_c R_L = \beta\, i_b R_L$$

so

$$V_o = \beta\,\frac{V_{\text{in}}}{R_{\text{in}}}\,R_L$$

voltage gain

$$A_v = \frac{V_o}{V_{\text{in}}} = \beta\,\frac{R_L}{R_{\text{in}}}$$

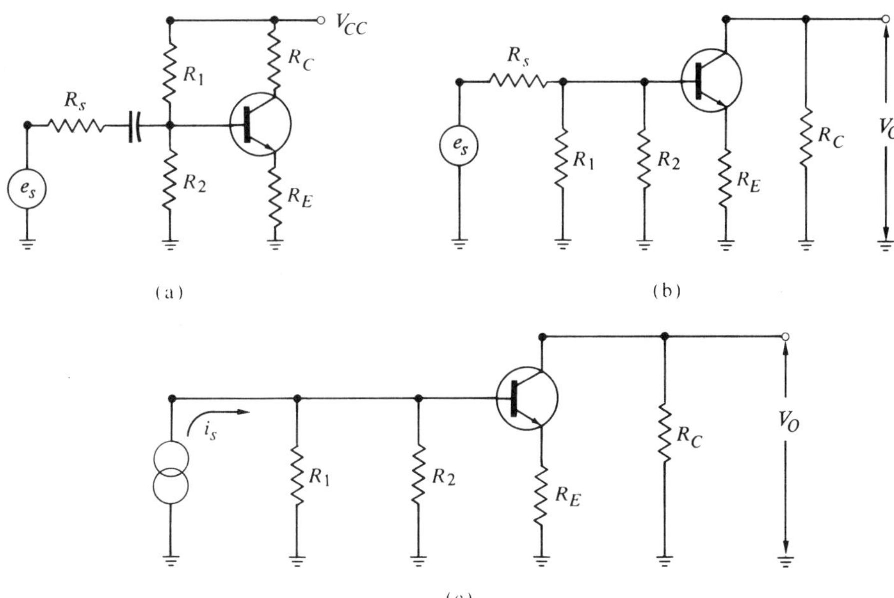

Fig. 2–35 (a) NPN common emitter amplifier with voltage source, (b) AC equivalent circuit, (c) AC equivalent circuit with high impedance current source.

Example 2–4. Given the amplifier of Fig. 2–35(a) with R_E = 100 ohms, I_B = 0.070 mA, and β = 100, specify R_L for a voltage gain of 10.

$$A_v = \beta \frac{R_L}{R_{\text{in}}}$$

or

$$R_L = \frac{A_v R_{\text{in}}}{\beta}$$

$$R_{BE} = \frac{25}{0.070} = 357\ \Omega$$

From Eq. (2.63),

$$R_{\text{in}} = (101)\,100 + 357 = 10{,}437\ \Omega \qquad \textbf{(2.64)}$$

$$R_L = \frac{10 \times 10{,}437}{100} = 1044\ \Omega$$

or

$$R_L = 1\ \text{k}\Omega \text{ nearest } 10\% \text{ standard value.}$$

Current gain can be determined from the circuit of Fig. 2–35(c) by applying a current source at the input.

$$i_b = i_s \frac{R_B}{R_B + R_{\text{in}}}$$

where $R_B = R_1 \| R_2$

$$i_c = \beta i_b$$

$$\text{current gain, } A_i = \frac{i_c}{i_s} = \frac{\beta i_b}{i_s} = \beta \frac{R_B}{R_B + R_{\text{in}}} \frac{i_s}{i_s}$$

$$A_i = \frac{\beta R_B}{R_B + R_{\text{in}}} \tag{2.65}$$

Example 2–5. In the amplifier of Fig. 2–35(c), $\beta = 100$, $I_{BQ} = 0.12$ mA, and $R_E = 100$ ohms. Specify R_B for $A_i = 35$.

$$R_{BE} = \frac{25}{0.12} = 208\ \Omega$$

$$R_{\text{in}} = (101)(100) + 208 = 10{,}308\ \Omega$$

Rearranging Eq. (2.65),

$$R_B = \frac{A_i R_{\text{in}}}{\beta - A_i}$$

$$= \frac{35 \times 10{,}308}{100 - 35} = 5560\ \Omega \tag{2.66}$$

Power gain is the product of voltage gain and current gain. Power gain is usually stated in decibels (dB) where

$$\text{dB} = 10 \log P_g$$

Example 2–6. An amplifier has a voltage gain = 15 and current gain = 85. What is the power gain in dB?

$$P_g = A_v A_i = 15 \times 85 = 1275$$

$$P_g = 10 \log 1275 = 31.1\ \text{dB.}$$

Input impedance is determined by the biasing resistors and input impedance of the transistor circuit.

In the circuit of Fig. 2–35(c)

$$Z_{\text{in}} = R_1 \| R_2 \| R_{\text{in}} = R_B \| R_{\text{in}} \tag{2.67}$$

Example 2–7. If $R_1 = 100\ \text{k}\Omega$, $R_2 = 50\ \text{k}\Omega$, $I_{BQ} = 0.1\ \text{mA}$, $\beta = 100$, in the circuit of Fig. 2–35(c), specify R_E for $Z_{\text{in}} = 15$ kilohms.

$$R_1 \| R_2 = R_B = 33\ \text{k}\Omega$$

$$R_{BE} = \frac{25}{0.1} = 250\ \Omega$$

$$R_{\text{in}} = (\beta + 1)R_E + R_{BE} = 101R_E + 250$$

Rearranging Eq. (2.67),

$$R_{\text{in}} = \frac{R_B Z_{\text{in}}}{R_B - Z_{\text{in}}} = \frac{33\ \text{k}\Omega \times 15\ \text{k}\Omega}{33\ \text{k}\Omega - 15\ \text{k}\Omega} = 27.5\ \text{k}\Omega$$

$$101R_E = 27.5\ \text{k}\Omega - 250, \quad R_E = 270\ \Omega$$

Output impedance of the amplifier is determined by the parallel combination of R_c and R_o where R_o is the collector-to-emitter impedance of the transistor. R_o is usually much larger than R_c so that $Z_o \approx R_c$.

The amplifier performance requirements are seldom specified as completely as we have just outlined them. Only the requirements important to the particular application will be specified. The designer will often impose further requirements in order to achieve a quality design based on previous experience.

Establishing the dc operating conditions requires that the Q point be selected and the bias circuit designed to establish the collector current and base current at the proper value. After the dc load line is drawn on the output characteristics [Figure 2–36(a)], it is useful to draw the dynamic transfer characteristics before selecting the Q point. Figure 2–36(b) shows the nonlinear transfer characteristics of the transistor.

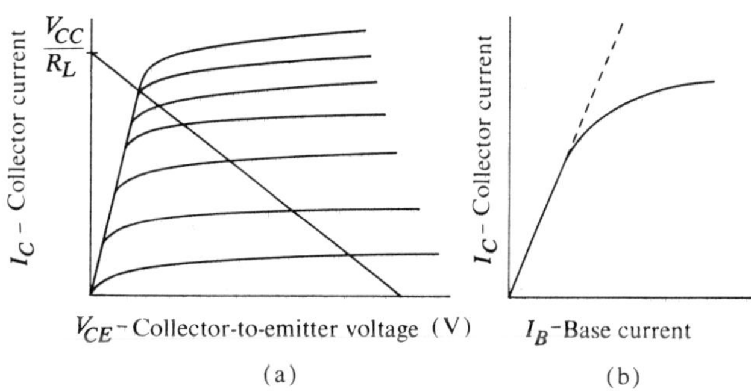

Fig. 2–36 (a) Transistor output characteristic, (b) dynamic transfer characteristic.

The dotted line represents true linearity. We can see how selection of the Q point affects the linearity of the amplifier by observing the two choices of Fig. 2–37. The same base current drive is applied at Q points A and B. At point A, the output (collector) current is sinusoidal. At point B, the output current is obviously distorted.

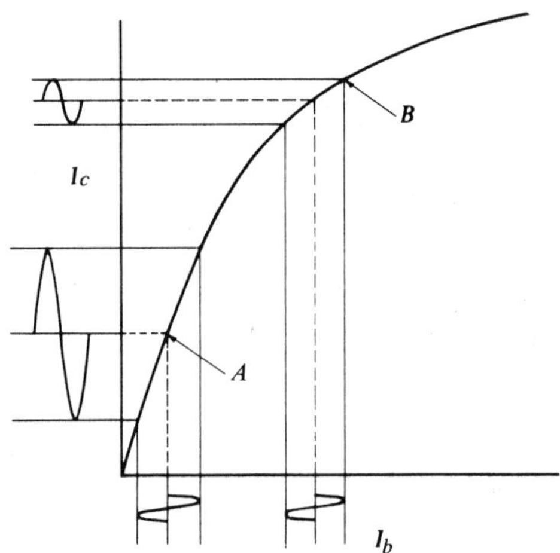

Fig. 2–37 Nonlinearity of transistor transfer characteristics.

The Q-point specifications are I_{CQ}, I_{BQ}, and V_{CEQ}. Let's now look to Fig. 2–38(a) and determine the values of R_1 and R_2 for a specific Q point. This is a good place to apply Thévenin's theorem. First, the biasing circuit is redrawn in Fig. 2–38(b). The Thévenin equivalent circuit is shown in Fig. 2–38(c) where

$$R_B = \frac{R_1 R_2}{R_1 + R_2} \tag{2.68}$$

$$V_B = \frac{R_2}{R_1 + R_2} V_{CC} \tag{2.69}$$

We can now replace the original biasing circuit with the thévenized equivalent [Fig. 2–38(d)]. Applying Kirchhoff's voltage law to the input circuit,

$$V_B = R_B I_{BQ} + V_{BE} + I_E R_E \tag{2.70}$$

The base-to-emitter voltage can be determined from manufacturers' specifications or estimated at 0.2 volts for

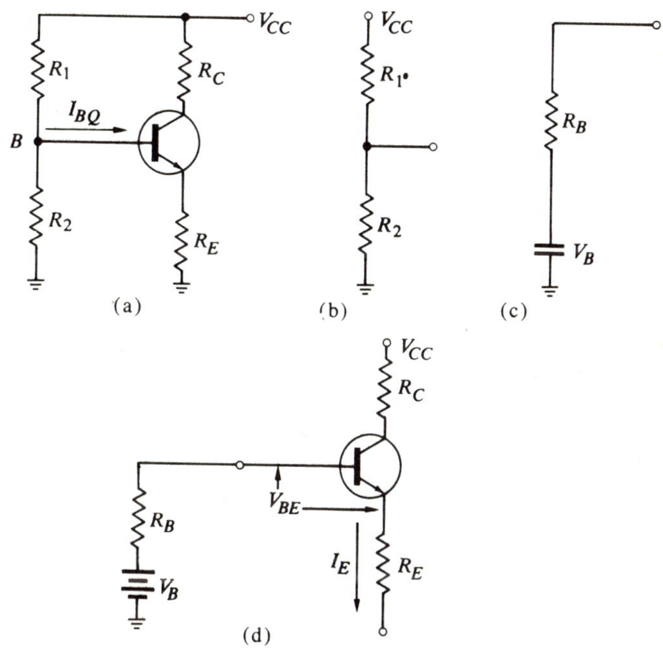

Fig. 2–38 (a) Common emitter amplifier, (b) voltage divider biasing network, (c) thévenized bias circuit, (c) common emitter amplifier with thévenized bias network.

germanium or 0.7 volts for silicon transistors. The emitter resistor, R_E, is usually chosen to be a small fraction of R_c. A good "rule of thumb" is $R_E \leq (R_C/10)$. There are still two unknowns in Eqs. (2.68) and (2.69), V_B and R_B. Our choice of R_B will be determined by the expected amount of variation in temperature of the environment in which the amplifier will be used, the variation in β of the transistors to be used, or variation in the bias supply. Analysis of the stability of the operating point indicates that the effects of variation in temperature β_1 and V_{CC}, will not be so critical if the following condition is met.

$$R_B \leq \frac{(\beta + 1) R_E}{10} \tag{2.71}$$

This value of R_B can now be substituted in Eq. (2.70) to solve for V_B. The values of R_1 and R_2 are determined from the solution of Eqs. (2.68) and (2.69).

$$R_1 = R_B \frac{V_{cc}}{V_B} \tag{2.72}$$

$$R_2 = \frac{R_B}{1 - \dfrac{V_B}{V_{cc}}} \tag{2.73}$$

We determined earlier (Eqs. 2.63 and 2.65) that the amplifier current gain and input impedance are both affected by R_B. It is the designer's function to adjust R_B to obtain maximum stability consistent with specified gain and impedance in the amplifier.

These examples will illustrate some of the approaches to transistor amplifier design.

Example 2-8. Use the 2N699B NPN silicon transistor in the amplifier of Fig. 2-39 to get 10 volts' peak ac output across the 10-kilohm load resistor.

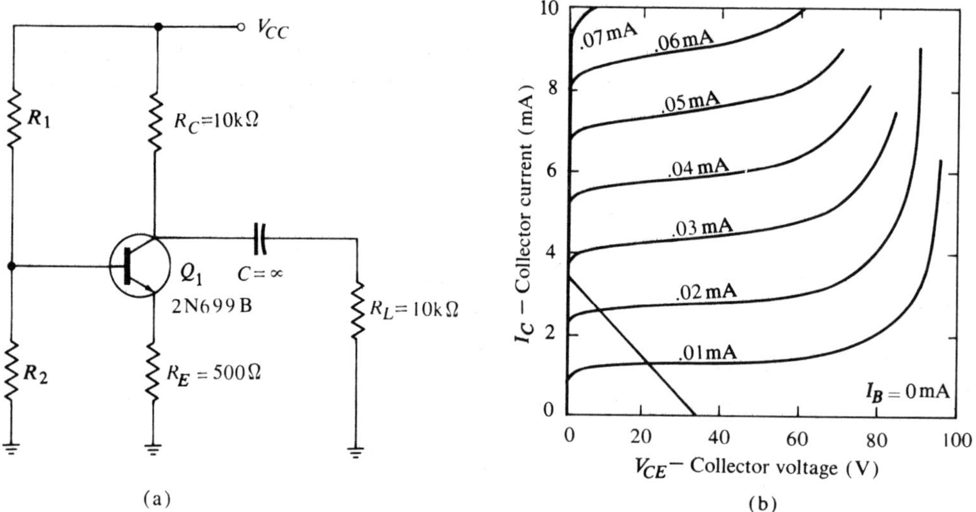

Fig. 2-39 (a) Transistor amplifier with capacitor coupled load, (b) output characteristics for 1Q.

First consideration must be given to the ac and dc load lines to determine where the Q point should be located. In Fig. 2-40, the collector current

$$I_{CEQ} = \frac{V_{CC} - V_{CEQ}}{R_C + R_E}$$

Also

$$I_{CEQ} = \frac{v_{ce}}{R_c \| R_L + R_E}$$

where v_{ce} is the collector-to-emitter voltage variation.

The quiescent collector-to-emitter voltage, V_{CEQ}, must be sufficiently large to prevent saturation distortion, or $V_{CEQ} > v_{ce} + V_{(sat)}$. The ac output voltage will depend on R_c and R_E as well as R_L, specifically

$$v_l = \frac{R_c \| R_L}{R_c \| R_L + R_E} v_{ce}$$

Putting in the actual values

$$v_l = \frac{10\text{ k}\Omega \| 10\text{ k}\Omega}{10\text{ k}\Omega \| 10\text{k}\Omega + 500} v_{ce} = \frac{5\text{ k}\Omega}{5.5\text{ k}\Omega} v_{ce}$$

or

$$v_{ce} \geq \frac{5.5\text{ k}\Omega}{5\text{ k}\Omega} v_l = 1.1 v_l = 11\text{ V}$$

Then

$$I_{CEQ} = \frac{11}{5\text{ k}\Omega + 500} = \frac{11}{5.5\text{ k}\Omega} = 2 \times 10^{-3}\text{ A}$$

Using $V_{CE(sat)} = 0.4$ V, $V_{CEQ} \geq 11.4$ V. Let $V_{CEQ} = 12$ V. Then

$$I_{CEQ} = 2\text{ mA} = \frac{V_{CC} - 12}{10\text{ k}\Omega + 500}$$

$$V_{CC} = 12 = (10.5\text{ k}\Omega)(2\text{ mA}) = 21\text{ V}$$

$$V_{CC} = 21 + 12 = 33\text{ V}$$

We have determined the important dc conditions based on performance requirements:

$$V_{CC} = 33\text{ V}$$

$$I_{CEQ} = 2.1\text{ mA}$$

$$V_{CEQ} = 12\text{ V}$$

Let's now go back to the output characteristics of Fig. 2–39(b) and draw the load lines. We see that $I_{BQ} = 0.018$ mA at the Q point,

$$\beta = \frac{I_{CQ}}{I_{BQ}} = \frac{2}{0.018} = 110$$

We will make $R_B = \dfrac{(\beta + 1)\, R_E}{10}$ for stability. Then $R_B = (111)50/10 = 5.5\text{ k}\Omega$, from Eq. (2.62)

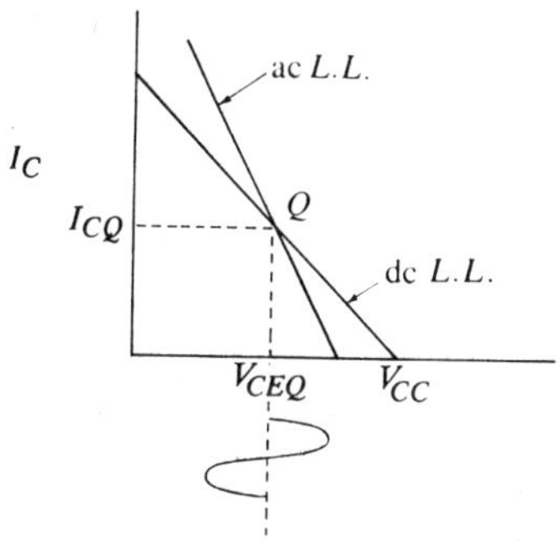

Fig. 2–40 Load line analysis of amplifier of Fig. 2–39.

$$V_B = 5.5 \times 10^3 \times 0.018 \times 10^{-3} + 0.7 + 2.018 \times 10^{-3} \times 500$$
$$= 1.81 \text{ V}$$

Then $$R_1 = R_B \frac{V_{CC}}{V_B} = 5.5 \times 10^3 \frac{33}{1.81} = 100 \text{ k}\Omega$$

$$R_2 = \frac{R_B}{1 - \dfrac{V_B}{V_{CC}}} = \frac{5.5 \times 10^3}{1 - 0.055} = 5.8 \text{ k}\Omega$$

The design is complete and we would now proceed to build a model and test the amplifier for performance and make any necessary adjustments.

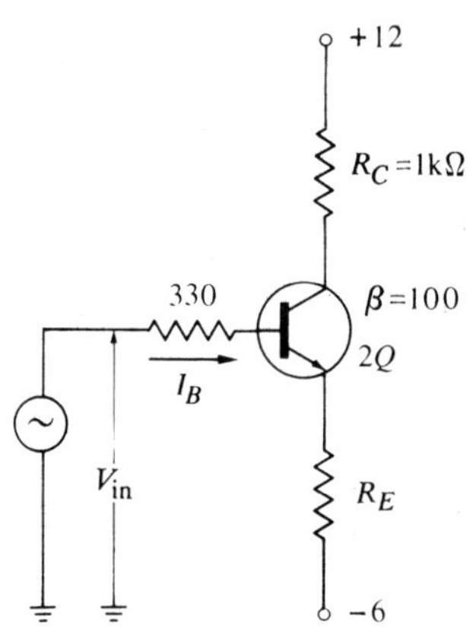

Fig. 2–41 Common emitter amplifier with dual supplies.

Example 2–9. Design the amplifier of Fig. 2–41 for an input impedance of 100 kilohms and determine the Q point.

Writing the summation of voltages in the input loop

$$V_{in} = I_b 330 + V_{be} + I_e R_E - 6$$
$$= I_b 330 + V_{be} + (\beta + 1) I_b R_E - 6$$
$$V_{in} = I_b [330 + (\beta + 1) R_E] + V_{be} - 6$$

or

$$V_{in} = I_{in} R_{in} + V_{be} - 6$$

where

$$R_{in} = 330 + (\beta + 1) R_E$$
$$100 \text{ k}\Omega = 330 + 101 R_E$$
$$R_E = \frac{100 \text{ k}\Omega}{101} \approx 1 \text{ k}\Omega$$

The Q point can be found by letting $V_{in} = 0$, then

$$I_{BQ} = \frac{6 - V_{BE}}{100 \text{ k}\Omega}$$
$$I_{BQ} = \frac{6 - 0.7}{100 \text{ k}\Omega} = 0.053 \text{ mA}$$
$$I_{CQ} = \beta I_{BQ} = 5.3 \text{ mA}$$

2–6 Field-Effect Transistor Circuits

FETs have the advantage over bipolar transistors when high-input impedance, low-noise small signal operation is required. There are three FET amplifier configurations analogous to the bipolar transistor amplifier configurations.

The common source amplifier of Fig. 2–42(a) is used most of the time. If the drain resistor is reduced to zero, the common drain, or source follower, of Fig. 2–42(b) is obtained. The source terminal can be used as the input resulting in the common gate configuration of Fig. 2–42(c). There is really only one FET amplifier. By changing the input and output connections, we form several configurations. We will deal with the common source FET amplifiers. The amplifiers of Fig. 2–42 used the N-channel JFET. The P-channel JFET or a MOSFET could very well have been used. Let's look at the JFET first.

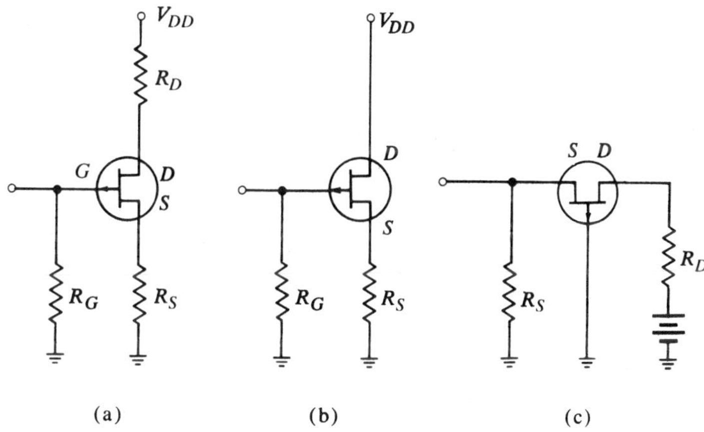

Fig. 2–42 JFET amplifiers: (a) common source, (b) common drain, (c) common gate.

The JFET is always biased in the depletion mode. Note that on the output characteristics of Fig. 2–43(a) V_{GS} is negative. Self biasing is usually provided in the common source amplifier by putting some resistance between the source and ground. Since there is negligible current flow in the gate lead, the gate is at ground potential, so $V_{GS} = -I_D R_S$. Writing the summation of dc voltages in the output circuit

$$V_{DD} = I_D(R_D + R_S) + V_{DS} \tag{2.74}$$

The dc biasing conditions can be determined from this equation. For example, let $V_{DD} = 20$ volts in the circuit of

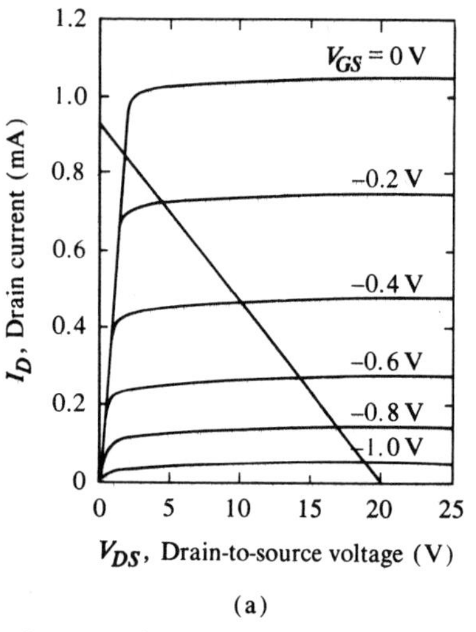

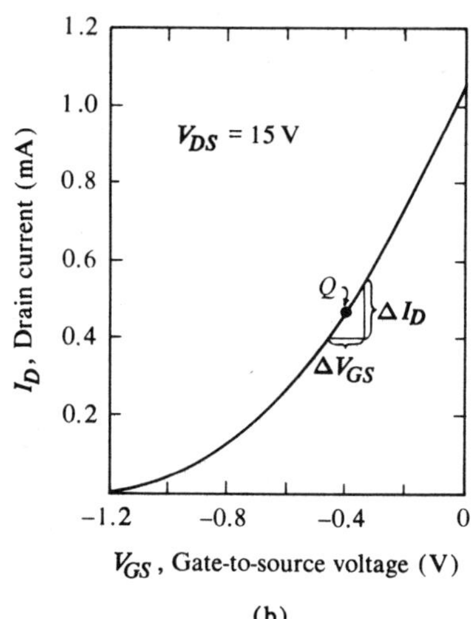

Courtesy of Motorola Inc.

Fig. 2–43 Commercial JFET characteristics: (a) output, (b) transfer.

Fig. 2–44. The FET is a 2N4220. It is desired to bias the amplifier at $V_{GS} = -0.4$

$$V_{DS} = 10 \text{ V}$$

I_{DQ} is read from Fig. 2–44(a) at 0.48 mA. Then $V_{GS} = -I_{DQ}R_S = -0.4$

$$R_S = \frac{0.4}{0.48 \times 10^{-3}} = 835 \ \Omega$$

From Eq. (2.74)

$$20 = 0.48 \times 10^{-3}(R_D + 835) + 10$$
$$R_D = 20 \text{ k}\Omega$$

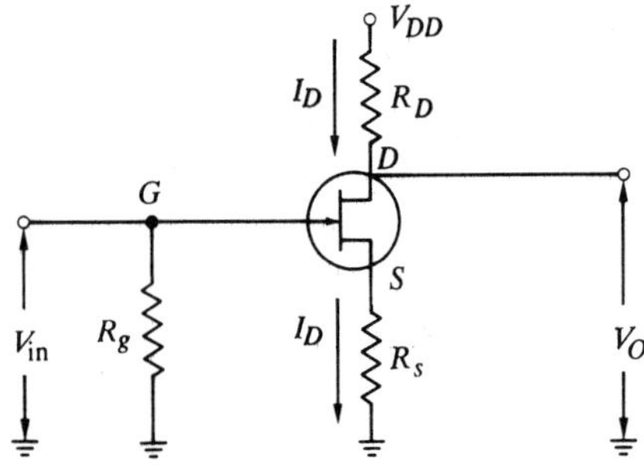

Fig. 2–44 Common source N-channel JFET amplifier.

The Q point is located on the transfer characteristics of Fig. 2–43(b). The slope of the transfer curve, $\Delta I_D/\Delta V_{GS}$, is called the forward transfer admittance, Y_{fs}. This important parameter is usually measured at $V_{GS} = 0$. Typical values range from 1000μS (micromhos) to 6000μS. At the Q point, $\Delta I_D = 0.15$, $\Delta V_{GS} = 0.11$, $Y_{fs} = 0.15/0.11 \times 10^{-3} = 1360\mu$S. At low frequencies where capacitance effects can be ignored $Y_{fs} = G_m$ or the "transconductance" of the FET. The voltage gain of the amplifier of Fig. 2–44 is given by the equation

$$A_v \approx \frac{g_m R_D}{1 + g_m R_S} \tag{2.75}$$

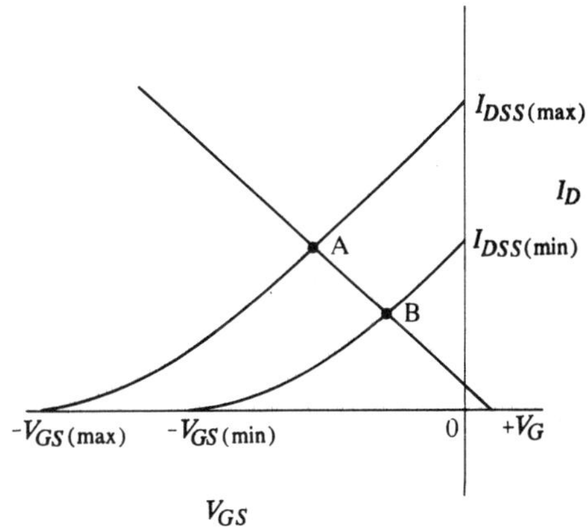

Fig. 2–45 JFET transfer characteristics and input load line.

FETs are subject to the same types of variation as bipolar transistors: manufacturing tolerances, temperature effects, and variation in bias supply. As you would expect, the designer must be concerned about the stability of the amplifier in spite of parameter variations. The transfer characteristics of Fig. 2–45 are typical of the range of variation you might expect for a given type JFET. Let's suppose that we would like to keep I_{DQ} between the points A and B. A line drawn through the two points specifies a source resistor with external gate bias. The biasing circuit of Fig. 2–46 will provide the external bias through the voltage divider network of R_1 and R_2.

The equation for the summation of dc voltages in the input loop is

$$V_G = V_{GS} + I_D R_S \tag{2.76}$$

or

$$I_D = \frac{V_G - V_{GS}}{R_S} = \frac{V_G}{R_S} - \frac{V_{GS}}{R_S} \tag{2.77}$$

This is the equation for the load line of Fig. 2–45 where $1/R_S$ is the slope of the line. Let's look at the design of a stable JFET amplifier. $V_{DD} = 20$ volts, I_{DQ} must be between 1.5 mA and 4.0 mA for the variation in transfer characteristics of Fig. 2–47. The biasing circuit of Fig. 2–46 will be used. Specify R_1, R_2, and R_s. From Fig. 2–47

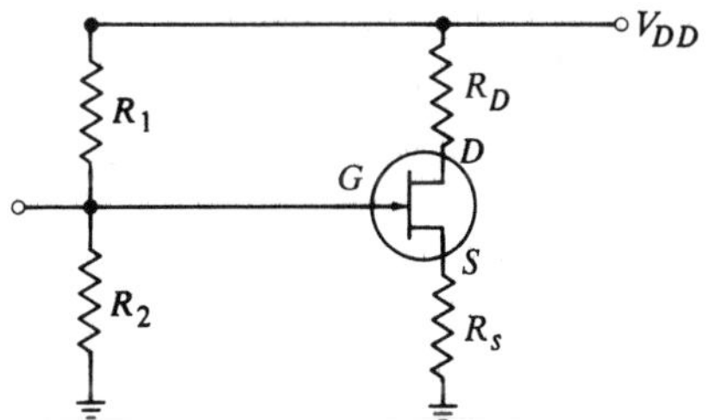

Fig. 2–46 Voltage divider biasing network for JFET amplifier.

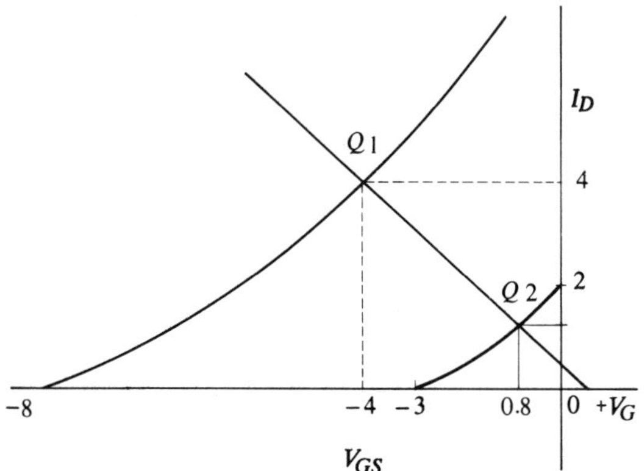

Fig. 2–47 Graphical analysis of JFET biasing.

$$\frac{10^3}{R_S} = \frac{4.0 - 1.5}{4.0 - 0.8}; \; R_S = 1.28 \text{ k}\Omega$$

and

$$-V_{GS_{Q1}} + V_G = -I_{DQ_1} R_S$$

or

$$\begin{aligned} V_G &= -I_{DQ_1} R_s + V_{GS_{Q1}} \\ &= 4.0 \times 10^{-3} \times 1.28 \times 10^3 \;\; + 4 \\ &= 1.1 \text{ V} \end{aligned}$$

Since there is no gate current flow, R_1 and R_2 must be selected to provide an open circuit voltage of 1.1 volts at the gate. Let $R_2 = 10 \text{ k}\Omega$, then

$$\frac{R_1}{10 \text{ k}\Omega} = \frac{18.9}{1.1}$$

$$R_1 = 172 \text{ k}\Omega$$

The design is complete except for R_D. R_D must be chosen from the output characteristics consistent with the Q point chosen.

The surface FET may be either the metal-oxide-silicon (MOS) type or the insulated gate (IG) type. The characteristics are so similar that we will concentrate on the MOSFET. Output and transfer characteristics for a typical N-channel MOSFET are shown in Fig. 2–48. The gate-to-source voltage may be negative (depletion node) or positive (enhance-

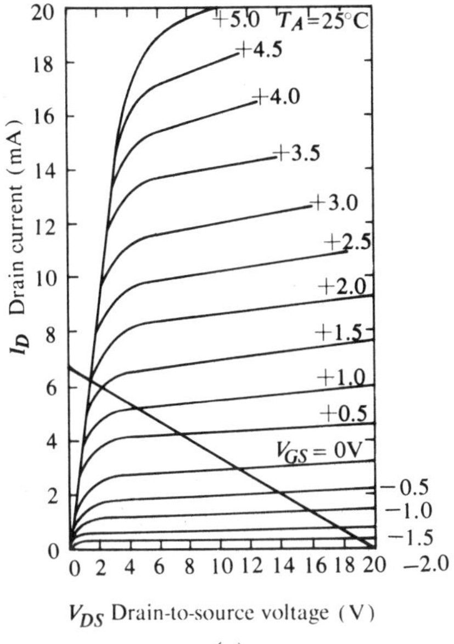

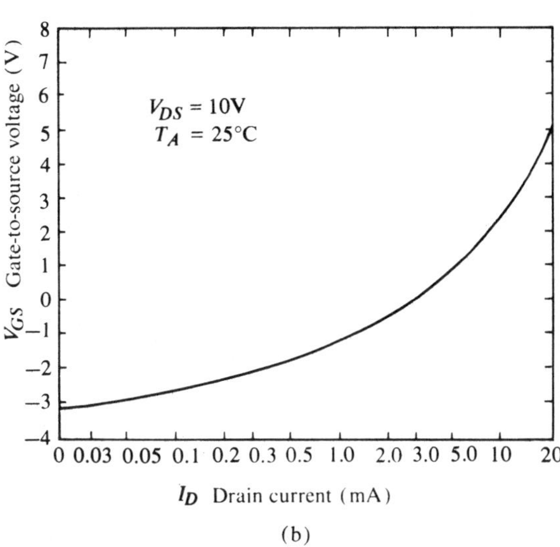

Courtesy of Motorola Inc.

Fig. 2–48 Characteristics for 2N3797 MOSFET: (a) output, (b) transfer.

ment node). Operation is usually in the enhancement node. The biasing circuit of Fig. 2–49 is often used. Let's look at an example of the design procedure.

Example 2–10. If V_{DD} = 20 V, R_D = 2 kΩ, and R_S = 1 kΩ in the circuit of Fig. 2–49, specify R_1 and R_2 to locate the Q point at V_{GS} = 1.0 V using the 2N3797 FET.

First the load line is drawn on the output characteristics of Fig. 2–48. The drain current at the Q point is 5.2 mA. The voltage dropped across R_S is then 5.2 mA × 1 kΩ = 5.2 V. Then

$$V_G = V_{GS} + V_{RS} = 1.0 + 5.2 = 6.2\text{ V}.$$

Let R_2 = 39 kΩ, then

$$\frac{6.2}{13.8} = \frac{39\text{ k}\Omega}{R_1}$$

$$R_1 = 87\text{ k}\Omega$$

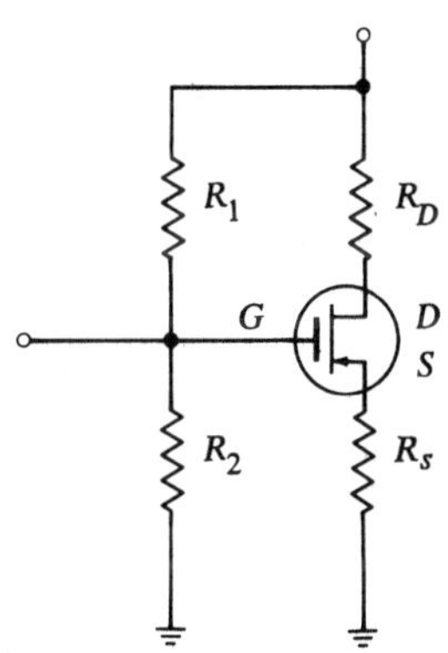

Fig. 2–49 Common source P-channel MOSFET amplifier.

The FET is being used at the front end of electronic devices where signal levels are low and very little noise can be introduced by the device.

2–7 Summary

We have discussed the design of semiconductor circuits using diodes, bipolar transistors, and FETs. Our design approach was based on the terminal characteristics and manufacturers' specifications. We realized that the final step must be to build a model of the circuit and test it for the required performance. In the next chapter, we will look at some commercial designs of semiconductor amplifiers in relay controls.

EXERCISES

1. When is a diode forward-biased?
2. Draw the circuit diagram of a series diode-resistor circuit with a dc battery connected for forward bias.
3. The V–I characteristics of a rectifier diode were measured in the lab. The data are shown below:

V_D (volts)	I_D (amps)
−2.00	2×10^{-6}
−1.00	2×10^{-6}
0.00	2×10^{-6}
0.05	2×10^{-6}
0.10	15×10^{-6}
0.20	100×10^{-6}
0.40	6×10^{-3}
0.60	32×10^{-3}
0.70	2
0.80	18
0.9	128

   Plot the V–I characteristics.

4. The diode of problem 3 is used in the circuit of Fig. 2–1. If $R_L = 50\ \Omega$ and $E = 5$ V, draw the dc load line.
5. In problem 4, $V(t) = 0.3 \sin \omega t$. What are I_D, i_d, and I_d?
6. The full-wave rectifier of Fig. 2–50 has an input voltage $V_{ac} = 30 \sin \omega t$. Assuming the load current is also sinusoidal, plot i_L if $R_L = 1$ kΩ and $1T$ is a center-tapped transformer. Ignore any voltage drop across $1D$ and $2D$ during conduction.
7. Specify the voltage and current ratings for $1D$ and $2D$ in problem 6.

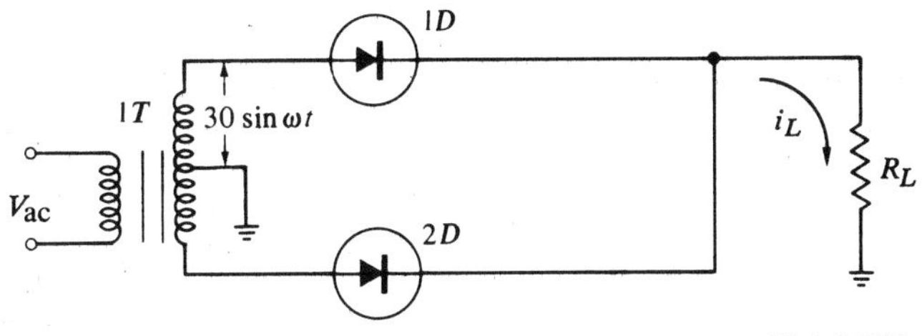

Fig. 2–50

8. The DRS-250 diode thermal resistance from junction to case is 0.22°C/W. The maximum junction temperature is 175°C. If the diode is dissipating 60 W, specify the maximum allowable case-to-air thermal resistance at 25°C ambient temperature.

9. If an insulating washer with a thermal resistance of 0.5°C/W is used between the diode and heat sink in problem 8, specify the maximum heat sink thermal resistance.

10. Specify the type and size heat sink you would use in problem 9 using the data of Fig. 2–12.

11. A rectifier diode is dissipating 25 W in a particular circuit application. The maximum allowable junction temperature is 175°C. The ambient temperature is 30°C. The thermal impedances are Θ_{JC} = 2.5°C/W and Θ_{CS} = 0.5°C/W.
    (a) Specify the minimum diode power rating if an infinite heat sink is used.
    (b) If Θ_{SA} = 2.5°C/W, at what temperature will the diode junction operate?

12. The load of Fig. 2–51 requires 100 mA to 200 mA. If E = 50 V, specify R_s, V_z, R_z, and P_z to maintain the voltage across the load between 29 and 31 V.

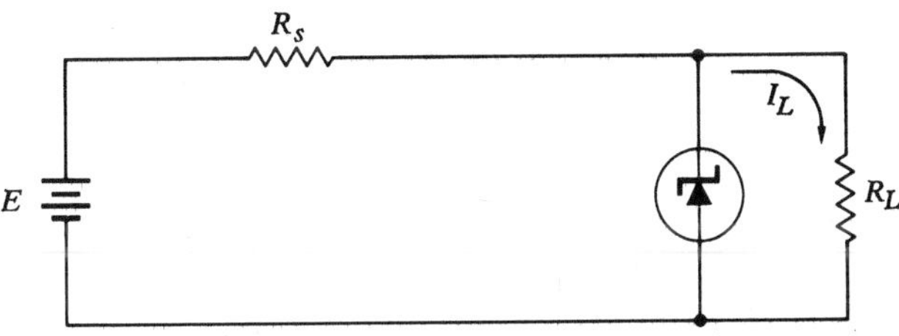

Fig. 2–51

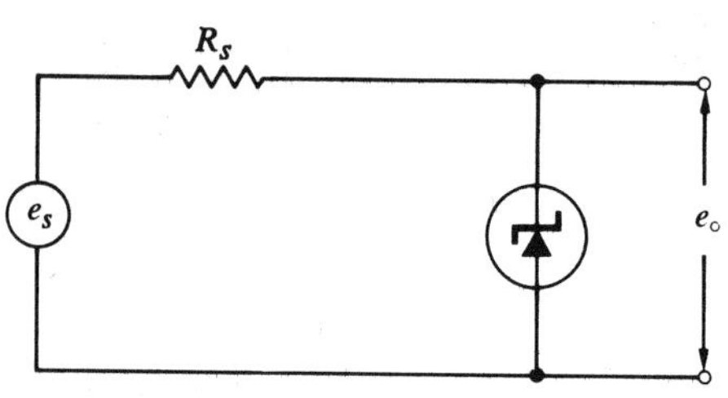

Fig. 2–52

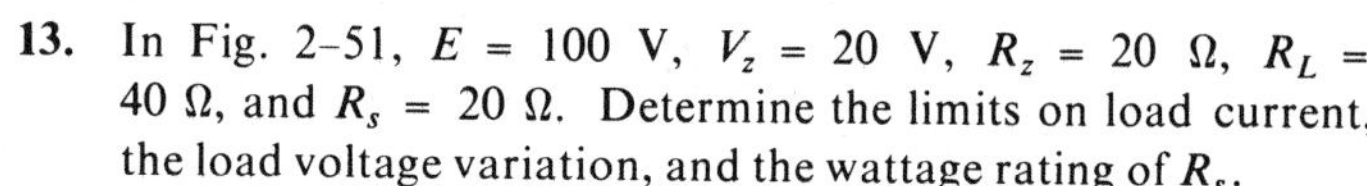

13. In Fig. 2–51, $E = 100$ V, $V_z = 20$ V, $R_z = 20\ \Omega$, $R_L = 40\ \Omega$, and $R_s = 20\ \Omega$. Determine the limits on load current, the load voltage variation, and the wattage rating of R_s.

14. The voltage source, E, in Fig. 2–51 varies between 80 V and 40 V. If $V_z = 20$ V, $R_z = 2.0\ \Omega$, $R_L = 500\ \Omega$, $R_s = 250\ \Omega$, determine the maximum load current when E is at maximum voltage. Determine the maximum load current when E is at minimum value.

15. A zener regulator is often used in ac circuits such as Fig. 2–52. If $e_s = 35 \sin \omega t$, $V_z = 14$ V, and $R_s = 1\ \text{k}\Omega$, sketch e_o.

16. Figure 2–53 shows the input rectangular pulse to an amplifier and the resulting output pulse.
    (a) What is the delay time?
    (b) What is the rise time?
    (c) What is the storage time?
    (d) What is the fall time?

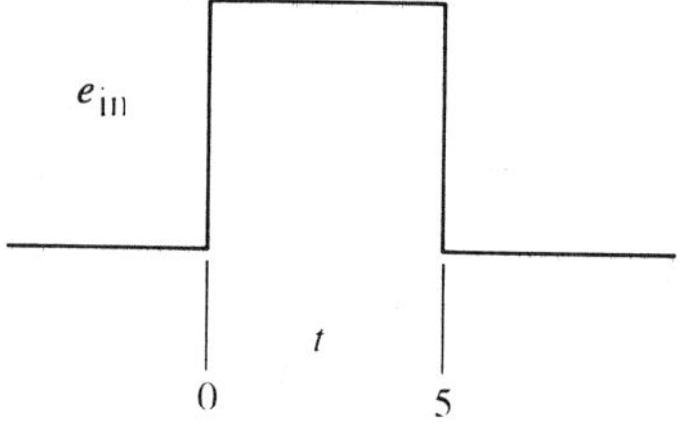

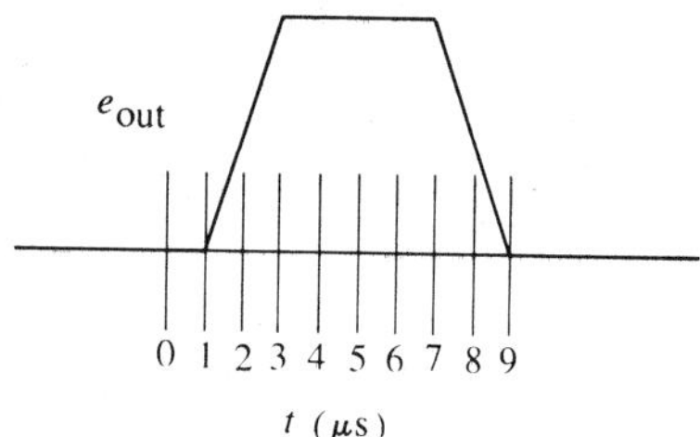

Fig. 2–53

17. Use the characteristics of Fig. 2–23 to design the switching amplifier of Fig. 2–54 to the following specifications:

$$V_{CC} = -60 \text{ V}$$
$$V_{BB} = +10 \text{ V}$$
$$V_{EB} = 1 \text{ V with } V_i = 0$$
$$I_L = 5 \text{ A in "on" state}$$
$$R_2 = 30\ \Omega$$
$$T_A = 80°\text{C}$$

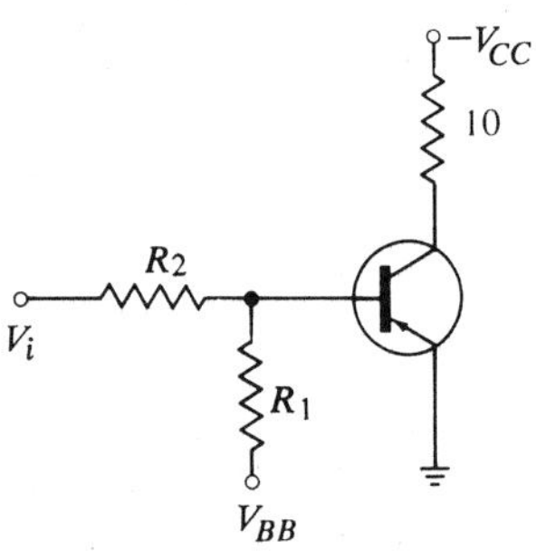

Fig. 2–54

18. Use the characteristics of Fig. 2–23 to design the switching amplifier of Fig. 2–54 to the following specifications:

$$V_{CC} = -80 \text{ V}$$
$$V_{BB} = +7 \text{ V}$$
$$V_{EB} = 2 \text{ V with } V_i = 0$$
$$I_L = 7.5 \text{ A in the "on" state}$$
$$R_1 = 150\ \Omega$$
$$T_A = 25°\text{C}$$

19. Determine the maximum value of R_E that will allow e_o to be 12 V peak-to-peak in Fig. 2–55. Assume $V_{ce\,(sat)} = 0.2$ V.

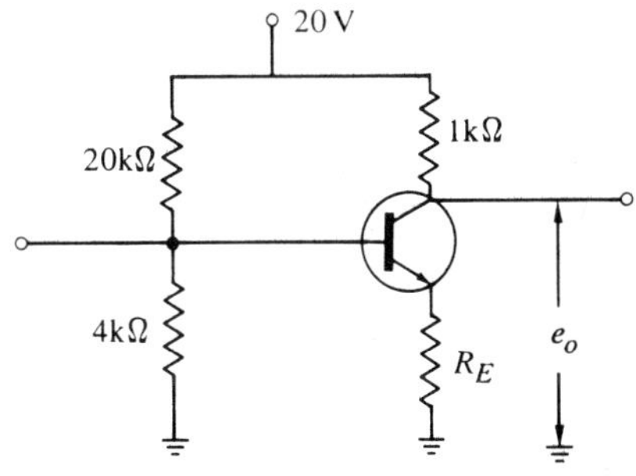

Fig. 2–55

20. Determine the maximum peak-to-peak ac output voltage of the amplifier in Fig. 2–56. Use the characteristics of the MPS 6518 in Appendix B.

21. In the amplifier of Fig. 2–56, $R_1 = 20$ kΩ, $R_2 = 8$ kΩ, and $\beta = 250$. Calculate I_{cQ}, V_{CEQ}, R_{be}, R_{in}, A_v, A_i.

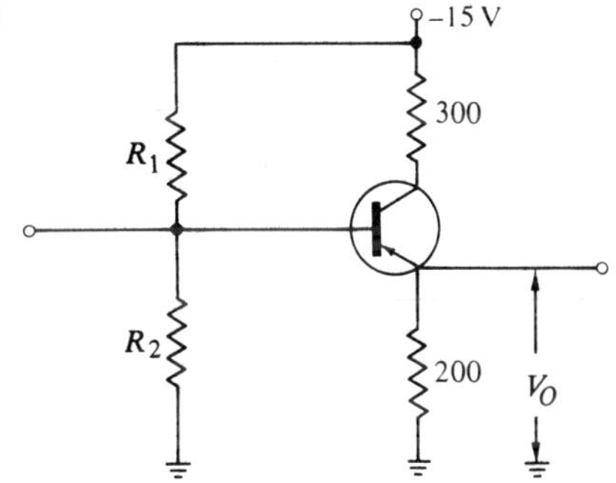

Fig. 2–56

22. In the amplifier of Fig. 2–57, $V_i = 10^{-2} \sin \omega t$. Use the bipolar transistor worksheet of Appendix B to determine:
    (a) R_B to bias at $V_{CEQ} = -6$ V
    (b) The peak-to-peak ac base current
    (c) The peak-to-peak ac current in the 300-Ω resistor

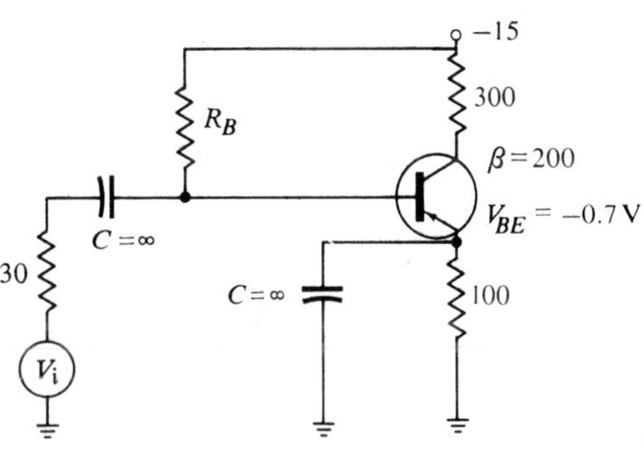

Fig. 2–57

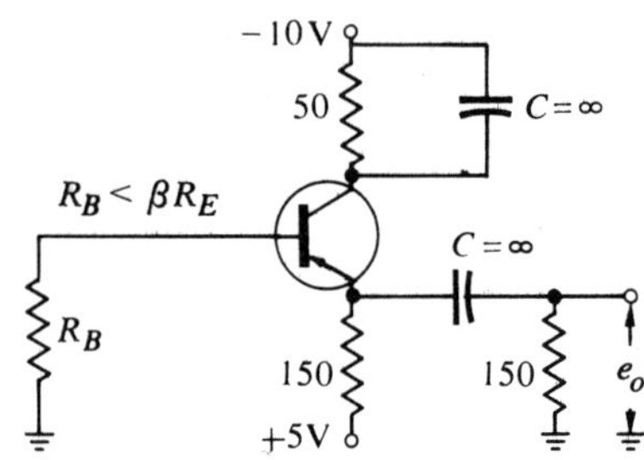

Fig. 2–58

23. Use the bipolar transistor worksheet of Appendix B to determine the Q point of the amplifier in Fig. 2–58. Determine the peak-to-peak ac current in the 150-Ω load resistor at maximum sinusoidal output.

24. What is the maximum sinusoidal output voltage of the amplifier of Fig. 2–59?

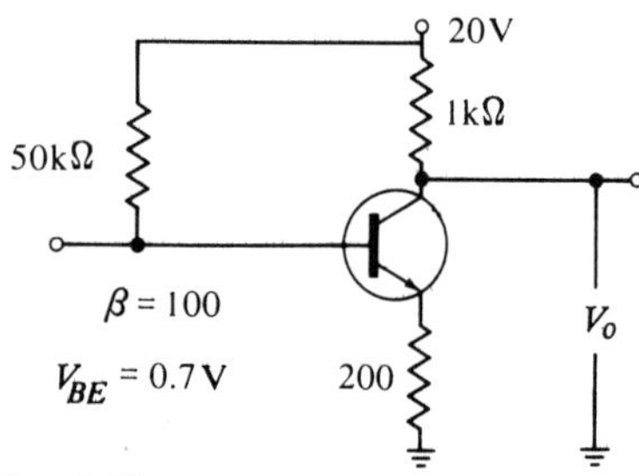

Fig. 2-59

25. Use the bipolar transistor worksheet of Appendix B to determine the values of R_1 and R_2 to bias the amplifier of Fig. 2–60 at $I_{CQ} = -20$ mA.

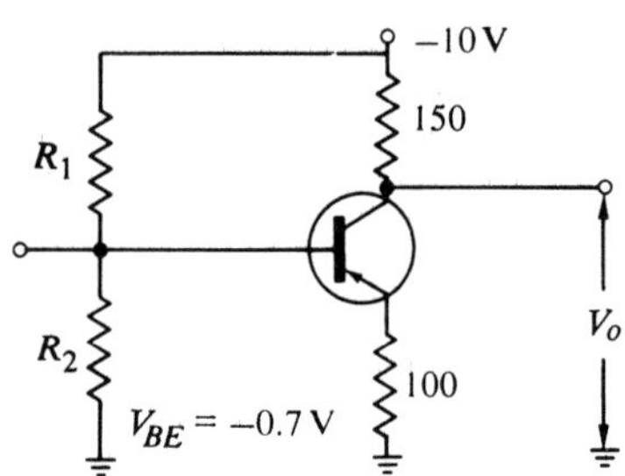

Fig. 2–60

26. The transistor used in Fig. 2–61 has an average $\beta = 100$.

(a) Determine I_{CQ} at $\beta = 50$.

(b) Find the percentage variation in I_c as β varies from 50 to 200.

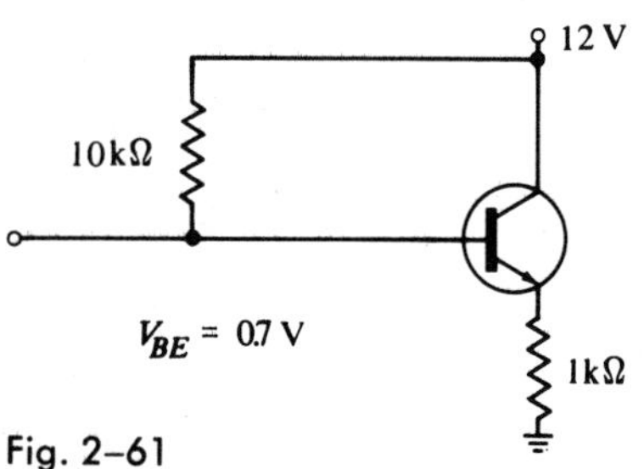

Fig. 2–61

27. Using the JFET characteristic worksheet of Appendix B, find R_1 and R_2 to bias the amplifier of Fig. 2–62 at $V_{DSQ} = 7.5$ V.

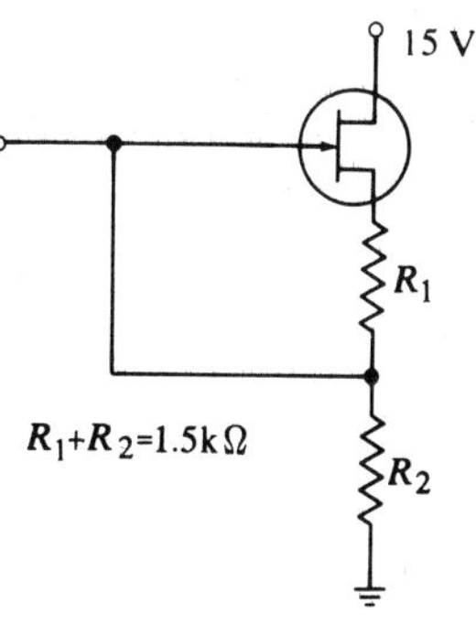

Fig. 2–62

28. Using the JFET characteristic worksheet of Appendix B, find R_1 and R_2 to bias the amplifier of Fig. 2–63 at

$$I_{DQ} = 3 \text{ mA}$$
$$V_{DQ} = 10 \text{ V}$$

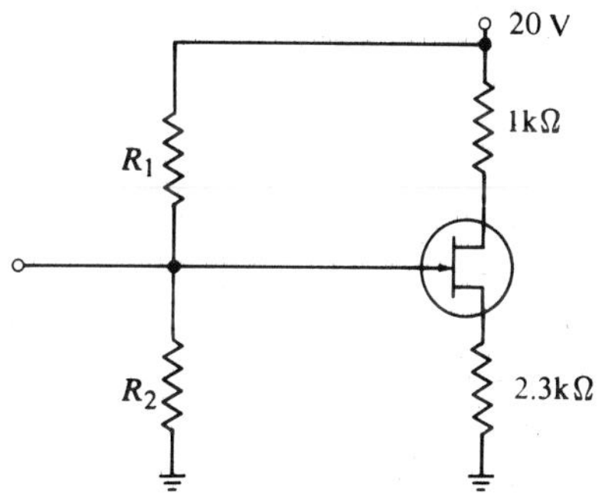

Fig. 2-63

3

Transistorized Industrial Control Relays

3–1 Introduction

One of the important applications of semiconductor devices in industry has been the sensing of nonelectrical variables such as light, heat, motion, material dimensions, polluting particles, or pressure.

Sensors are used to convert the nonelectrical variables to electrical outputs. You have probably seen some of these sensors such as photoelectric transistors, thermistors, strain gauges, thermocouples, and microphones used in many applications. The sensors do not have sufficient electrical output to handle loads such as relays, motors, solenoids, or audio speakers. Electronic amplifiers are used to boost the electrical output to a power level sufficient to handle load requirements. It is often more economical to use an electromechanical relay at the output of the amplifier and control the load current flow through the relay contacts. We will look at some commercial relay amplifiers as one illustration of the use of semiconductor devices in industrial electronics.

3–2 A Transistorized Photoelectric Control

Photoelectric controls are used in industry for automatic door openers, package sorting, counting, detection, part inspection, and identification. A commercial transistorized photoelectric control that can be used in these applications is the General Electric 3S7505 PG520. Figure 3–1 shows the control chassis. The simplified schematic is shown in Fig. 3–2. The light sensor for this control is a silicon photovoltaic cell, 1*PD*. The amount of current flow through the cell varies with the intensity of the light beam received. A preamplifier raises the signal to a level sufficient to drive the output amplifier. The control relay is energized by the output amplifier. The load is controlled by the normally open or normally closed high-current relay contacts. Let's see how the circuit operates with and without the light beam present.

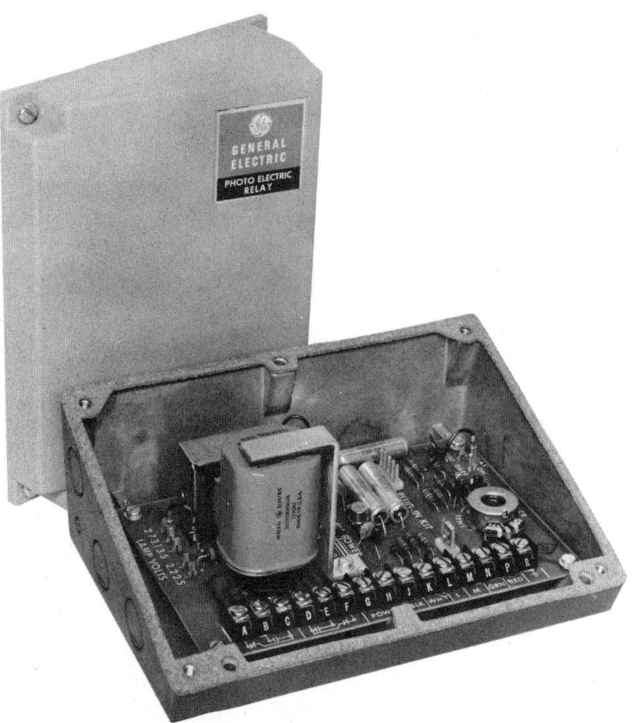

Courtesy of General Electric Company

Fig. 3–1 Commercial transistorized photoelectric relay control.

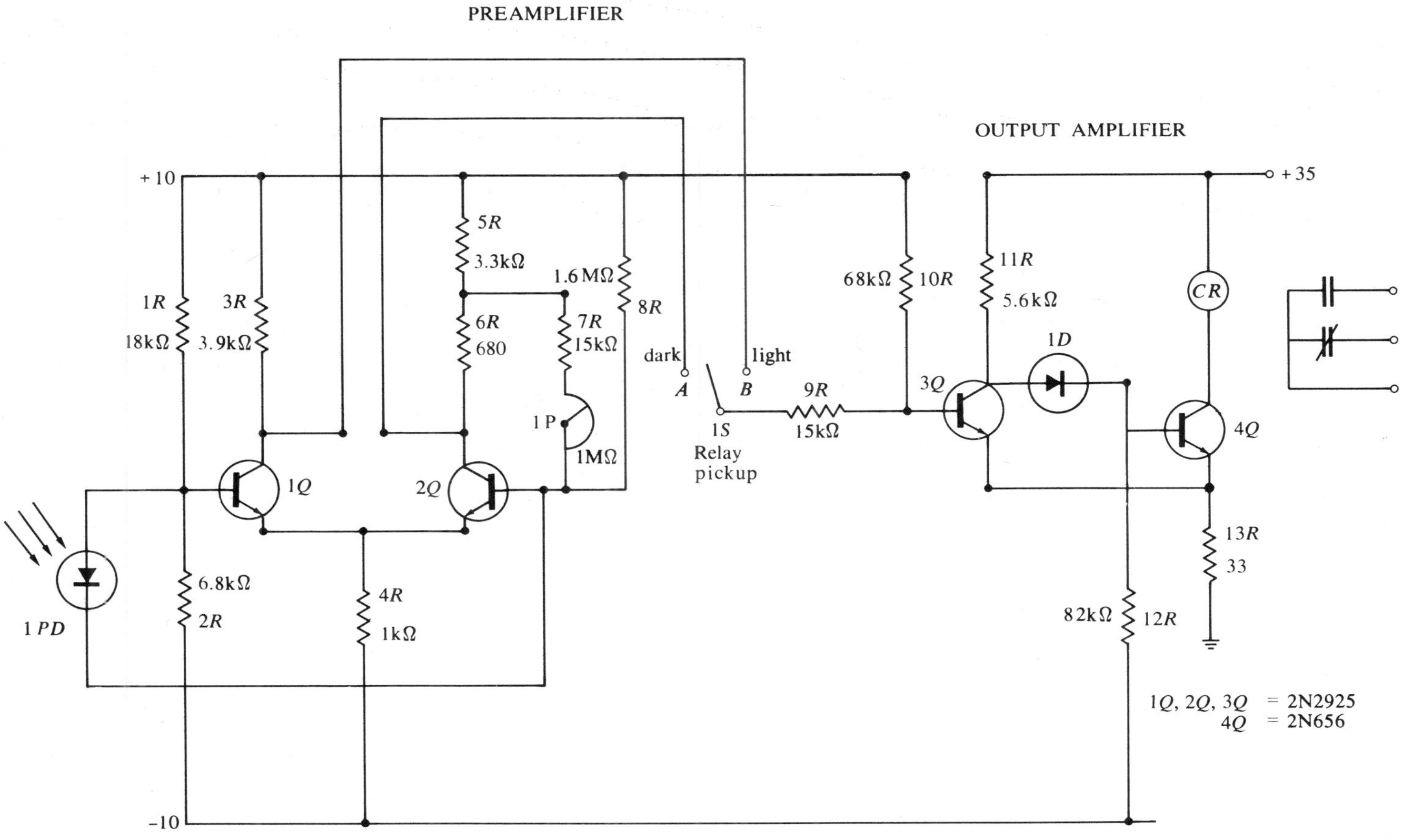

Courtesy of General Electric Company

Fig. 3–2 Typical elementary schematic for photoelectric amplifier.

Transistors $1Q$ and $2Q$ form a difference amplifier. The output voltage at the collector of $2Q$ is proportional to the difference between the voltage at the base of $1Q$ to ground and the voltage at the base of $2Q$ to ground. In equation form:

$$V_{C_2} = k(V_{B_1} - V_{B_2}) \tag{3.1}$$

and

$$V_{C_1} = k(V_{B_2} - V_{B_1}) \tag{3.2}$$

When the light beam is not present (dark), $2Q$ is biased on by current flow through $7R$, $8R$, and $1P$. Transistor $1Q$ is biased off by voltage divider $1R$, $2R$, and the voltage across $4R$ due to the emitter current of $2Q$. Under these bias conditions V_{B_2} is more positive than V_{B_1} so the voltage at the collector of $1Q$ is positive and the voltage at the collector of $2Q$ is negative with respect to ground. The voltage levels can be controlled by adjusting pot $1P$. Switch, S, allows us to choose whether the output relay will be energized or de-energized under "dark" conditions. Let's choose to put the switch in position B so that the relay is de-energized under "dark" conditions and will "pick-up" under "light" conditions. Resistor $10R$ is selected so that $3Q$ is based near cutoff with switch S open. When the switch is thrown into position B applying a positive voltage at its base, $3Q$ is driven into saturation. The current flowing through $11R$ is shunted to ground through $3Q$ driving $4Q$ to cutoff. Since $4Q$ is at cutoff, no current flows through the relay coil and the coil is de-energized.

When the light beam is focused onto the photo cell, the current through the photovoltaic cell increases and current flow into the base of $1Q$ is increased forcing $1Q$ into conduction. The base current flow into $2Q$ is reduced and $2Q$ is driven toward cutoff. The voltage V_{B_1} now becomes more positive than the voltage V_{B_2}. From Eqs. (3.1) and (3.2) the voltage at the collector of $2Q$ goes positive with respect to ground while the voltage at the collector of $1Q$ goes negative with respect to ground. Since we chose to put switch S in position B, the base of $3Q$ is driven below ground potential forcing $3Q$ into cutoff. Current flow through $11R$ now flows through $1D$ and the base of $4Q$, biasing $4Q$ near saturation. Current flow through the collector circuit of $4Q$ energizes the relay coil and the relay is picked up. The relay action would be just the opposite if we had chosen to put switch S in position A.

3–3 A LED Photoelectric Control

The photoelectric control of the previous section is designed to detect a continuous beam of light. The sensitivity of the device is affected by changes in ambient lighting. A new approach to photoelectric controls is the General Electric 3S7505 PS 800 LED Photoelectric Control. The light emitting diode (LED) is one of the miracles of modern semiconductor technology. This P-N junction diode emits light when current is passed through it. General Electric engineers have designed a transmitter using the LED as the transmitting device. The LED transmission is modulated and picked up by an ac receiver circuit so that ambient lighting conditions do not interfere with the photoelectric control. The LED photoelectric control is shown in Fig. 3-3. The simplified schematic is shown in Fig. 3–4. A free-running flip-flop (see Sec. 6-3-2 for a detailed discussion of the flip-flop) operating as an on-off switch applies a

Courtesy of General Electric Company

Fig. 3–3 Commercial LED photoelectric control.

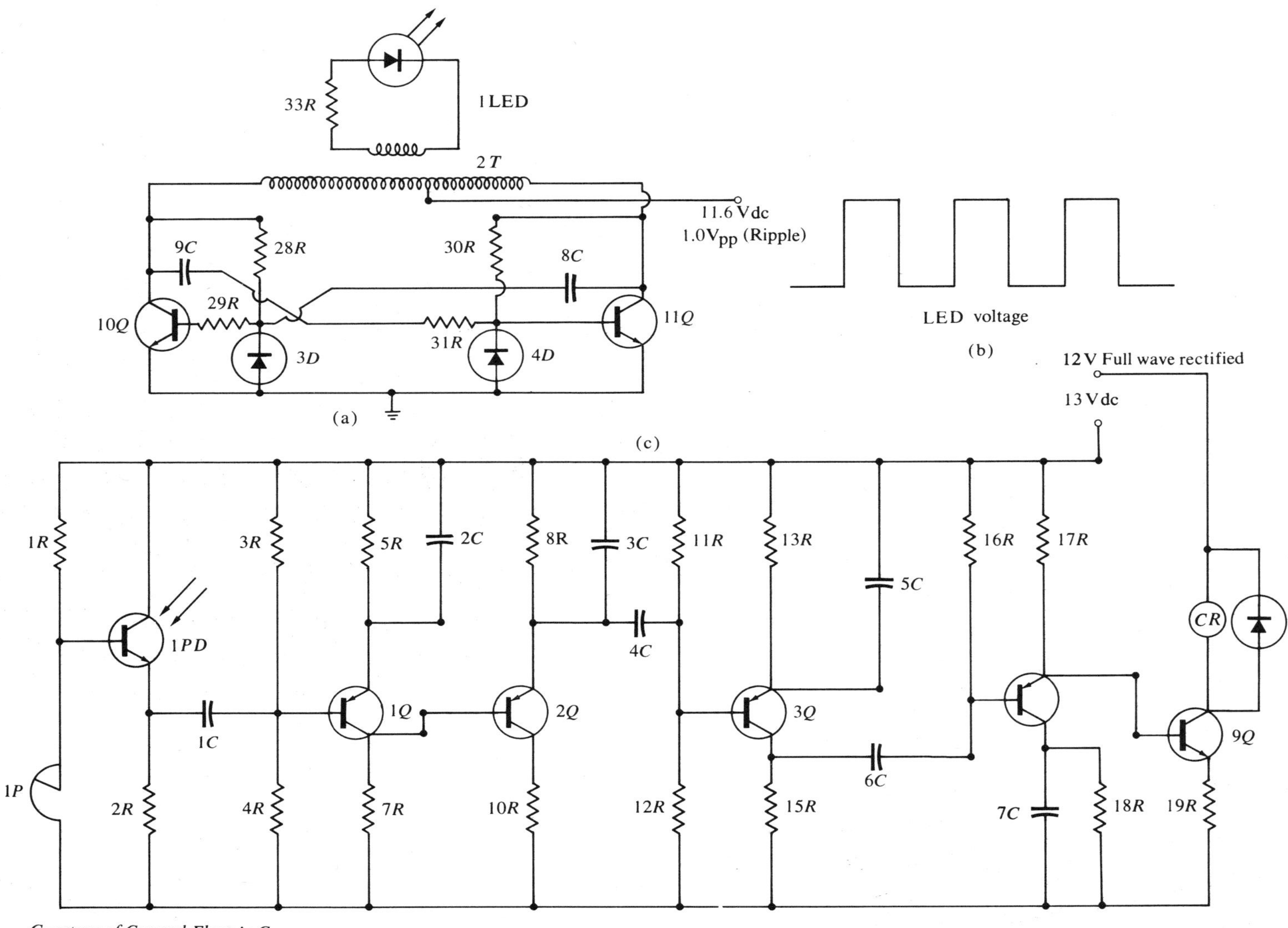

Courtesy of General Electric Company

Fig. 3–4 Typical LED photoelectric control: (a) transmitter, (b) LED voltage waveform, (c) simplified receiver.

chopped voltage to the LED. The chopper is shown in Fig. 3–4(a) and the resulting voltage at the secondary of $2T$ is shown in Fig. 3–4(b). The light emitted from the diode is transmitted directly or reflected back into the receiver shown in Fig. 3–4(c). A phototransistor is used to detect the chopped light beam. The phototransistor is a bipolar junction transistor (see Appendix A) whose output current is controlled by the intensity of light shining on the surface of the window. The sensitivity of the phototransistor amplifier is controlled by adjusting the pot, $1P$. The output of the phototransistor is passed through a multistage transistor amplifier. Bypass and coupling capacitors (which we will look at in the next chapter) are used to shape the frequency response of the transistor amplifier so that only the high-frequency chopped signal from the LED passes through to $9Q$ and energizes the control relay. We have looked at a simplified schematic of this photoelectric control. The actual circuit has several more amplifier stages and options that would not add to our understanding of the basic design.

3–4 A Resistance-Sensitive Relay

Sensing of a change in resistance can be an indication of a break in a piece of thread or wire, the presence or absence of a metallic object, or a change in the level of a liquid. The change in resistance can be used to operate a relay. The General Electric 3S7511 RS 570 of Fig. 3–5 is designed to convert the change in current due to a change in resistance into a relay action. The basic schematic is shown in Fig. 3–6. The external resistance to be monitored, R_X, is connected at terminals L and M. The circuit will detect a decrease in the value of the external resistor. Potentiometer $1P$ is adjusted so that the base of $1Q$ is more positive than its emitter. Current flow through $1P$, $5R$, and $8R$ drives $1Q$ into saturation. Current flow through $7R$ is shunted through the collector-to-emitter circuit of $1Q$. Since no current flows into the base of $2Q$, the collector current of $2Q$ is near zero and the relay is unenergized. When the value of R_X is decreased, some of the current flowing through $1P$ is shunted through R_X, thus reducing the biasing current flow through the base of $1Q$. Transistor $1Q$ is forced to cutoff. Current flow through $7R$ is now diverted through the base to emitter of $2Q$ driving $2Q$ into saturation. The collector

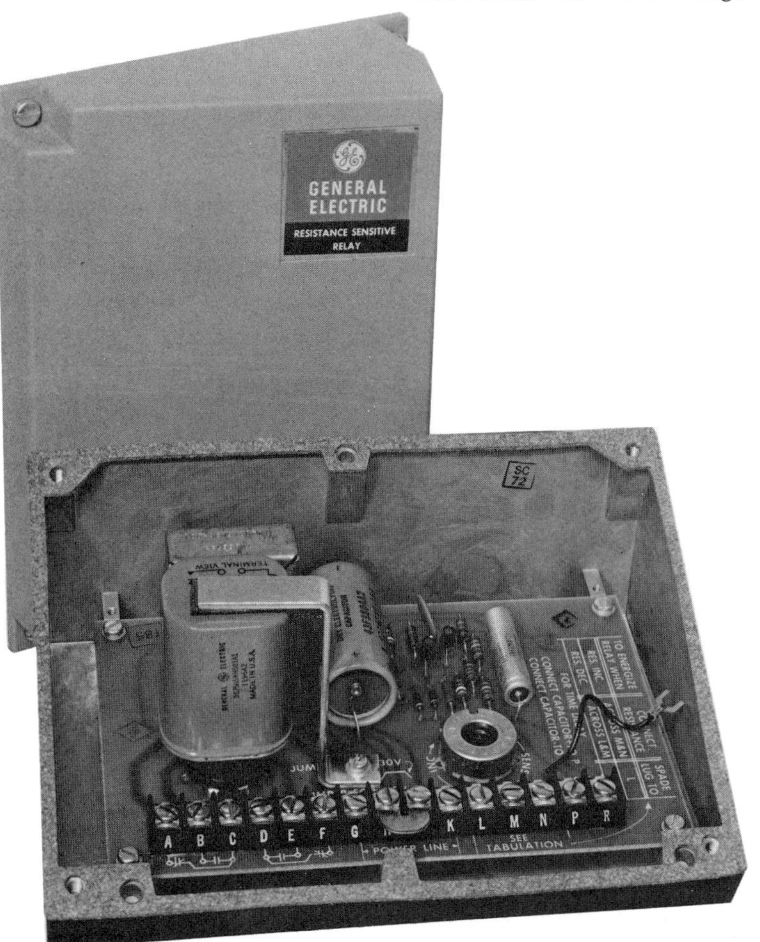

Courtesy of General Electric Company

Fig. 3–5 Commercial resistance-sensitive relay control.

current of $2Q$ energizes the relay coil opening the normally closed (NC) contact and closing the normally open (NO) contact. Diodes $2D$ and $3D$ along with capacitors $2C$ and $3C$ rectify and filter the secondary voltage of transformer $1T$ to provide the +20 and −20 V dc (see Sec. 12–2). The sensitivity pot, $1P$, is adjusted to match the relay amplifier to the expected change in resistance to be detected. The sensitivity pot and external resistor may be interchanged in the circuit to energize the relay when the external resistance, R_X, increases.

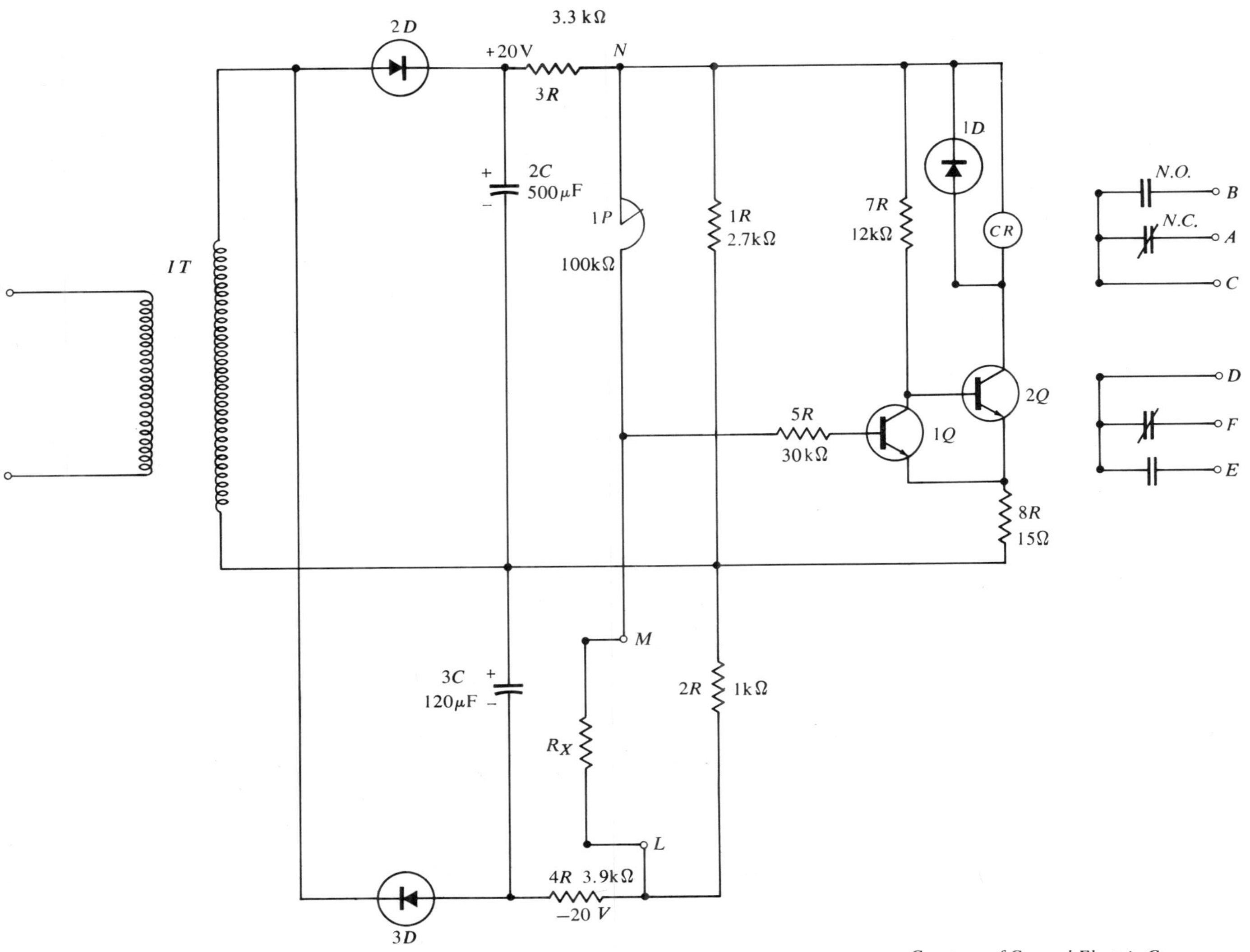

Courtesy of General Electric Company

Fig. 3–6 Simplified schematic of typical relay control.

Figure 3–7 shows a typical application of the resistance-sensitive relay as suggested by the manufacturer. It is necessary to pump liquid from the tank repeatedly as in the case of a sump pump. Motor maintenance will be reduced if the pump is not operated any more than necessary. The resistance-sensitive relay may be used as a two-point liquid level control to maintain the level between two preset points. The conductivity of the liquid as contrasted to the conductivity of the air in the tank will provide the necessary resistance change to energize the relay. The motor starter relay coil, M, is connected to the ac line through normally closed contacts of the output relay. In case of failure of

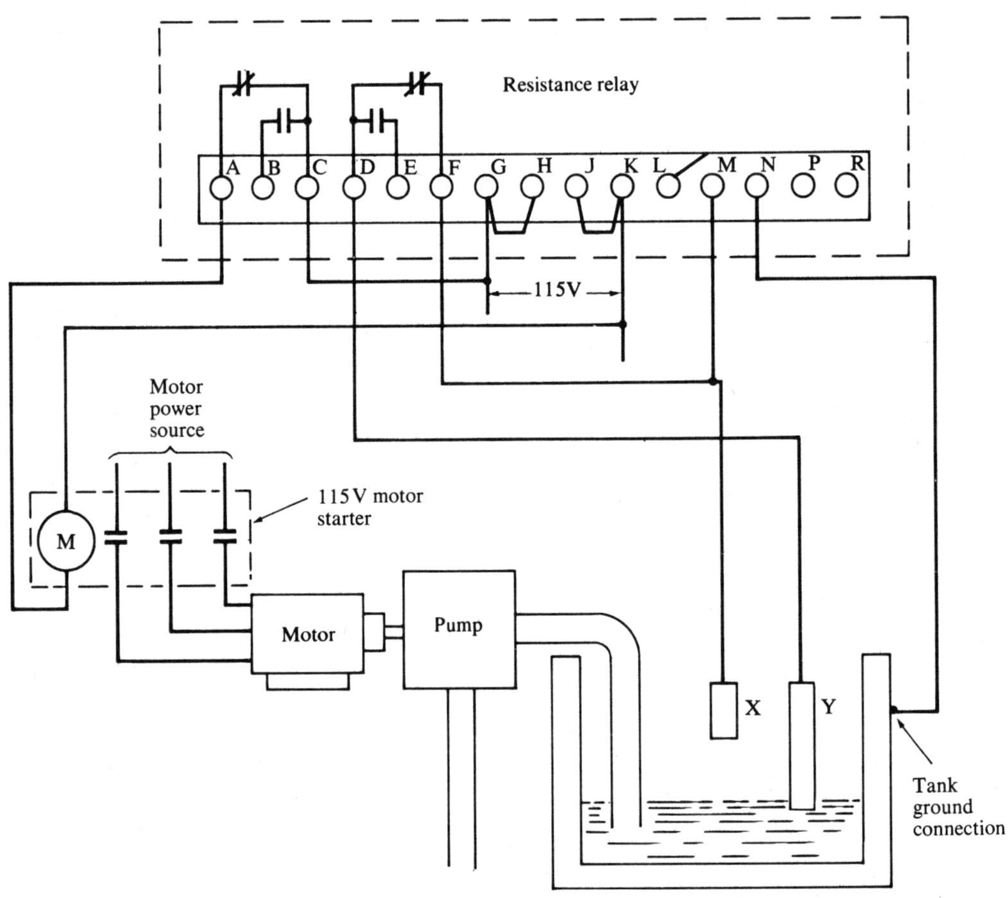

Courtesy of General Electric Company

Fig. 3–7 Relay used in two-point liquid-level control.

the resistance-sensitive relay, the pump will run continuously so that the tank will not overflow. The resistance-sensitive relay is wired to energize the relay on an increase in resistance. When the liquid level falls below the level of the metal probe *Y*, the high resistance between the tank and air is sensed at the *M* to *N* terminals of Fig. 3–6. The pot 1*P* is connected between terminals *L* and *M*. Sensing of the high resistance causes the output relay to be energized, shutting off the motor and removing the *Y* probe from the circuit. The relay action will now depend on the resistance between the tank and the metal probe *X*. When the liquid level reaches the *X* probe, a low resistance is sensed at the *M* to *N* terminals of the relay amplifier and the output relay is de-energized, turning on the pump motor to empty the tank. If the motor starter relay connection is moved from terminal *A* to terminal *B*, the same circuit will operate to keep the tank filled if the pump input and output connections are reversed. The pump will operate when the liquid level drops below the *Y* probe and shut off when the liquid level reaches the *X* probe.

3–5 Summary

The transistorized industrial control relays that we have looked at in this chapter represent the combination of electrical and mechanical components in simple control circuits. These electromechanical controls are sensitive enough to respond to transducer input signals and switch several amperes of current in the load circuit. Though simple and inexpensive, they are quite adequate for many applications.

EXERCISES

1. Why is an amplifier used with most sensors?
2. What is the advantage of using an electromechanical relay in transistor amplifier circuits?
3. List some applications of photoelectric controls.
4. With the photoelectric relay of Fig. 3–2 connected to pickup on light, what would happen if the photodiode were shunted with a 100-kilohm resistor? Explain why.

5. What would happen if the 15-kilohm resistor at the base of $3Q$ in Fig. 3–2 were connected to the -10-V supply instead of being connected to switch $S1$?

6. Why is the 35-V supply necessary in Fig. 3–2?

7. What is the use of potentiometer $1P$ in Fig. 3–2?

8. Why is the voltage applied to the LED of Fig. 3–4 chopped instead of continuous?

9. Should potentiometer $1P$ of Fig. 3–4 be rotated clockwise or counterclockwise to increase the sensitivity of the photoelectric relay? Explain why.

10. How would you modify the circuit of Fig. 3–6 to pick up the relay on an increase in the resistance of $R1$? Explain why your method would work.

11. Why is diode $1D$ placed across the relay coil in Fig. 3–6?

12. Redraw the liquid level control diagram of Fig. 3–7 to keep the tank filled rather than empty using the resistance-sensitive relay.

4

Passive Time Delay Elements

4–1 Introduction

A critical variable that the industrial electronics designer is trying to control is timing. Timing most often involves fixed amounts of delay between the occurrence of events in a circuit or system. Capacitors and inductors are the passive elements used to achieve time delays. In this chapter, we will become familiar with the behavior of inductors and capacitors in circuits and look at some applications in industrial electronics.

4–2 Energy Storage in Capacitors

One of the most important elements in the design of industrial electronic circuits is the capacitor. Figure 4–1 is an example of capacitors that you may have seen or will see frequently in the future. Capacitors are constructed in many different ways. Some factors that influence the construction are the value of capacitance required, temperature rating, voltage rating, cost, frequency of operation, physical limitations, or reliability requirements.

Fig. 4–1 Industrial capacitors.

Capacitors are used in industrial electronic circuits for temporary storage of electrical energy. The energy is stored in the form of electrical charge on the conductors of the capacitor. Figures 4–2 to 4–4 help in understanding the concept of charge on capacitors. If a brand new parallel plate capacitor were taken from the shelf and connected in the circuit of Fig. 4–2, no current would flow in the circuit. That is because metallic conductors in their normal state are electrically neutral; there is as much positive charge as there is negative charge so that they cancel each other. (See Appendix A for a more complete discussion of the electrical properties of metals.) If a battery of voltage V is now connected in the circuit by closing switch S in Fig. 4–3, the positive side of the battery will draw electrons from the conductor on the right. Electrons carry negative charges away from the conductor so that the number of positive charges is now greater than the number of negative charges on the conductor. The conductor on the right is now positively charged. A common symbol for electrical charge is Q. A net positive charge is indicated by Q_+. The electrons leaving the conductor on the right flow through the battery and resistor until they reach the conductor on the left. Electrons coming to the conductor on the left bring negative charges so that there are now more negative charges than positive charges on the conductor on the left. The conductor on the left is now negatively charged. While electrons are flowing from the plate on the right to the plate on the left to charge the capacitor, the current will flow from left to right in the circuit. This is consistent with the definition of *conventional current flow* which we will use throughout this text. Current flow in the circuit will stop when the charge on the capacitor reaches its final value. We can now open switch S and the capacitor will maintain its charge until we are ready to use it. Let's look at the amount of energy stored in the capacitor and how we determine how much charge it holds. Given the voltage of the battery, V, and the value of capacitance, C, the energy stored in the capacitor can be calculated from Eq. (4.1)

$$W = \frac{1}{2} CV^2 \qquad \textbf{(4.1)}$$

where

W = energy in joules
C = capacitance in farads
V = voltage in volts

The charge stored in the capacitor can be calculated from the equation:

$$Q = CV \qquad \textbf{(4.2)}$$

where

Q = charge in coulombs
V = voltage in volts
C = capacitance in farads

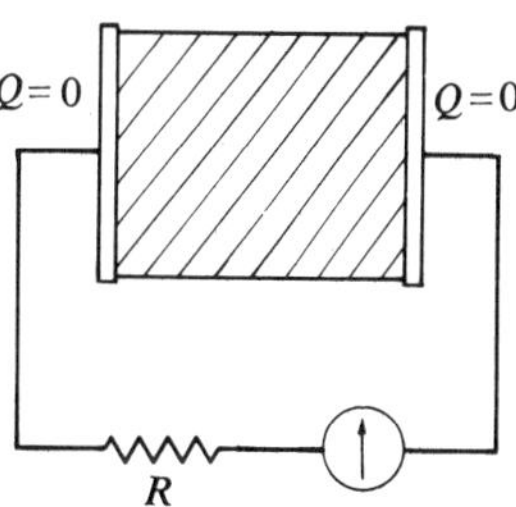

Fig. 4–2 Uncharged capacitor (no electron flow).

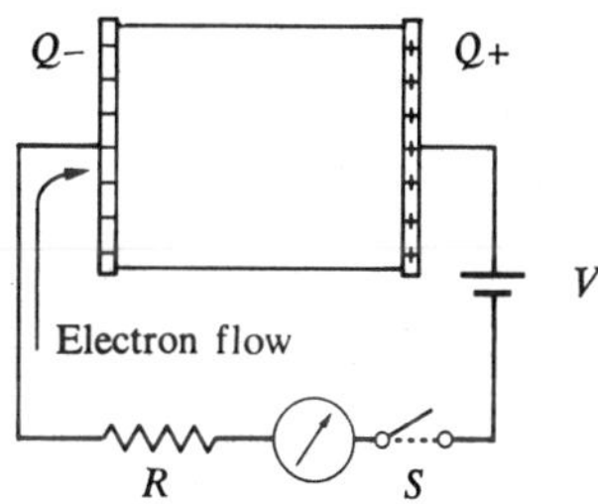

Fig. 4–3 A charged capacitor.

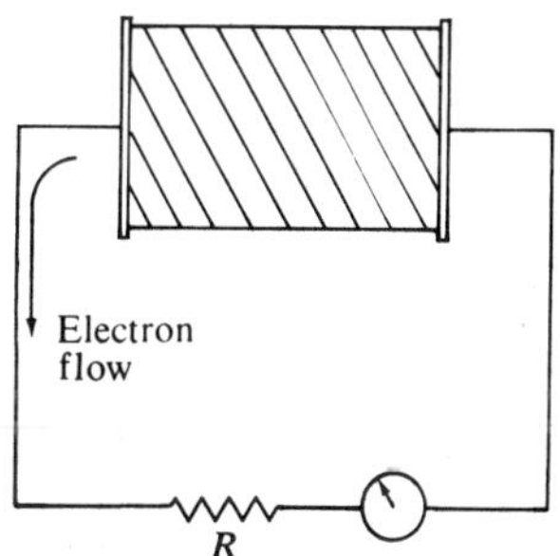

Fig. 4–4 A capacitor discharging.

Equation (4.2) is sometimes rearranged as $C = Q/V$, and used to define capacitance as charge per volt.

The charge can be removed from the capacitor by allowing the capacitor to deliver its stored energy to resistor R in Fig. 4–4. Notice that the battery has been removed from the circuit so that the excess electrons stored on the left plate are allowed to flow through the resistor and ammeter back to the right plate. These electrons will cancel the net positive charge on the right plate, restoring the circuit to its "initial condition" of Fig. 4–2. While electrons are flowing from the left plate to the right plate, "conventional current" flows from right to left. This is exactly opposite the current flow for "charging the capacitor" and is referred to as "discharging." When the net charge on both plates is zero, the current flow ceases.

Let's look at an example of the design of a simple capacitive circuit for charge storage.

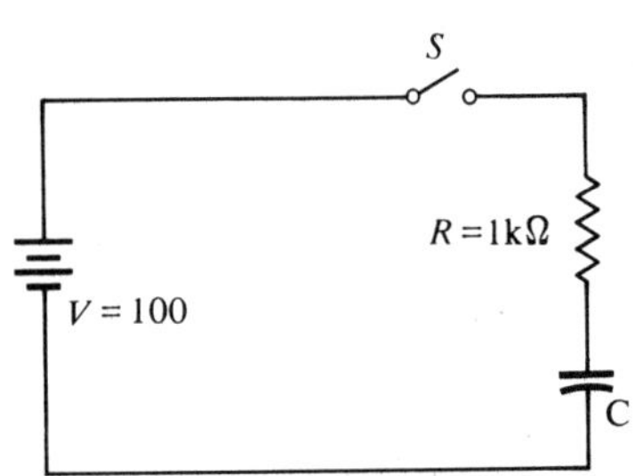

Fig. 4–5 RC series circuit.

Example 4–1. It is required to store 20×10^{-6} coulombs on capacitor C when switch S is closed in the circuit of Fig. 4–5. Specify the capacitor and calculate the energy stored after charging. Using Eq. (4.2)

$$Q = CV$$

$$20 \times 10^{-6} = C \times 100$$

$$C = \frac{20 \times 10^{-6}}{100} = 0.2 \times 10^{-6} = 0.2\ \mu\text{F}$$

$$\text{Working voltage*} = 100\ \text{V}$$

A 0.2-μF capacitor with a working voltage rating of at least 100 V should be specified. Using Eq. (4.2), $W = 1/2\ CV^2$

$$\begin{aligned} W &= (\tfrac{1}{2})(0.2 \times 10^{-6})(100)^2 \\ &= (\tfrac{1}{2})(0.2 \times 0.0^{6})(10^{4}) \\ &= (\tfrac{1}{2})(0.2 \times 10^{-2})(0.1 \times 10^{-2}) \\ &= 0.001\ \text{J} \end{aligned}$$

4–3 Capacitors in DC Circuits

We have discussed the concept of energy storage in a capacitor and the method of charge and discharge associated with it. In order to establish the concept firmly in

*Sum of dc-peak ac voltage seen by the capacitor.

our minds, we simplified the discussion by emphasizing the steady-state conditions. The steady-state conditions were in Fig. 4–2 with no charge on the conductors, in Fig. 4–3 after current flow had stopped and both conductors were charged, and in Fig. 4–4 after current flow had stopped and charge on both conductors was zero. The currents flowing in Figs. 4–3 and 4–4 are commonly called the *transient currents*. Transient means that they are only temporary and will cease to exist within a reasonable time. This current flow and the voltage associated with it are quite predictable and can be described mathematically. Many industrial control applications of the capacitors make use of the predictable transient current. Let's take a closer look at charge and discharge cycles of capacitors in dc circuits.

Consider the circuit of Fig. 4–6(a) with no charge on capacitor C and switch S in the open position. This is the *initial condition* or "starting condition" of the circuit.

In Fig. 4–6(b), switch S is moved to position 1 at time $t = 0$. It is common practice to start observing transient currents or voltages at $t = 0$. This establishes a reference for discussing behavior of the circuit any time after the switch is closed. As soon as the switch is put in position 1, current starts to flow in the circuit charging capacitor C. We will discuss the transient voltages and currents related to charge transfer in the capacitive circuit.

The mathematical relationships between charge, voltage, and current for a capacitor are:

$$Q = \int_0^t i(t)\,dt \tag{4.3}$$

$$V = Q/C = 1/C \int_0^t i(t)\,dt \tag{4.4}$$

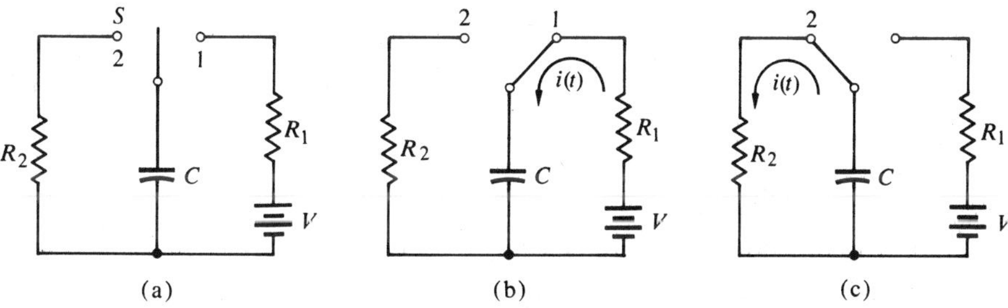

Fig. 4–6 DC resistor capacitor circuit with one voltage source.

Figure 4–7 is a plot of the transient current flow in the circuit of Fig. 4–6(b). Maximum current flows at the instant the switch is put in position 1. The value of the maximum current is V/R_1. Current flow decreases as time passes. At time $t = R_1C$ (ohms × farads = seconds), the current is down to 37% of its maximum value; at $t = 2R_1C$, the current is down to 14% of its maximum value; at $t = 3R_1C$, the current is down to 5% of its maximum value. It will take an infinite length of time for the current to actually decrease to zero, but after $t = 5RC$ the amount of current flow is usually negligible. The product $R_1 \times C$ is called the *time constant* of the circuit. The Greek letter tau (τ) is commonly used, or $\tau_1 = R_1C$. If R_1 is in megohms and C is in microfarads, then τ_1 will be in seconds.

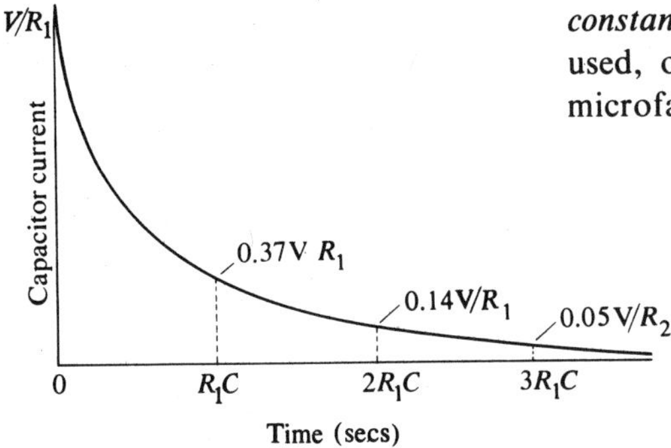

Fig. 4–7 Transient current in RC circuit while charging.

Figure 4–8 is a plot of the transient voltage on the capacitor C in the circuit of Fig. 4.6(b). The capacitor voltage starts at 0 when the switch is put in position 1. Capacitor voltage increases with increasing time. The maximum value of capacitor voltage is the battery voltage V. At $t = R_1C$, capacitor voltage is at 63% of maximum value; at $t = 2R_1C$, capacitor voltage is at 86% of maximum value; at $t = 3R_1C$, capacitor voltage is at 95% of maximum value. It will take

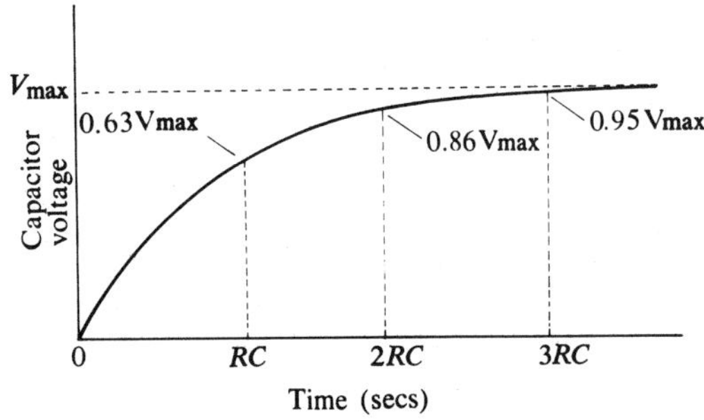

Fig. 4–8 Capacitor voltage vs time.

an infinite length of time for capacitor voltage to reach its maximum value V, but after $t = 5RC$, the actual value is usually close enough to the maximum value that the difference is negligible.

Figure 4–9 is a combined plot of i/I_{max} and V_c/V vs t/τ for values of t between 0 and 5τ. This plot will be very handy for more exact analysis of transient conditions in an RC circuit as we shall see later in this chapter.

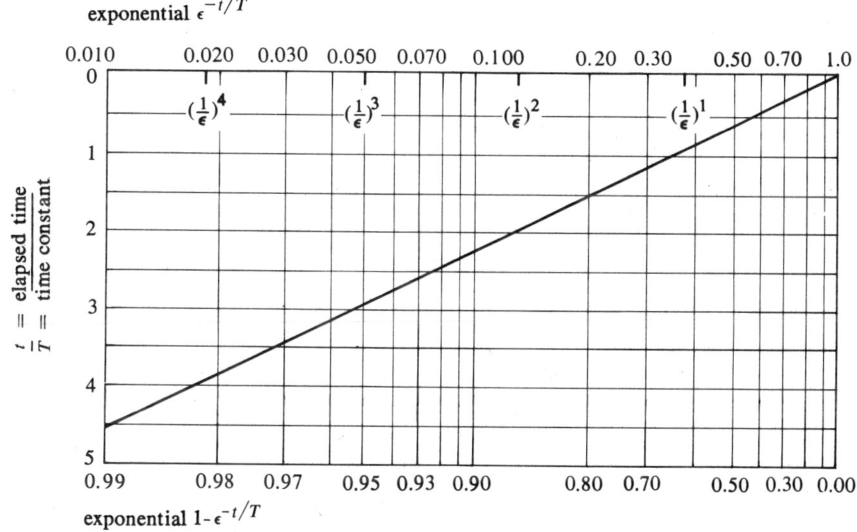

Fig. 4–9 Exponential functions $\epsilon^{-t/T}$ and $1 - \epsilon^{-t/T}$ applied to transient in *R-C* and *L-R* circuits. (From "Reference Data for Radio Engineers," I.T.T., 4th Ed., p. 153.)

The reader with an advanced mathematical background will be interested in the more rigorous mathematical treatment of the transient circuit analysis. Let's examine a brief treatment of the circuit in Fig. 4–6(b) immediately after the switch is put into position 1. Kirchhoff's voltage law requires that the sum of voltages in the closed loop must be zero, or

$$R_1 i(t) + V_c(t) - V = 0 \tag{4.5}$$

$$V = R_1 i(t) + V_c(t) \tag{4.6}$$

Substituting Eq. (4.4) into Eq. (4.6) we get

$$V = R_1 i(t) + 1/C \int_0^t i(t)\,dt \tag{4.7}$$

The solution of Eq. (4.7), using the initial conditions already stated, is:

$$i(t) = \frac{V}{R}\exp\left(\frac{-t}{R_1 C}\right) \tag{4.8}$$

Substituting Eq. (4.8) into Eq. (4.6) we get

$$V_c(t) = V - V\exp\left(\frac{-t}{R_1 C}\right) = V\left(1 - \exp\frac{-t}{R_1 C}\right) \tag{4.9}$$

The term $\exp(-t/R_1 C)$ is called an exponential with the exponent being $-t/R_1 C$. The form of the exponential is e^X Values of e^X for both positive and negative values of X can be found in Table 1 of Appendix C. Turn to Table 1 and notice the value of e^{-X} when $X = 0$. You will find that value is 1. Then look at $t = 0$ in Eq. (4.8). If $t = 0$, then $\exp(-t/R_1 C) = \exp(0) = 1$; then $i(t) = (V/R_1)(1) = V/R_1$. This is the same value we read from the plot of Fig. 4–7. Figure 4–7 is a graphical representation of Eq. (4.8). We could use either method to find the value of the transient current in Fig. 4–6(b).

Let's look at an example to which we can apply some of our knowledge about the transient behavior of capacitive circuits.

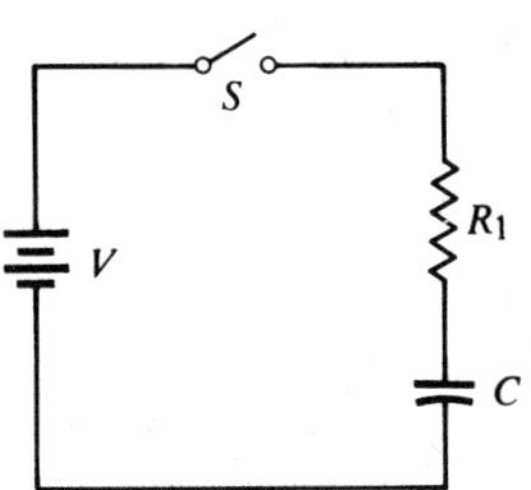

Fig. 4–10 Single-loop RC circuit.

Example 4–2. The circuit of Fig. 4–10 is used in the laboratory to demonstrate transient conditions in an RC circuit. The voltage source V is set at 10 volts, R_1 is a 1-megohm resistor, and C is a 5-microfarad capacitor with a working voltage rating of 25 volts. How much current will be flowing in the circuit at 5, 10, and 50 seconds after switch S is closed?

A good starting point in RC circuits is the calculation of the time constant, $\tau = R_1 C$. In the circuit given, $\tau = 1$ megohm $\times$ 5 microfarads or $\tau = 5$ seconds. The times at which we must determine the current flow are

$$t = 5\text{ s} = 1 \times \tau \quad \text{or} \quad t/\tau = 1$$
$$t = 10\text{ s} = 2 \times \tau \quad \text{or} \quad t/\tau = 2$$
$$t = 50\text{ s} = 10 \times \tau \quad \text{or} \quad t/\tau = 10$$

Establish the association between time and time constant firmly in your mind. From Fig. 4–7, we are reminded that the current will be maximum at $t = 0$, $I_{max} = V/R_1 = 10\text{ V}/1\text{ M}\Omega$ or $I_{max} = 10 \times 10^{-6}\text{ A} = 10\ \mu\text{A}$. Using the plot of Fig. 4–9, when $t = 5$ s, $t/\tau = 1$,

$$\begin{aligned} i/I_{max} &= 0.368, \quad \text{or} \quad i = 0.368 I_{max} \\ &= 0.368 \times 10\ \mu\text{A} \\ &= 3.68\ \mu\text{A} \end{aligned}$$

at $t = 10$ s, $t/\tau = 2$

$$i/I_{max} = 0.136$$

$$i = 0.136 I_{max} = 1.35\ \mu\text{A}$$

at $t = 50$ s, $t/\tau = 10$

$$i \approx 0$$

We can get the answers just as well by using Eq. (4.8):

$$i(t) = \frac{V}{R_1} \exp(-t/R_1 C) \qquad \textbf{(4.8)}$$

putting in the actual values of V, R, and C at $t = 5$ s

$$i(t) = 10 \exp(-5/5)$$

$$i(t) = 3.68\ \mu\text{A}$$

putting in the values of V, R, and C at $t = 10$ s

$$i(t) = 10 \exp(-10/5)$$

$$i(t) = 1.36\ \mu\text{A}$$

putting in the values of V, R, and C at $t = 50$ s

$$i(t) = 10 \exp(-50/5)$$

$$i(t) \approx 0$$

Consider the circuit of Fig. 4–6(b) after more than five time constants have elapsed. The capacitor is now changed to voltage V and current flow is reduced to zero. If switch S is now open, as in Fig. 4–6(a), capacitor C will remain charged indefinitely. We can now talk about the energy stored in the capacitor. From Eq. (4.1), the energy stored in the capacitor is $1/2\ CV^2$ joules. This energy is ready to be delivered to a load in the form of *discharge current and voltage*. The capacitor has now become a *source*. We will view it as a current source or voltage source depending upon its application in industrial electronics circuits. Let's look at Fig. 4–6(c) with the switch, S, in position 2 and see how the capacitor discharges, delivering its energy to a resistor,

R_2. Figure 4–11 is a plot of current flow when switch S is in position 2. Notice that since we used positive current flowing into the capacitor in Fig. 4–6(b), we reverse the sign of the capacitor in Fig. 4–6(c); therefore, the current in Fig. 4–11 is always negative. Think about this carefully and don't be confused by it.

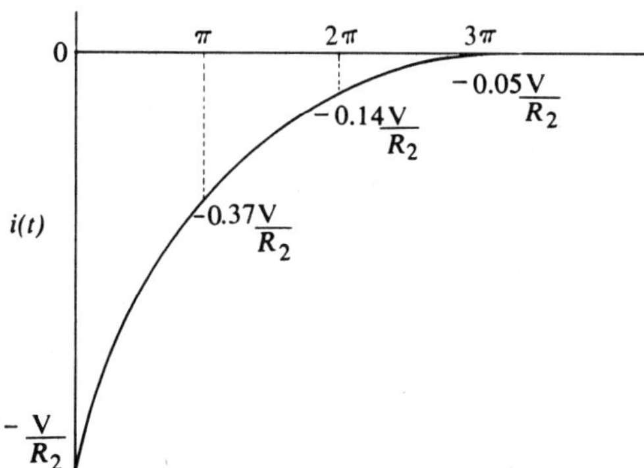

Fig. 4–11 Transient current in *RC* circuit while discharging.

Notice that current flow is maximum at $t = 0$, when the switch S is first put into position 2. Figure 4–12 is a plot of the voltage across the capacitor after the switch S is put into position 2. You will no doubt have noticed the similarity between transient currents and voltage waveforms in capacitor charge and discharge circuits. It is this similarity that often leads to confusion in later stages of industrial electronic design. Let's note these similarities and plant them firmly in the back of our mind.

(1) The amount of current flow is the same in the charge or discharge circuits if the number of time constants is the same and there is no voltage source in the discharge circuit.

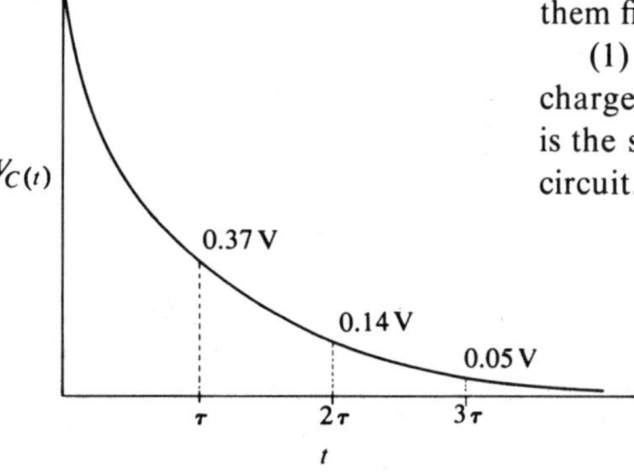

Fig. 4–12 Transient voltage in *RC* circuit while discharging.

(2) The ratio $i(t)/I_{max}$ is the same as the ratio $v(t)/V$ in the discharge circuit.

(3) The voltage across the capacitor in the charge circuit is the source voltage minus the product of $i(t)$ and R_1.

The plot of Fig. 4–9 is sufficient to handle all of the problems you will have with RC circuits using dc sources.

The reader with an advanced mathematics background will notice that Fig. 4–11 is a plot of the equation $i(t) = (V/R_2)\exp(-t/\tau_2)$ or

$$i(t) = -I_{max}\exp\left(\frac{-t}{\tau_2}\right) \tag{4.10}$$

Notice the time constant, τ_2, is now R_2C. The voltage across the capacitor is then given by

$$V(t) = V\exp\left(\frac{-t}{\tau_2}\right) \tag{4.11}$$

Let's look at an example to which we can apply our knowledge about RC discharge voltages.

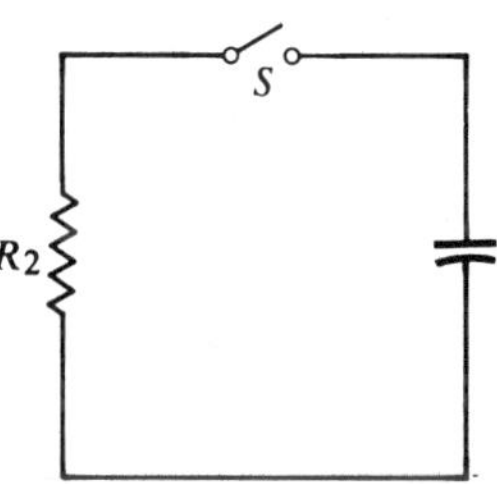

Fig. 4–13 Capacitor discharge circuit.

Example 4–3. The capacitor of Fig. 4–13 is 10 μF. Before switch S was closed, 5×10^{-4} joules of energy were stored in the capacitor. The resistor R = 1 megohm. Find the voltage across the capacitor at 10, 20, and 60 seconds after switch S was closed.

The time constant in this circuit is $R_2 \times C = 10^6 \times 10 \times 10^{-6} = 10$ seconds. From Eq. (4.1)

$$5 \times 10^{-4} = \tfrac{1}{2}CV^2 = \tfrac{1}{2} \times 10 \times 10^{-6} \times V^2$$

$$V^2 = 100, \quad V = 10\text{ V}$$

The times that we are interested in are:

$$t = 10\text{ s} = \tau_2 \quad \text{or} \quad t/\tau_2 = 1$$

$$t = 20\text{ s} = 2\tau_2 \quad \text{or} \quad t/\tau_2 = 2$$

$$t = 60\text{ s} = 6\tau_2 \quad \text{or} \quad t/\tau_2 = 6$$

Remembering that the ratio $V(t)/V = i(t)/I_{max}$ in the discharge circuit and using the plot of Fig. 4–8:

at t = 10 s $V(t)/V = 0.368$, $V(t) = 3.68$ V

at t = 20 s $V(t)/V = 0.136$, $V(t) = 1.36$ V

at t = 60 s $V(t)/V = 0.0$, $V(t) \approx 0$

We have just looked at the transient analysis of RC circuits using a dc voltage source. This might seem like a trivial analysis at first, but, as we will show later, most of your design problems can be reduced to a simple *RC* circuit with a single voltage source. Solve as many of the review problems as your time allows.

4–4 Capacitors as Current and Voltage Sources

We mentioned earlier that charged capacitors can deliver their stored energy to a load in the form of discharge current and voltage. When looked at from the load, the capacitor appears to be the source from which power is drawn. If we are controlling the load current, the capacitor looks like a current source. If we are controlling the load voltage, the capacitor looks like a voltage source. Since a capacitor cannot generate power internally, it can be viewed as a source only if it has some way of recharging itself. The model of Fig. 4–14 should help us to keep this in mind.

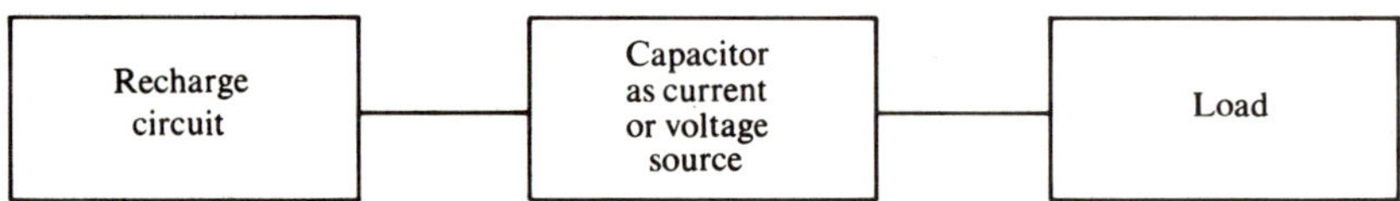

Fig. 4–14 Block diagram of capacitor as voltage or current source.

Many applications of capacitors in industrial electronics circuits can be simplified by separating the circuit components into these three categories. Keep this model in mind whenever capacitors are discussed.

Let's look at the circuit of Fig. 4–15 in terms of our model in Fig. 4–14. When switch S is in position 1, capacitor C charges. R_1 and V form the capacitor recharge portion of the model.

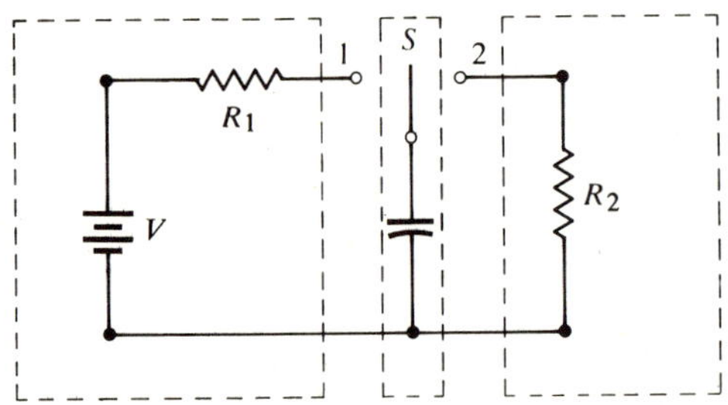

Fig. 4–15 Capacitor as a voltage source.

We will control the current flow into the load so capacitor C will be used as a current source. When switch S is in position 2, capacitor C delivers current to the load R_2. We must now face the problem of describing the capacitor as a current source. What kind of current source is it? Since the current involved is the capacitor discharge current, the current depends on the value of the load resistance R_2. We could describe the capacitor as a current source in terms of Figure 4–15. Unfortunately, designers are often required to make quick estimates when plots such as Figure 4–9 or exponential tables are not available. Figure 4–16 is an approximate description of a capacitor as a current source that is easy to remember. You can make quick calculations of sufficient accuracy for practical design. This is a piece-wise linear approximation to the actual discharge current of Fig. 4–10. The key numbers to remember are 37% at one time constant, 14% at two time constants, and 5% at three time constants.

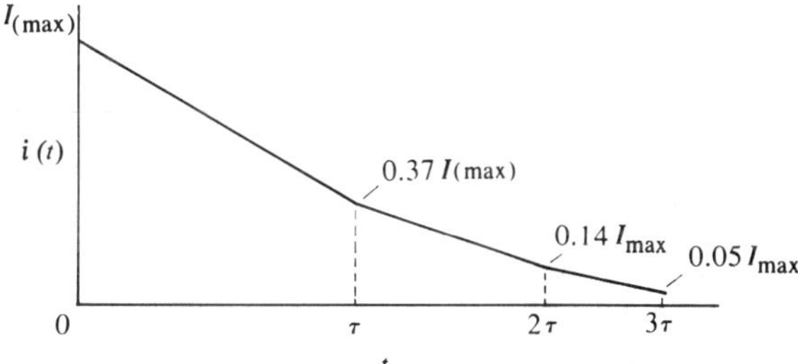

Fig. 4–16 Piecewise linear approximation of capacitor discharge current.

It is convenient for design purposes to talk about the average current delivered to the load during the first time constant.

$$
\begin{aligned}
I_{av} &= 1/RC \int_0^{RC} I_{max} \exp\left(\frac{-t}{RC}\right) dt \\
&= -I_{max} \int_0^{RC} \left[\exp\left(\frac{-t}{RC}\right)\right]\left(\frac{-dt}{RC}\right) \qquad (4.12)
\end{aligned}
$$

This equation is the same as

$$
A \int_0^{RC} \exp(-X)\, dx,
$$

if we let $X = t/RC$.

From the table of integrals (Appendix C), we find the answers to be

$$I_{max} \exp\left(\frac{-t}{RC}\right)\Bigg]_0^{RC}$$

Therefore

$$\begin{aligned} I_{av} &= -I_{max}\left[\exp\left(-\frac{RC}{RC}\right) - \exp\left(\frac{-0}{RC}\right)\right] \\ &= -I_{max}[\exp(-1) - 1] \\ &= I_{max}\, 0.63 \qquad \textbf{(4.13)} \end{aligned}$$

$$I_{av} = 0.63 I_{max} = 0.63 \frac{V_{max}}{R}$$

Let's look at a sample problem involving the capacitor as a current source.

Example 4–4. In the circuit of Fig. 4–15, it is required to deliver an average of 10 milliamps to the 200-ohm load resistor during the first time constant. The R_2C time constant must be 2 microseconds. Specify the value of C and V and calculate the amount of charge transferred during the first time constant.

From Eq. 4.13,

$$I_{av} = 0.63 \frac{V}{R}$$

or

$$10 \times 10^{-3}\,\text{A} = 0.63 \frac{V}{200}$$

$$V = \frac{200 \times 10^{-2}}{0.63} = \frac{2}{0.63} = 3.17\ \text{V}$$

$$\tau = R_2C = 2\,\mu\text{s} = 2 \times 10^{-6}\,\text{s}$$

$$200C = 2 \times 10^{-6}, \quad C = 0.01 \times 10^{-6}\,\text{F} = 10 \times 10^{-9}\,\text{F}$$

Since charge = current × time

$$\left(\text{coulombs} = \frac{\text{coulombs}}{\text{second}} \times \text{seconds}\right)$$

The amount of charge transferred in the first time constant is the product of average current and the circuit time constant.

$$\begin{aligned} Q = I_{av} \times \tau &= 10 \times 10^{-3} \times 2 \times 10^{-6} \\ &= 20 \times 10^{-9}\,\text{C} \end{aligned}$$

The circuit of Fig. 4–15 might also be used to control the voltage drop across load resistor R_2. If this were the case, the capacitor, C, would be acting as a voltage source. The source voltage would be the discharge voltage of C through R_2. We have already described this waveform in Fig. 4–12. Again we would like a simplified approximation for design use. Figure 4–17 is a sufficiently accurate piecewise linear approximation to Fig. 4–10. The key numbers to remember are 37% at one time constant, 14% at two time constants, and 5% at three time constants.

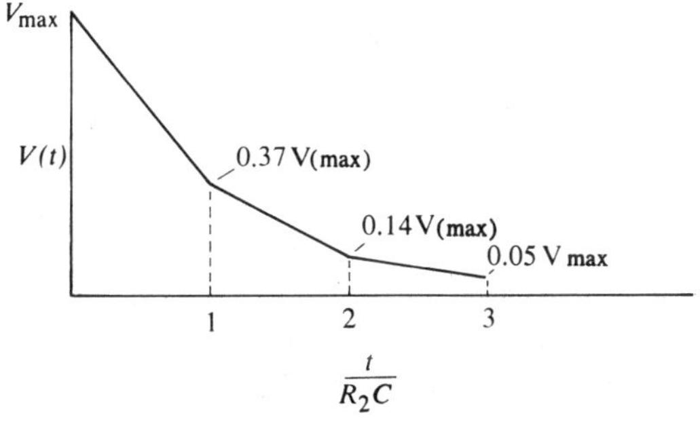

Fig. 4–17 Capacitor discharge linear approximation.

It is also necessary to specify the recharge circuit in many applications. Figure 4–18 is a piecewise linear approximation to the charging voltage waveform of Fig. 4–8.

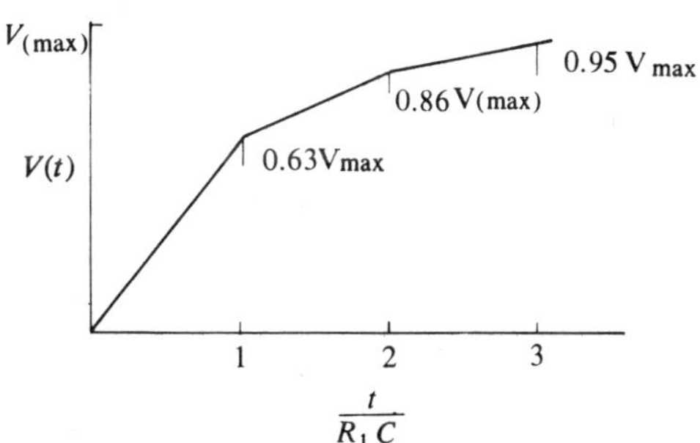

Fig. 4–18 Linear approximation of capacitor charge voltage.

The key numbers to remember are 63% at one time constant, 86% at two time constants, and 95% at three time constants. Let's make a couple of calculations using the linear approximations to capacitor voltage.

Example 4–5. In Example 4–4 it is required that the capacitor be recharged to 86% of its maximum voltage in 7 microseconds. Specify the value of R_1.

The value of C is already established at 10 millimicrofarads. Using the linear approximation of Fig. 4–18, the capacitor will be charged to 86% of its maximum value in two time constants so that $2R_1C$ should be 7 microseconds.

$$2R_1C = 7 \times 10^{-6}$$

$$2R_1 \times 10 \times 10^{-9} = 7 \times 10^{-6}$$

$$R_1 = \frac{7 \times 10^{-6}}{20 \times 10^{-9}}$$

$$R_1 = 0.35 \times 10^3 = 350\ \Omega$$

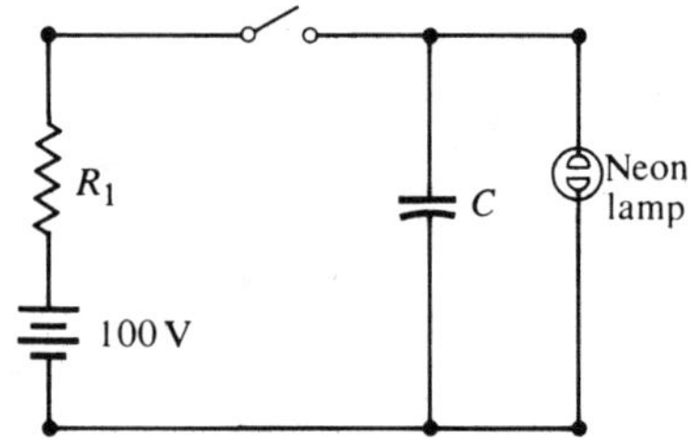

Fig. 4–19 Application of capacitor in active circuit.

Example 4–6. In the circuit of Fig. 4–19, 85 volts are required to cause the neon lamp to start conducting. If the product of R_1C is 30 seconds, at what time will the lamp start conducting if switch S is closed at time $t = 0$?

R_1C is the circuit time constant = 30 seconds. The capacitor charge voltage is 100. It is required that the capacitor charge to 85 volts or 85% of its maximum voltage to cause the lamp to conduct. Remembering that the capacitor will reach 86% of its maximum voltage in 2 time constants, it will take approximately 2 × 30 = 60 seconds for the capacitor voltage to reach 85 volts. The lamp will start to conduct at 60 seconds.

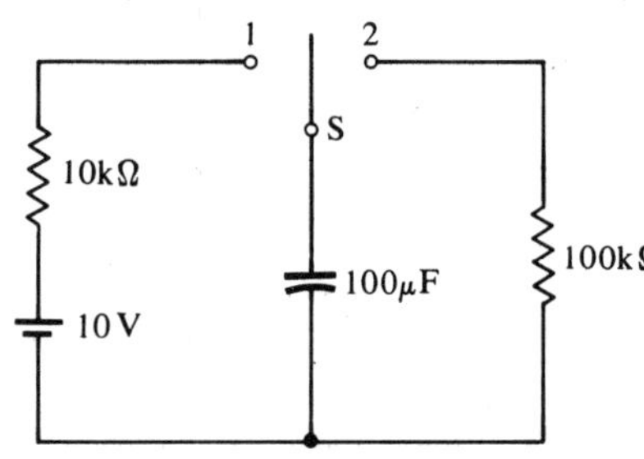

Fig. 4–20 DC *RC* circuit.

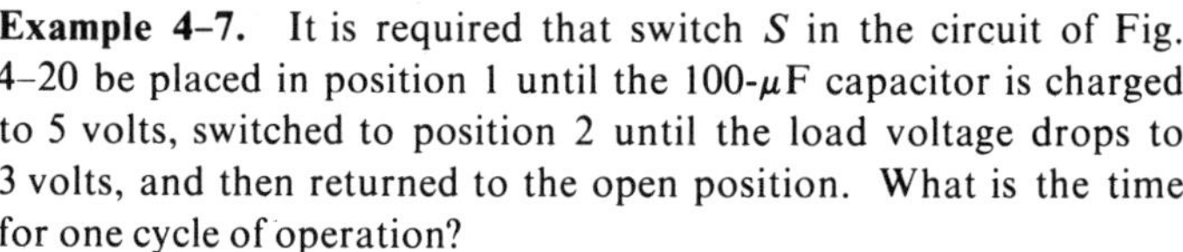

Example 4–7. It is required that switch S in the circuit of Fig. 4–20 be placed in position 1 until the 100-μF capacitor is charged to 5 volts, switched to position 2 until the load voltage drops to 3 volts, and then returned to the open position. What is the time for one cycle of operation?

Solution: The charging circuit time constant is 10^4 ohms × 100 × 10^{-6} farads = 1 second. The discharging circuit time constant is 10^{-5} ohms × 100 × 10^{-6} farads = 10 seconds. It is a good idea to make a sketch of the waveform in switching circuits such as this. Figure 4–21 tells us that at $t = 0$ the capacitor voltage

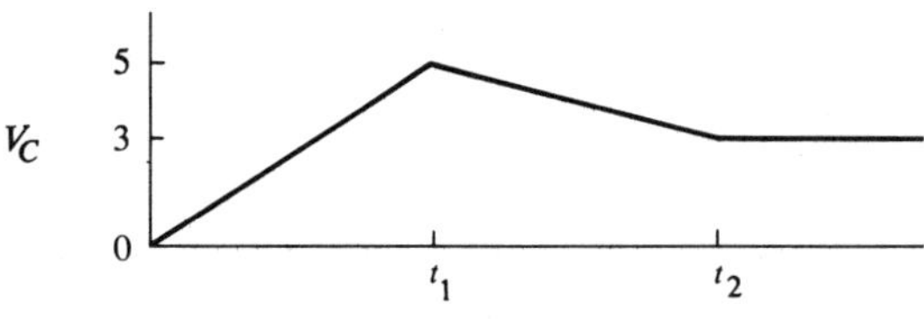

Fig. 4–21 Sketch of capacitor voltage of Example 4–7.

will start to rise as the switch is placed in position 1; at t, the capacitor voltage starts to drop off as the switch is placed in position 2; and at t_2 the capacitor voltage remains steady as the switch is opened.

The time for a cycle of operation is $t_1 + t_2$.

From Fig. 4–18, the capacitor will charge to one half of its maximum voltage (5 volts = 1/2 × 10 volts) in 0.74 time constants. From Fig. 4–17, the capacitor will discharge to 3/5 of its maximum voltage (3 volts = 3/5 × 5 volts) in 0.574 time constants. The time for a cycle is 0.74 × 1 s + 0.574 × 10 s = 0.74 s + 5.74 s = 6.48 s.

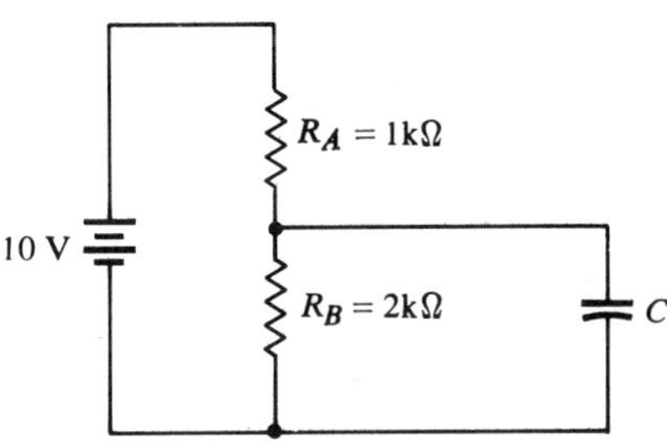

Fig. 4–22 Capacitor charge circuit.

4–5 Multiple-Loop DC Capacitive Circuits

The capacitive circuits we discussed in Section 4–3 involved a single capacitor-resistor-voltage source combination at each point in the analysis. Life is not always so simple. What if more than one resistor or voltage source is involved in the analysis? How does the designer handle these problems? Most designers will try to simplify the circuit and reduce it to the point where reasonably accurate estimates can be made and rely on final adjustments of the prototype components to meet performance specifications. We will look at three situations in which this approach is successfully used for design purposes.

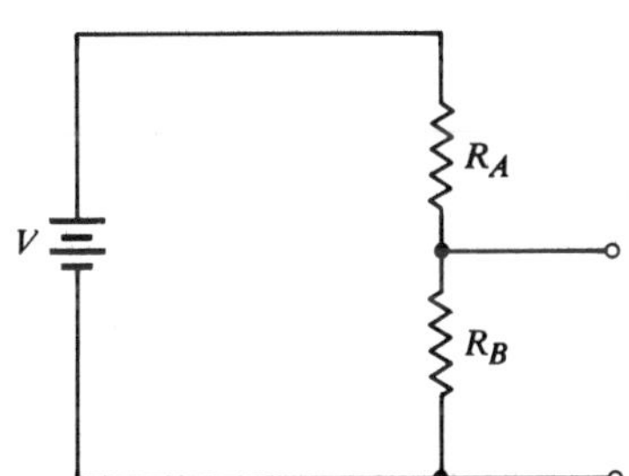

Fig. 4–23 Equivalent circuit analysis of Fig. 4–22.

The circuit of Fig. 4–22 is used to charge capacitor C from battery V. What should the value of the capacitor C be if it is desired that the capacitor voltage reach 85% of its maximum value in 0.02 μs?

We know that the capacitor will reach 85% of its maximum charging voltage in 2 time constants. Our first problem is to determine what value of resistance to use in the RC time constant. The following simple rules can be followed in circuits of this type:

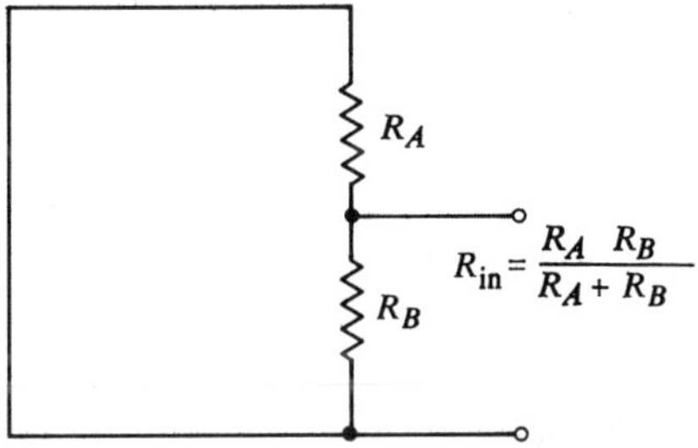

Fig. 4–24 Circuit for finding equivalent resistance in Fig. 4–22.

(1) Redraw the circuit without the capacitor, using the points at which the capacitor was connected as the input terminals (Fig. 4–23).

(2) Replace all voltage sources in the redrawn circuit with short circuits.

(3) Determine the equivalent resistance as seen at the new input terminals (Fig. 4–24).

(4) Use the equivalent resistance in the RC time constant calculation.

The application of these rules gives the solution to our problem.

$$R_{in} = \frac{R_A R_B}{R_A + R_B} = \frac{1\,\text{k}\Omega \times 2\,\text{k}\Omega}{1\,\text{k}\Omega + 2\,\text{k}\Omega} = \frac{2}{3}\,\text{k}\Omega$$

$$R_{in}C = 0.01 \times 10^{-6}\,\text{s}$$

$$0.67 \times 10^{3}C = 0.01 \times 10^{-6}\,\text{s}$$

$$C = \frac{0.01 \times 10^{-6}}{0.67 \times 10^{3}} = 0.015 \times 10^{-9}\,\text{F}$$

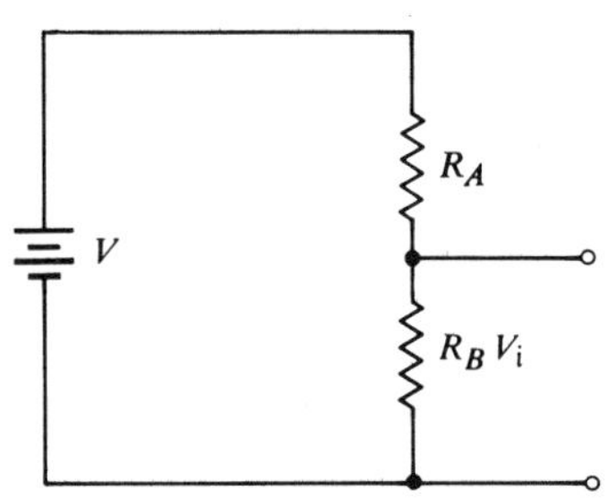

Fig. 4–25 Circuit for finding V_i in Fig. 4–22.

Suppose we want to estimate the value of the capacitor voltage at 0.02 μs. We know that the capacitor will be charged to 85% of its maximum charging voltage. Then all that remains is to determine the maximum charging voltage.

The simple rules to follow for this type of circuit are:

(1) Redraw the circuit without the capacitor, using the points at which the capacitor was connected as the input terminals (Fig. 4–23).

(2) Determine the voltage that would be measured at the new input terminals using a good high impedance voltmeter (Fig. 4–25).

(3) Use this new input voltage as the maximum value to which a capacitor will charge when placed at the input terminals.

Applying the rules to our problem,

$$V_{in} = \frac{VR_B}{R_A + R_B} = 10\,\frac{2\,\text{k}\Omega}{2\,\text{k}\Omega + 1\,\text{k}\Omega} = 10\,\frac{2}{3} = 6.7\,\text{V}$$

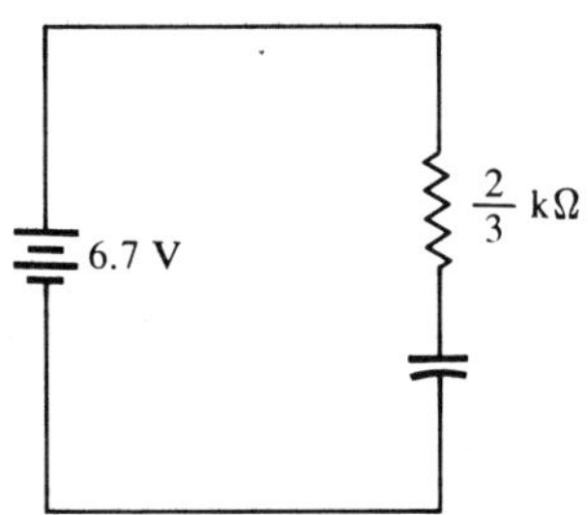

Fig. 4–26 Equivalent circuit of Fig. 4–22.

The voltage on the capacitor C at $t = 0.02$ μs will be $0.85 \times 6.7 = 5.7$ V.

The circuit of Fig. 4–22 can be redrawn as a single-loop circuit as in Fig. 4–26 where R_{in} is the charging resistance and V_{in} is the maximum charging voltage. You might recognize this as the Thévenin equivalent circuit from your previous experience with circuits.

A more complicated circuit is shown in Fig. 4–27. If there is no initial charge on the capacitor C, find the voltage on the capacitor after the switch is put in position A for 1.5 seconds. Here is a good opportunity to use the rules for simplification. If the two voltage sources are short-circuited, the circuit seen by the capacitor with the switch in position A is shown in Fig. 4–28. Further combination of resistors leads to the circuit of Fig. 4–29.

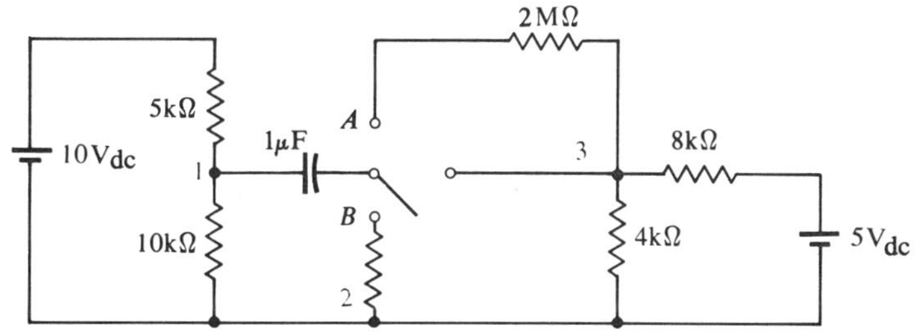

Fig. 4–27 Multiple-loop DC *RC* circuit.

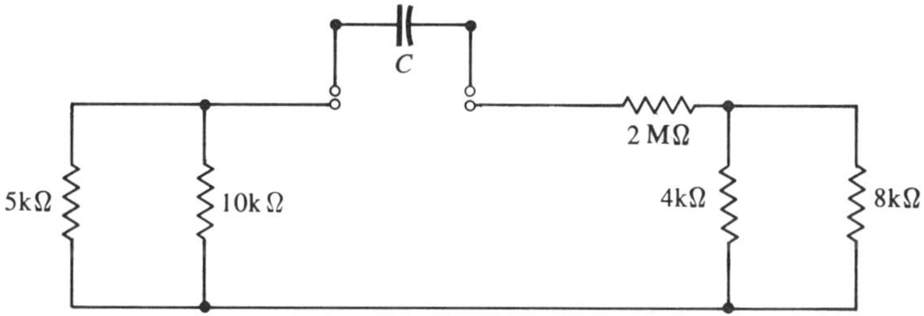

Fig. 4–28 Circuit of Fig. 4–27 with voltage sources short-circuited.

The two-megohm resistor is so large compared to the other two that, for all practical purposes, the equivalent resistance can be estimated at 2 megohms.

The voltage between points (1) and (2) in Fig. 4–27 is:

$$10 \frac{10\ \text{k}\Omega}{5\ \text{k}\Omega + 10\ \text{k}\Omega} = \frac{20}{3} = +6.67\ \text{V}$$

The voltage between points (3) and (2) in Fig. 4–27 is:

$$5 \times \frac{4\ \text{k}\Omega}{8\ \text{k}\Omega + 4\ \text{k}\Omega} = \frac{5}{3} = +1.67\ \text{V}$$

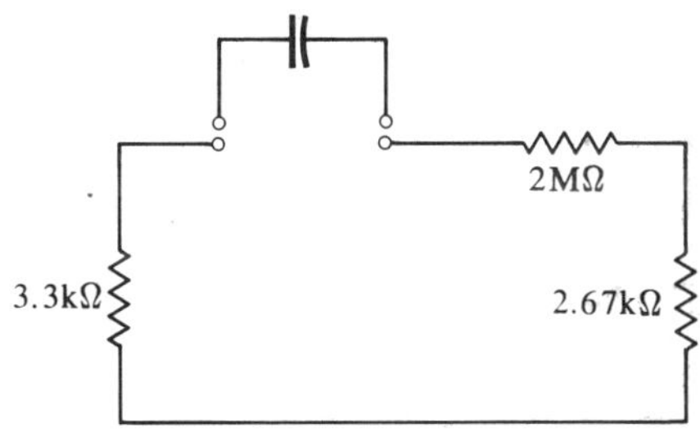

Fig. 4–29 Simplified circuit of Fig. 4–28.

The voltage that would be measured between points (1) and (3) would then be 6.67 – 1.67 = +5 V. Since there is no current flowing in the 2-megohm resistor connected at point (3), the voltage measured between points (1) and (*A*) would also be +5 V. This is the charging voltage to be used in the simplified *RC* circuits. The complete simplified circuit is shown in Fig. 4–30.

The charging time constant is $RC = \tau = 2$ megohms × 1 μ farad = 2 seconds. The maximum voltage that the capacitor will charge to is 5 volts.

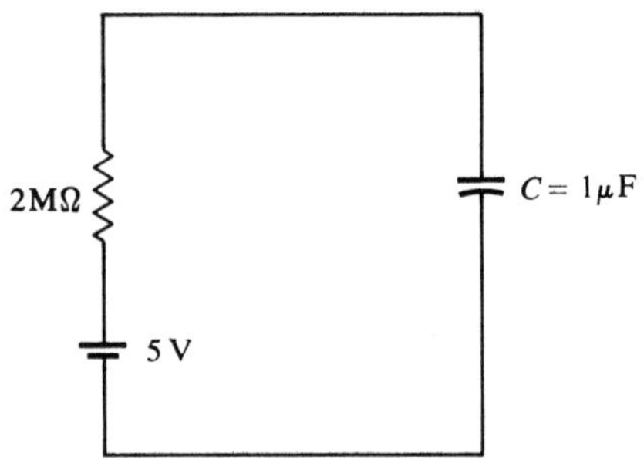

Fig. 4–30 Single-loop equivalent circuit in switch position *A*.

We would like to know what the voltage on the capacitor will be at $t = 1.5$ seconds. Then

$$\frac{t}{\tau} = \frac{1.5}{2} = 0.75, \quad \frac{V_C}{V_{max}} = 0.53.$$

$$V_C = 0.53\,V_{max} = 0.53 \times 5 = 2.65 \text{ V}$$

Keep this simplification technique in mind when you are faced with seemingly complicated *RC* circuits. It will save you many hours of grief.

4–6 The Capacitor in AC Circuits

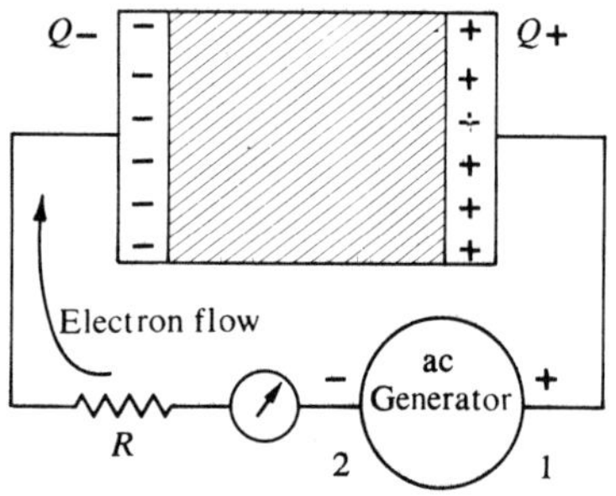

Fig. 4–31 Circuit of Fig. 4–3 with DC battery replaced by an AC generator.

Let's go back to Fig. 4–3 and replace the battery with an ac generator (Fig. 4–31). The output of the ac generator is the standard sine wave (Fig. 4–32). The mathematical expression is $V = V_{max} \sin \omega t$ where $\omega = 2\pi/T$, T = the time for one cycle of the sine wave.

The charges on Q_- and Q_+ in Fig. 4–31 are shown at the instant when terminal 1 of the generator is at a higher voltage than terminal 2 (point 1, Fig. 4–32). Electrons are flowing from Q_+ to Q_-. Consider what will happen one-half cycle later when the voltage at terminal 1 is lower than the voltage at terminal 2 (point 2, Fig. 4–32). Electrons will try to rush in the opposite direction to reverse the charge on the capacitor plates.

One-half cycle later, the circuit will return to the condition of Fig. 4–32. The capacitor will not be allowed to remain fixed but will tend to follow the output of the signal generator. Current in the circuit will flow in the proper direction to allow the capacitor voltage to follow the generator voltage. Figure 4–33 is a sketch of capacitor voltage and current.

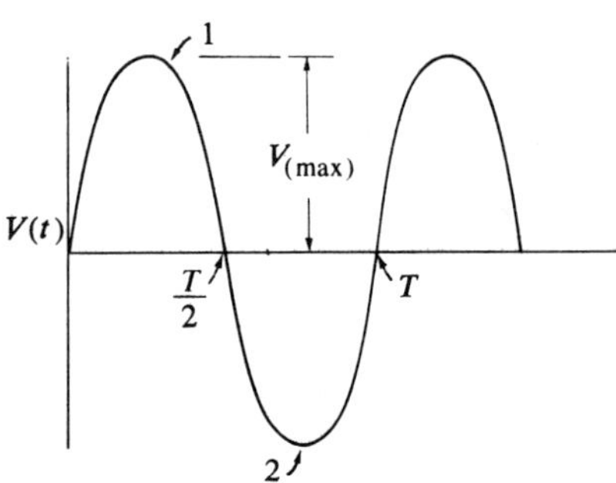

Fig. 4–32 Instantaneous sine wave voltage.

The voltage across a capacitor cannot change instantly; current must flow to build up the charge ahead of the voltage. The current waveform is always one-fourth of a period ahead of the voltage waveform. This represents a 90° phase difference between V_c and i_c, with the current leading the voltage.

There is no phase shift between the current flowing in a resistor and the voltage drop across the same resistor.

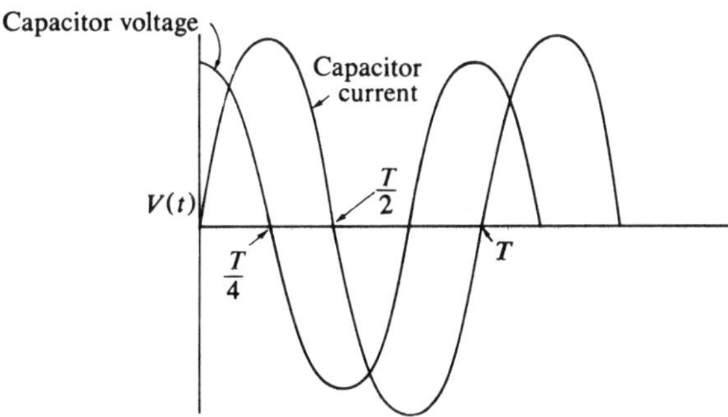

Fig. 4–33 Capacitor voltage and current vs time.

Therefore, the voltage across R in Fig. 4–34 will be in phase with the current $i(t)$ and lead the voltage across the capacitor by 90°. We can make a phasor diagram showing the 90° phase relationship of V_c to V_R (Fig. 4–35). Using Kirchhoff's voltage law (KVL) in Fig. 4–34, $V_i = V_C + V_R$. Figure 4–36 is the combination of the phasors tip to tail. KVL requires that the three phasors form a closed loop when connected in this form.

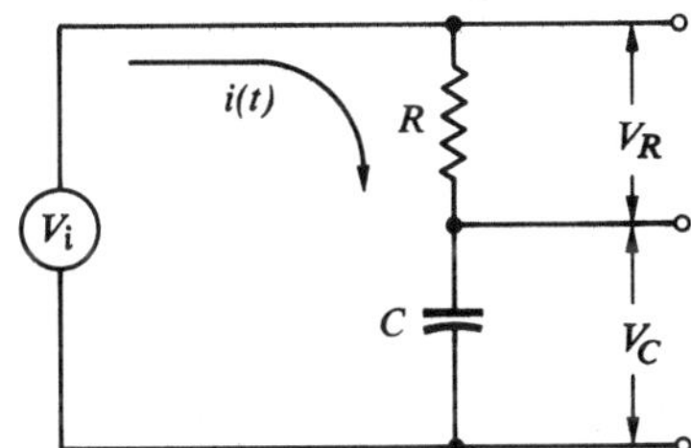

Fig. 4–34 AC resistor capacitor circuit.

The angle between V_R and V_i is the impedance angle or power factor angle that you learned about in your basic circuits course. This angle is commonly designated by the Greek letter theta, Θ. Since the closed triangle of Fig. 4–36 is a right triangle, the angle between V_C and V_i is 90° − Θ. We can see in Fig. 4–35 that V_c lags behind V_i. This is a very important relationship in industrial electronics. The circuit of Fig. 4–34 allows us to apply a sine wave input to the RC circuit and get a sine wave output across the capacitor that lags the input waveform by some angle 90° − Θ. This is called an RC phase shift circuit. We'll see more of it later.

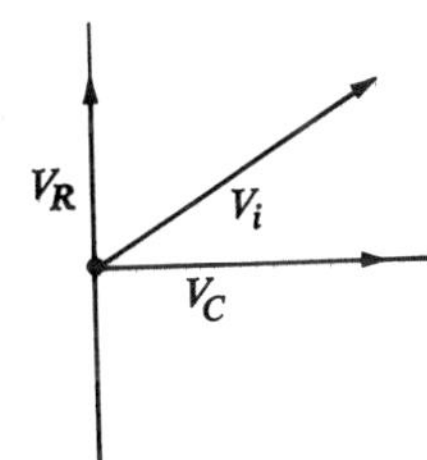

Fig. 4–35 Phasor plot of voltages in circuit of Fig. 4–34.

Example 4–8. Let's assume that we had used an ac voltmeter to measure the voltages across V_R, V_C, and V_i in the circuit of Fig. 4–34. We found V_R = 3 volts, V_C = 4 volts, and V_i = 5 volts. We could now use the closed triangle of Fig. 4–36 to calculate the angle of phase shift between V_C and V_i.

Observing that

$$\tan \Theta = \frac{V_C}{V_R} = \frac{4}{3} = 1.33$$

$$\Theta = 53.2°$$

The angle of phase shift between V_c and V_i = 90° − Θ = 36.8° with V_c lagging.

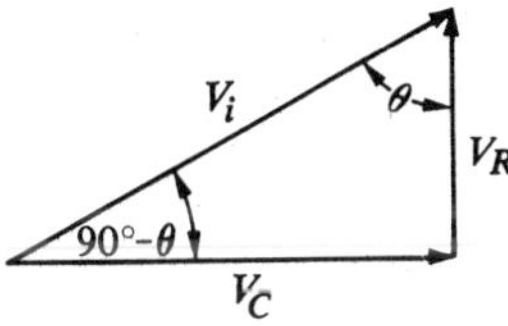

Fig. 4–36 Closed-loop vector diagram of Fig. 4–35.

4–7 Energy Storage in Inductors

The inductor, or coil, is often used in industrial electronics circuits to store energy and return it to the circuit after a period of delay. Unlike the capacitor, the energy is not stored in the inductor itself. The energy is stored in the magnetic field surrounding the inductor. If the coil is detached from the circuit, the field collapses and the stored energy is released. The inductor, L, in the circuit of Fig. 4–37(a) has no current flowing through it so there is no magnetic field around it. When switch S is closed, current flows through the coil and a magnetic field is built up as shown in Fig. 4–37(b). If we wait until the current reaches its final value, V/R, then the energy stored in the magnetic field of the coil is given by Eq. (4.14):

$$W = \frac{1}{2} LI^2 \tag{4.14}$$

where

W = energy in joules
L = inductance in henrys
I = current in amperes

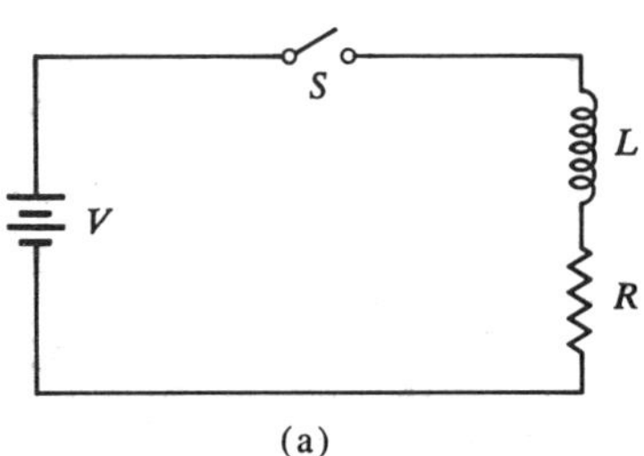

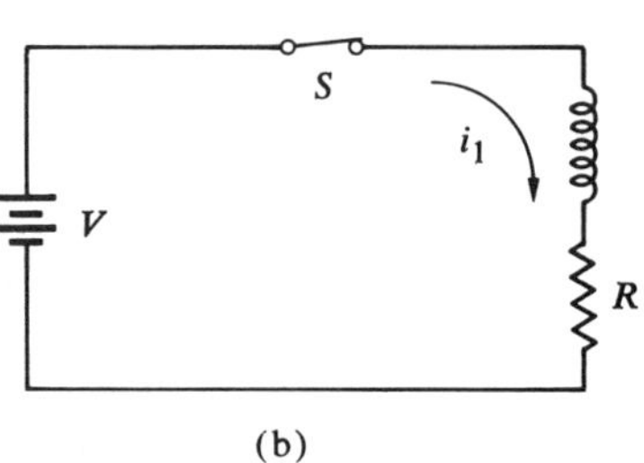

Fig. 4–37 (a) *RL* circuit with no current flow, (b) *RL* circuit with current flow and field around coil.

Let's look at the design of a simple *RL* circuit.

Example 4–9. In the circuit of Fig. 4–37, if V = 10 volts, R = 1 kilohm, and it is required to store 10^{-7} joules in the inductor, specify the value of L.

$$I = \frac{10}{1\ \text{k}\Omega} = 0.01\ \text{A}$$

From (4.14)

$$10^{-7} = \frac{1}{2} L(0.01)^2$$

$$10^{-7} = \frac{1}{2} L\, 10^{-4}$$

$$L = 2 \times 10^{-3}\ \text{H}$$

4–8 Inductors in DC Circuits

In our discussion of energy stored in inductors, our calculations were based on the steady-state current, or the final value after all transient effects have disappeared. Let's take a closer look at current flow during the period when the current is building up to its final value. Figure 4–38 is a plot of the current flow in the circuit of Fig. 4–37(b) when switch S is closed at $t = 0$. You will notice that this curve is very similar to Fig. 4–8, the transient voltage across the capacitor. The time constant, τ, in the circuit of Fig. 4–37 is L/R. At $t = L/R$, the current has reached 63% of its final value. At $t = 2L/R$, the current has reached 86% of its final value. At $t = 3L/R$, the current has reached 95% of its final value. It will actually take an infinite amount of time for the current to reach its final value, but after 5 time constants it is sufficiently close that the error is negligible.

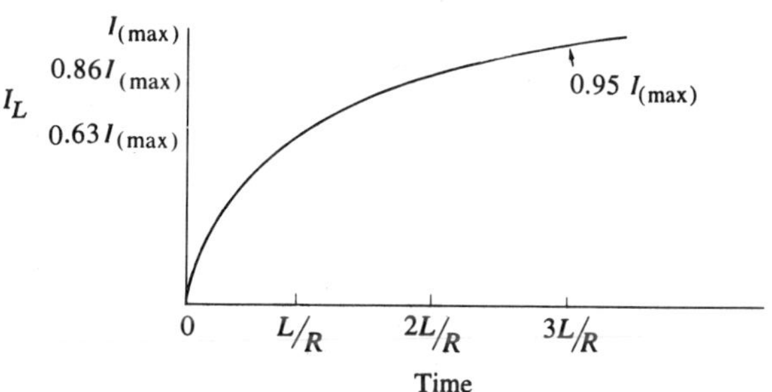

Fig. 4–38 Transient current in the circuit of Fig. 4–37(b).

We have said nothing yet about voltages in the *R-L* circuit. The circuit equations reveal quite simply the transient voltages in the circuit of Fig. 4–37(b). Writing the summation of voltages around the loop

$$V = L\frac{di}{dt} + iR \qquad (4.15)$$

using the initial condition that

$$i = 0 \quad \text{at } t = 0$$

the solution of Eq. (4.15) is

$$i_L = \frac{V}{R}(1 - e^{-tR/L}) \qquad (4.16)$$

$$i_L = I_{max}(1 - e^{-t/\tau}) \qquad (4.17)$$

where $I_{max} = V/R$ and $\tau = L/R$.

Equation (4.16) describes the graphical plot of Fig. 4–38. The voltage across the coil

$$V_L = V - i_L R \qquad (4.18)$$

Substituting (4.17) into (4.18)

$$V_L = Ve^{-t/\tau} \qquad (4.19)$$

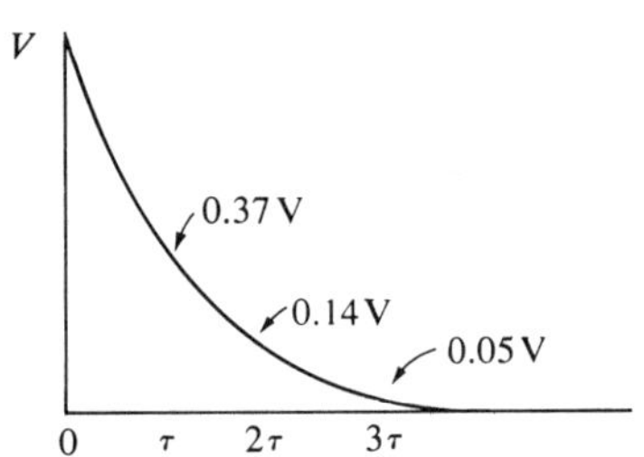

Fig. 4–39 Transient voltage across the coil in Fig. 4–37(b).

Figure 4–39 is a plot of the voltage across the coil if switch *S* is closed at $t = 0$. The voltage across the coil is the same as the battery voltage, *V*, at $t = 0$. $V_L = 37\%$ of battery voltage at one time constant, 14% at 2 time constants, and 5% of battery voltage at 3 time constants.

On observing Figs. 4–38 and 4–39, we notice that the voltage across the coil jumps instantaneously to the value of the battery voltage when switch *S* is closed but the current flow through the coil cannot change instantaneously. The maximum rate of change in current occurs immediately after $t = 0$. From Eq. (4.16) and the initial condition $i_L = 0$ at $t = 0$,

$$\left(\frac{di}{dt}\right)_{max} = \frac{V}{L} \qquad (4.20)$$

The inductor, or coil, tends to delay the rise of current flow in the *R-L* circuit. This property make coils useful as *current smoothing elements*. Let's assume that switch *S* has been closed long enough for the current in Fig. 4–37(b) to reach its maximum value. What happens when the switch is opened again? Since the current flow in the coil cannot change instantly, at the moment the switch is opened, Kirchhoff's voltage law requires that:

$$V = L\frac{di}{dt} + I_{max}R + V_{switch} \qquad \textbf{(4.21)}$$

Since the resistance of the air, R_{air}, is very large, $V_{switch} = I_{max}R_{air}$ is a very large voltage. Let's put in some numbers to dramatize the values involved.

Example 4–10. Let V = 10 volts, R = 1 kilohm, and R_{air} = 1 megohm. Calculate the voltage drop across the switch when switch S is opened.

$$I_{max} = \frac{V}{R} = \frac{10}{10^3} = 0.01\text{ A}$$

$$V_{switch} = I_{max}R_{air} = 10^2\text{ A} \times 10^6\ \Omega$$

$$= 10{,}000\text{ V}$$

From Eq. (4.21), the voltage across the coil, $V_L = L(di/dt)$, must be equal to the switching voltage to make the equation balance since $I_{max}R = V$. Then the voltage across the coil in Example #10 will also be 10,000 volts. This is the *counter-emf* produced by the coil according to Lenz's law of magnetic induction. Obviously, we cannot allow 10,000 volts to develop across the switch or across the coil. Either the switch will arc or the insulation of the coil will break down. Two methods of preventing excessive voltages in the circuit are shown in Fig. 4–40. In Fig. 4–40(a) an RC circuit is placed across the switch. The energy returned to the circuit

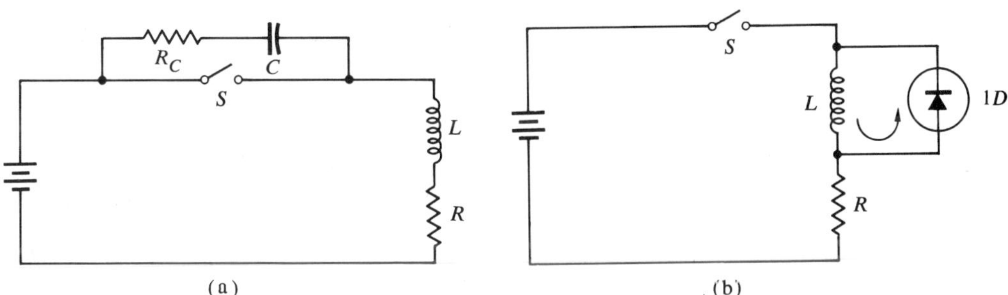

Fig. 4–40 Methods of reducing switching voltage in ***RL*** circuit (a) capacitor, and (b) freewheeling diode.

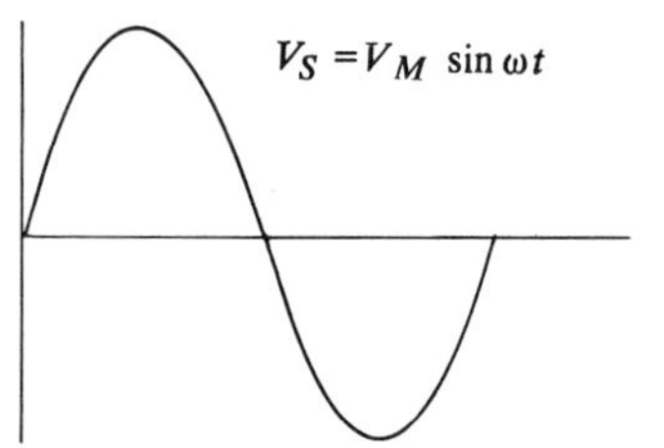

Fig. 4–41 AC source to replace battery in Fig. 4–37.

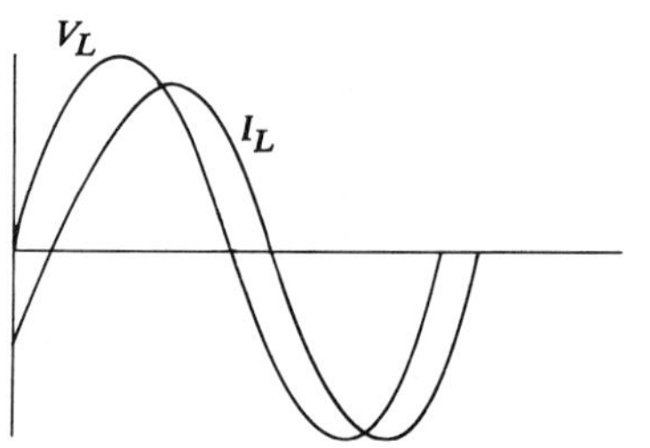

Fig. 4–42 Voltage across coil and coil current in AC *RL* circuit.

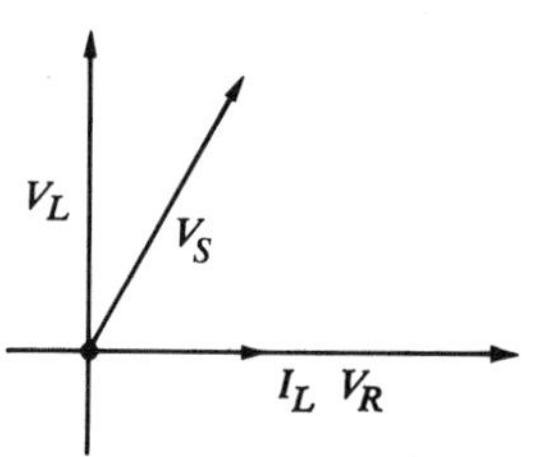

Fig. 4–43 Phasor diagram of waveforms in Figs. 4–41 and 4–42.

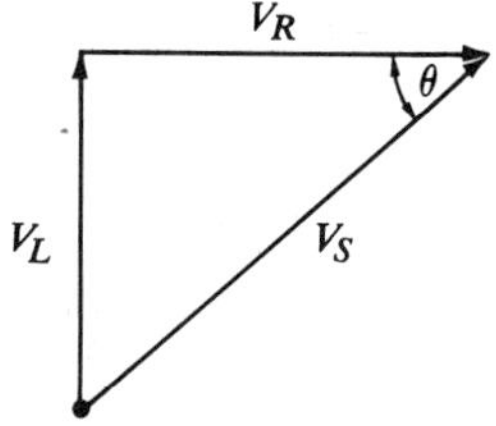

Fig. 4–44 Vector diagram of the phasors in Fig. 4–43.

due to the collapsing magnetic field of the coil is temporarily stored in the capacitor and later returned to the resistors in the circuit and dissipated in the form of heat. In Fig. 4–40(b) the current flow is through the freewheeling diode *D*, instead of through the high resistance of the air at the switch so that excessive switch voltage is avoided.

4–9 Inductors in AC Circuits

Let's go back to Fig. 4–37(b) and replace the battery with an ac source. The output of the ac source is the standard sine wave of Fig. 4–41. Since the voltage across the coil can change, there is no time lag between source voltage and coil voltage. The current through the coil cannot change instantly. Time is required for the magnetic field to build up and collapse. The current through the coil will lag 90° behind the voltage across the coil (Fig. 4–42). The voltage across the resistor, *R*, will be in phase with the current through the coil. The phase relationships between V_S, V_L, and V_R are shown in the phasor diagram of Fig. 4–43. The summation of voltages around the loop of Fig. 4–37(b) with V replaced by the ac source V_S is

$$V_S = V_L + V_R \tag{4.22}$$

If the phasors of Fig. 4–43 for each of the voltages are placed tip to tail, then Eq. (4.22) requires that the phasors form a closed loop as in Fig. 4–44. The angle between V_R and V_S is the impedance angle or *power factor angle* of the circuit. This angle is commonly called theta, Θ. Notice that V_R lags behind V_S. This is an important property. We can put a sine wave signal, V_S, into the circuit and get out a sine wave across the resistor that lags behind it. This is the *R-L* phase shift circuit. We'll deal with this circuit in more detail later. Let's look at a sample calculation of the phase shift in an *R-L* circuit.

Example 4–11. Let's assume that we had used an ac voltmeter to measure the voltages across V_R, V_L, and V_S in Fig. 4–44. We found that $V_R = 3$ volts, $V_L = 4$ volts, and $V_S = 5$ volts. Calculate the angle of phase lag between V_R and V_S.

The angle of phase lag is Θ. From Fig. 4–44

$$\text{tau } \Theta = \frac{V_L}{V_R} = \frac{4}{3} = 1.33$$

$$\Theta = 53.2$$

The angle of phase lag is 53.2°.

4–10 Applications of Passive Time Delay Elements

Some applications of capacitors and inductors in industrial electronics circuits are shown in Figs. 4–45 to 4–49. Let's look at each one. The circuit of Fig. 4–45 is an *R-L-C* delay line. The delay line is used in digital electronics circuits to prevent races and oscillations. When the input is at 2 volts, all capacitors are fully charged. When the input goes to ground, *C*, discharges through 1*L* and 1*R*, 2*C* discharges through 2*L*, 1*L*, and *R*, each capacitor discharges in succession so that the ground potential travels down the line at some velocity, v_o. Additional *LC* sections are added to achieve the desired delay.

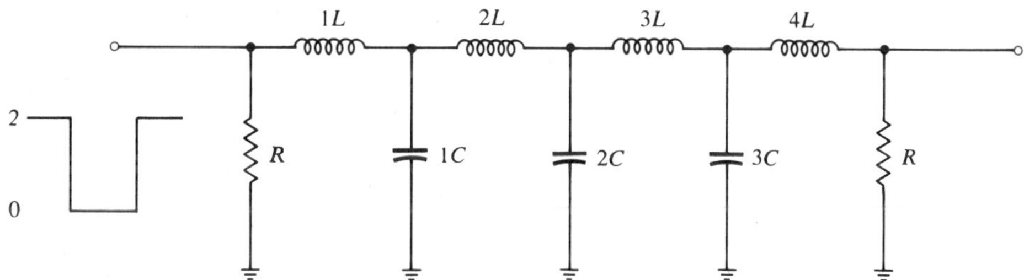

Fig. 4–45 Passive delay line.

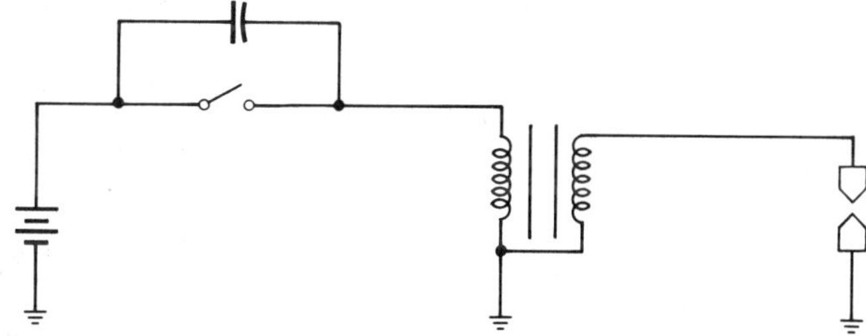

Fig. 4–46 Induction sparking circuit.

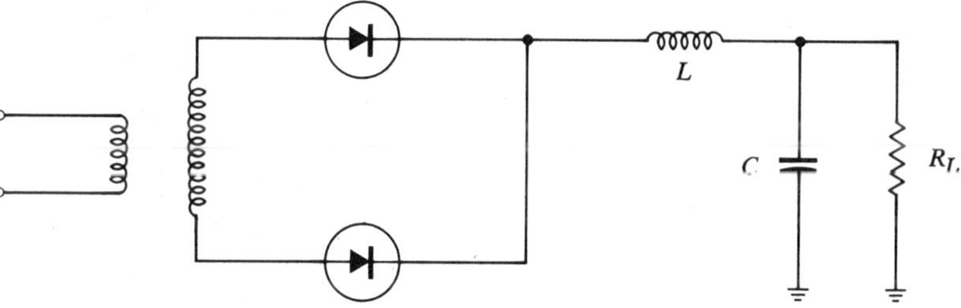

Fig. 4–47 LC filter at output of diode rectifier circuit.

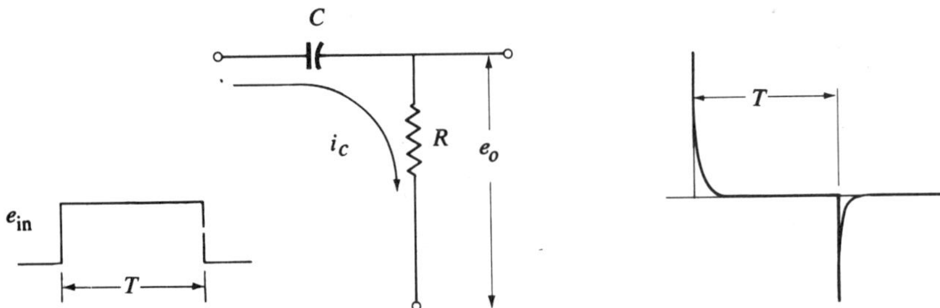

Fig. 4–48 *RC* differentiating circuit for pulse shaping.

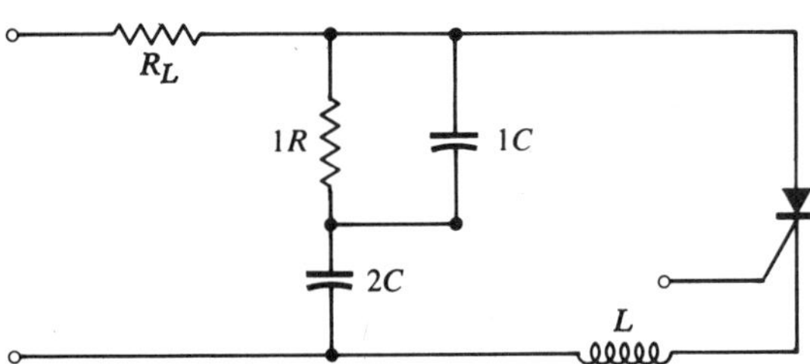

Fig. 4–49 *LC* filter to suppress high-frequency switching noise in *SCR* control circuit.

The circuit of Fig. 4–46 is used to develop a high-voltage spark. When the switch is closed, energy is stored in the magnetic field of the coil. Opening the switch allows the primary coil voltage to increase; the high voltage is applied through a step-up transformer to the electrodes where the spark is developed between two electrodes. This is the operating principle of the automotive ignition system.

The *LC* filter of Fig. 4–47 is used at the output of the full-wave rectifier dc power supply. The inductor smooths out ripple in the load voltage.

The circuit of Fig. 4–48 is an *RC* pulse-shaping circuit. A square wave pulse in provides a positive and negative spike out. The time constant, *RC*, must be much less than the width of the input pulse.

When an SCR or TRIAC is used to switch on current in a resistive load, high-frequency noise is generated. This noise is radiated or conducted along the power line and sometimes interferes with other equipment. The *LC* combination in Fig. 4–49 is used to suppress the noise output from the switching device. The value of *L* and *C* depends on the frequency and amount of suppression required. Low-pass filter design techniques are used.

4–11 Summary

Capacitors and coils are critical elements in industrial electronics because they affect timing and frequency performance. We have looked at coils and capacitors as temporary energy storage devices. This property is very important; you will see it used many times. We also evaluated the transient voltages and currents in *RL* and *RC* circuits with dc sources, the phase shift properties of *RL* and *RC* circuits with ac sources, and some simple applications. I have found that many of the problems in the design and analysis of industrial electronics circuits are due to misunderstanding of the application of passive time delay elements. We will discuss them many times as they appear in later chapters.

EXERCISES

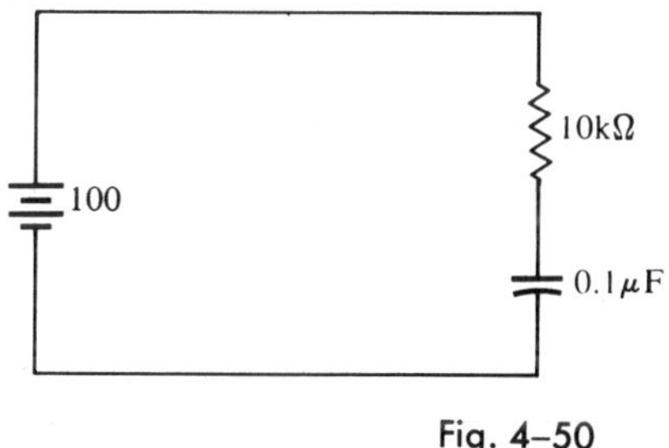

Fig. 4–50

1. How much energy is stored in the capacitor of Fig. 4–50 when it is fully charged?
2. What is the time constant in the circuit of Fig. 4–50?
3. The capacitor of Fig. 4–51 is completely discharged before switch *S* is closed.
   (a) Make a plot of $V_C(t)$ after the switch is closed.
   (b) Make a plot of $i(t)$ after the switch is closed.

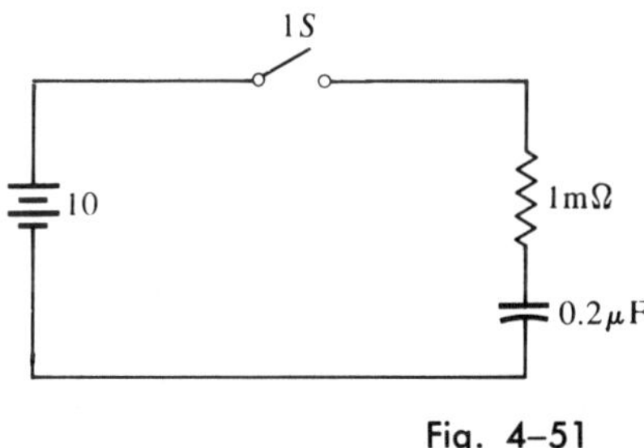

Fig. 4–51

4. The capacitor of Fig. 4–52 is initially uncharged. The voltage on the capacitor reaches 40 volts 0.5 seconds after switch S_2 is closed. Find the value of the battery voltage, *V*.
5. The capacitor of Fig. 4–53 is initially uncharged. Switch S_3 is placed in position 1 at $t = 0$. At $t = 0.1$ seconds, S_3 is instantly moved to position 2. At $t = 0.3$, S_3 is opened.
   (a) What is the maximum voltage on the capacitor?
   (b) At what time does the capacitor voltage pass through zero?
   (c) What is the final voltage on the capacitor?
   (d) Make a sketch of $V_C(t)$ versus time.

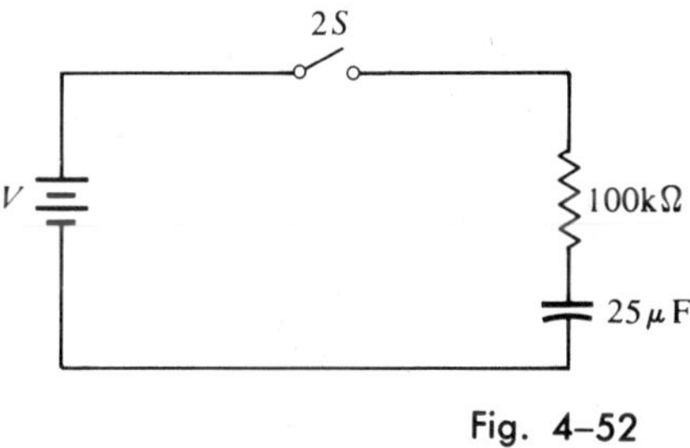

Fig. 4–52

6. The capacitor of Fig. 4–54 is initially uncharged. At $t = 0$, switch S_4 is placed in position 1. At $t - 3$ milliseconds, S_4 is moved instantly to position 2.
   (a) What is the maximum voltage on the capacitor?
   (b) What is the voltage on the capacitor at $t = 5$ milliseconds?

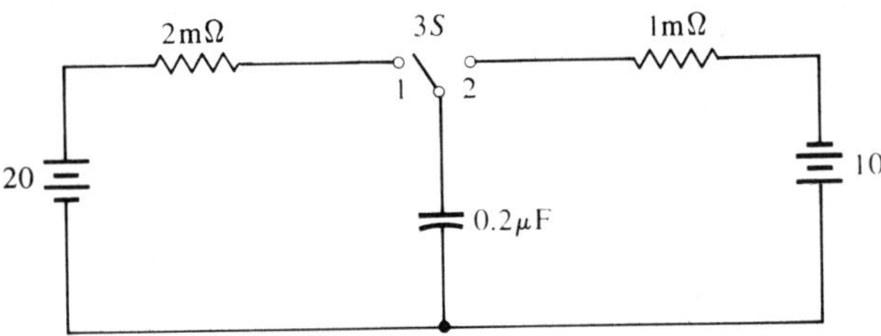

Fig. 4–53

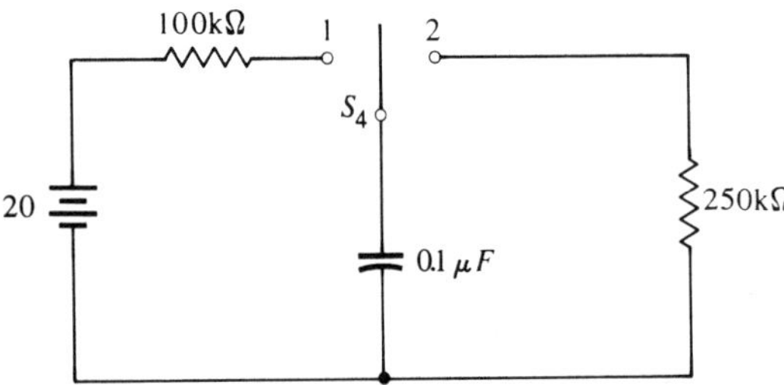

Fig. 4–54

7. Draw the equivalent circuit for charging the capacitor of Fig. 4–55.

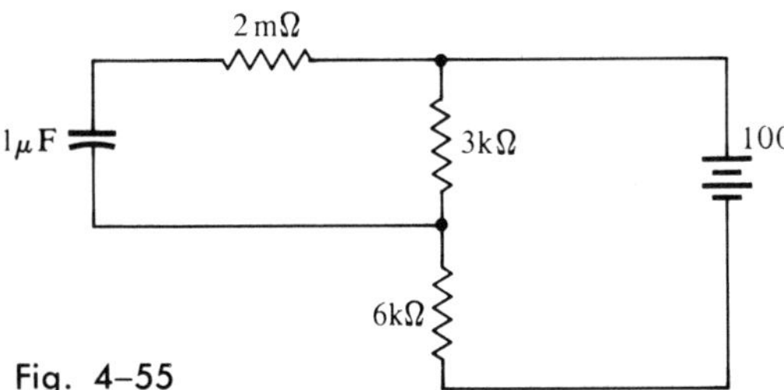

Fig. 4–55

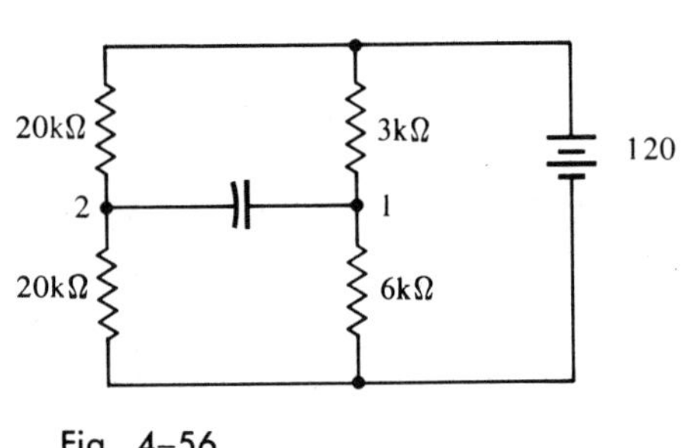

Fig. 4–56

8. Draw the equivalent circuit for charging the capacitor of Fig. 4–56.

9. The capacitor of Fig. 4–57 is charged in position 1 for 30 seconds. The switch is rotated clockwise, remaining in each position for 2.0 seconds.
   (a) Make a plot of the voltage on the capacitor for one complete rotation of the switch.
   (b) Make a plot of the capacitor voltage waveform you would observe on an oscilloscope if the vertical amplifier leads were placed across the capacitor. The scope has an input impedance of 1 megohm.

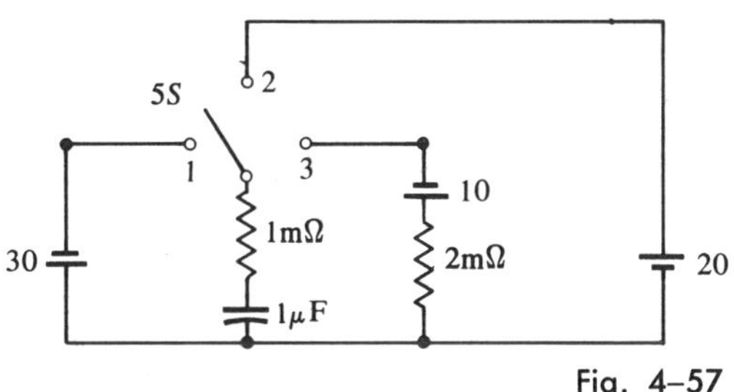

Fig. 4–57

10. The capacitor of Fig. 4–58 is initially uncharged. At $t = 0$, switch S_1 is put in position A and S_2 is put in position D. At the end of 5 seconds, the switches are placed in the opposite positions.

(a) Draw the equivalent circuit for charging the capacitor for each pair of switch settings.

(b) Plot the capacitor voltage for the first ten seconds. Use the polarity indicated.

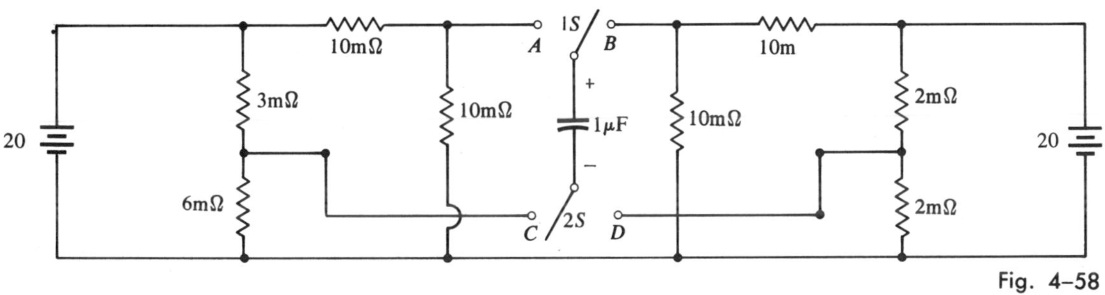

Fig. 4–58

11. Starting with switch S in Fig. 4–59 in the open position and no charge on the capacitor at $t = 0$

(a) Move S to position 1 and find the capacitor voltage at $t = 2$ seconds.

(b) Move S instantly to position 2 and determine the capacitor voltage 3 seconds later.

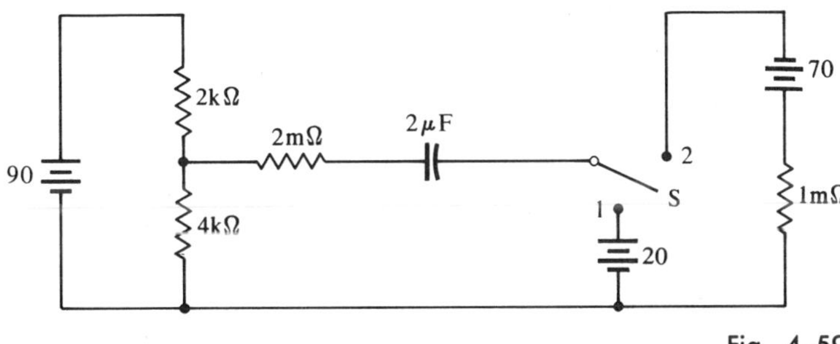

Fig. 4–59

12. Switch S_A of Fig. 4–60 is in the open position. The voltage on C is zero. At $t = 0$, the switch is placed in position 1. At $t = 250$ milliseconds, S_A is placed in position 2. At $t = 0.5$ seconds, S_A is opened.

(a) Plot the capacitor voltage for the first 500 milliseconds after S_A is placed in position 1 using $C = 0.1\ \mu F$.

(b) Repeat part A if $C = 0.05\ \mu F$.

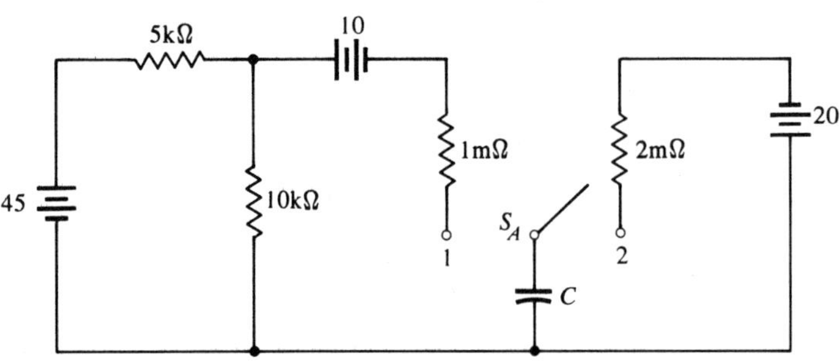

Fig. 4–60

13. Capacitor C in Fig. 4–61 is uncharged with switch S_B open. At $t = 0$, S_B is put in position 2. Plot the voltage on the capacitor for a complete clockwise rotation of the switch if it is left in each position for one time constant of the charging circuit in each position.

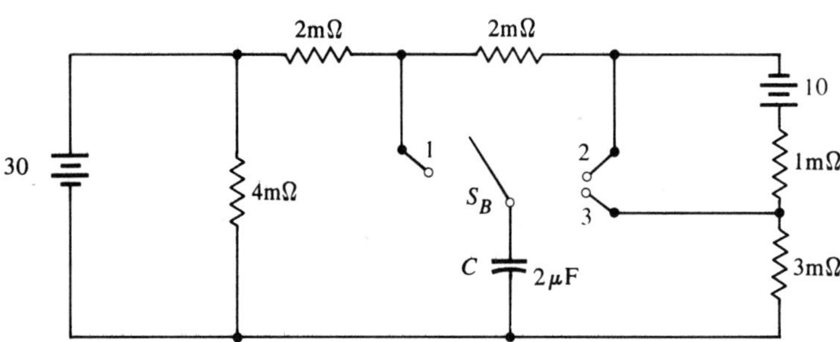

Fig. 4–61

14. The passive time delay circuit of Fig. 4–62 is combined with a transistor amplifier.

(a) Determine the value of R_B.

(b) If switch S_c is in position 1 until the capacitor is fully charged, determine the base current 0.5 milliseconds after S_c is placed in position 2.

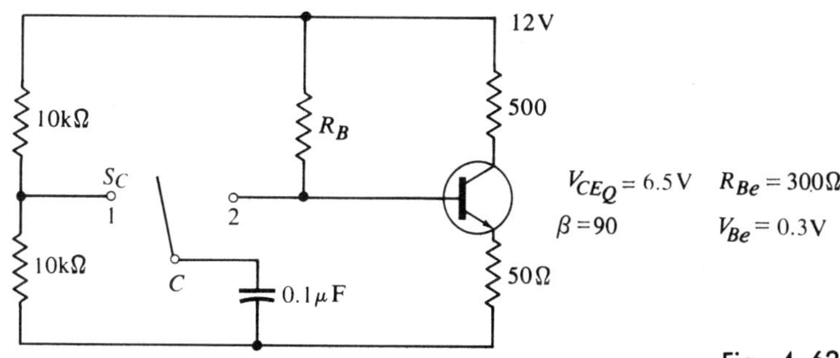

Fig. 4–62

15. The ac voltage source in Fig. 4–63 is $V_i = 70 \sin 377t$.
(a) Make the vector diagram showing the angle Θ.
(b) What is the angle of phase lag between the capacitor voltage and V_i?

16. The ac voltage source of Fig. 4–64 is $V = 100 \sin 100t$. Determine the amplitude and phase lag of the voltage across the capacitor.

17. How much energy is stored in the field of the coil of Fig. 4–65?

18. What is the time constant of the inductive circuit of 4–65?

19. How long after switch S_D is closed does it take for the current in the circuit of Fig. 4–66 to reach 40 mA?

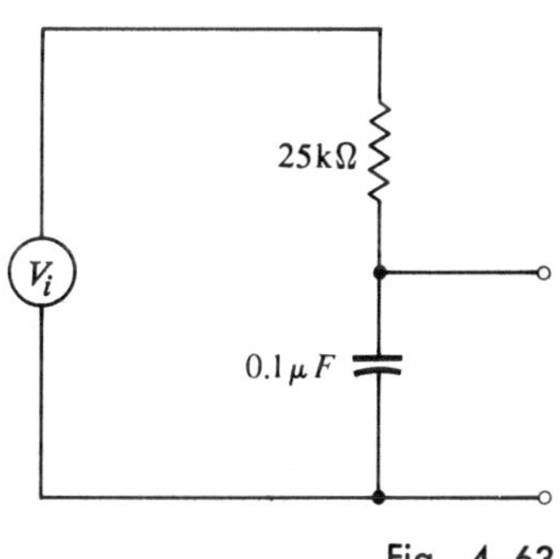

Fig. 4–63

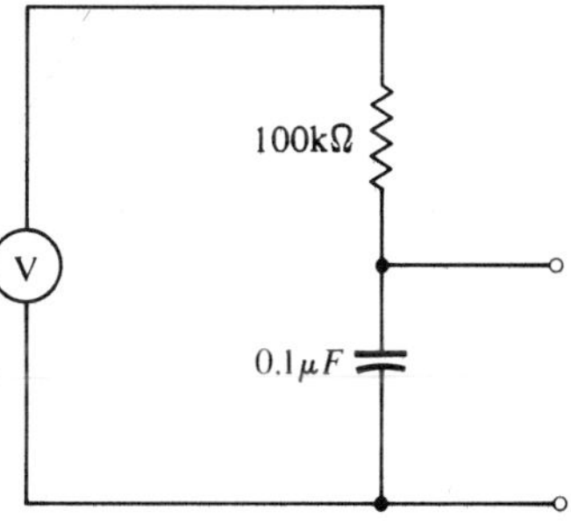

Fig. 4–64

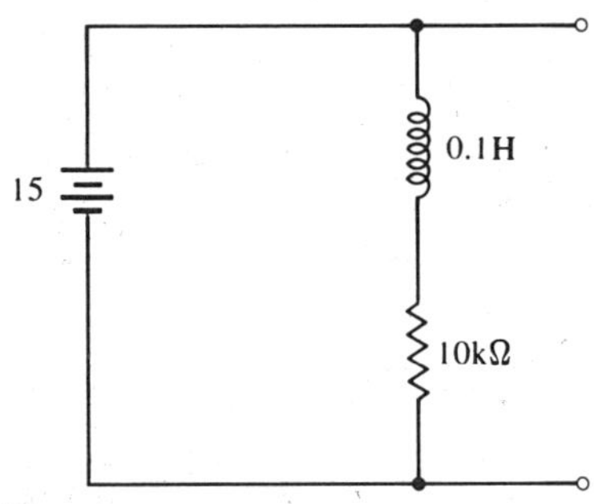

Fig. 4–65

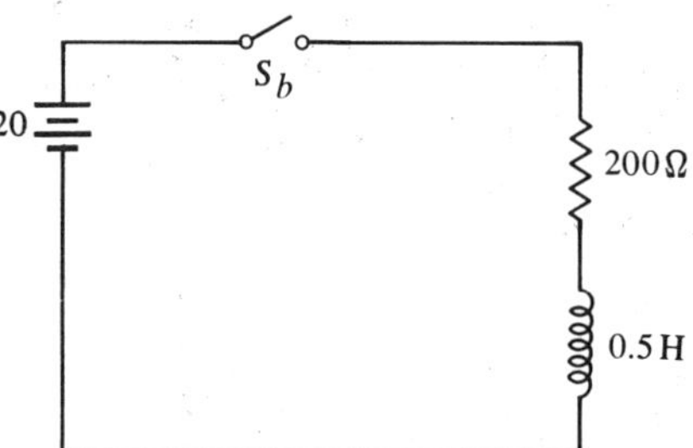

Fig. 4–66

5

PNPN Control Circuits

5–1 Introduction

We discussed the terminal characteristics of PNPN switching devices in Chapter 1. The SCR and TRIAC are the most frequently used of these devices. The design of PNPN circuits relies heavily on capacitors and inductors for timing and control. We will use the terminal characteristics of PNPN devices and the passive time delay elements to design control circuits for industrial electronics applications.

5–2 SCR Circuits

Figure 5–1 is an illustration of the advantage of the SCR in industrial control circuits. The normal rectifier diode in the circuit of Fig. 5–1(a) provides a half-wave rectified current to the load. The amount of power delivered to the fixed load resistor remains constant unless the source voltage is changed. If the diode is replaced by an SCR, then the

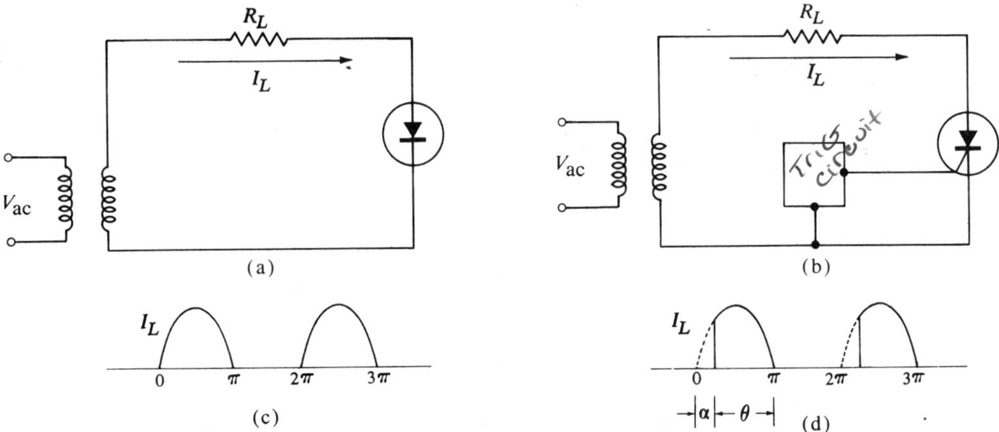

Fig. 5–1 (a) Rectifier diode in half-wave control circuit, (b) SCR in half-wave control circuit, (c) load current in circuit (a), (d) load current in circuit (b).

amount of power delivered to the fixed load resistor can be controlled by varying the firing angle of the SCR. The SCR then conducts during the remainder of the 180° while the anode is more positive than the cathode. The firing angle (α) and the conduction angle (θ) are shown in Fig. 5–1(d). In summary, in Fig. 5–1(b), the SCR blocks the flow of current in the circuit until a triggering signal is applied at the gate terminal. The trigger signal turns on the SCR; the load draws current during the period of conduction. The SCR is turned off when the anode voltage falls below the cathode voltage. The five major phases of operation of the circuit of Fig. 5–1(b) provide a good checklist for the specification of the SCR. They are:

(a) blocking
(b) triggering
(c) turn-on
(d) conduction
(e) turn-off

Let us look at each one of these important phases separately.

5–2–1 Blocking

We saw, in Fig. 1–28, that the SCR will block the flow of current in either direction between the anode and cathode if the applied voltage is kept within the proper limits. It is these *proper voltage limits* that must be specified in the design of SCR circuits. Specifications are usually stated at the maximum operating temperature. Specified voltage limits are:

(1) Repetitive Peak Reverse Voltage—V_{RRM}—This represents the maximum reverse voltage that can be repeatedly applied to the SCR with the gate circuit open and still block current flow.

(2) Nonrepetitive Peak Reverse Voltage—V_{RSM}—This represents the maximum transient reverse voltage level that should be applied to the SCR under reverse blocking conditions with the gate circuit open.

(3) Peak Forward Blocking Voltage—V_{FOM}—This represents the maximum instantaneous value of positive anode voltage guaranteed by the manufacturer to maintain the blocking capability of the SCR with the gate circuit open. If this voltage value is exceeded, even by transients, the SCR might go into conduction.

(4) Peak Forward Voltage—*PFV*—This represents the maximum instantaneous positive voltage that should ever be applied at the anode under any conditions. If this voltage is exceeded, even under transient conditions, then the SCR might be permanently damaged.

(5) Peak Forward Blocking Voltage with a Gate Resistor—V_{DRM}—This represents the maximum instantaneous value of positive anode voltage guaranteed by the manufacturer to maintain the blocking capability of the SCR with a specific gate-to-cathode resistor. This technique is often used to increase the forward blocking voltage of low current SCRs.

(6) Forward Breakover Voltage—V_{BRO}—This represents the maximum positive anode-to-cathode voltage that can be applied without causing the SCR to go into the conducting state with the gate circuit open.

The diagram of Fig. 5–2(a) will help to explain the SCR blocking voltage specifications for the center-tapped full-wave rectifier of Fig. 5–2(b) using SCRs in place of rectifier diodes. The maximum positive voltage at the anode of either SCR will be 170 volts. V_{FOM} is then specified at greater than 170 volts, say 200 volts. With an external gate-to-cathode resistor, the forward blocking voltage might be

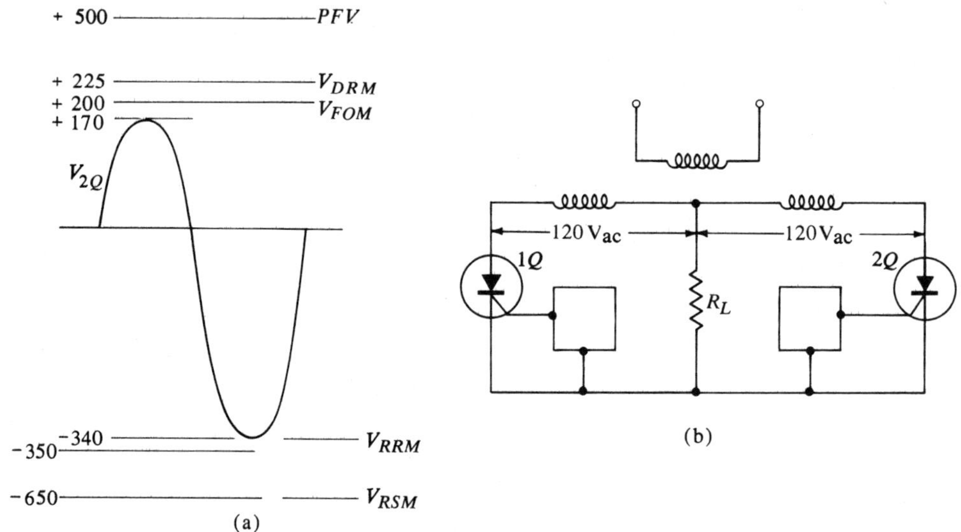

Fig. 5–2 (a) Maximum SCR anode voltage specifications, (b) center-tapped full-wave SCR rectifier.

increased to 225 volts, V_{DRM}. A *PFV* of 500 volts would allow for 330 volt line transients in addition to the normal 170 volts without damage to the SCR. The normal reverse voltage applied to either SCR in this circuit will be 2 × 170, or 340 volts. V_{RRM} is specified at more than 340 volts, say 350 volts. Then V_{RSM} should be specified to handle 330 volt transients without damage to the SCR. Let V_{RSM} be 650 volts.

Most SCRs are made so that V_{FOM} and V_{RRM} are equal. In this rectifier application, you would probably specify V_{FOM} and V_{RRM} at 350 volts.

5–2–2 Triggering

An SCR may be triggered into conduction by exceeding the anode forward blocking voltage, V_{FOM} or V_{DRM}, by exceeding the maximum allowable temperature, by excessive rate of change of anode voltage, *dv/dt*, or by applying a signal at the gate terminal. Of these methods, nearly all triggering is done with a gate signal. We shall discuss gate triggering as the intentional mode of turning the SCR on. "Turn-on" by the other methods is usually unintentional and prevented if possible.

I have found that successful gate triggering depends on the satisfaction of three conditions:

(1) trigger current and voltage must be within the trigger zone;

(2) gate dissipation should be minimized;
(3) trigger signal must be properly timed.

Let us look at these conditions and the factors that influence them.

The gate-triggering circuit is usually designed after the SCR has been selected. The manufacturer will provide the data from which you can determine the trigger zone.

Figure 5–3 is the gate triggering data provided by one SCR manufacturer. The trigger zone is defined by limiting

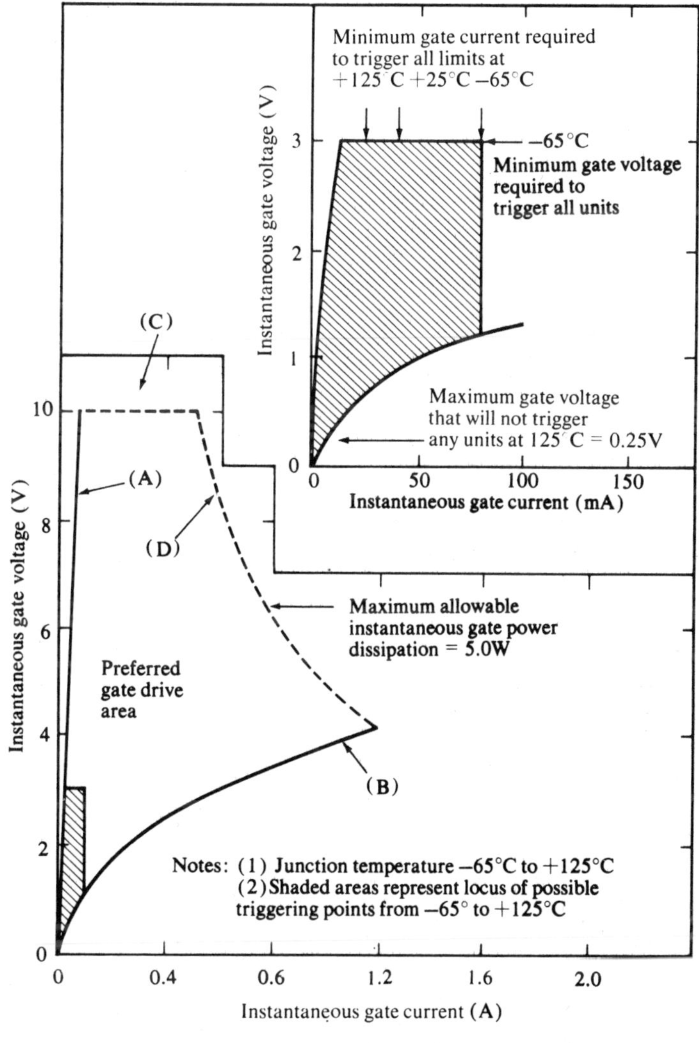

Courtesy of General Electric Semiconductor Dept., Syracuse, N.Y.

Fig. 5–3 SCR gate characteristics.

diode characteristic curves (A and B), maximum power dissipation hyperbola (D), and maximum gate voltage (C). The shaded area in the lower left-hand corner represents the combinations of gate voltages and gate currents that will not guarantee triggering of this particular SCR. Similar data must be provided for each type of SCR.

After the trigger zone is identified, the triggering source must be designed to operate within that zone. This is best done by simple load line analysis.

The Thévenized trigger source is shown in Fig. 5–4. R_g is the equivalent source impedance and V_g is the open-circuit source voltage. The source load line is shown in Fig. 5–5 for three different values of source resistance with $V_g = 10$ volts.

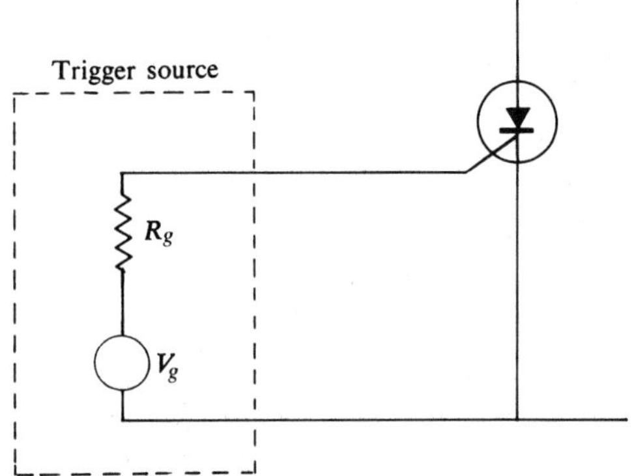

Fig. 5–4 SCR gate triggering source.

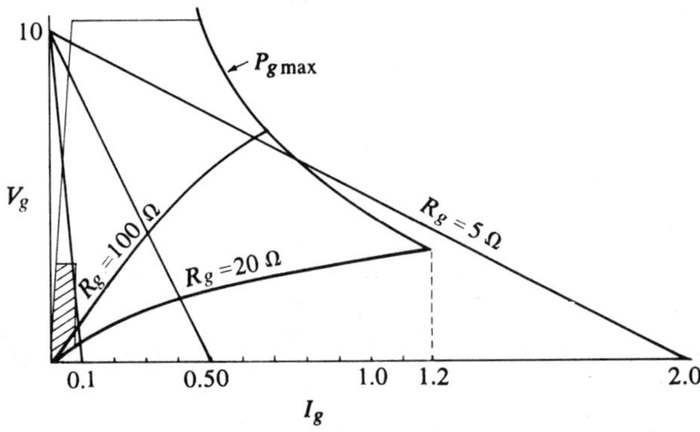

Fig. 5–5 Gate source load lines.

We notice that R_g = 100 ohms is too large because it passes through the shaded area. R_g = 5 ohms is too small because it intersects the maximum power dissipation curve. R_g = 20 ohms is a reasonable value of source impedance for this particular SCR. You will find that most SCRs require low-impedance trigger sources.

The gate-to-cathode junction of the SCR is essentially a P–N junction. Whenever voltage is applied at the junction such that a forward bias is achieved, current flows across the junction and power is dissipated in the form of heat. Gate power dissipation can be minimized by using a pulse triggering source rather than a continuous signal. Pulses of 500-microsecond duration are sufficient to trigger most SCRs. You should refer to the manufacturer's specification to be sure the pulse width is sufficient for a given SCR. We will discuss the design of pulse circuits that can be used as trigger sources later in the chapter.

The pulses from the triggering source must be timed so that they cause the SCR to "turn on" at the proper time. This is called *synchronization* of the triggering pulse. Figure 5–6 shows the relationship between the anode voltage and triggering pulses in an ac-controlled SCR circuit. The technique for synchronizing the gate triggering source with the anode voltage depends on the specific design of the source. We will get into the details of synchronization when we discuss the design of pulse circuits later in this chapter.

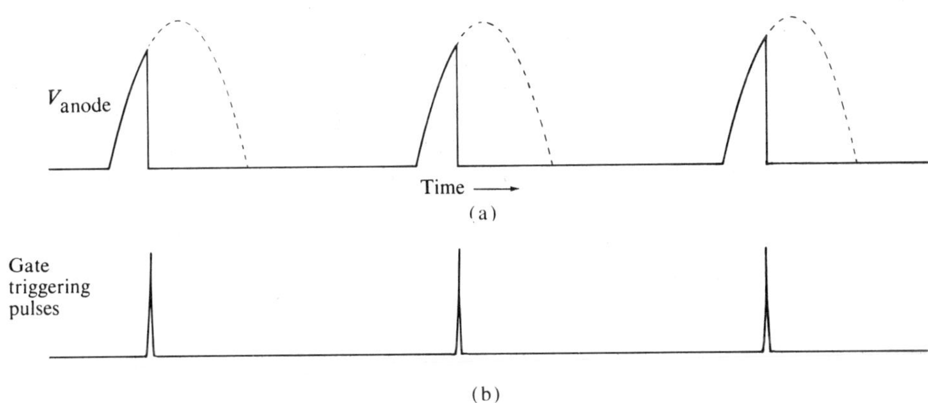

Fig. 5–6 (a) SCR anode voltage for controlled triggering, (b) triggering pulses.

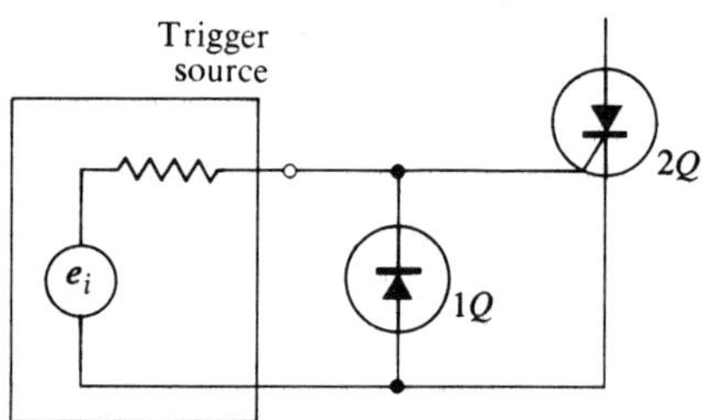

Fig. 5–7 Clamping the gate reverse bias with a rectifier diode.

One precaution that must be taken in applying pulse sources at the gate of the SCR is that the reverse bias limit of the gate is not exceeded. Most manufacturers will specify the maximum allowable negative voltage between the gate and cathode. A common technique for limiting the negative voltage at the gate is to clamp the gate to the forward drop across a diode. This technique is illustrated in Fig. 5–7. If a negative voltage is applied at the gate, diode 1*Q* conducts and limits the gate voltage to approximately –0.7 volts.

5–2–3 Turn-on of SCRs

"Turn-on" occurs when an SCR switches from the forward blocking state to conduction. The turn-on may be intentionally caused by gate triggering or may be the result of transient voltages in the anode circuit. In either case, the designer must be sure that the current flow in the anode circuit does not increase too rapidly or the SCR might be damaged. The maximum *di*/*dt* is usually stated by the manufacturer in amps per microsecond. The inductance inherent in many industrial loads, such as motors and transformers, tends to limit the rate of change of current in the anode circuit depending on the time constant *L*/*R*. If the load is highly resistive, then switching an SCR at high voltages could result in excessive *di*/*dt* due to the extremely short rise time of the SCR (typically 1 or 2 microseconds). It is sometimes necessary to add external inductance in the load circuit as in Fig. 5–8. Let us look at a sample calculation of the amount of inductance required in the load circuit to protect the SCR.

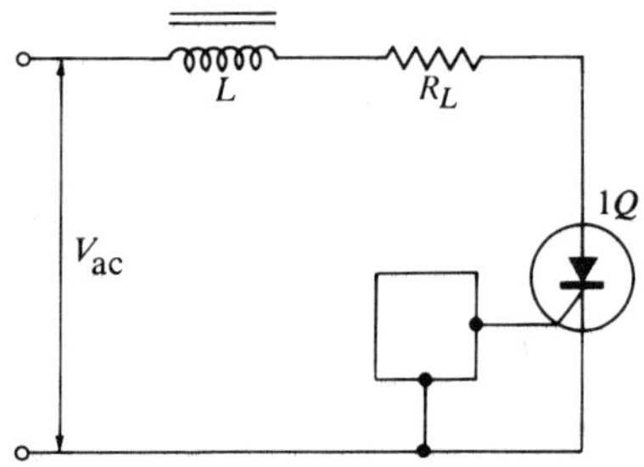

Fig. 5–8 External inductance in the anode circuit to limit *di*/*dt*.

Example 5–1. In the circuit of Fig. 5–8, if V_{ac} = 120 V rms, R_L = 50 Ω, and 1*Q* is a G.E. C106B1 SCR where *di*/*dt* maximum is 50A/μs, determine the value of *L* to protect the SCR.

The maximum *di*/*dt* would occur if the SCR were switched on when the voltage was at its peak.

$$V_{peak} = \sqrt{2}\, V_{rms} = \sqrt{2} \times 120 = 170\text{ V} \tag{5.1}$$

From Eq. (4.20)

$$\left(\frac{di}{dt}\right)_{max} = \frac{V}{L}$$

then

$$L = V/\left(\frac{di}{dt}\right)_{max} \tag{5.2}$$

It would not be wise to design at the specified maximum; some allowance should be made for variations in line voltage and tolerance on the value of L. Let us use $(di/dt)_{max} = 40$ A/μs or 80% of the specified value. Then $L = 170/40 \times 10^{-6} = 4.25\ \mu$H.

Unintentional turn-on of SCRs may be caused by rapidly increasing forward voltage in the anode circuit. Manufacturers will specify the maximum dv/dt a given SCR will withstand without going into conduction. This specification is at a given operating temperature, usually the maximum operating temperature of the device. Since the SCR is a junction semiconductor device, it is natural to characterize its dv/dt limitation in terms of a capacitive charge curve. Figure 5–9 is a graphical display of how dv/dt limits are specified. The piecewise linear approximation (shown in dotted lines) of Fig. 4–18 is used to define dv/dt. Between $t = 0$ and $t = \tau_1$

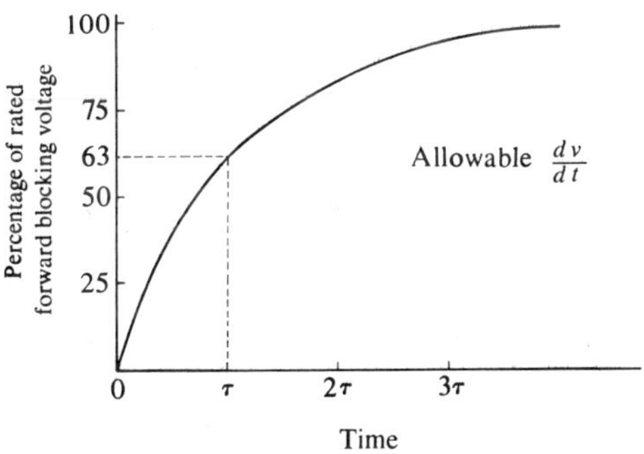

Fig. 5–9 Characterization of dv/dt anode limitations of SCRs.

$$\frac{dv}{dt} = \frac{0.63\ V_{FOM}}{\tau} \tag{5.3}$$

An example of the interpretation of dv/dt specification and its application to an SCR circuit will probably increase our understanding of this important part of the design of reliable SCR controls.

Example 5–2. The SCR in the circuit of Fig. 5–10(a) is a G.E. C106B1 with a specified dv/dt limit of 20 V/μs and $V_{DRM} = 200$ V. A step input of 100 V will be applied to the anode before gate

triggering occurs. Determine the earliest time at which the anode voltage may reach 100 V without causing turn-on of the SCR.
From Eq. (5.3)

$$\tau = \frac{0.63 \times 200}{20} \times 10^{-6} \qquad \textbf{(5.4)}$$

$$= 6.3\ \mu s$$

The required 100 V is 50% of the rated forward blocking voltage. From Fig. 5–9, 50% of rated forward blocking voltage occurs at 0.69 τ. The earliest time at which the anode is allowed to reach 100 V is 0.69 × 6.30 μs or 4.35 μs.

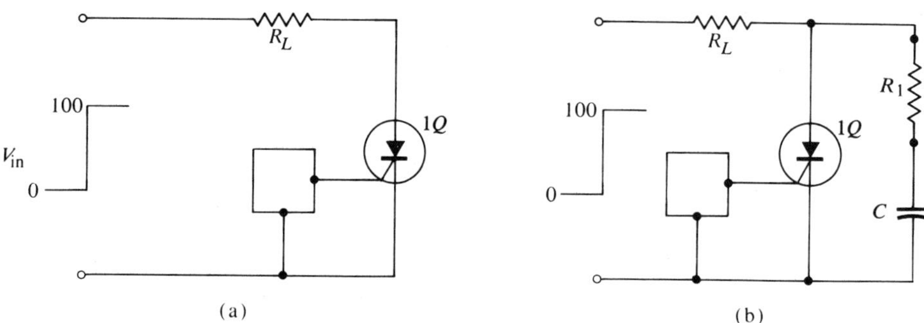

Fig. 5–10 (a) Step input to SCR circuit without *dv/dt* protection, (b) *RC* circuit to limit *dv/dt*.

Since a step input implies an instantaneous change in voltage, the SCR in Fig. 5–10(a) would turn-on as soon as the anode voltage was applied. Figure 5–10(b) shows a technique for limiting *dv/dt* at the anode of the SCR. The $R_L C$ time constant limits the rate of rise anode voltage. The resistor R_1 limits the rate of discharge current in the SCR due to the capacitor.

Let us look at calculation of the values of R_1 and C.

Example 5–3. The SCR in the circuit of Fig. 5–10(b) is the G.E. C106B1 (same as Example 5–2). If $R_L = 150\ \Omega$, and *dv/dt* is 50 V/μs, determine the values of R_1 and C to be used. From Example 5–2, $\tau = 6.3\ \mu s$. Then:

$$\tau = R_L C = 6.3 \times 10^{-6}$$

$$C = \frac{6.3 \times 10^{-6}}{R_L} = \frac{6.3}{150} \times 10^{-6} = 0.042\ \mu F \qquad \textbf{(5.6)}$$

The closest standard value of 0.047 μF, 200 V, is specified for C.

From Eq. (4.10), for the discharging capacitor,

$$\frac{di}{dt} = \frac{V}{R^2C} \exp\left(-\frac{t}{RC}\right) \tag{5.7}$$

$$\left(\frac{di}{dt}\right)_{max} \text{ occurs at } t = 0$$

$$\left(\frac{di}{dt}\right)_{max} = \frac{V}{R^2C} \tag{5.8}$$

or

$$R_1 = \sqrt{\frac{V}{\left(\frac{di}{dt}\right)_{max} C}} \tag{5.9}$$

$$R_1 = \sqrt{\frac{100}{50} \times \frac{1}{.047}} = 6.5\ \Omega \tag{5.10}$$

This says that almost any small resistor will be satisfactory. In order to limit the maximum current in the capacitor, let R_1 = 100 Ω. From Fig. 4-11, $I_{max} = V/R_1 = 100/100 = 1$ A, which is reasonable. Figure 5-11 shows the complete circuit.

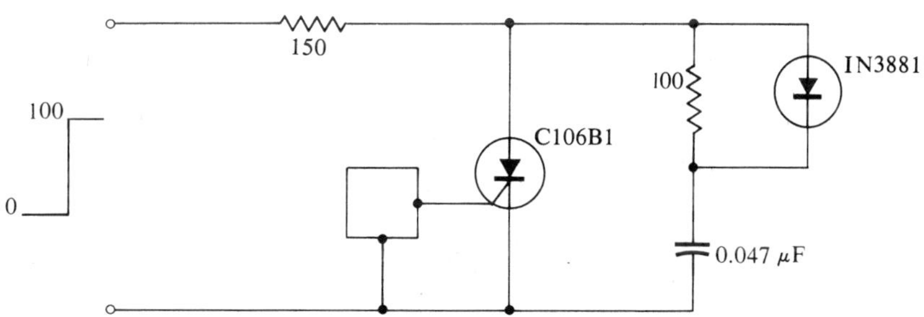

Fig. 5-11 Complete circuit of Example 5-3.

5-2-4 SCR in Conduction

Once an SCR is "triggered" and turn-on occurs, the anode voltage drops to approximately 1 volt and anode current is limited by the external resistance in the anode circuit. This is called the conduction mode of operation. The angle at which triggering occurs is called the firing angle, α. The remaining portion of the conduction cycle is called the conduction angle, θ. The sum of the firing and conduction angles is 180° in a resistive circuit. If the load is highly inductive, such as a motor or welding transformer, then the sum of the two angles might exceed 180° due to the inductive carry-over current of the load. Figure 5–12 is an illustration of the firing and conduction angles for a resistive load and an inductive load.

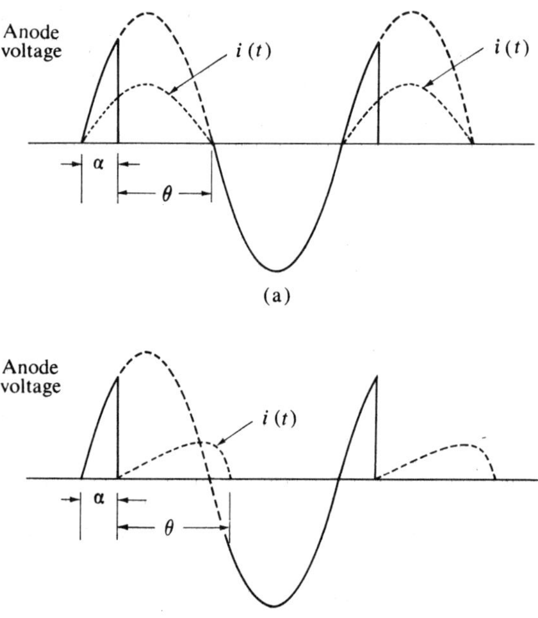

Fig. 5–12 SCR anode voltage and current to (a) resistive load; (b) inductive load.

Manufacturers rate SCRs by the maximum allowable rms current in a resistive circuit with a conduction angle of 180° operating from a sinusoidal voltage source. An 8-amp SCR will conduct 8 amps half-wave rms sinusoidal current. Figure 5–13 is a graphical presentation of the allowable forward anode current for an SCR rated at 8 A. The SCR case must not exceed the maximum allowable temperature to prevent overheating of the junction. The manufacturer also specifies this maximum allowable junction temperature.

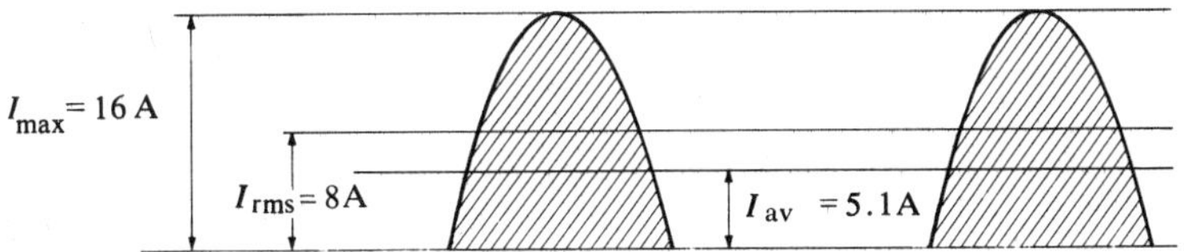

Fig. 5–13 Maximum allowable currents in an 8-A SCR.

We are interested in operating the SCR at conduction angles less than 180°. We would like to know the maximum allowable case temperatures for angles of conduction between 0° and 180°. Graphical data is usually provided from which this information can be obtained. Figure 5–14(a) is the form in which this data is provided. Notice that the manufacturers use "average forward current" rather than "rms forward current." The reason is that

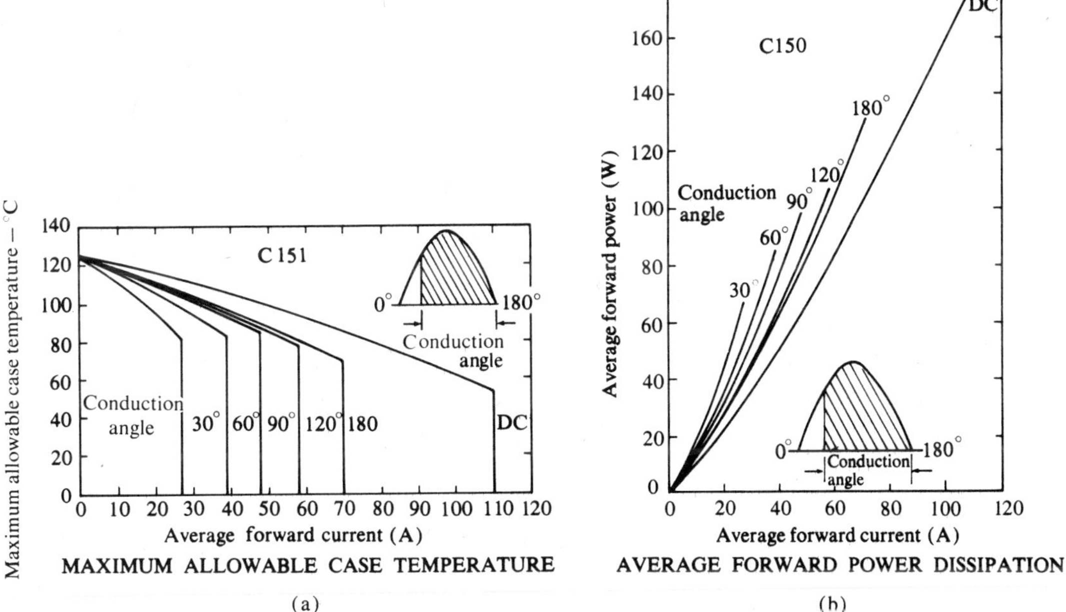

Courtesy of General Electric Semiconductor Dept., Syracuse, N.Y.

Fig. 5–14 Current rating and power dissipation for a typical SCR.

"average current" can be measured with an ordinary dc ammeter; rms current measurement would require a more expensive "true rms ammeter" due to the nonsinusoidal current waveform. Figure 5–15 can be used to convert rms current to average current in the SCR circuit. The data of Fig. 5–14(b) relate the power dissipated in the SCR to the average forward current for different conduction angles. This data includes power dissipated in the gate circuit, dissipated power due to the small leakage current that flows when the SCR is in the blocking state, and switching losses.

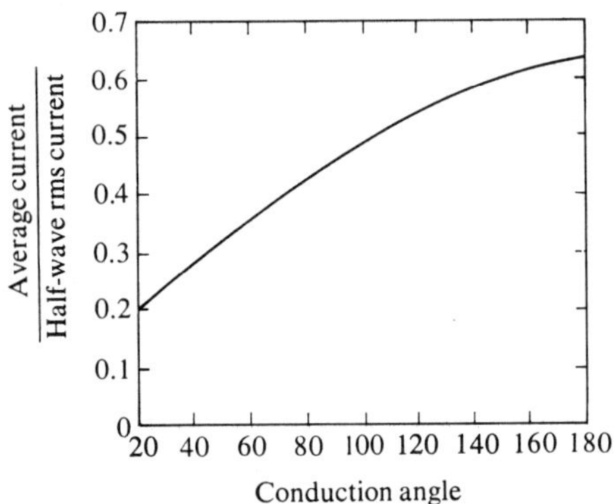

Fig. 5–15 Ratio of average current to half-wave rms current vs conduction angle.

Let us look at the interpretation of the data of Fig. 5–14. From Fig. 5–14(a), when the conduction angle is 180°, the maximum allowable average forward current is 70 amperes. Using the conversion curve of Fig. 5–15,

$$\frac{I_{av}}{I_{rms}} = \frac{\text{average current}}{\text{half-wave rms current}} = 0.636$$

at 180° conduction angle. For this SCR,

$$I_{rms} = \frac{70}{0.636} = 110\text{ A}$$

This is a 110-A SCR. Notice that the maximum allowable dc forward current is the same as the maximum allowable half-wave rms current, in this case 110 amperes. Operating at 110 amperes rms requires that the case temperature be kept at 70°C or less to prevent the junction temperature from exceeding 125°C. From Fig. 5–14(b), the power being dissipated at 110 amperes rms is 130 watts. The temperature drop between the case of the SCR and ambient temperature of 25°C is 70°C − 25°C = 45°C. Then the total thermal impedance between the case and air must not exceed 45°C/130W = 0.346°C/W.

It requires an excellent heat sink to keep the thermal resistance at such a low level! Let us look at an example of how this particular SCR might be used at conduction angles less than 180°.

Example 5–4. The 110-A SCR whose ratings are shown in Fig. 5–14 will be used in a resistive circuit. It is required that the average load current will be 45 A at 90° conduction angle. Determine the rms current being delivered to the load and specify the heat sink thermal impedance for operation at 25°C ambient temperature.

The ratio of rms forward current to average forward current for a specific conduction angle is called the *current form factor*,

F_θ. A plot of F_θ as a function of conduction angle is shown in Fig. 5-16. At 90° conduction angle, $F = I_{rms}/I_{av} = 2.2$. The rms current being delivered to the load is $2.2 \times 45 = 99$ A. From Fig. 5-14(b), the power dissipation is 95 W. From Fig. 5-14(a), the maximum allowable case temperature at the operating point specified is 90°C.

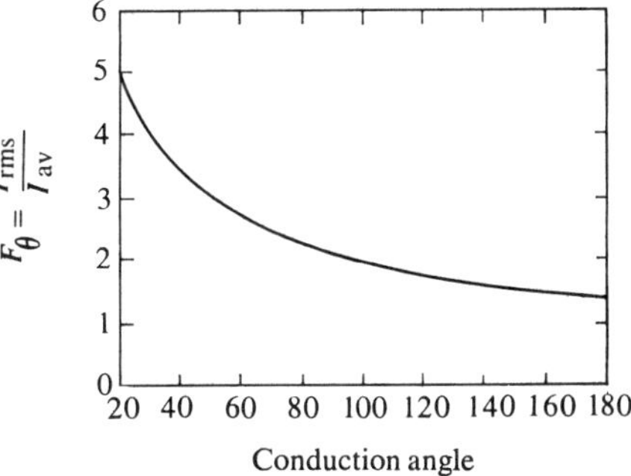

Fig. 5-16 Current form factor vs conduction angle.

From the circuit of Fig. 2-10(b),

$$\Delta T = T_c - T_A = P_d \theta_{CA} \tag{5.11}$$

The thermal impedance between the case of the SCR and air must be less than

$$\frac{T_c - T_a}{P_d} = \frac{90°\text{C} - 25°\text{C}}{95\text{ W}} = 0.685°\text{C/W} \tag{5.12}$$

Manufacturers' ratings are for resistive loads or unity power factor. Most industrial loads are not unity power factor but are inductive with lagging power factors. Since unity power factor loads place the heaviest demand on the SCR in terms of power dissipation, heat sinking, and current limitations, ratings such as those of Fig. 5-14(a) and (b) are slightly conservative for most applications that you will encounter.

5-2-5 Turn-off of SCRs

In order to effectively control the power delivered to the load in an SCR circuit, we must be able to turn off the SCR or switch it from the *conducting state* to the *forward blocking state*. The gate has no control over the SCR once it goes into conduction. Turn-off must be achieved in the anode-to-cathode circuit. There are three ways in which turn-off, or *commutation* as it is commonly called, can be achieved:

(1) reverse the anode-to-cathode voltage;
(2) reduce the anode current below the holding level;
(3) force current through the anode circuit in the reverse direction.

When a sinusoidal voltage source is used, turnoff of the SCR occurs automatically at the end of each positive half cycle of applied voltage. When a dc or unidirectional voltage source is used, the anode current must be interrupted or a passive energy storage element is used to attempt to force current through the anode circuit in the reverse direction, which reverses the anode voltage. Since the SCR is a PNPN junction semiconductor structure, a minimum time is required for the charges to reverse at the junction after conduction has been interrupted. This time is called the *turn-*

off time of the SCR. The manufacturer will specify the minimum turnoff time for each SCR under specified operating conditions. If forward voltage is applied to the anode before the turnoff time has expired, the SCR will go into conduction without a gate trigger signal. The turnoff time for SCRs is typically 10 to 100 μs.

We noted in Fig. 1–28 that each SCR has a specified holding current. This is the minimum value of anode current required to keep the device in full conduction. If the anode current is reduced below the holding current level, then turnoff will occur. Two simple ways to reduce anode current are shown in Fig. 5–17. We could use a series switch to interrupt current flow or a low-impedance shunt switch to divert the flow of current from the anode.

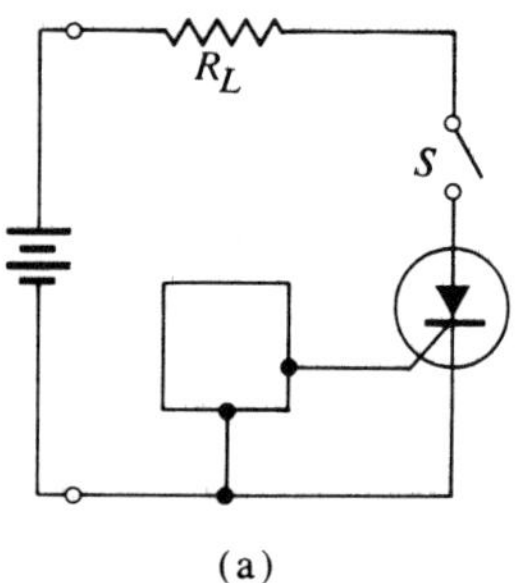

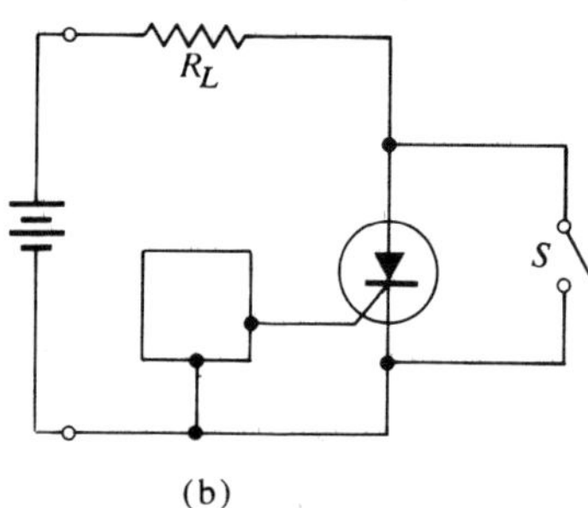

Fig. 5–17 (a) Series interruption of anode current for commutation, (b) shunting of anode for commutation.

The most common method of turn-off of the SCR in a dc circuit is self-commutation. Figure 5–18 is a diagram of the very popular Jones circuit used for self-commutation in dc SCR controls. 1*Q* is the load conducting SCR. 2*Q* is the turnoff SCR. 3*Q* is the commutating diode. 1*L* and 2*L* are the primary and secondary windings on the charging transformer. When 1*Q* is triggered at the gate, current flows through 1*Q*, 1*L*, and the load R_L. The current flow through 1*L* induces a voltage across 2*L*, which forward biases diode 3*Q*. Current flows through 2*L*, 3*Q*, and 1*Q* to charge capacitor *C* with the polarity shown. After the capacitor is fully charged, current flow in the charging circuit decreases to zero and the load continues to draw current from the dc source. When $2Q$ is triggered at its gate, the voltage on capacitor C is applied across $1Q$. This negative anode-to-cathode potential forces $1Q$ to turn off. The capacitor then charges through $2Q$, $1L$, and R_L opposite to the polarity shown in Fig. 5–18. When $1Q$ is triggered at its gate again, the circuit operation repeats. This type of commutation is used in commercial electric vehicles for speed control. We will look at this application in more detail later.

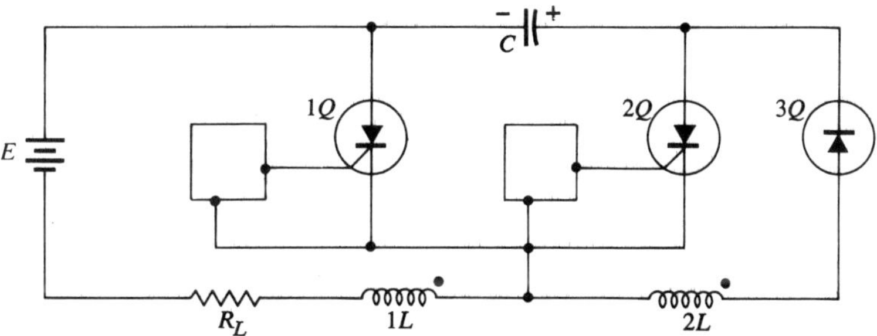

Fig. 5–18 Jones circuit for self-commutation.

Another very useful commutation technique is shown in Fig. 5–19. $1Q$ and $2Q$ are both load conducting SCRs. When neither SCR is conducting, there is virtually no charge on the capacitor. If a triggering pulse is applied to the gate of $1Q$, current flows through $1R$, R_L, and $1Q$. Current also flows through $2R$, C, and $1Q$ to charge the capacitor with the polarity shown in Fig. 5–19. When a gate triggering pulse is applied at $2Q$, $2Q$ conducts—dropping its anode voltage to approximately one volt. The capacitor voltage is then placed across the anode to cathode of $1Q$.

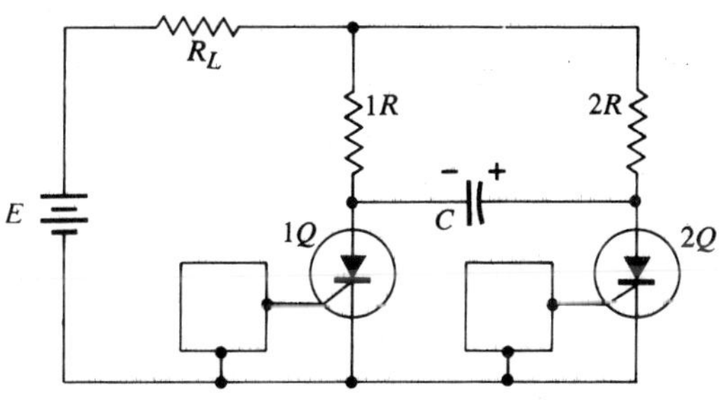

Fig. 5–19 Commutation with two load conducting SCRs.

This reverse anode-to-cathode voltage causes $1Q$ to turnoff. The capacitor discharge path is then $1R$, R_L, $2Q$, and E. Current flows through $1R$ and $2Q$ to charge the capacitor to the polarity opposite that shown in Fig. 5–19. The circuit is now ready for a gate triggering signal at $1Q$ and the cycle repeats. This type of commutation is readily adaptable to digital brushless motor controls.

5–2–6 Inverse Parallel Operation of SCRs

When an SCR is used in an ac control circuit, current flows through the load only when anode voltage is positive with respect to the cathode. At maximum conduction angle, current only flows half the time. This is adequate for some dc loads; ac loads require that current flows during both half cycles. Two SCRs are used in a back-to-back or *inverse parallel* configuration to control high current ac loads.

Figure 5–20 shows the inverse parallel connection of two SCRs and the full-wave current flow. Each SCR has its own gate trigger circuit so that the firing angles are independently controlled. The conduction angles of the SCRs are usually the same so that load current is equally shared by the SCRs. This provides a symmetrical load current with an average value of zero. Any nonsymmetry in the load current results in a dc current component which might cause saturation or excessive heat dissipation in inductive loads such as transformers.

When selecting SCRs for inverse parallel operation, remember that the mean squared load current is the sum of the mean squared current flow in each SCR, or

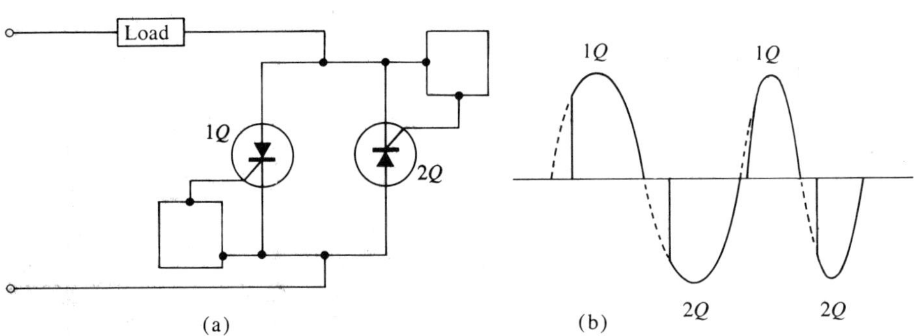

Fig. 5–20 (a) Inverse parallel operation of SCRs in an ac circuit, (b) SCR current flow with a purely resistive load.

$$[I_{\text{rms}}]^2_{\text{load}} = [I_{\text{rms}}]^2_{Q_1} + [I_{\text{rms}}]^2_{Q_2} \tag{5.13}$$

For symmetrical operation,

$$[I_{\text{rms}}]_{Q_1} = [I_{\text{rms}}]_{Q_2} = \frac{[I_{\text{rms}}]_{\text{load}}}{\sqrt{2}} \tag{5.14}$$

Special care must be exercised to prevent undesired triggering due to dv/dt transients in inverse parallel operation with an inductive load. Figure 5–21 shows the waveforms in a full-wave SCR control with a highly inductive load. Notice the inductive carry-over of current flow in $1Q$ extends beyond the 180° point and into the portion of the applied voltage waveform when $2Q$ is forward-biased. During the early portion of the time when $2Q$ is forward-biased, its anode voltage is held at approximately 1 volt by $1Q$. At the

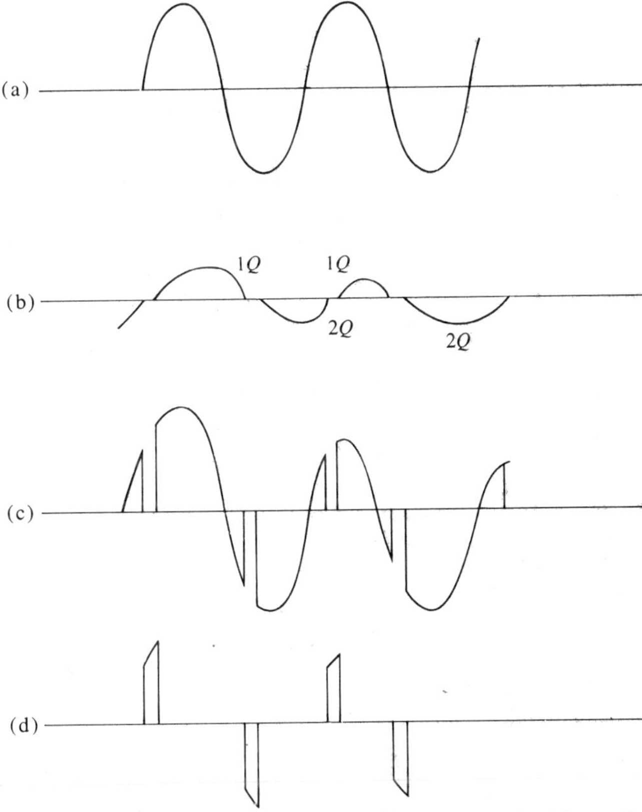

Fig. 5–21 Wavefore in the circuit of Fig. 5–20(a) with an inductive load: (a) source voltage, (b) SCR current flow, (c) load voltage, (d) SCR voltage.

point of commutation of 1Q, the voltage on 2Q suddenly rises to the value of the source voltage. It is at this point that the value of dv/dt will cause 2Q to turn on if a protective circuit, such as we discussed in section 5–2–3, is not included in the design of the control.

5–3 TRIAC Circuits

We noticed in section 5–2 that the SCR is only capable of half-wave rectification. In order to achieve ac full-wave control, two SCRs are used in back-to-back or inverse parallel connection. It would seem natural then to design a semiconductor device that could replace the inverse parallel SCR configuration. The TRIAC performs that function. The term TRIAC is derived from triode ac switch. TRIACs are limited to relatively low-current application. As of 1972, TRIACs with conduction capability of more than 50 amperes (rms) were not available commercially.

Figure 5–22(a) is a typical TRIAC control. The load current waveform of Fig. 5–22(c) is the same as Fig. 5–20(b) for inverse parallel SCR control. The firing angle, α, and the conduction angle, Θ, are shown in Fig. 5–22(c). The total conduction angle is the sum of Θ_1 and Θ_2.

The design of TRIAC switching circuits is basically the same as for SCR circuits. We will note the important exceptions in each of the design phases.

5–3–1 Blocking

The TRIAC will block voltage of either polarity so long as the magnitude of the voltage is less than the forward

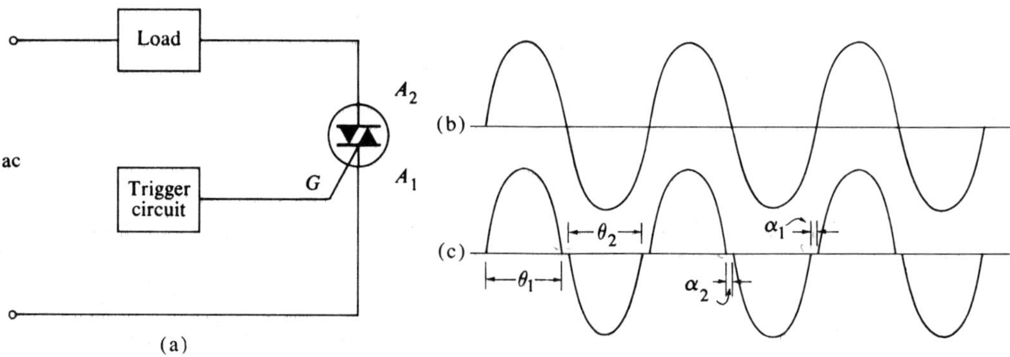

Fig. 5–22 (a) Typical TRIAC control circuit, (b) source voltage, (c) load current with a resistive load.

breakover limitation. Figure 5–23 shows that the anode current is limited to a very low leakage level, usually less than a few milliamperes under forward blocking conditions. There is no reverse voltage specification for the TRIAC. V_{FOM}, PFV, V_{DRM}, and V_{BR} are defined exactly the same as SCR specifications of section 5–2–1.

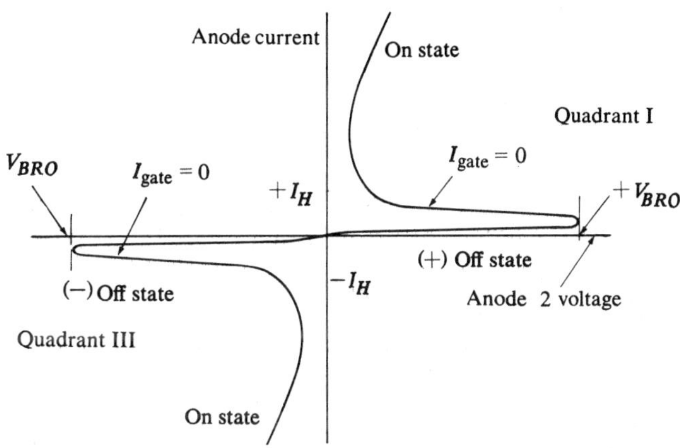

Fig. 5–23 TRIAC anode characteristics.

5–3–2 Triggering

The difference between SCR and TRIAC triggering is in the polarity of gate triggering signals. The SCR is triggered by applying a positive gate signal while the anode voltage is positive. The TRIAC can be triggered under four gate conditions:

(1) I+: A_2 positive, V_G positive
(2) I−: A_2 positive, V_G negative
(3) III+: A_2 negative, V_G positive
(4) III−: A_2 negative, V_G negative

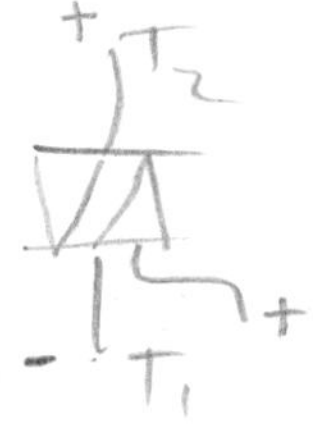

Figure 5–24 shows the manufacturer's gate trigger current requirements for a sensitive, low-current TRIAC. Notice that the gate current required to trigger in the I− and III+ modes is higher than the gate current required to trigger in the I+ and III− modes under the same conditions. This is typical of TRIAC triggering requirements. The significance of this property is that given a symmetrical gate triggering source, either I+ and III− and I− and III+ modes should be used in order to get symmetrical triggering. Other combinations will result in α_1 being different than α_2 and nonsymmetrical load current.

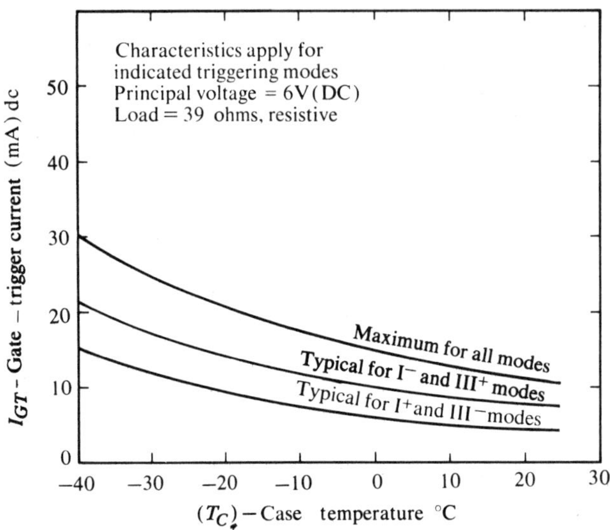

Courtesy of R.C.A. Solid State Div.

Fig. 5–24 Gate trigger current requirements for a commercial TRIAC.

5–3–3 TRIACs in Conduction

TRIACs are rated according to the maximum allowable full-wave rms current in a resistive circuit with a total conduction angle of 360° operating from a sinusoidal voltage source. The rated current is usually stated at maximum operating temperature. Figure 5–25 represents typical data supplied by manufacturers with each TRIAC. Maximum

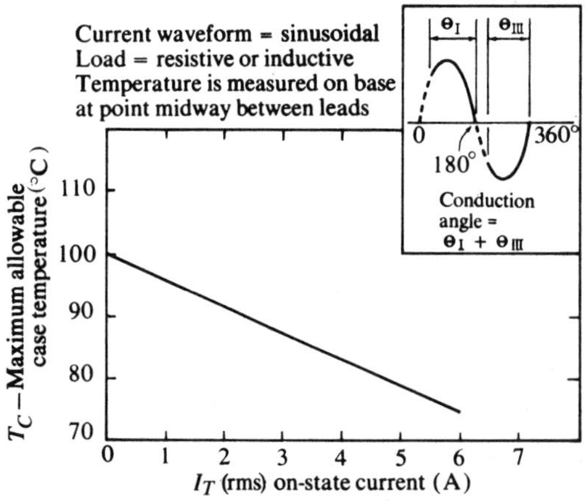

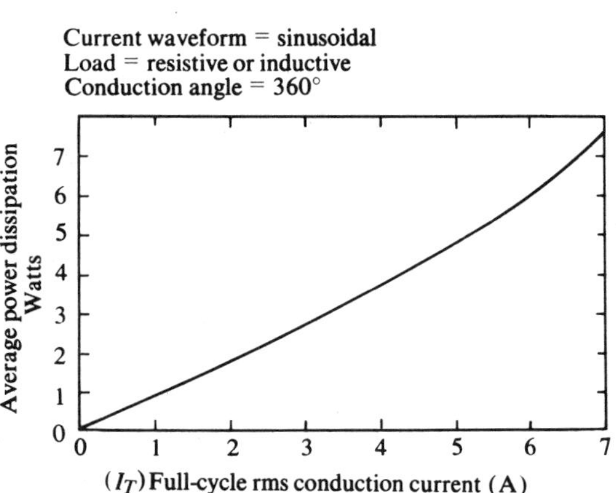

Courtesy of R.C.A. Solid State Div.

Fig. 5–25 Current ratings and power dissipation for a commercial TRIAC.

allowable case temperature and power dissipation are specified for full-wave rms current levels up to the rated current level. Proper heatsinking must be provided to prevent overheating of the semiconductor junctions and permanent damage to the TRIAC.

5–4 Silicon Controlled Switches

SCRs and TRIACs are the workhorses of industrial electronics. They are used for direct control of watts or kilowatts of power. Other PNPN devices, or thyristors, are used in electronic circuits when milliwatts of power are to be controlled. The silicon-controlled switch is a commonly used low-power PNPN switch.

The SCS is essentially a miniature SCR with leads attached to all four semiconductor layers. The additional lead gives the designer access to the anode gate as well as the cathode gate. The increased accessibility to the PNPN semiconductor allows us more flexibility in turn-on and turn-off methods than the conventional SCR. We will look at some examples later in this chapter.

Figure 5–26 shows the anode characteristics for the SCS. You will notice the similarity to the SCR anode characteristics. The SCS can be triggered or turned off by a control signal at either gate. Figure 5–27 shows the basic methods of gate turn-on and gate turn-off. Gate triggering of the SCS is highly sensitive. Less than one microamp and one volt are required for triggering at the cathode gate. Less than one milliamp and one volt are required to trigger at the anode gate. This makes the SCS desirable for sensing circuits and low-power timing circuits.

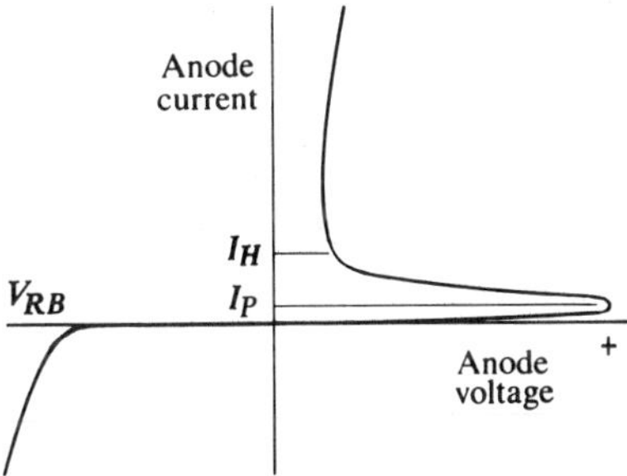

Fig. 5–26 SCS anode characteristics.

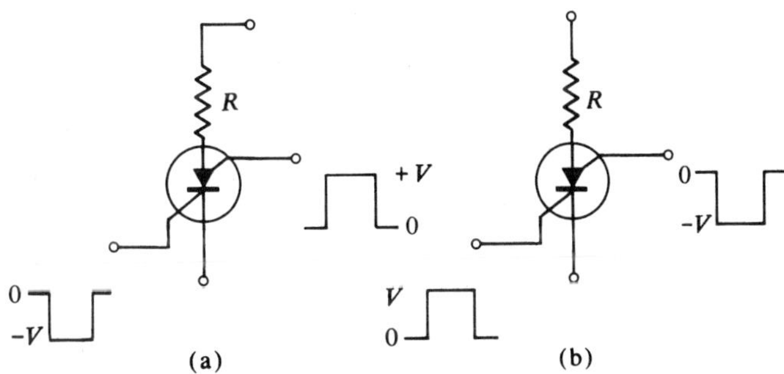

Fig. 5–27 (a) SCS gate turn-off methods, (b) SCS gate turn-on methods.

Let us look at the SCS in a few industrial electronic circuits.

Example 5–5. The circuit of Fig. 5–28(a) is a resistance-sensitive alarm actuator. The resistor R_S might be sensitive to light, temperature, or radiation. The voltage at the cathode gate of the SCS

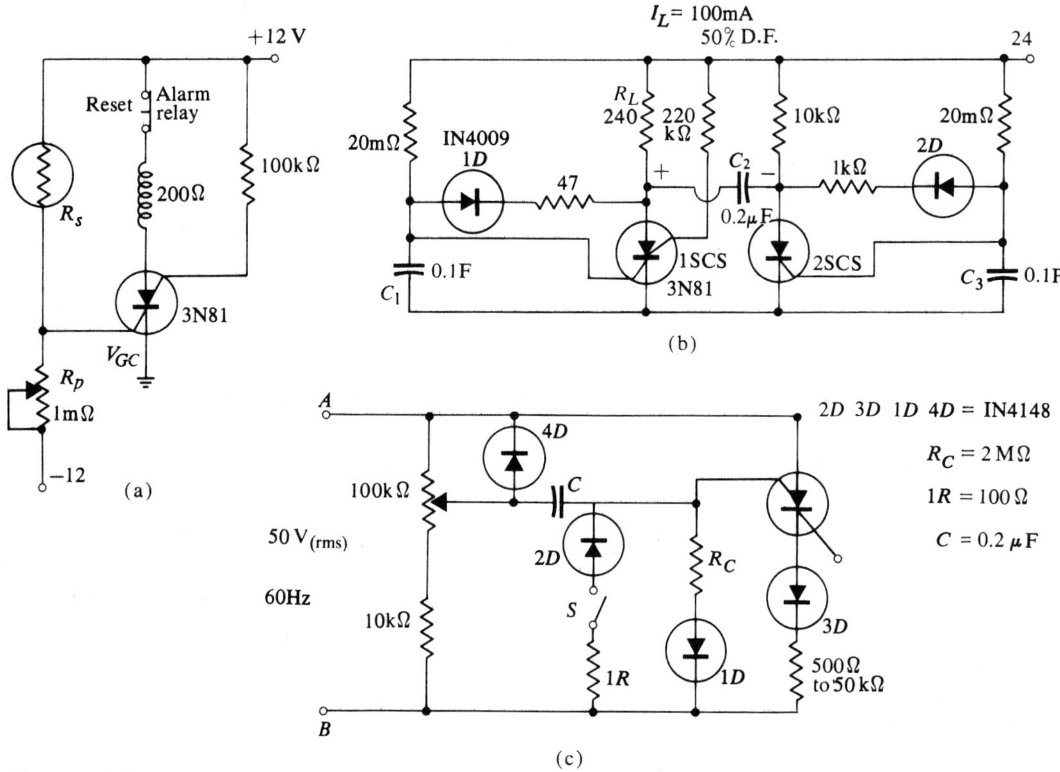

Courtesy of General Electric Semiconductor Dept., Syracuse, N.Y.

Fig. 5–28 SCS circuits: (a) alarm circuit, (b) low-frequency oscillator-flasher, (c) ac time delay relay.

will depend on the voltage divider circuit of R_s , R_p, and the two dc supplies. From Fig. 5–29:

$$I_s = \frac{12 - (-12)}{R_p + R_s} \quad \textbf{(5.15)}$$

Then:

$$V_{GC} = 24 - V_{RS} - 12$$
$$= 12 - V_{RS} \quad \textbf{(5.16)}$$

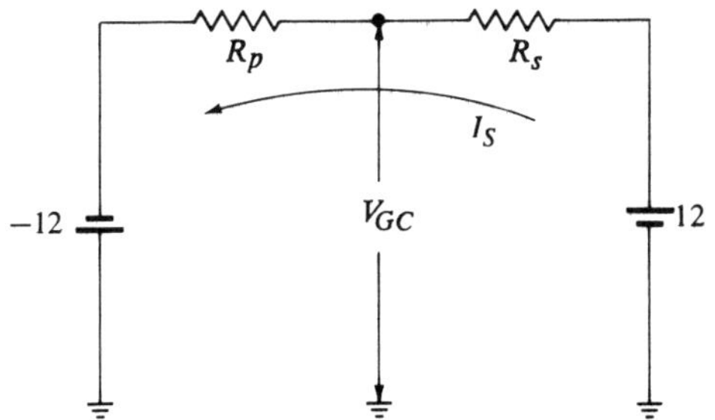

Fig. 5–29 Cathode gate voltage source of Fig. 5–28(a).

$$V_{GC} = 12 - R_s I_s = 12 - \frac{24 R_s}{R_p + R_s} \quad (5.17)$$

$$V_{GC} = 12 \frac{(R_p - R_s)}{R_p + R_s} \quad (5.18)$$

When $R_s = R_p$, $V_{GC} = 0$ and the SCS is off. When $R_s < R_p$, $V_{GC} > 0$, the SCS is on. When $R_s > R_p$, $V_{GC} < 0$, the SCS is off. The alarm will be actuated if R_s falls below the resistance of the potentiometer. The 100-kΩ resistor in the anode gate reduces the effect of dv/dt transients when the manual reset is operated.

Example 5–6. The circuit of Fig. 5–28(b) is a low-frequency oscillator without the usual large electrolytic capacitors. 1SCS is the load conducting switch. 2SCS provides the commutation to interrupt current flow through 1SCS. When the 24-volt dc source is connected to the circuit, 2SCS switches on, capacitor C_2 will charge to 24 volts through R_L with a time constant $\tau_1 = 48$ μs, at the same time C_1 will be charging through the 20-mΩ resistor to approximately one volt required to trigger 1SCS. When 1SCS is turned on, capacitor C_2 sees a 24-volt drop at the anode of 1SCS. Since the voltage across the capacitor cannot change instantly, the voltage at the anode of 2SCS must also drop 24 volts. This drop in anode voltage reduces the anode current in 2SCS below the holding current required for conduction and 2SCS turns off. (Notice that this action is very similar to the forced commutation of Fig. 5–19.) At the point of commutation of 2SCS, the simplified circuit seen by the capacitor C_2 is shown in Fig. 5–30(a). C_2 will discharge, current I_1 will flow through capacitor C_3, diode 2D, and the 1-kΩ resistor with the time constant

$$\tau_2 = \frac{C_2 C_3}{C_2 + C_3} \times 1\text{k}\Omega$$

$$\tau_2 = 0.067 \times 10^{-6} \times 10^3$$

$$\tau_2 = 0.067 \times 10^{-3} = 67\mu\text{s}$$

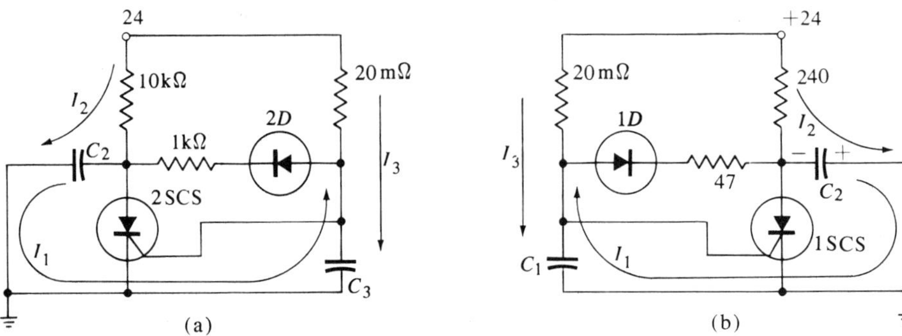

Fig. 5–30 Current flow in oscillator circuit of Fig. 5–28(b).

Capacitor C_3 will be charged negatively at the gate of 2SCS due to the direction of current flow I_1. Capacitor C_2 will be charged to 24 volts by current flow I_2 through the 10-kΩ resistor with the time constant $\tau_3 = 0.2 \times 10^{-6} \times 10 \times 10^3 = 2 \times 10^{-3} =$ 2 ms. Capacitor C_3 will be charged toward 24 volts by current flow I_3 through the 20-MΩ resistor with the time constant $\tau_4 = 0.1 \times 10^{-6} \times 20 \times 10^6 = 2$ seconds. When the voltage at the gate of 2SCS reaches approximately one volt, 2SCS will be turned on. The anode voltage of 2SCS will drop 24 volts.

Since the voltage across the capacitor C_2 cannot change instantly, the anode of 1SCS will be pulled down 24 volts. The anode current of 1SCS will drop below the required holding current and 1SCS will turn off. Capacitor C_2 will then discharge through C_1 with the time constant

$$\tau_5 = \frac{C_1 C_2}{C_1 + C_2} \times 47 = 0.067 \times 10^{-6} \times 0.047 \times 10^3$$

$$= 3.3\ \mu s$$

Capacitor C_1 will be negatively charged at the gate of 1SCS due to current flow I_1 as shown in the simplified circuit of Fig. 5–30(b). C_2 will then charge to 24 volts at the anode of 1SCS by current flow I_2 with the time constant

$$\tau_6 = 0.2 \times 10^{-6} \times 240 = 48\ \mu s$$

Capacitor C_1 will charge toward 24 volts by the current flow I_3 with the time constant $\tau_4 = 0.1 \times 10^{-6} \times 20 \times 10^6 = 2$ seconds. When the voltage at the gate of 1SCS reaches approximately one volt, 1SCS will turn on and the cycle will repeat. Notice that τ_4 is much longer than any of the other time constants in this circuit. The value of the gate capacitor and its charging resistor, RC, determines the frequency of the oscillator.

The action of this oscillator is similar to the free-running multivibrator that you will look at in detail in the next chapters. We have gone over this circuit in detail because it includes features that you will recognize in many industrial electronics circuits. Go over this entire example again.

Example 5–7. The circuit of Fig. 5–28(c) uses the anode gate to control the SCS in an ac time delay. The load is in the cathode lead in this application.

The switch is normally closed. When point A is positive with respect to ground, capacitor C charges through the 100-kΩ potentiometer, R_c and $1D$. The time constant $\tau = 0.2 \times 2 = 0.4$ s is so long compared to the period of the ac source voltage that the capacitor voltage changes less than a volt during a positive half cycle. When point A is negative, capacitor C charges through $1R$, $2D$, and $4D$ with a time constant $\tau = 0.2 \times 100 \times 10^{-6} = 0.02$ ms. The capacitor charges to the peak source voltage during a negative half cycle of source voltage. Since the capacitor is charged positively at the anode gate of the SCS with respect to the anode, the SCS remains in the blocking state. Now to actuate the circuit, the switch S is opened. There is no path for current flow when point A is negative so current flows through the capacitor only during the positive half cycle of source voltage. The number of cycles required to reverse the voltage on the capacitor depends on R_c, C, and the setting of the look potentiometer. When the anode gate voltage goes negative with respect to the anode, the SCS is turned on and load current flows for each positive half cycle. The delay is measured from the time the switch is opened to the time that load current begins to flow.

5–5 The Programmable Unijunction Transistor

The programmable unijunction transistor, PUT, is a PNPN silicon switch very similar to the SCS. Major differences are that the PUT operates at very low current levels, usually less than one microampere, and only the anode gate is available for controlling the triggering of the PUT. Figure 5–31 shows the anode-to-cathode characteristics of a

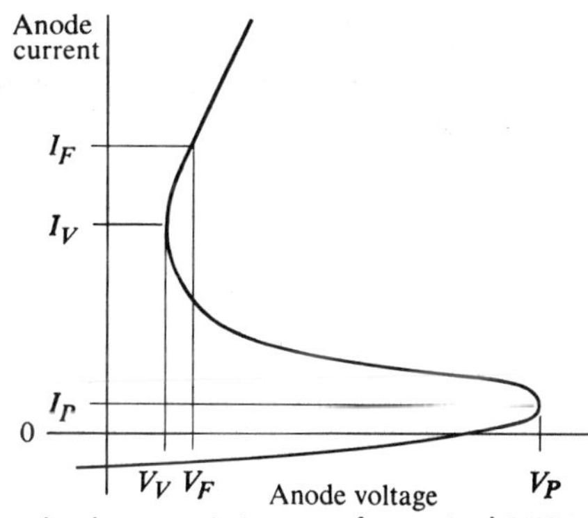

Fig. 5–31 Anode characteristic curve for typical PUT.

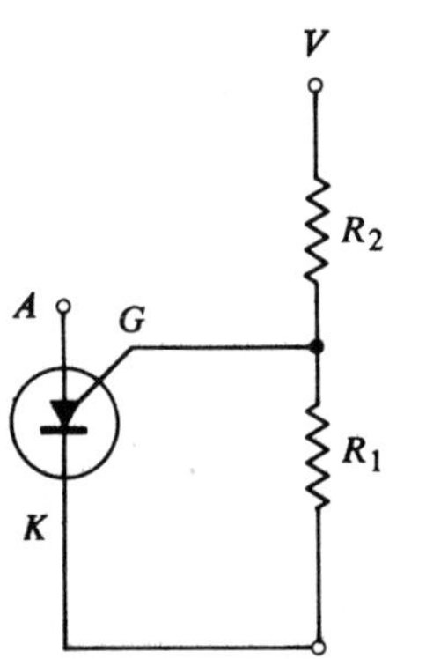

Fig. 5–32 Gate biasing circuit for PUT.

typical PUT. The anode current is very low when the anode voltage is less than the voltage required to cause the transistor to switch on. Once conduction occurs, the anode current is limited only by external resistance in the anode circuit. Figure 5–32 shows the usual gate biasing circuit for the PUT. The peak point current, I_p, and the valley point current, I_v, of Fig. 5–31 can be varied by the proper choice of R_1 and R_2. The peak point voltage, $V_p = [R_1 V/(R_1 + R_2)] +$ 0.5 V at 25°C. Figure 5–33 shows how I_p and I_v vary as the Thévenin equivalent biasing resistance is changed.

Let us look at some examples of the PUT in action.

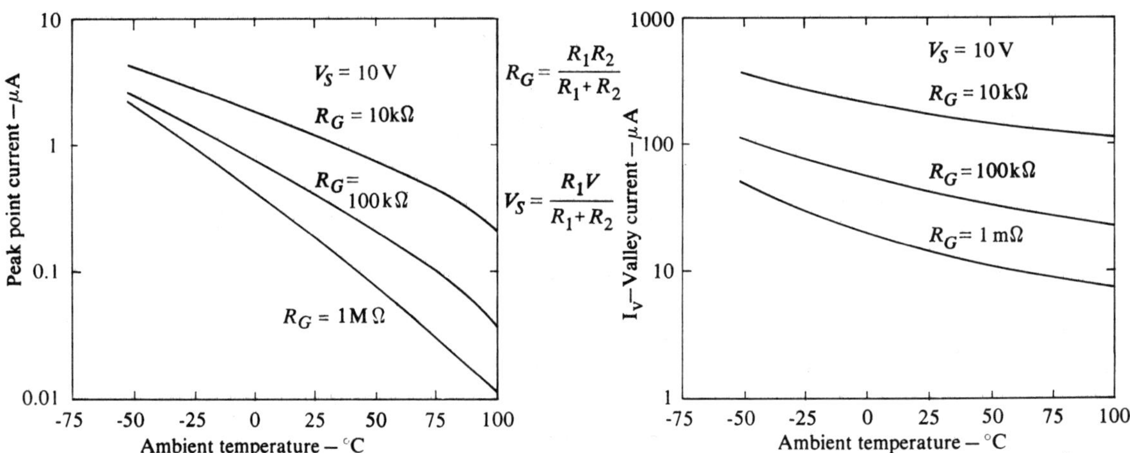

Courtesy of General Electric Semiconductor Dept., Syracuse, N.Y.

Fig. 5–33 Control characteristics for programmable control of commercial PUT.

Example 5–8. The PUT of Fig. 5–34 is used in the gate triggering circuit for SCR 2*Q*. The PUT characteristics are those of Fig. 5–33. Determine the values of *R* and *C* to trigger the SCR 4 ms after the switch is closed. Use $T_A = 25°C$.

$$R_g = \frac{R_1 R_2}{R_1 + R_2} = \frac{39\ \text{k}\Omega \times 22\ \text{k}\Omega}{39\ \text{k}\Omega + 22\ \text{k}\Omega} = 14\ \text{k}\Omega$$

$$V_s = \frac{R_1}{R_1 + R_2} V = \frac{22\ \text{k}\Omega}{22\ \text{k}\Omega + 39\ \text{k}\Omega} \times 40 = 14.4\ \text{V}$$

Then

$$V_p = V_s + 0.5 = 14.9\ \text{V}$$

From Fig. 5–33,

$$I_v = 150\ \mu\text{A}$$

The maximum value of R that will maintain conduction is

$$\frac{40}{150 \times 10^{-6}} = 266\ \text{k}\Omega$$

let $R = 220\ \text{k}\Omega$, for $I_F > I_v$

$$\frac{V_p}{V} = \frac{14.9}{40} = 0.37$$

From Fig. 4–9, $V_c(t)/V = 0.37$ at $t/\tau = 0.53$

Then

$$\frac{t}{0.53} = \frac{4 \times 10^{-3}}{0.53} = 7.5\ \text{ms}$$

$$\tau = RC, \quad C = \frac{\tau}{R} = \frac{7.5 \times 10^{-3}}{220 \times 10^3} = 0.034\ \mu\text{F}$$

use $C = 0.033\ \mu\text{F}$.

The small value of valley current allowed us to use a large resistor and small capacitor in the RC circuit. This is one of the main advantages of the PUT, especially in applications where the time delay is very long.

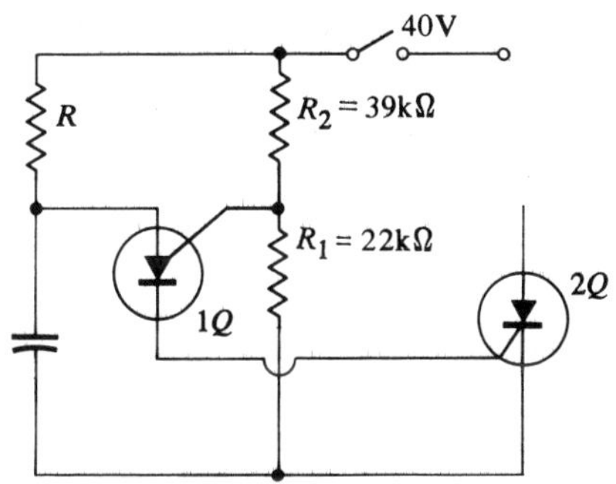

Fig. 5–34 Gate triggering circuit for SCR using a PUT.

5–6 The Shockley Diode

The final PNPN device we will discuss is the two-terminal diode thyristor commonly called the Shockley diode. The Shockley diode construction is basically the same as the PUT. The major differences are that there are no gate terminals on the diode and the current conduction capability is higher than the PUT. Figure 5–35 shows the V-I characteristics of the Shockley diode.

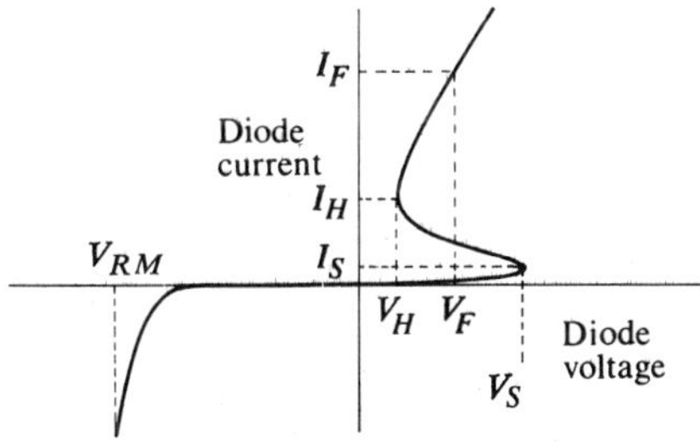

Fig. 5–35 *V-I* characteristics for the Shockley diode.

Example 5–9. Figure 5–36 shows the Shockley diode in a simple switching application. The Shockley diode, Q_1, is chosen so that V_s is greater than 40 volts and I must be at least as great as the

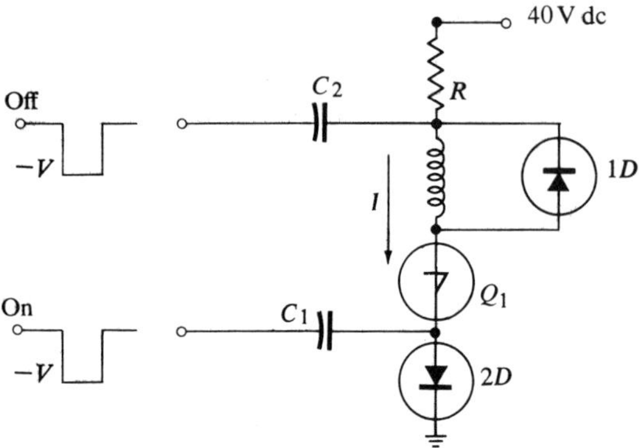

Fig. 5–36 Relay driver circuit using Shockley diode.

holding current. A negative pulse applied at C_1 drives the cathode potential of the Shockley diode below ground so that the anode-to-cathode potential exceeds V_s and the diode switches on. The Shockley diode will continue to conduct until a negative pulse is applied at C_2. The current flow through R will be diverted through the capacitor C_2. The loss of current flow through the Shockley diode will switch it back to the off state. $2D$ is the freewheeling diode similar to that of Fig. 4–40(b).

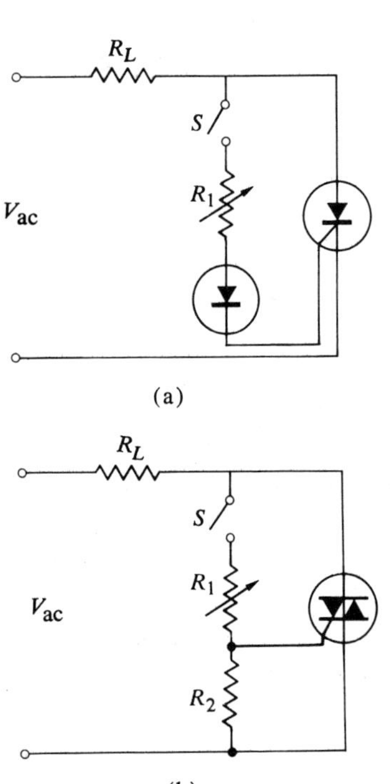

Fig. 5–37 Simplest SCR and TRIAC triggering circuits.

5–7 Triggering Circuits for SCRs and TRIACs

We have shown boxes in place of the actual gate triggering circuits in all of our SCR and TRIAC switching circuits up to this point. The reason was that we wanted to devote all of our attention to the power control devices themselves. Let us take a look inside the boxes and see what kinds of circuits are used for triggering these solid-state power control devices.

Figure 5–37 shows what are probably the simplest triggering circuits for the SCR and TRIAC. The resistor R_1 is adjusted to limit the gate current to a value less than the manufacturer's maximum allowable level that we discussed in section 5–2–2 and 5–3–2. When the SCR or TRIAC goes into conduction, the voltage across the triggering resistor is dropped to approximately one volt and the gate is protected from excess voltage. This type of triggering might be considered for on-off switching, but more sophisticated phase control would require the switch, S, to be closed at a specific point on the ac source voltage waveform. Most triggering circuits use the negative resistance portion of the characteristics of an active device to discharge a capacitor through the gate circuit. These pulse triggering techniques limit power dissipation in the gate circuit thus extending the useful life of the thyristor. A diode is often placed in the gate circuit of the SCR as in Fig. 5–37(a) to protect the gate-to-cathode junction from damage due to reverse bias.

5–7–1 The Relaxation Oscillator

Most triggering circuits for SCR and TRIAC switching use the basic relaxation oscillator as a pulse generator. Any semiconductor device whose V–I characteristics include a negative resistance portion may be used. Devices commonly used include the UJT, PUT, SCS, DIAC, and Shockley diode. Figure 5–38 shows the UJT in a basic relaxation oscillator. When switch S is closed, capacitor C_1 charges toward the voltage V by current flow I_1 through R_1. The UJT will switch on at the point A in Fig. 5–39 where the capacitor voltage reaches the peak point voltage. The capacitor will then discharge through the emitter-to-base 1 circuit of the UJT producing a voltage pulse output across R_{B_1}. The amplitude of the output pulse will be

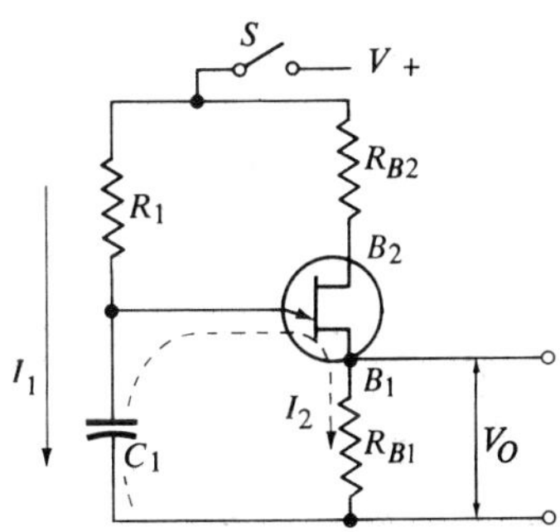

Fig. 5–38 Basic UJT relaxation oscillator.

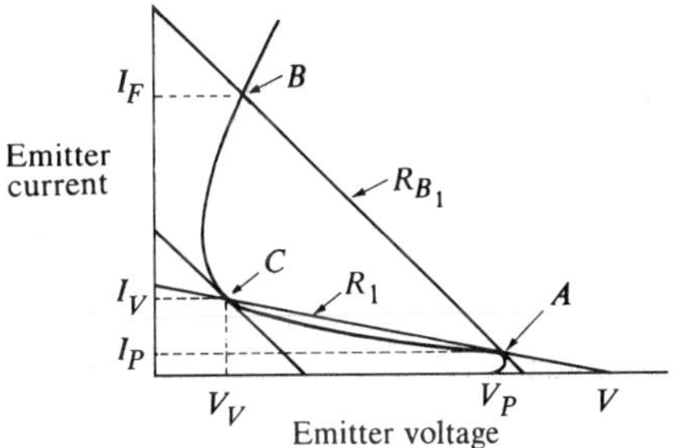

Fig. 5–39 Load lines for oscillator of Fig. 5–38.

$$V_0 \approx V_P - V_{E(\text{sat})} \tag{5.19}$$

where $V_{E(\text{sat})}$ is the saturation voltage at the emitter-to-base 1 terminals of the UJT. When the emitter current decays to the level of the UJT valley point current, point C of Fig. 5–39, the UJT switches off and the capacitor starts to charge again. The cycle repeats as long as switch S remains closed.

There are some conditions that must be met in the design of the relaxation oscillator as a triggering circuit. They are:

(1) The voltage V must be large enough to switch the UJT on, or

$$V \geq V_P + I_p R_1 \tag{5.20}$$

(2) The load line for R_1 must intersect the UJT characteristic curve in the negative resistance region.

$$R_1 > \frac{V_P - V_v}{I_v - I_P} \tag{5.21}$$

(3) The capacitor discharge time constant should be greater than the turn-on time of the UJT so that saturation resistance will not affect the R_{B_1} load line. If we make $R_{B_1} \times C$ at least 10 times T_{on}, then

$$C > 10 \frac{t_{\text{on}}}{R_{B_1}} \tag{5.22}$$

(4) In order to avoid premature triggering, R_{B_1} should be chosen so that:

$$V_{GT(\max)} > \frac{R_{B_1} V_{B_2}}{R_{BB} + R_{B_1}} \tag{5.23}$$

$$(I_F - I_V) R_{B_1} > V_{GT\,\min} \tag{5.24}$$

where $V_{GT\max}$ is the maximum gate voltage that will not trigger the SCR or TRIAC and $V_{GT\,\min}$ is the minimum gate voltage that will trigger the device. These are rather general conditions. They might even be called rules-of-thumb, due to the lack of precise values of device parameters such as t_{on}, I_p, V_P, I_s, V_v, and I_v, but reasonable approximations will result in a workable design with little effort. These same conditions apply to other semiconductor devices used in relaxation oscillators if the analogous device parameters are used.

The UJT, like most semiconductor devices, is sensitive to temperature variations. The resistor R_{B_2} in base two lead provides thermal stability for the oscillator. A precise formulation of the value to be used is quite complex due to the number of parameters involved. Motorola engineers report that an empirically derived formula

$$R_{B_2} = 0.015\, V R_{BB}\, \eta, \qquad \begin{aligned} \eta &= \text{UJT standoff ratio} \\ R_{BB} &= \text{internal } B_1 \text{ to } B_2 \text{ resistance} \end{aligned} \tag{5.25}$$

results in near optimum performance for stabilizing the frequency of the oscillator.

Let us look at the design of the relaxation oscillator of Fig. 5–40.

Example 5–10. The SCR of Fig. 5–40 has $V_{GT_{\max}} = 0.40$ V, $I_{GT(\min)} = 20$ mA and $I_{F(\max)} = 2$ A at 25°C. For the UJT, $\eta = 0.65$, $R_{BB} = 7\text{k}\,\Omega$, $I_P = 5.0\,\mu\text{A}$, $I_v = 6$ mA, and $V_v \approx 2$ V. Design the relaxation oscillator circuit.

The zener diode regulator will clamp the voltage at point A to 20 volts during the positive half cycle of anode voltage (see chapter 3, section 2). From Equation (5.25)

$$R_{B_2} = 0.015 \times 20 \times 7 \times 10^3 \times 0.65 = 1360\ \Omega$$

use 1.2 kΩ

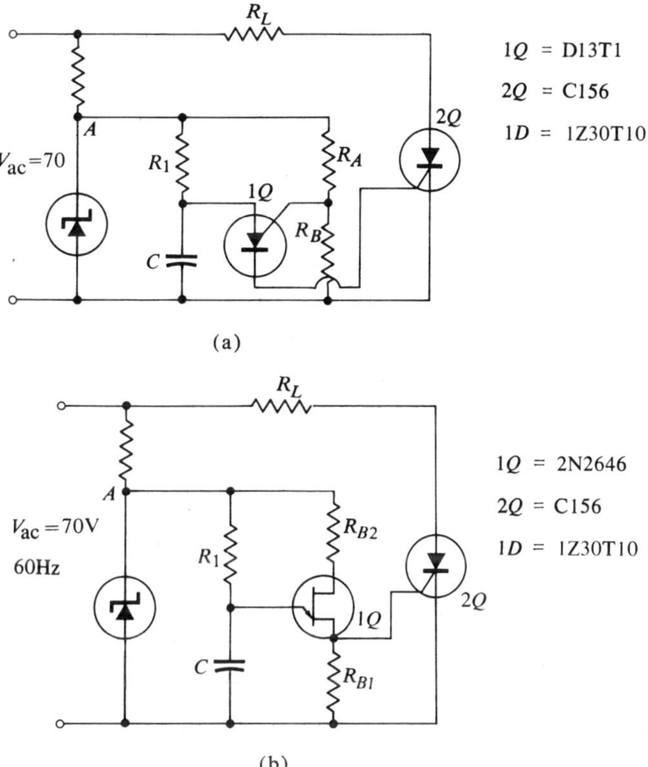

Fig. 5–40 (a) PUT relaxation oscillator to trigger SCR, (b) UJT relaxation oscillator to trigger SCR.

Allowing for the voltage drop across R_{B_2},

$$V_{B2} = \frac{R_{BB}}{R_{BB} + R_{B2}} V = \frac{7\text{k}\Omega}{7\text{k}\Omega + 1.2\text{k}\Omega} 20 = 17 \text{ V}$$

The value of R_{B_1} can be determined from Eqs. (5.25) and (5.24):

$$0.40 > \frac{R_{B_1} 17}{7\text{k}\Omega + R_{B_1}}$$

or

$$R_{B_1} < 169\ \Omega$$

If we choose to trigger the SCR at 3.0 volts, 0.1 amps, then from Eq. (5.24):

$$R_{B_1} > \frac{3.0}{(100 - 6)\text{ mA}} = 31\ \Omega$$

use 33 Ω

Then the UJT will switch on when

$$V_P = \eta V_{B_2} + 0.5 = 11.5\ \text{V}$$

from Eq. (5.20)

$$R_1 < \frac{V - V_P}{I_P} = \frac{20 - 11.5}{5} \times 10^6 = 1.7\ \text{M}\Omega$$

from Eq. (5.21)

$$R_1 > \frac{V_P - V_v}{I_v - I_P} \approx \frac{11.5 - 2}{6} \times 10^3 = 1.6\ \text{k}\Omega$$

from Eq. (5.22)

$$C > \frac{10\,t_{on}}{R_{B_1}} = \frac{10 \times 10^{-6}}{33} = 0.3\ \mu\text{F}$$

use 0.5 μF

for $t_{on} = 1\ \mu\text{s}$

The value of R_1 will be determined by the frequency of oscillation required. Figure 5-41 shows the waveform at point A in the UJT oscillator circuit. The period of oscillation can be found from Eq. (4.9),

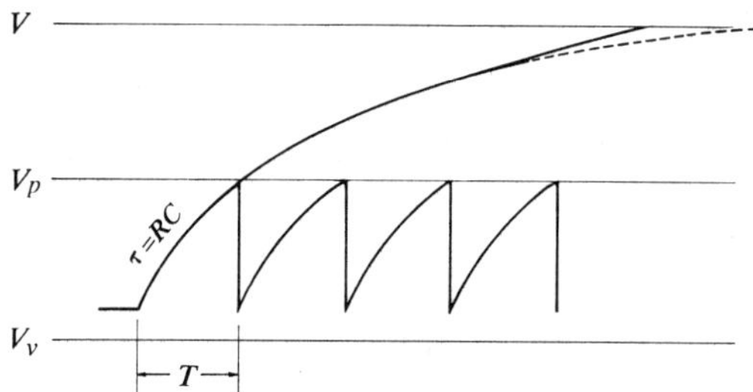

Fig. 5-41 Voltage waveform at point *A* in Fig. 5-40.

$$V_P = (V - V_v)\left[1 - \exp\left(-\frac{t}{R_1 C}\right)\right]$$

Since $V \gg V_v$ in most applications

$$V_P \approx V\left[1 - \exp\left(-\frac{T}{R_1 C}\right)\right] \tag{5.26}$$

This equation can be rearranged to

$$T = R_1 C \ln \frac{V}{V - V_P} \tag{5.27}$$

if $R_1 = 10\ \text{k}\Omega$ in Fig. 5–40.

$$\text{Then } T = 10 \times 10^3 \times 0.5 \times 10^{-6} \ln \frac{20}{20 - 11.5}$$

$$= 5 \ln 2.35 = 4.28 \text{ ms}$$

5–7–2 PNPN Devices in SCR and TRIAC Triggering Circuits

Figure 5–42 shows an example of the DIAC, SCS, and Shockley Diode in SCR and TRIAC triggering circuits. The DIAC of Fig. 5–42(a) is designed specifically for triggering the TRIAC. The DIAC has a breakover voltage of approximately 35 volts and exhibits negative resistance after breakover. It is used in a relaxation oscillator circuit to trigger the TRIAC on both positive and negative half cycles of the voltage at anode 2.

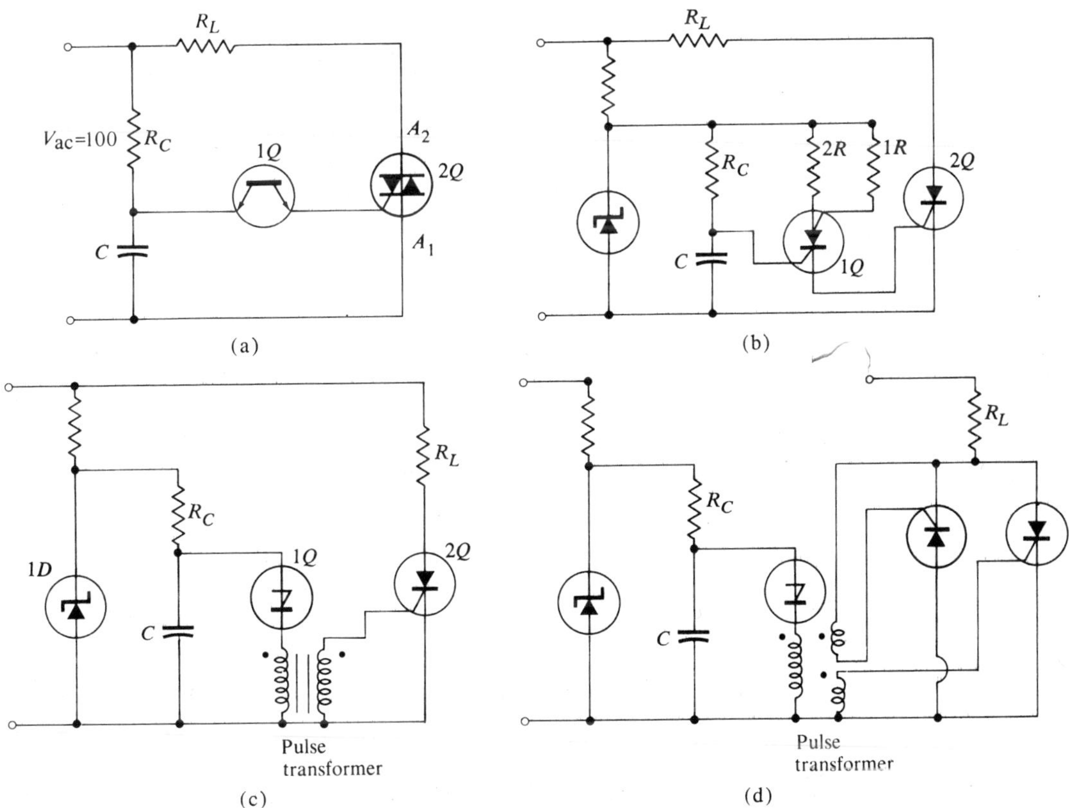

Fig. 5–42 PNPN devices in SCR and TRIAC triggering circuits: (a) DIAC, (b) SCS, (c) Shockley diode, (d) Shockley diode.

The SCS of Fig. 5–42(b) is used in the same relaxation oscillator as the UJT of Section 5–7–1. The anode resistor R_1 is made large enough so that the anode current is below the required holding current for conduction. The SCS only conducts while the capacitor is discharging through the gate circuit.

The Shockley diode is also used in a relaxation oscillator configuration to trigger the SCR. A pulse transformer may be used to boost the low current output of the diode as shown in Fig. 5–42(c). The design of the Shockley diode oscillator is subject to the same conditions as the UJT oscillator of the previous section. Two SCRs in inverse parallel may be triggered from the Shockley diode circuit by using a dual secondary transformer as in Fig. 5–42(d).

5–8 Summary

In this chapter we have seen how the terminal characteristics of PNPN devices can be applied to the design of control circuits. SCRs, TRIACs, UJTs, and low-current PNPN switching devices are very common in industrial electronics systems. In the chapters that follow, we will see how the basic circuits of this chapter are used in timing, sequencing, power control, and speed control of industrial equipment and processes.

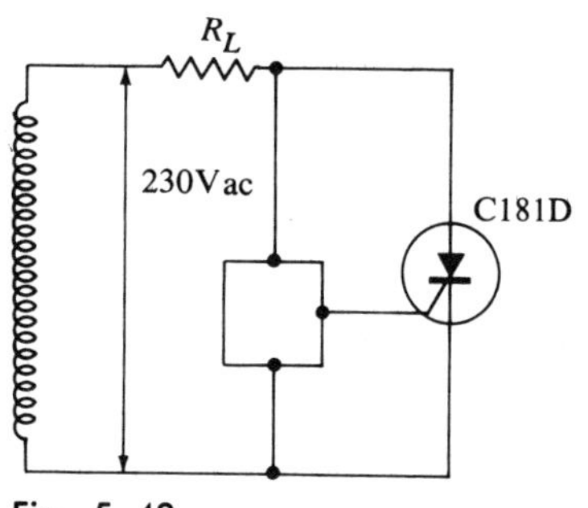

Fig. 5–43

EXERCISES

1. What is the main advantage of using an SCR in place of a normal rectifier diode?
2. State four ways to trigger an SCR into conduction.
3. The SCR of Fig. 5–43 is used in a rectifier circuit with a purely resistive load. Using the characteristics of Appendix B, how much inductance must be added in the load circuit to protect the SCR from di/dt damage?
4. Determine the RC circuit that should be placed across the SCR anode to cathode to protect against dv/dt damage if R_L is a 5-Ω resistor in Fig. 5–43.
5. The 2N3653 SCR is used in the circuit of Fig. 5–44. Using the characteristics of Appendix B, determine if the dv/dt, di/dt, or rms current limitations have been exceeded.

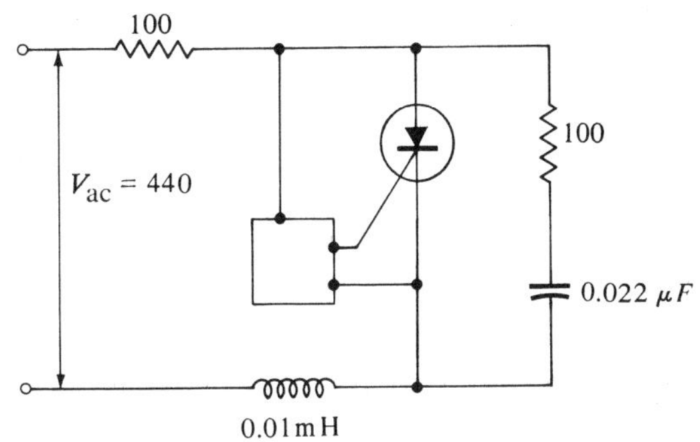

Fig. 5–44

6. The dc ammeter of Fig. 5–45 indicates that 27 A are being drawn by the resistive load.
   (a) Determine I_{rms} if $\alpha = 75°$.
   (b) Determine R_L if V_{rms} is 230 V and $\alpha = 45°$.
   (c) Determine P_L if $\alpha = 110°$.

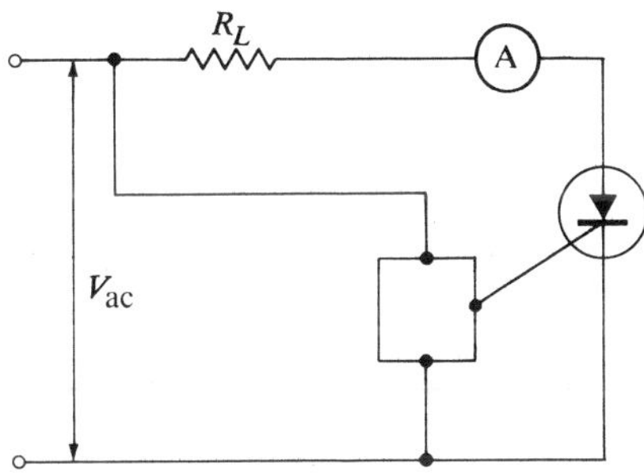

Fig. 5–45

7. The 20-Ω resistive load of Fig. 5–46 requires 1 kW of power. If V_{ac} = 230 V, determine:
   (a) Average current if $\alpha = 90°$.
   (b) The conduction angle, θ.
   (c) Voltage read by the dc voltmeter.
   (d) $1R$, $1C$, and $1L$ if $dv/dt = 100$ V/μs, $di/dt = 40$ A/us, $V_{FOM} = 600$ V.
   (e) The SCR current ratings.

8. The resistive load of Fig. 5–47 requires 800 W of power. If each SCR is sharing the load equally, how much rms current is flowing through each? If $\alpha = 80°$ for each SCR, specify the value of V_{av}.

9. What are the specifications for a TRIAC to replace the SCRs of problem 8?

10. What are the four modes of operation of a TRIAC? Explain with a diagram.

11. Use the characteristics of Appendix B to determine the maximum value of V_{ac} that can be used without damaging the TRIAC of Fig. 5–48.

12. If the load of Fig. 5–48 is only dissipating half the maximum power that the source can supply:
    (a) What is the maximum allowable case temperature?
    (b) What is the maximum allowable θ_{CA}?

13. Repeat problem 12 using the 2N5568 TRIAC.

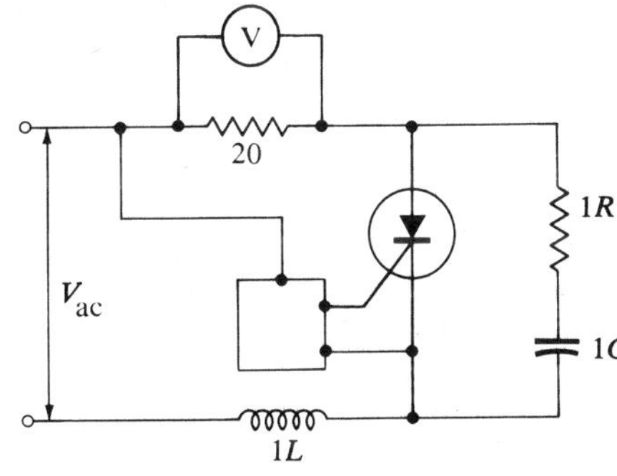

Fig. 5–46

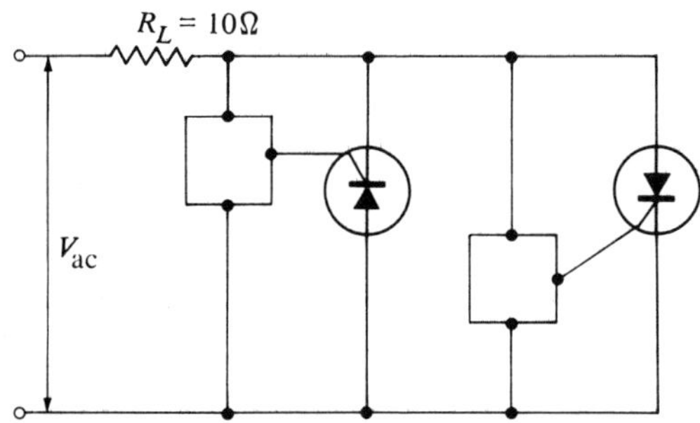

Fig. 5–47

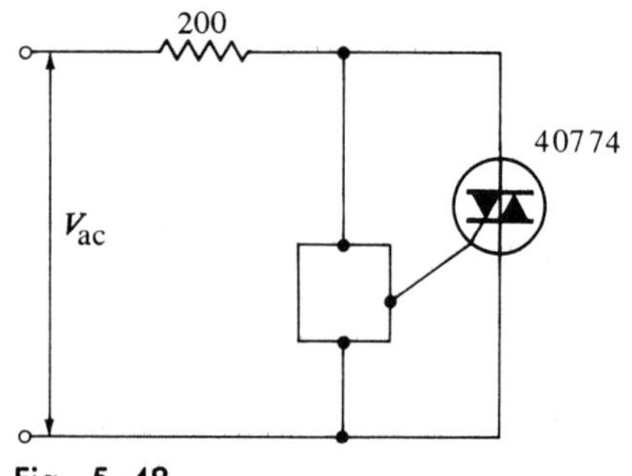

Fig. 5–48

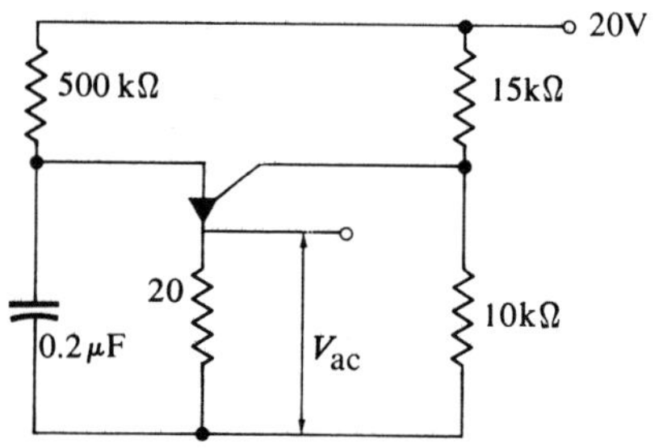

Fig. 5–49

14. How would you make the SCS alarm circuit of Fig. 5–28(a) sensitive to an increase in the resistance of R_S?

15. Make a plot of the voltage of capacitor $1C$ of Fig. 5–28(b) for two complete cycles of oscillation.

16. How does the SCS differ from the SCR in its terminal characteristics?

17. Why must the capacitor of Fig. 5–28(c) be charged negatively at the anode gate before the SCS will conduct?

18. Why is the PUT considered programmable?

19. The 2N6027 PUT of Appendix B is used in the circuit of Fig. 5–49.

(a) What is the frequency of oscillation of 25°C?
(b) What is the frequency of oscillation at –50°C?
(c) What is the output pulse amplitude at 25°C?

20. In Fig. 5–49, $1R$ is changed to 15 kΩ, $2R$ is changed to 10kΩ. Repeat problem 19 using these new values.

21. What is the meaning of the "intrinsic standoff ratio" of a UJT?

22. Determine the percent variation in η for the 2N2646 UJT of Appendix B.

23. The SCR of Fig. 5–50 requires 2.5 V to guarantee triggering and definitely will not trigger on less than 300 mV.

(a) Specify R_{B_1} and R_{B_2}.
(b) If C_1 = 0.1 uF, specify R_1 for an oscillation frequency of 200 Hz.

24. The IN 5411 DIAC of Appendix B is used in the circuit of Fig. 5–42(a) to trigger the 2N5568 TRIAC. If R_C = 15kΩ, determine the values of C to trigger the TRIAC at 45° on each half cycle using the minimum rated breakover voltage of the DIAC.

25. If the DIAC of problem 24 requires the maximum rated breakover voltage, calculate the percentage error in the triggering angle.

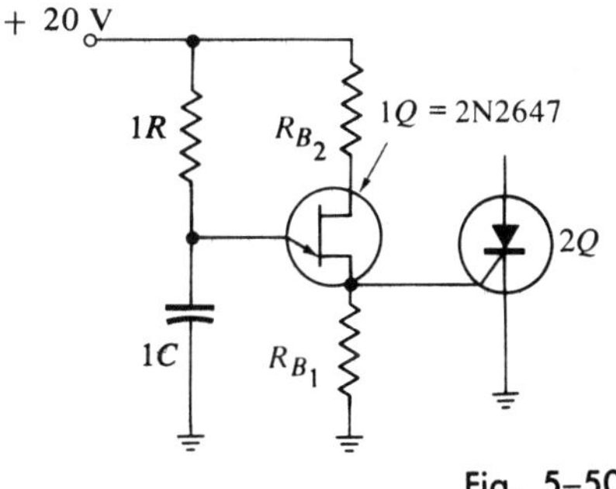

Fig. 5-50

6

Timing and Active Time Delays

6–1 Introduction

Many of the applications of electronics in industry are in the circuitry for control of the timing of devices or operations. Control requirements may be as simple as a time delay between the starting of two motors or as complicated as the precision timing of pulses in a digital computer. A variety of circuit techniques is used to cover the wide range of requirements. For purposes of simplification, industrial devices or operations may be classified as "event-oriented" or "time-oriented." Event-oriented operations generally require time delay between the occurrence of specific events. Electronic time delays are usually adequate to control the timing of the events. Time-oriented operations generally require that events occur at a specific time regardless of other events. The control of timing of such operations requires a clock or *real time* pulse generator. We will look at the design of electronic circuitry for both time delays and real time control in this chapter.

6–2 Analog Time Delays

Some electronic time delays are operated from a continuous or analog source such as ac line voltage or a dc battery. We will call these "analog" time delays. Other electronic time delays are designed specifically for use in digital applications where a digital pulse at the input is delayed and then occurs at the output. We will call these digital time delays.

6–2–1 The AC Time Delay

The simplest electronic time delay uses an *RC* circuit and one or more semiconductor devices and operates at ac or line voltage. Let's look at the very common SCR ac time delay. Figure 6–1 is typical of a circuit that you might find in an ac time delay. This particular circuit controls the actuation of an electromechanical relay. The load being controlled may be placed in series with the normally open or normally closed relay contacts.

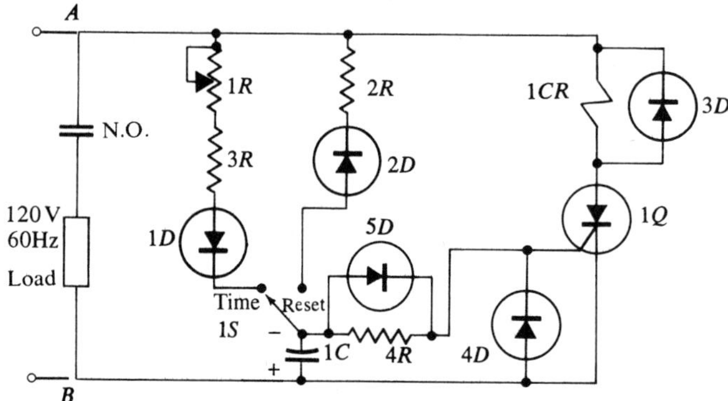

Fig. 6–1 An AC time delay relay using an SCR.

The control switch, 1*S*, may be placed in the *reset* position or the *time* position. With 1*S* in the reset position, the load of Fig. 6–1 is disconnected from the ac line voltage and capacitor 1*C* is charged negative with respect to the cathode of the SCR due to the rectified current flow through 2*D* and 2*R*. The peak forward blocking voltage of the SCR is greater than the peak ac line voltage so that almost no current flows through the relay coil 1*CR*. The charge on capacitor 1*C* might be as low as −150 volts. Diode 4*D* protects the gate of the SCR by limiting the gate-to-cathode reverse bias to −0.7 volts. The resistor, 4*R*, limits the current flow through 4*D* and the SCR gate.

When switch 1*S* is placed in the time position, the charge path of capacitor 1*C* is now through 1*D*, 1*R*, and 3*R*. Current rectification is now such that 1*C* starts to charge positive with respect to the cathode of the SCR as shown in Fig. 6–2. The time constant $\tau = (1R + 3R)1C$ is usually much longer than the period of the ac line voltages so that it takes many ac cycles between the point *M* where 1*S* is switched to the point *N* where the SCR fires. The number of cycles between *M* and *N* is the amount of time delay. The positive charge on 1*C* is limited to +1 to +2 volts by the forward drop of 5*D* and the gate-to-cathode forward voltage drop. Diode 5*D* bypasses the resistor 4*R* so that sufficient current will flow in the SCR gate circuit for triggering. When the SCR is triggered, current flow in the relay coil 1*CR* activates the relay and the normally open contacts in the load circuit are closed. Current flows through the load at the end of the time delay. This action is called *delay on*. If the normally open contacts had been used in the load circuit, the action would be *delay off*. Diode 3*D* provides the freewheeling action to smooth the current flow in the inductive relay coil.

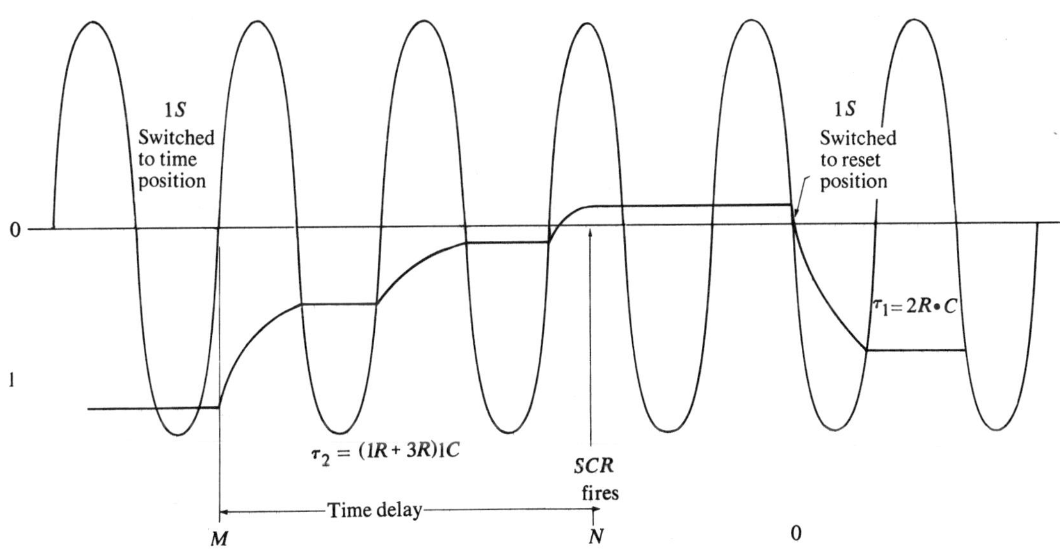

Fig. 6–2 Voltage on capacitor of Fig. 6–1.

Example 6–1. Let's look at the design requirements for a time delay to meet the following specifications:

(a) Line voltage = 120 V_{ac}, 60 Hz
(b) Load current = 10 A_{max}
(c) Time delay = 0.1 to 1 minute

First a relay must be chosen with a normally closed contact current capability of 10 A_{ac}. Since the relay will operate from the rectified line voltage a general purpose ac relay such as the Guardian 900-2C will be sufficient. The 1N 5059, 1-A 200 *PRV* diode is satisfactory for the freewheeling action required of 3*D*.

The General Electric C 106 B SCR with I_F = 2 A, V_{DRM} = 200 V is satisfactory for 1*Q*. Diode 2*D* must have a *PRV* of at least 340 V; the 1N5060, 1-A, 400-V diode is sufficient. The 1N5059 diode can also be used for 1*D*, 4*D*, and 5*D*. Resistor 4*R* must be large enough to limit current flow through 4*D* and the SCR gate when the capacitor is charged negatively. A 1-MΩ resistor will be used. The values 1*R*, 2*R*, 3*R*, and 1*C* are determined by the amount of time delay required. As you can see from Fig. 6–2, calculating the exact amount of time delay is no easy task due to the discontinuous charge waveform. Common practice is to make some design estimates, build the circuit, and make necessary changes. Good design estimates can be made by assuming that the source voltages are dc as in Fig. 6–3. The time constant τ_1 =

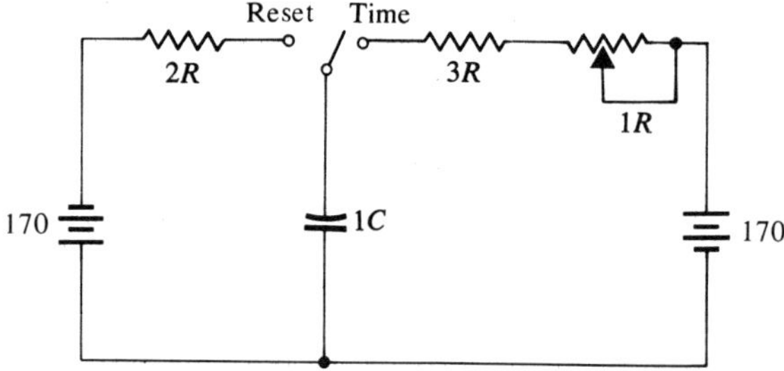

Fig. 6–3 DC circuit for estimating values of *RC* components in AC time delay.

RC should be approximately one period of the ac source voltage for quick resetting.

$$\frac{1}{60} = RC \tag{6.1}$$

if

$$1C = 100\,\mu\text{F}$$

then

$$2R = \frac{1}{60 \times 100 \times 10^{-6}} = 160\ \Omega, \quad \text{use } 180\ \Omega \tag{6.2}$$

The charge on $1C$ is going from -170 to $+170$ V in Fig. 6–3. The voltage will become positive when

$$V_c(t) = \frac{\Delta V}{2}$$

From Fig. 4–8,

$$\frac{V_c}{\Delta V} = \frac{1}{2}$$

when

$$\frac{t}{\tau} \approx 0.53$$

then

$$\tau_2 = \frac{t}{0.53} = (1R + 3R) \times 1C \tag{6.3}$$

where t is the required time delay, or

$$R_1 = \frac{t}{0.53C_1} - R_3 \tag{6.4}$$

let

$$3R = 100\ \Omega$$
$$1R \approx 1\ \text{M}\Omega \quad \text{for} \quad t = 60\ \text{s}$$
$$1R \approx 1\ \text{k}\Omega \quad \text{for} \quad t = 6\ \text{s}$$

Use a 1-MΩ pot for $1R$ and adjust for the exact amount of time delay.

6–2–2 DC Time Delays

The general design of dc time delays includes a passive *RC* circuit (see chapter 4) and an active semiconductor device. The building block of many dc time delays is the UJT relaxation oscillator that we discussed in chapter 5. Figure 6–4 is the basic circuit used in the UJT time delay. You will notice the circuit is the same as that of Fig. 5–40 except that the source voltage is now dc. Let's review the operation of the relaxation oscillator time delay and discuss its design problems. When switch $1S$ is closed, the dc supply voltage is applied to the SCR, 2Q, through the coil of the control relay, *CR*. The peak forward breakover voltage of the SCR is selected at a value greater than the dc source voltage so that almost no current flows through the coil. The zener regulation circuit (see section 3–2), consisting of

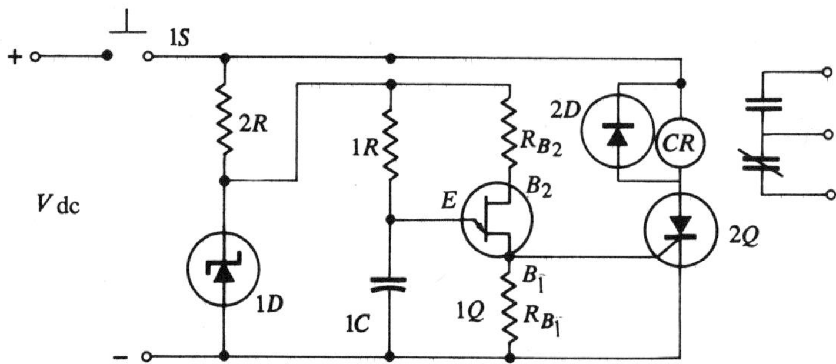

Fig. 6–4 Basic UJT DC time delay.

$2R$ and $1D$, provides the voltage for operation of the oscillator. Current flow through resistor $1R$ charges capacitor $1C$ toward the zener voltage of $1D$, (see Fig. 5–41). When the voltage of $1C$ reaches the peak point voltage of the UJT, the pulse output at B_1 triggers the SCR, supplying the current required to energize the control relay. The load is connected in series with the contacts of the relay and might be powered from another source independent of V_{dc}.

Example 6–2. Let's suppose that we are designing the time delay of Fig. 6–4 for a 10-A load and a 0.1- to 1.0-minute delay operating from a 24-V dc source. We would choose a dc relay with 10-A contact capacity, such as the Potter and Brumfield KA series. A 1-A, 50-V diode such as the G.E. A14F should be placed across the coil to reduce the high voltage transients due to current switching of the SCR. The G.E. C106F, 2-A, 50-V SCR will be sufficient for $2Q$.

We discussed the design of zener regulators in section 2–2. The design of the relaxation oscillator follows the same procedure as example 5–10. It will be left as an exercise for you to go over the details of these designs. It is important to note that one of the limitations of the circuit is that $1R < 3.7\ \text{M}\Omega$. In order to get a 1-minute time delay:

$$1C \approx \frac{T}{1R} = \frac{60\ \text{s}}{3.7\ \text{M}\Omega} = 16.2\ \mu\text{F}$$

A 10-minute time delay would require a 162-μF capacitor. A one-hour time delay would require 972 μF, and so on.

We see that long time delays would require large and expensive capacitors. The alternative is to make $1R$ larger, but this requires larger dc source voltages to maintain the peak current required by the UJT. Figure 6–5 shows some of the techniques used to extend the time delay without expensive capacitors.

In Fig. 6–5(a), the charging resistor is replaced by a junction transistor. The amount of charge current is controlled by the biasing resistors 1*R*, 2*R*, and 3*R*. A constant current source of higher impedance is achieved by using a FET in place of the charging resistor as in Fig. 6–5(b). The amount of charge current is controlled by the choice of source resistor 1*R*. The PUT operates at much lower peak point current than the UJT. This characteristic makes it more desirable for time delays of long duration. Figure 6–5(c) is a typical design configuration. The biasing resistors, 2*R* and 3*R*, are usually greater than a megohm to reduce the peak point current requirement.

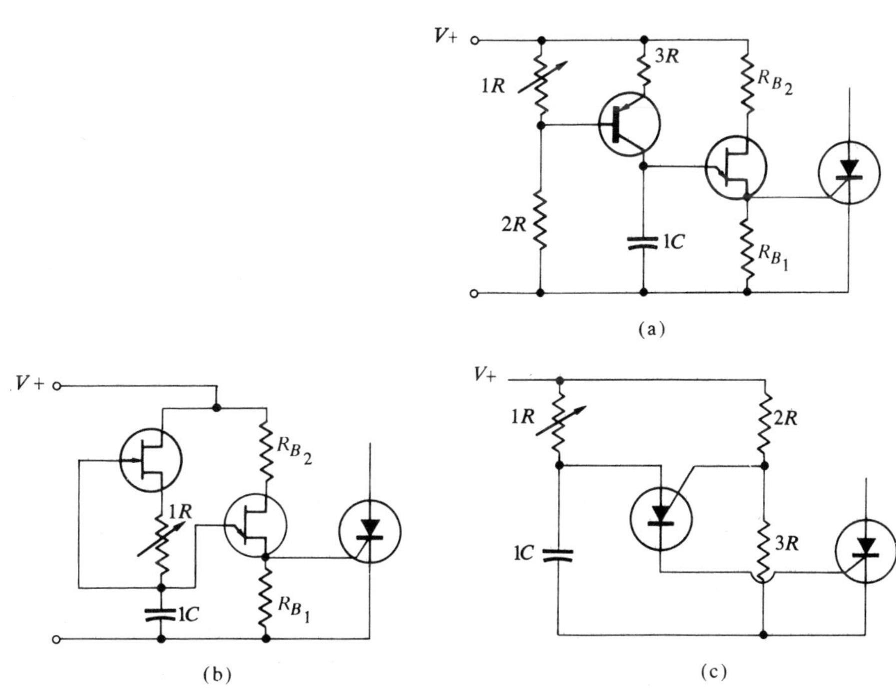

Courtesy of Motorola Inc.

Fig. 6–5 Techniques used to achieve long-duration time delay. (a) Transistor controlled UJT time delay. (b) FET constant current UJT time delay. (c) PUT time delay.

6–3 Timing Sources

Time-oriented devices and operations in industrial electronics require a time base from which events may be synchronized. Generally an electronic timing source is required. The source generates output signals at a fixed frequency. The output signals may be reshaped for triggering gates, counted for interval timing, or delayed through frequency division. We will look at the most widely used techniques for generating timing pulses and discuss their design criteria.

6–3–1 The Schmitt Trigger

The Schmitt trigger is used as a shaping circuit in the generation of timing signals because of its simplicity and general application. It can be used to convert sinusoidal line voltages into rectangular timing signals. As seen in Fig. 6–6, the circuit operates as a regenerative switch.

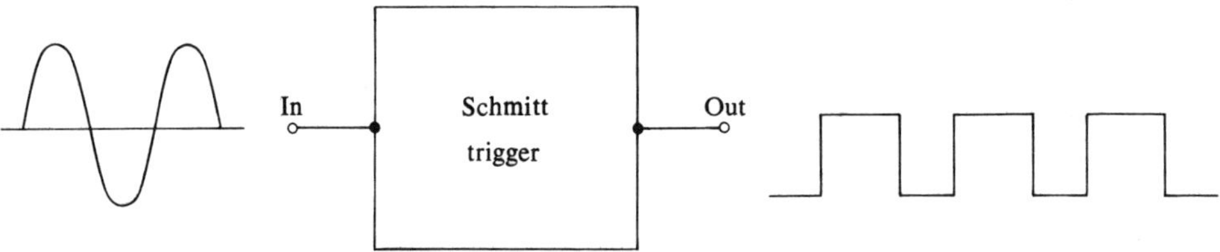

Fig. 6–6 Schmitt trigger used to generate rectangular timing signals.

Whenever the input signal is above a predetermined threshold level, the output is switched *on*. When the input signal falls below the threshold level, the output switches *off*. The Schmitt trigger is sometimes used as a squaring circuit, voltage comparator, level detector, or as a peak sensing amplifier. Figure 6–7 shows the basic Schmitt trigger circuit. Transistors $1Q$ and $2Q$ are connected in a high-gain positive feedback configuration. Resistor R_4 provides the feedback that makes the output response very sensitive to the level at the input. With no signal applied at the input, $2Q$ is biased in the on state so the output voltage is very low. The voltage

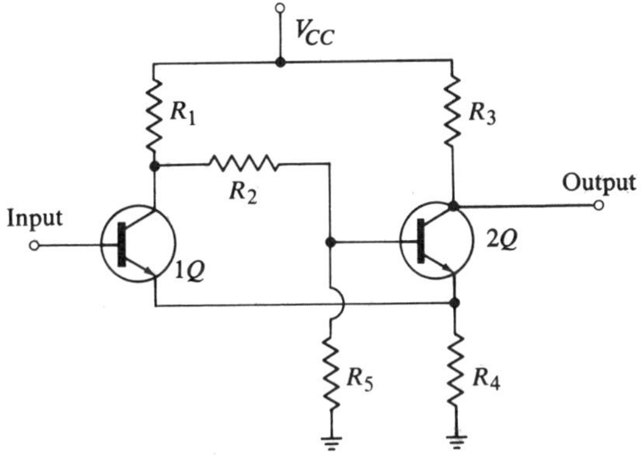

Fig. 6–7 Basic Schmitt trigger circuit.

at the input must be more positive than the voltage across R_4 plus the turn-on base-to-emitter voltage of $1Q$ to switch $2Q$ to the off state. When $1Q$ is turned on two things occur: the base current flow to $2Q$ through $2R$ is diverted through the collector-to-emitter circuit of $1Q$ and the voltage drop across R_4 is increased. Both of these actions tend to force $2Q$ into the off state so that very fast switching is achieved. The input switching can be determined by considering the partial circuit of Fig. 6–8(a) and the Thévenin equivalent circuit of Fig. 6–8(b)(see Sec. 2–4).

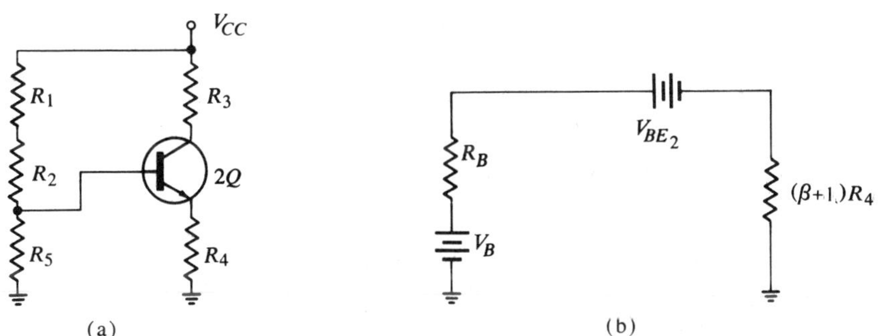

Fig. 6–8 (a) Circuit for determining trigger voltage of Schmitt trigger, (b) Thévenin equivalent base biasing circuit.

From Eqs. (2.68) and (2.69) and Fig. 2–38,

$$R_B = \frac{(R_1 + R_2)\,R_5}{R_1 + R_2 + R_5} \tag{6.5}$$

$$V_B = \frac{R_5}{R_1 + R_2 + R_5}\,V_{cc}$$

Then

$$V_{B_2} = V_{BE_2} + I_{B_2}[R_B + (\beta + 1)\,R_4] \tag{6.6}$$

or

$$I_{B_2} = \frac{V_{B_2} - V_{BE_2}}{R_B + (\beta + 1)\,R_4} \tag{6.7}$$

The voltage across R_4 is $R_4 I_{E_2}$ or $R_4 I_{B_2}(\beta + 1)$. Then the required turn-on voltage V_{B_1} is given by Eq. (6.8):

$$V_s = V_{B_1} = V_{BES} + (\beta + 1)\,R_4\left[\frac{V_{B_2} - V_{BE_2}}{R_B + (\beta + 1)\,R_4}\right] \tag{6.8}$$

where V_{BES} is the base-to-emitter switching voltage. For silicon transistors,

$$V_{BE} \approx 0.6 \text{ V}$$

$$V_{BES} \approx 0.5 \text{ V}$$

so that $V_{B_1} = V_s$ can be selected by the proper choice of biasing resistors and transistor gain, β.

Let's look at a sample design of a Schmitt trigger as a timing source.

Example 6–3. The circuit of Fig. 6–7 is chosen as a timing source to operate from a transformer whose output is 3 volts peak, 60 Hz. The transistors 1*Q* and 2*Q* are identical, silicon with $\beta = 100$. Bias the trigger and specify its output.

Let's select the trigger point at the 1-V level of the sinusoidal input signal. Then from Eq. (6.8):

$$V_S = V_{BES} + R_4(\beta + 1)\,I_{B_2} = V_{BES} + R_4 I_{E_2} \tag{6.9}$$

Then:

$$R_4 I_{E_2} = 1 - V_{BES} = 0.5 \text{ V} \tag{6.10}$$

In order to keep 2*Q* away from saturation let the output "on" voltage be 1 V. Then

$$V_{CE_2}\,\text{“on”} = 0.5 \text{ V}$$

Let $R_4 = 1\ \text{k}\Omega$; then from Eq. (6.10)

$$I_{E_2} = \frac{0.5 \text{ V}}{1\ \text{k}\Omega} = 0.5 \text{ mA}$$

Let $V_{cc} = 6$ V; then the voltage across $3R$ when $2Q$ is on will be 5 V.

$$R_3 = \frac{5\text{ V}}{0.5\text{ mA}} = 10\text{ k}\Omega$$

If we make $R_B \ll (\beta + 1)R_4$, then from Eq. (6.7):

$$R_4 I_{E_2} \approx V_{B_2} - V_{BE_2}$$

or

$$V_{B_2} = R_4 I_{E_2} + V_{BE_2} = 0.5 + 0.6 = 1.1\text{ V}$$

Make $R_B = 10\text{ k}\Omega$. Then

$$R_1 + R_2 = \frac{10\text{ k}\Omega \times 6}{1.1} = 54.5\text{ k}\Omega$$

$$R_5 = \frac{10\text{ k}\Omega}{1 - \dfrac{1.1}{6}} = 12.7\text{ k}\Omega \quad \text{use } 12\text{ k}\Omega$$

We must still decide on the values of R_1 and R_2. R_1 must be large enough to keep $1Q$ from going into saturation in the "on" state. The maximum emitter current in $1Q$ when $2Q$ is off will be

$$I_{E_1} = \frac{3 - 0.6}{1\text{ k}\Omega} = 2.4\text{ mA} \tag{6.11}$$

If we limit V_{CE} "on" to 1.0 V minimum, then

$$R_1 \geq \frac{6 - 3.4}{2.4\text{ mA}} = 1.1\text{ k}\Omega$$

Use 1.8 kΩ. Then $R_2 \approx 52.5\text{ k}\Omega$. Use 47 kΩ and 4.7 kΩ in series.

The completed circuit design is shown in Fig. 6–9. The output voltage is a rectangular waveform at levels of +1 and +6 volts as shown.

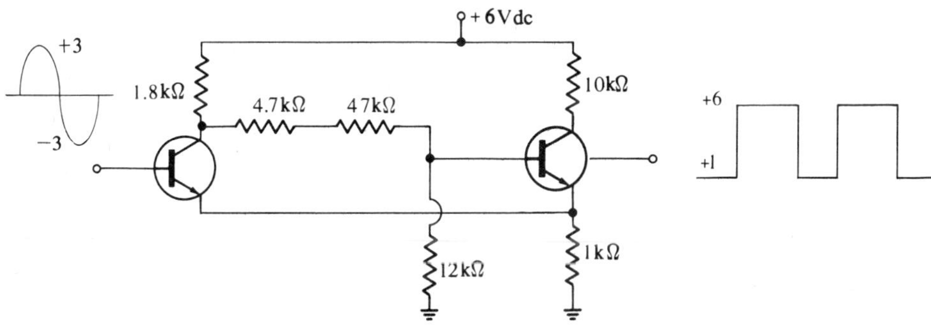

Fig. 6–9 Schmitt trigger design of Example 6–3.

6–3–2 The Astable Multivibrator

The astable multivibrator is a regenerative amplifier that requires no external input signal and produces a rectangular output similar to the Schmitt trigger (Fig. 6–10). A basic understanding of the design of the astable multivibrator lays the groundwork for understanding a whole family of digital circuits including the monostable multivibrator, the bistable multivibrator (or flip-flop), counters, decoders, and gating circuits. Figure 6–11 is the basic transistor astable multivibrator.

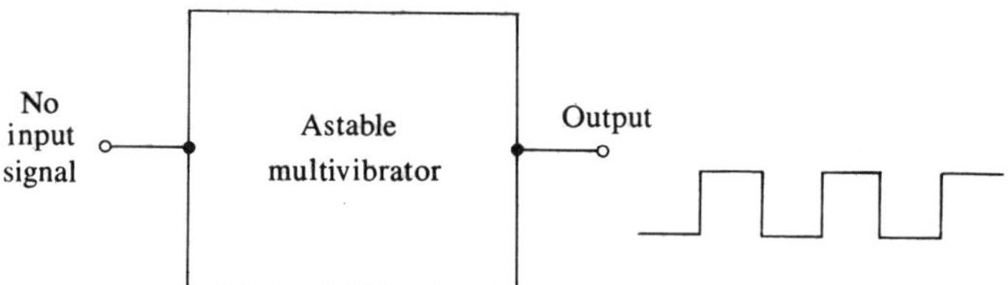

Fig. 6–10 The astable or free-running multivibrator.

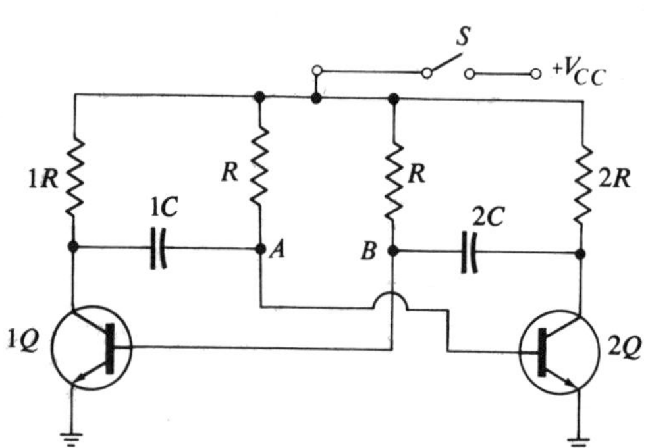

Courtesy of ITT Semiconductor Div.

Fig. 6–11 Basic transistor astable multivibrator.

Let's look at the switching action that causes the circuit to operate. The values of $1R$ and $2R$ are chosen so that when switch S is closed both $1Q$ and $2Q$ try to go into saturation. Since the two transistors are not identical, one will switch on faster than the other. Suppose $1Q$ switches on first; then the voltage at the collector of $1Q$ will drop from V_{cc} towards ground level, since the voltage across capacitor $1C$ cannot charge instantly. The voltage at the base of $2Q$ will drop by the same amount, driving $2Q$ into cutoff. Capacitor $1C$ will then charge through R and $1Q$ at the time constant $1RC$. When the voltage at point A goes positive, $2Q$ switches on. The voltage at the collector of $2Q$ drops from V_{cc} toward ground level; since the voltage across $2C$ cannot change instantly the voltage at the base of $1Q$ drops by the same amount forcing $1Q$ into cutoff. Capacitor $2C$ then charges up through R and $2Q$ at the time constant $2RC$. When the voltage at point B goes positive, $1Q$ switches on and the cycle is repeated. *It is important here to grasp the concept of the internal switching action of this circuit.* The capacitors act as positive feedback elements. For example, when $1Q$ switches on, $2Q$ switches off; the increasing voltage at the collector of $2Q$ is fed through capacitor $2C$ to aid in switching $1Q$ on even faster. The result is that rapid switching can be achieved. Microsecond switching times are easily accomplished.

The design of astable transistor multivibrators has been dealt with extensively in the literature for years. With your background from chapters 2 and 4, you would have no trouble designing your own circuit. While this bipolar transistor circuit is a very good one for discussing the concept of regenerative switching, you will not find it being used in industry very much. Other semiconductor devices are being used, such as the Shockley diodes of Fig. 6–12 (see chapter 5, section 6). The basic concept is the same. The voltage V is chosen to be greater than the breakover voltage, V_s, of

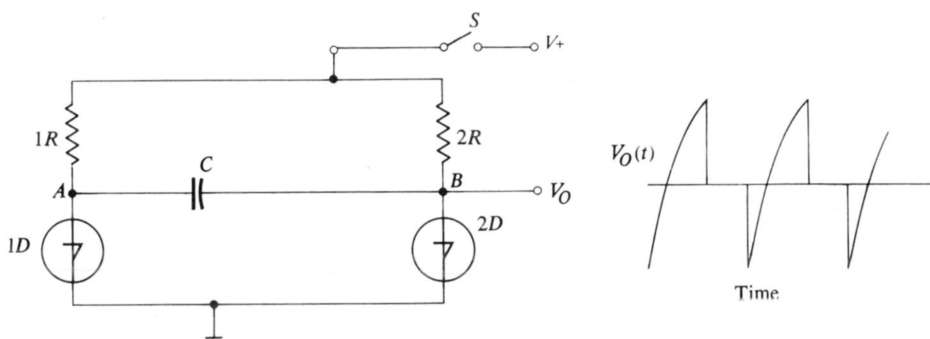

Courtesy of ITT Semiconductor Div.

Fig. 6–12 Shockley diode basic astable multivibrator.

either diode. Resistors $1R$ and $2R$ are chosen so that both diodes will operate in the conduction mode. When switch S is closed, one of the diodes will switch faster than the other. Let's assume that $1D$ is faster so that it goes into conduction. The voltage at point A will drop rapidly to ground level, since the voltage across capacitor C cannot change instantly. The voltage at point B will drop by the same amount forcing $2D$ into the off condition. Capacitor C will now charge through $2R$ and $1D$ at the time constant $2RC$. When the voltage at point B reaches the diode breakover level, $2D$ will switch into conduction. Since the voltage across capacitor C cannot change instantly, the voltage at point A drops by the same amount forcing $1D$ into cutoff. Capacitor C will now charge through $1R$ and $2D$ at the time constant $1RC$. When the voltage at A reaches the diode breakover voltage, $1D$ goes into conduction and the cycle repeats. The output taken from A or B is not rectangular and often requires shaping before applying to the circuit to be timed (see Fig. 6–13). Let's look at a design example of the Shockley diode astable multivibrator.

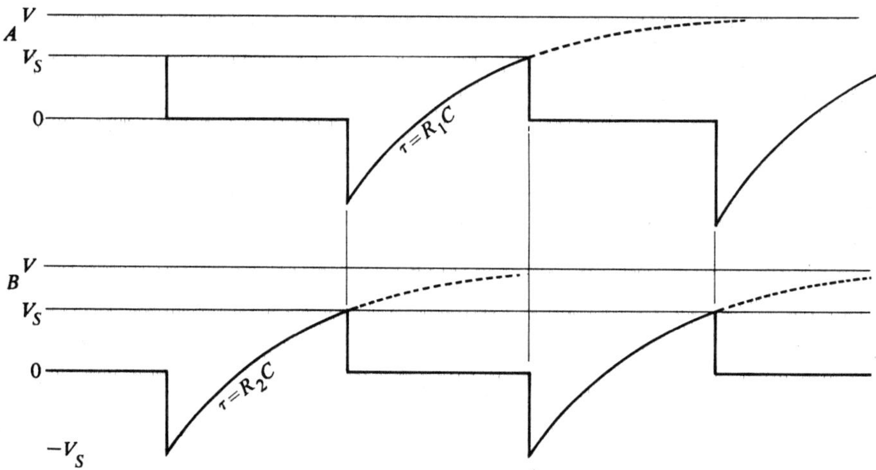

Fig. 6–13 Voltage waveforms in the circuit of Fig. 6–12.

Example 6–4. It is desired to design a free-running multivibrator to operate at 200 kHz. The 1N3841 diode will be used; the "on" time of each diode should be the same. Specify V, R_1, R_2, and C in the circuit of Fig. 6–12.

The specifications of the 1N3841 at 25°C as listed by ITT are:

$$V_S = 30 \text{ V} \pm 4$$
$$I_h = 14 - 50 \text{ mA}$$
$$C_D = 60 \text{ pF}$$

Reverse breakdown voltage = 18 V minimum.

The circuit of Fig. 6–12 will be used for this design. The switch S will not be necessary as voltage will be applied directly to the multivibrator.

The voltage, V_1, must be greater than V_s. Let V = 40 V. Then V/R must be greater than I_h. Since the maximum value of I_h is 50 mA, make V/R_1 = 100 mA. R_1 = 40/100 mA = 400 Ω, use 390 Ω. Let $R_2 = R_1$ = 390 Ω. The value of C will be determined by the frequency of the timer. At f = 200 kHz, T = 1/(200 × 103) = 0.5 × 10^{-5} = 4 μs. The on time of each diode is $T/2$. Then t_{on} = 2.5 μs.

From Fig. 6–13, the capacitor must charge from $-V_s$ to $+V_s$ in 2.5 μs.

$$V_{c(\text{max})} = V + V_s = 70$$
$$V_c(t) = 2V_s = 60$$

Then

$$\frac{V_c(t)}{V_{c\text{max}}} = \frac{60}{70} = 0.86$$

From Fig. 4–9,

$$\frac{t}{\tau} \approx 2$$

Then

$$\tau = R_1 C = \frac{2.5}{2} \times 10^{-6}$$

$$C = \frac{1.25}{R_1} \times 10^{-6} = 0.0031 \times 10^{-6}$$

use 3,300 pF

There is one other problem to be solved in this circuit. Ideally, the reverse voltage applied to the diodes during switching could be as much as 34 V! One solution would be to put a signal diode in the cathode-to-ground circuit of each diode. The resistance of the diode will change the time constants so an adjustment must be made in the values of R_1 and R_2. Build the circuit and make necessary adjustments to meet the design specifications. Remember, the proof of a good industrial electronic design is performance.

6–3–3 Oscillators

Quite often the source of timing signals in electronic systems is an oscillator. Oscillators may be sinusoidal or nonsinusoidal. Nonsinusoidal oscillators are usually relaxation oscillators (see section 5–7). The astable multivibrator that we discussed in the last section is classified as a nonsinusoidal oscillator. The sinusoidal oscillator is usually an amplifier with a tuned circuit for a load with positive feedback between the output and input circuits. Figure 6–14 is a block diagram of the typical sinusoidal oscillator. We have already looked at the design of transistor amplifiers in

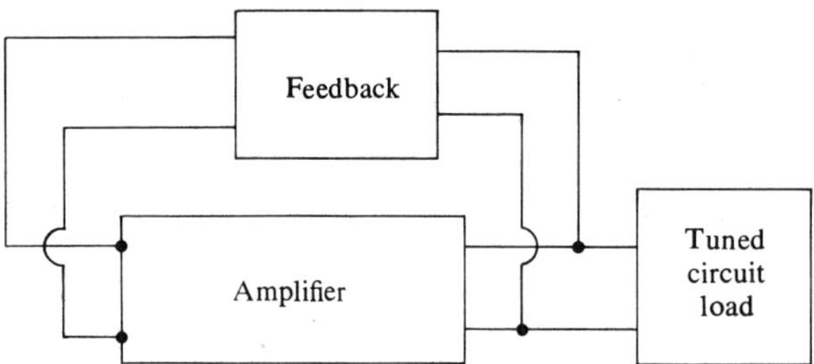

Fig. 6–14 Block diagram of a sinusoidal oscillator.

chapter 2. Let's look briefly at the tuned circuit. An ideal parallel tuned circuit is shown in Fig. 6–15(a). The ac impedance of the capacitor is $1/j\omega c$. The reactance of the coil is $j\omega L$. The impedance seen at terminals AB is

$$Z_{AB} = \frac{j\omega L\left(\dfrac{1}{j\omega c}\right)}{j\omega L + \dfrac{1}{j\omega c}} \tag{6.12}$$

or

$$Z_{AB} = \frac{L/C}{j\left(\omega L - \dfrac{1}{\omega C}\right)} \tag{6.13}$$

When $\omega L = 1/\omega C$, the denominator of Eq. (6.13) will be zero; then $Z_{AB} = \infty$ (infinity). If we applied an ac voltage at the frequency $f = \omega/2\pi$ where $\omega = \sqrt{1/LC}$, no current would flow into the terminals A–B. This condition is known as resonance and the circuit is tuned at the frequency $f_r = 1/2\pi\sqrt{LC}$. If a small amount of energy is supplied to start the circuit resonating, current will continue to flow in the

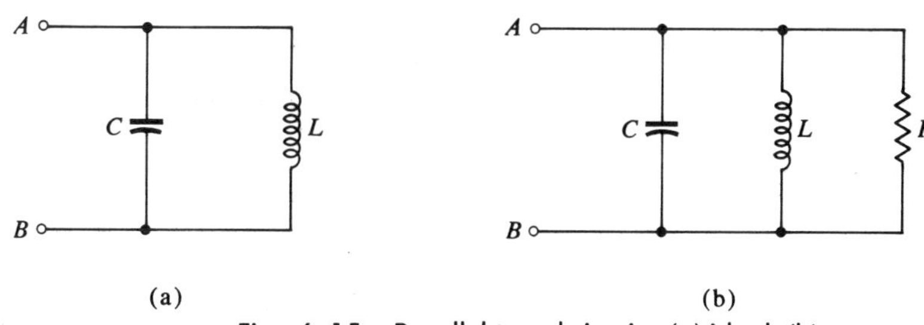

Fig. 6–15 Parallel tuned circuits: (a) ideal, (b) practical.

loop between L and C. This ideal condition is never attained in actual practice.

The circuit of Fig. 6–15(b) includes a resistor in parallel to account for losses in the coil, capacitor, and external load. The magnitude of the impedance seen at terminals A-B is shown in Fig. 6–16. Notice that the impedance at f_1 and f_2 is one-half times the impedance at resonance. The power dissipated in the resistor R at f_1 and f_2 is one-half times the power dissipated at resonance. The distance between the half-power points, f_2 and f_1, is called the bandwidth of the tuned circuit.

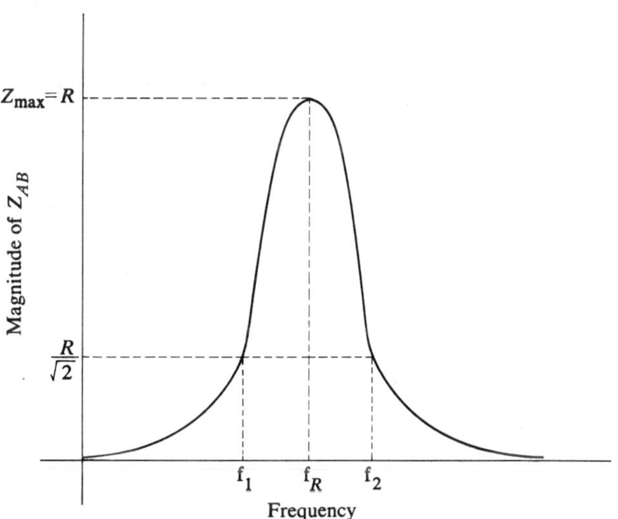

Fig. 6–16 Terminal impedance of practical parallel tuned circuit of Fig. 6–15(b).

Quite often it is important to know the relationship between the bandwidth and resonant frequency. The ratio of f_r to f_2-f_1 defines the quality factor or "Q" of the circuit.

$$Q = \frac{f_r}{f_2 - f_1} \approx \frac{R}{2\pi f L} \tag{6.14}$$

The higher the "Q" the more nearly perfect the circuit will be. The lower the Q the more energy there is being dissipated in the tuned circuit. This energy must be supplied by an external source if the circuit is to continue oscillating. The amplifier supplies this energy in a sinusoidal oscillator.

Let's look at a simple analysis of a parallel tuned circuit.

Example 6–5. In the circuit of Fig. 6–15(b), $L = 10^3$ H, $C =$ 0.10 μF, $R = 10^4\Omega$. What are Q, f_r, and Z_{AB} at resonance? We can calculate the resonant frequency from the equation

$$f_r = \frac{1}{2\pi\sqrt{LC}}$$

then $$f_r = \frac{1}{2\pi\sqrt{10^{-3} \times 10^{-7}}} = \frac{1}{2\pi\sqrt{10^{-10}}} = 15.9 \text{ k Hz}$$

$$Q = \frac{R}{2\pi f_r L} = \frac{10^7}{2\pi(15.9 \times 10^3)} = \frac{10^4}{100} = 100$$

$$Z_{AB} = \text{at resonance} = R = 10^4\Omega$$

We used the equation for f_r derived for the ideal LC-tuned circuit. As long as the circuit Q is greater than 10, the resonant frequency of the actual circuit will be very close to the ideal value.

Figure 6–17 is the circuit diagram of the parallel LC tuned circuit applied to a transistor amplifier to make an oscillator. The capacitor is split into a series combination

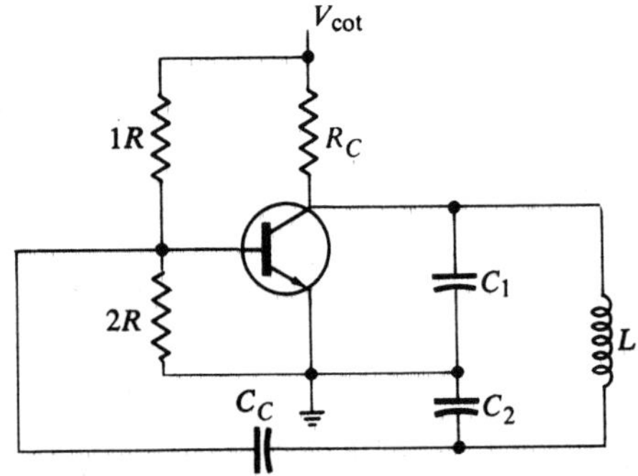

Fig. 6–17 Colpitts sinusoidal oscillator.

so that C_2 can be used to feed back a portion of the output signal to the base of the transistor R_c. $1R$ and $2R$ are the biasing resistors for the amplifier. The resonant frequency is given by $1/2\pi\sqrt{LC}$ where $C = C_1C_2/(C_1 + C_2)$. It has been proven mathematically that the condition for oscillation in this circuit is that $\beta > C_1/C_2$. The Colpitts oscillator is only one of many sinusoidal oscillators that have been developed and well documented in the literature. You will find elements that perform the block diagram functions of Fig. 6–14 in each design.

The output of sinusoidal oscillators is not always directly applicable to the devices to be timed, especially in digital systems. Shaping circuits, such as Schmitt triggers, are often used between the timing source and the devices to be timed.

6–4 Digital Time Delays

Quite often in digital electronic controls for welders, automatic machines, or sequencing operations it is necessary to introduce time delay. If the precision requirements are not severe, synchronized ac or dc time delays may be used. More precise delay times require that the time delay operate from the same time base as the system timing. We will look at the technique of frequency division to provide digital time delay and the extension of this technique to counting circuits.

6–4–1 Frequency Division

Let's assume we are designing a digital control circuit operating from a timing source at a frequency of 20 kHz. At a point in the design, we must introduce a one-half second delay. As shown in Fig. 6–18, by dividing the pulse rate by

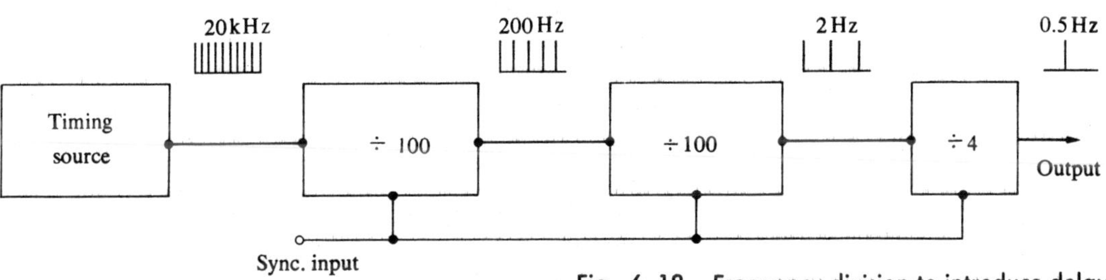

Fig. 6–18 Frequency division to introduce delay.

100 we could reduce the output to 200 Hz; dividing the rate by 100 again would reduce the output to 2 Hz: further division by 4 would reduce the output to one pulse each one-half second. By applying a synchronizing pulse at the beginning of the time interval, the first output pulse would signal the end of one-half second time delay.

Frequency division circuit designers use a large number of semiconductor devices. Figure 6–19 is the circuit diagram for a 100:1 divider circuit using cascaded UJT relaxation oscillators. Positive pulses applied at the emitter input of the first oscillator produce outputs at 5 kHz. Current pulses generated by the first oscillator produce negative voltage pulses across the 120-Ω resistors. The second oscillator generates pulses at the rate of 1 kHz. The final oscillator generates the output pulses at base one at a rate of 200 Hz.

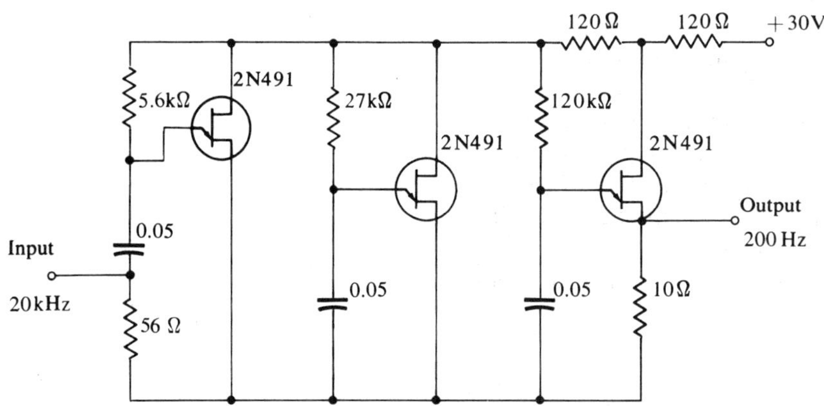

Courtesy of General Electric Semiconductor Dept., Syracuse, N.Y.

Fig. 6–19 UJT frequency divider circuit (100:1).

One of the most common frequency division circuits is the bistable multivibrator or flip-flop. We looked at the astable multivibrator in section 6–3–2. The astable multivibrator of Fig. 6–12 can be modified so that both diodes are normally in stable states. Figure 6–20 shows a typical design of the flip-flop using PNPN diodes. Both diodes are chosen so that V_s is greater than V_{dc}. When switch S is closed both diodes remain in the nonconduction or "off" condition. Application of negative pulses at the cathodes of $1D$ and $2D$ will cause the voltage across the diodes to

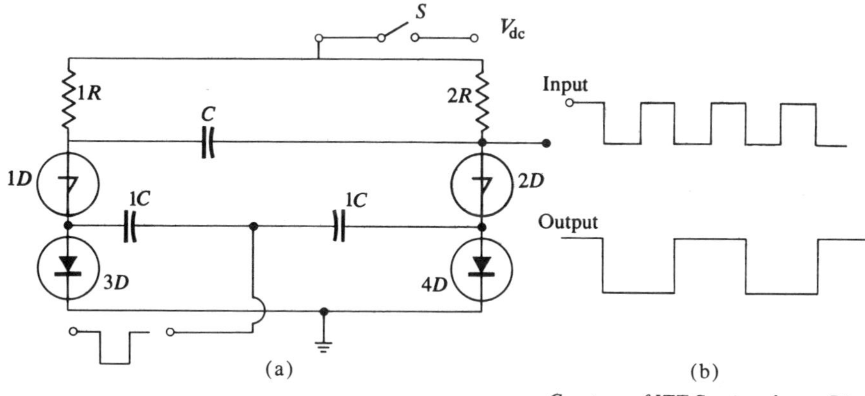

Courtesy of ITT Semiconductor Div.

Fig. 6–20 Bistable multivibrator using PNPN diodes.

exceed V_S. Both diodes will try to go into conduction. Since one of the diodes will switch on faster than the other, one diode will remain in the off state. Let's assume 1*D* goes into conduction first. Capacitor *C* will then charge through 2*R* and 3*D* to the voltage V_{dc}. Since V_{dc} is less than V_s, diode 2*D* will still remain in the off state. The circuit will remain stable until a second pulse is applied at the input. The second pulse will turn 2*D* on and 1*D* off. The output pulse rate is one-half the rate of the input pulses. The bistable multivibrator is an automatic two-to-one frequency divider.

6–4–2 Digital Counters

While straight frequency division may be adequate for some digital delay applications, it is difficult to vary the amount of delay after the circuit is designed. A more flexible technique is to combine bistable flip-flops in a counting circuit. By sampling the output of each flip-flop, we can determine how many input pulses have occurred at any time. Figure 6–21 is a block diagram of three flip-flops cascaded to form a counting circuit. Each flip-flop would contain the circuit of Fig. 6–20(a). Each flip-flop is forced into one of its stable states by applying a separate *reset* pulse. Let's assume that 2*D* is in the conducting state for each flip-flop; then Fig. 6–22 shows the voltage levels after each input pulse. The first negative-going input pulse switches 2*D* off and 1*D* on in flip-flop number one. The output of flip-flop number one is used to trigger flip-flop number two. Since positive-

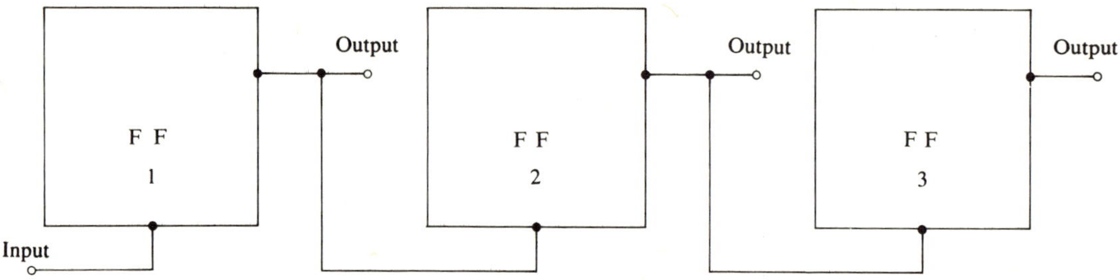

Courtesy of General Electric Company

Fig. 6–21 Counting circuit using three flip-flops in cascade.

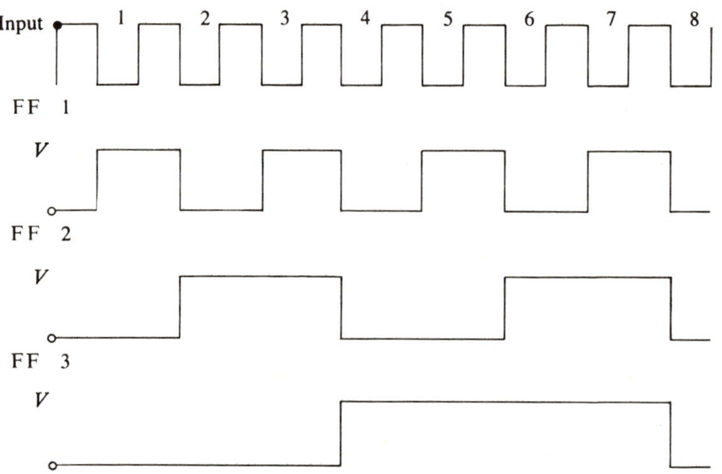

Fig. 6–22 Output voltages for counting circuit of Fig. 6–21.

going pulses do not change the state of the flip-flops, the output of number two and number three circuits remains the same. Notice that after the first input pulse, output number one is at voltage *V*; outputs number two and three are at zero volts.

We can follow through the first eight input pulses and make a table of the outputs after each input. This is shown in Table 6–1. This counting circuit resets itself on the eighth impulse so the maximum delay would be eight pulses. The maximum time delay would be 7 × *T* where *T* is the period of the input pulses. Additional delay would require the addition of more flip-flops. How will we know when the proper amount of time delay has occurred? We would have to build another circuit to indicate each combination of outputs in Table 6–1. Such a circuit is often called a decoder. Figure 6–23 shows that the outputs of each flip-flop in the counter are fed into the decoding circuit. The decoder then has one output for each amount of delay. We have simplified the operation of the counter in order to isolate the basic concept from the details surrounding the complete circuit. We will discuss counters and decoders in more detail later. Let's grasp the basic concept and retain it for future applications.

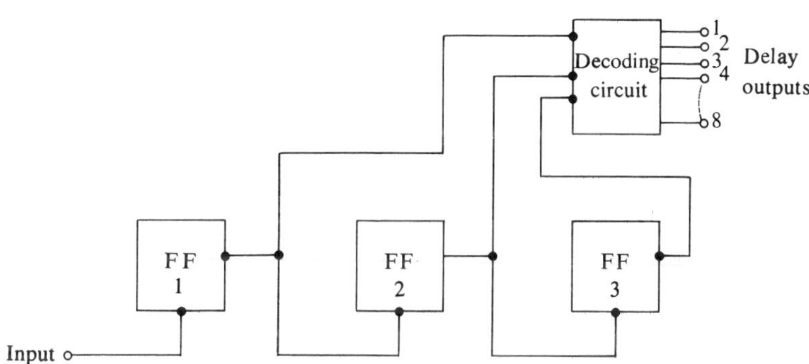

Fig. 6–23 Counting circuit with decoding for selection of time delay.

TABLE 6–1

Table of outputs for the first eight input pulses for the counter of Fig. 6–2

Input Pulse	Output One	Output Two	Output Three
Reset	0	0	0
1	V	0	0
2	0	V	0
3	V	V	0
4	0	0	V
5	V	0	V
6	0	V	V
7	V	V	V
8	0	0	0

6–5 Applications of Time Delays and Timers

Our theoretical discussion of timers and time delays will be aided by a close look at two industrial devices, an analog time delay used in small controls and a digital time delay used in welding controls.

6–5–1 A 60-Second Analog Timer

Simple analog time delays find many applications in industrial control situations. A semiconductor electronic time delay typical of the kind used in industry is the General Electric 3S7504 TM 560 shown in Fig. 6–24. The electronic circuitry operates an electromechanical relay to control the load. As much as 60 seconds' time delay can be introduced between the time the initiation switch is actuated and the

Courtesy of General Electric Company

Fig. 6–24 An industrial time delay for small controls.

time the relay contacts are actuated. The schematic of the 3S704 TM 560 time delay is shown in Fig. 6–25. Let's look at how it operates. The initiation switch, 1*S*, is in the closed position in the normal or ready condition. Diode 4*D* is forward-biased and conducting. Diode 5*D* is reverse-biased and cut off. Current flows through 4*D* and 100 *R* to charge

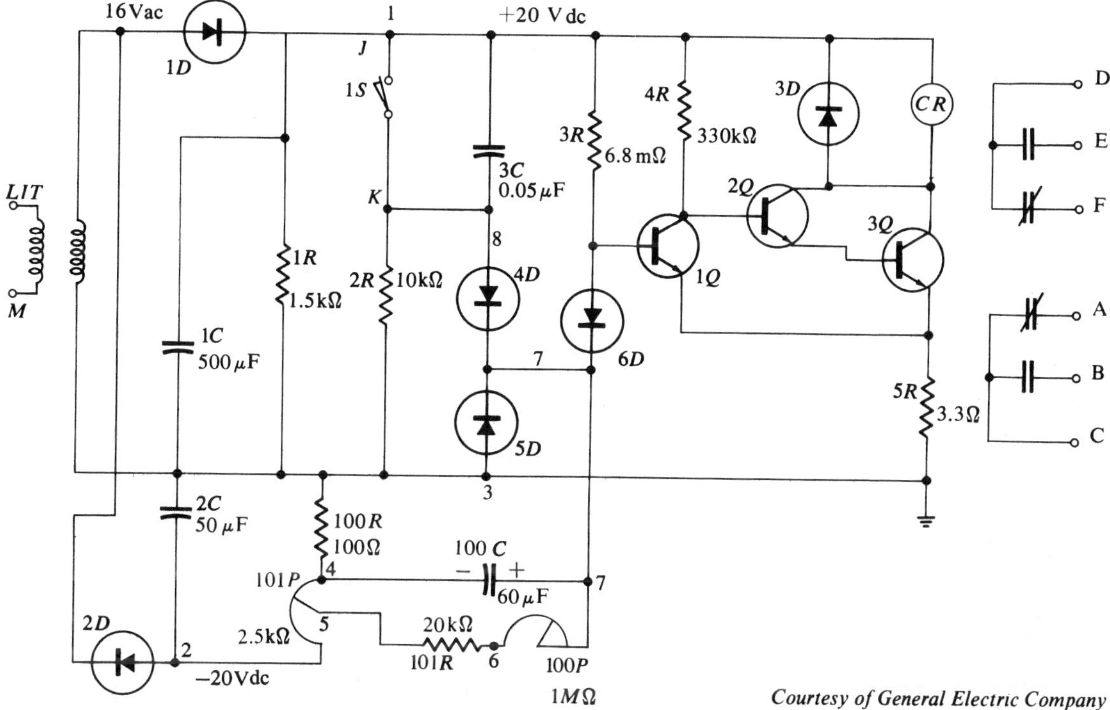

Courtesy of General Electric Company

Fig. 6–25 Typical time delay circuit (simplified).

capacitor 100*C* to approximately 20 volts. The cathode of diode 6*D* is clamped at 19.5 volts by 4*D* so no current flows through 6*D*. Transistor 1*Q* is biased into conduction by current flow through 3*R* and 5*R*. Current flow through 4*R* is shunted through the saturated collector-to-emitter circuit of 1*Q* so the "Darlington pair," 2*Q* and 3*Q*, is at cutoff. No current flows through the relay coil, *CR*, so the relay is not energized. The time delay is started by opening the initiation switch, 1*S*. Diode 4*D* will now be reverse-biased by the voltage on 100*C* so it stops conducting. Diode 5*D* is still reverse-biased by the voltage on 100*C*. The anode of diode 6*D* is held near ground potential by 1*Q*; the cathode of 6*D* is held at 20 volts by the charge on 100*C* so the base of 1*Q* is still isolated from the circuit of 100*C* by the high reverse resistance of 6*D*. As soon as the switch is opened, current flows through 100*R*, 100*P*, 101*R*, and 101*P* to reverse the charge on 100*C*. Figure 6–26 shows the voltage on the

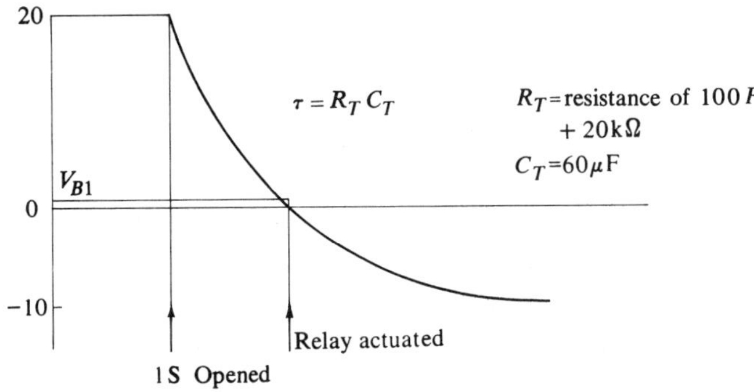

Fig. 6–26 Voltages in circuit of Fig. 6–25.

capacitor before and after 1S is opened. Capacitor 100C will charge to the voltage between points 4 and 5 at a rate determined by the setting of pot 100P. When the capacitor voltage goes through zero volts, diode 6D becomes forward-biased and conducts. The voltage at the base of 1Q goes to ground and 1Q is forced into cutoff. Current flow through 4R biases 2Q and 3Q into saturation. The current flow through the collector of 3Q energizes the relay coil. The voltage curve of Fig. 6–26 assumes that the voltage from 4 to 5 in Fig. 6–25 was 10 volts. The time delay can be calculated from:

$$V(t) = V_{\text{exp}} \left(- \frac{t}{\tau} \right) \tag{4.11}$$

Where $V = 30$ V

$\tau = (20\ \text{k}\Omega + R_{100P})\ 60\ \mu\text{F}$

If 100P is set at 0.5 mΩ, then:

$$\tau \approx 0.5 \times 10^6 \times 60 \times 10^{-6} = 30\ \text{s} \tag{6.15}$$

$$20 = 30\, e^{-t/30} \tag{6.16}$$

Solving for t

$$t = 30 \ln \frac{30}{20} = 12\ \text{s} \tag{6.17}$$

The circuit is reset by recharging capacitor 100C to 20 volts when the initiation switch is closed again.

Figure 6–27 shows the connection for a possible industrial application of this electronic time delay. Two motors must be operated in sequence. One motor must be started when 2*S* is closed, and run for 30 seconds. At the end of 30 seconds, the first motor must be shut off and the second motor must be started. The second motor must continue to run until 2*S* is reopened. Switch 1*S* should be placed in the closed position before 2*S* is actuated. The 3S704 ET 560 time delay can be used to energize the motor control relays 1*M* and 2*M* to start the first and second motors respectively.

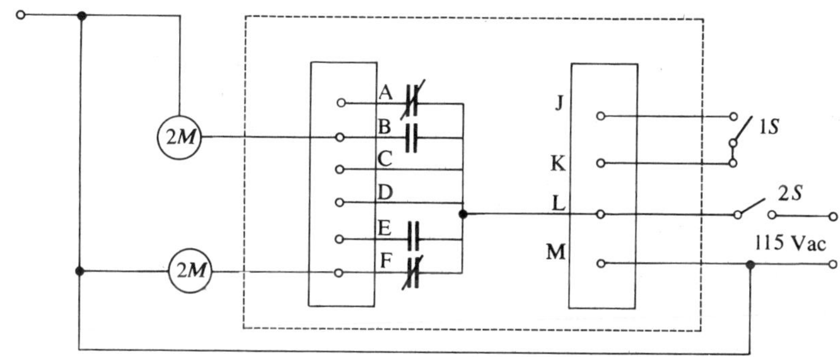

Fig. 6–27 Electronic timer used to operate two motors.

6–5–2 A Binary Welding Timer

The binary counter is often used to control the timing of sequential machines and processes. Let's look at how a binary timer is used to determine when a certain number of cycles of welding current has been applied to a work piece in a resistance welding machine.

The Robotron Mod 3095 resistance welding machine uses a binary counter to determine when the proper amount of current has passed through a resistance weld. A simplified schematic of the counter is shown in Fig. 6–28. The output voltage of each flip-flop is shown in Fig. 6–29(a) as the input pulses are applied to FF(flip-flop) #1. All outputs return to the starting voltage level after 16 input pulses. This is a scale of 16 counter. For this particular machine, the amount of welding current must be easily adjustable by the machine operator. It is more convenient if counting is done in the familiar decimal system. The scale of 16 counter

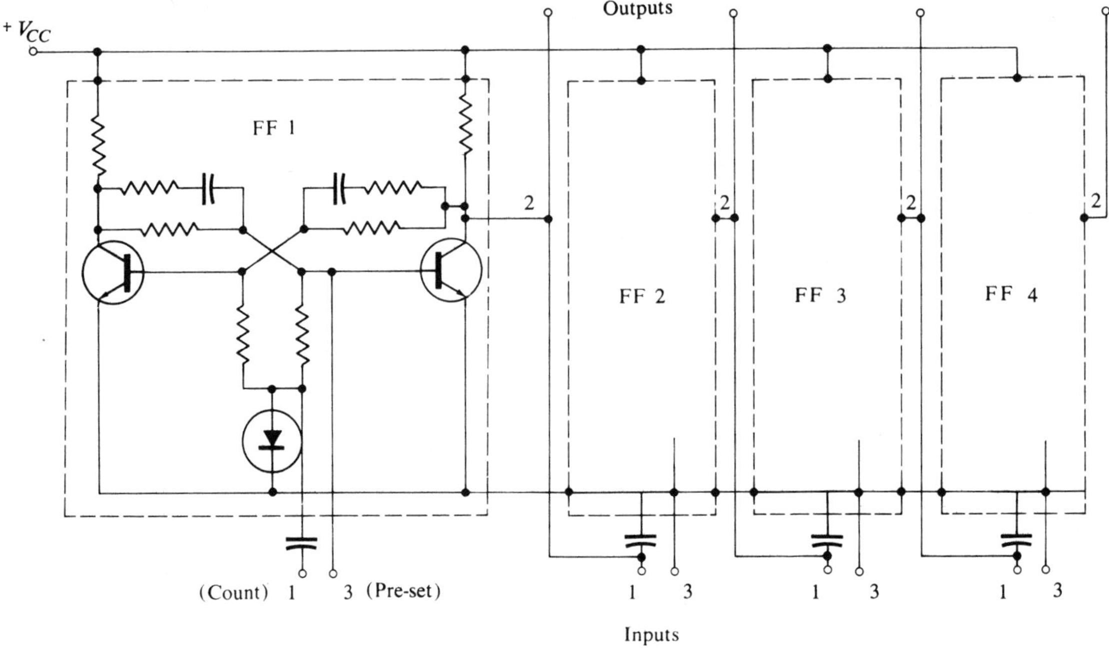

Fig. 6–28 Simplified schematic of binary counter of MOD 3095 welder.

can be converted to a scale of 10 counter by modifying the inputs to FF #4 as shown in Fig. 6–30. For the first seven input pulses, FF #4 is not affected. On the count of eight both inputs to FF #4 are pulsed and the output switches. At the same time, the input to FF #3 is clamped at ground potential by the saturated collector to emitter of transistor 2*Q* in FF #4. When the ninth input pulse is applied at FF #1, FF #1 is switched but all other flip-flops remain in the same state. The tenth input pulse causes FF #4 to switch and all output voltages are at the starting potential again. Figure 6–29(b) shows the output voltages of the scale of 10 counter.

One way to use a binary counter is to start with all outputs at zero and detect when the proper count has been reached by observing the decoded output. Another way is to preset the counter at the complement of the desired count and determine when all outputs reach zero. For example, if we wished to detect the count of six, we would preset the binary counter at four—six pulses later the output voltages would all be zero. The flip-flops are preset by applying a

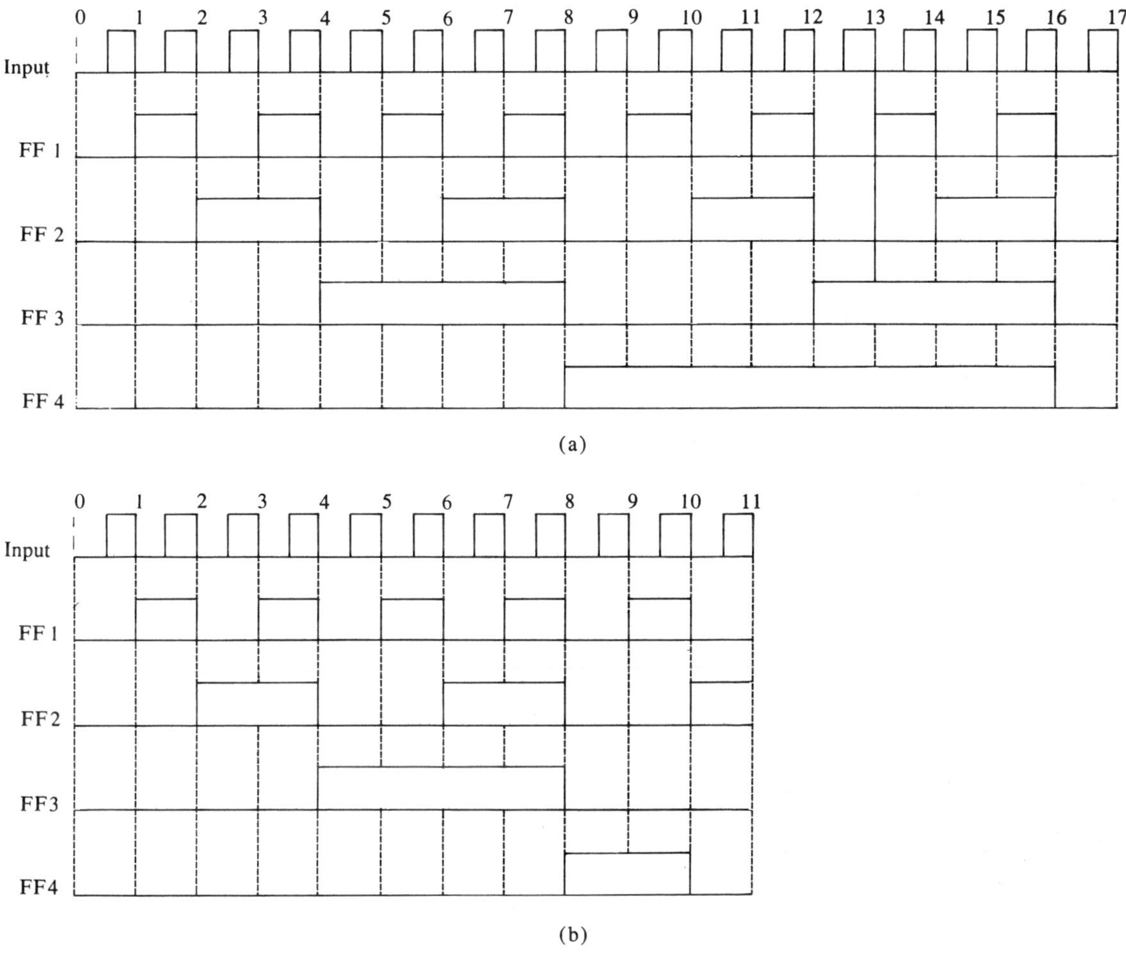

Courtesy of Robotron Corp.

Fig. 6–29 (a) Output voltages of flip-flops in Fig. 6–28, (b) output voltages of flip-flops in Fig. 6–28 after conversion to scale of 10.

negative pulse at the preset input terminal 3 of Fig. 6–28. The counter is preset to the proper count by operating the external selector switches on the welder control panel shown in Fig. 6–31. The selector switch chooses the proper flip-flops to be preset through the use of the resistance matrix of Fig. 6–32. The counter is "programmed" by selecting the proper preset network. The output is taken from FF #4. A

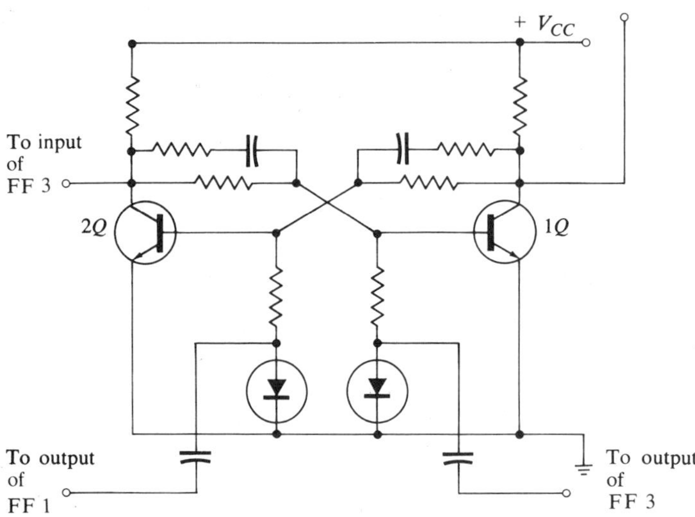

Fig. 6–30 Modification of FF#4 to make scale of 10 counter in Fig. 6–28.

Courtesy of Robotron Corp.

Fig. 6–31 Welder control panel.

capacitor at the output of FF #4 provides a negative pulse when the proper number of inputs has occurred. Figure 6–33 shows the printed circuit board on which the binary timer is built in the MOD 3095A welder. A simplified block diagram of the welding machine control showing the application of the binary timer is in Fig. 6–34.

6–6 Summary

The time delays that we looked at in this chapter may be used to control a very simple function such as the start of a motor or they may be part of a very complex electronic system such as a weld controller. We will see more applications in later chapters. We looked at some of the circuitry for generating digital timing signals. The understanding of the techniques applied will serve you well in your experiences with today's complex electronic digital systems.

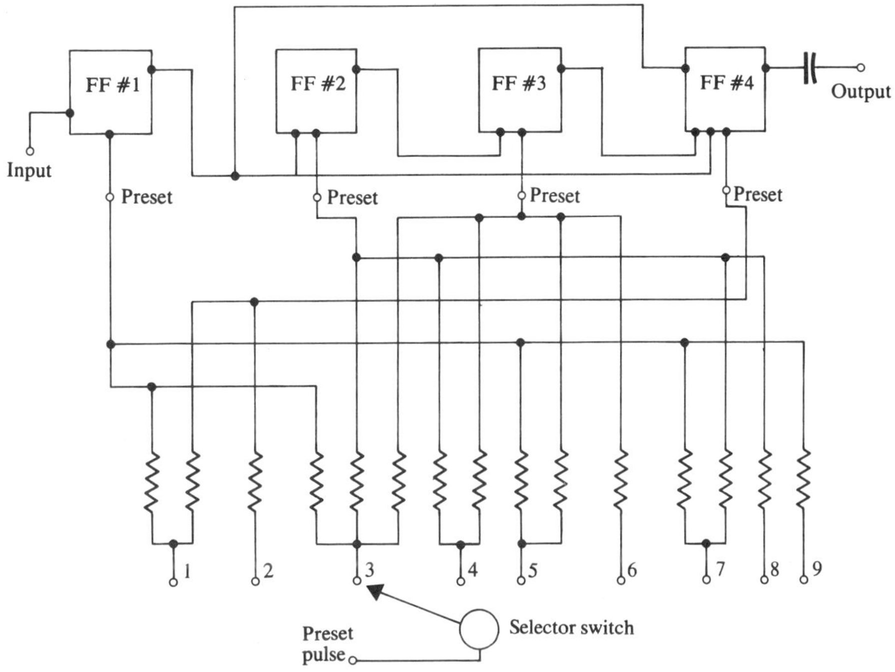

Fig. 6–32 Resistance network for preset selection.

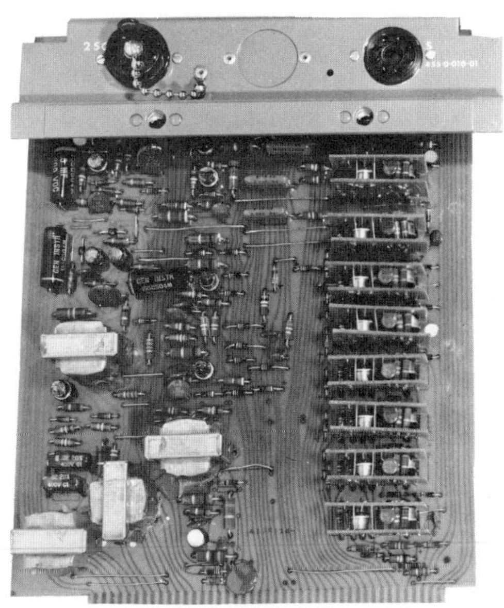

Courtesy of Robotron Corp.

Fig. 6–33 Binary counter for MOD 3095 welder.

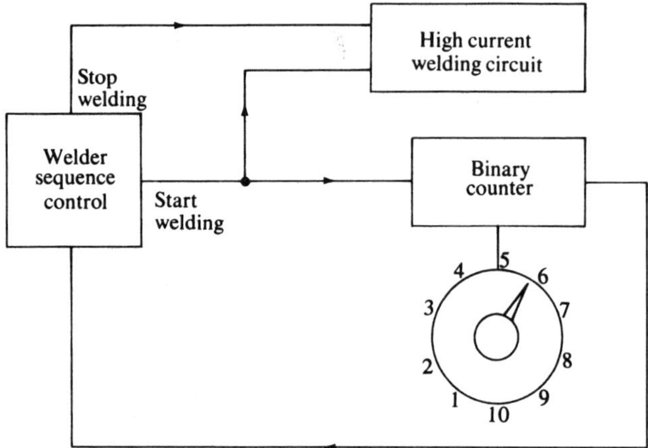

Fig. 6–34 Block diagram of binary counter in welding machine circuit.

EXERCISES

1. What is the meaning of "analog time delay" as compared to "digital time delay"?

2. In the time delay circuit of Fig. 6–4, $1D$ is a 20-V, 1-W zener diode, V_{dc} is 40 V, and $1Q$ is the 2N 2646 UJT.
   (a) Specify R_2 for a maximum output current of 40 mA.
   (b) What is the maximum value of $1R$ that can be used with the 2N 2646?
   (c) Calculate $1C$ for 90 minutes of time delay in the operation of the relay.

3. The UJT of problem 2 is replaced by the PUT of Fig. 6–5(c). Using the same zener regulator and $R_2 = R_3 = 20\ \text{k}\Omega$:
   (a) Determine the maximum value of R_1 if the D13T1 PUT is used.
   (b) Calculate the value of $1C$ required to get 90 minutes' time delay.

4. What is the basic function of a Schmitt trigger?

5. In the design of example 6–3, the trigger point is chosen at 1.5 V instead of 1 V. If all other design decisions remain the same, how will the final circuit of Fig. 6–9 be changed?

6. In the Schmitt trigger design of example 6–3, V_{CC} is decreased to 4 V. If all other design decisions remain the same, how is the final circuit of Fig. 6–9 changed?

7. What are some of the timing circuits that can be developed from the basic astable multivibrator?

8. Plot the voltage across 1C and 2C in Fig. 6–11.

9. Calculate the value of R, 1C, and 2C for an oscillation frequency of 10 kHz in Fig. 6–11.

10. Design the oscillator of example 6–4 at a frequency of 100 kHz using the IN 3841 diode, V = 35 V, and I_H = 25 mA.

11. In the circuit of Fig. 6–35, L = 0.5 μH, C = 0.022 μF, and $R = 10^4\ \Omega$.
    (a) Find f_r.
    (b) What is the Q?
    (c) Find the impedance at resonance.

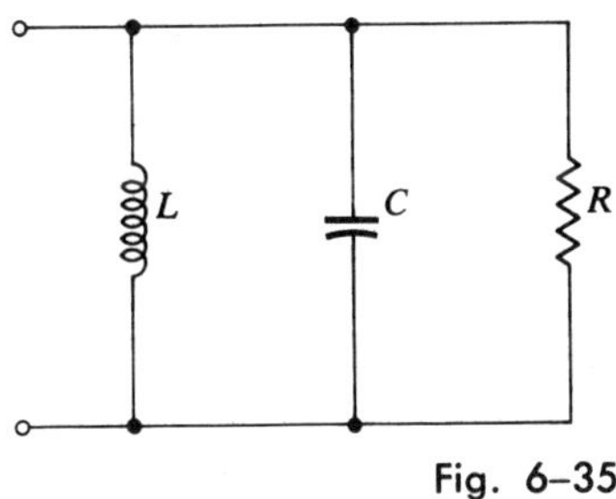

Fig. 6–35

12. Explain the operation of the bistable flip-flop of Fig. 6–20(a) if diode 2D is chosen to have a forward breakover voltage less than the dc voltage applied.

13. In the electronic timer circuit of Fig. 6–25, how would the relay perform if diode 5D was shorted? Open-circuited?

14. Why is the Darlington combination of transistors used in the timer circuit of Fig. 6–25?

7

Phase Shift Controls

7–1 Introduction

The basic purpose of industrial electronic controls is to regulate the transfer of energy from a source to a load. It may be a weld control to control the conversion of electrical energy to heat; it may be a motor control to control the conversion of electrical energy to mechanical force; or it may be a safety alarm to convert electrical energy to sound. If the energy transfer is at a constant rate, then the control may be as simple as an on–off switch. Quite often it is necessary to adjust the rate of energy transfer to control the output, such as speed of a motor or loudness of an alarm. The most convenient way to control the rate of energy transfer from an ac source is to control the portion of each cycle that current is allowed to flow into the load. This is accomplished in SCR and TRIAC circuits by controlling the phase angle at which the thyristor is turned on during each cycle of the ac voltage. The technique is called *phase control*. Control of the triggering of the thyristor may be delayed by using reactive ac circuits or by using digital triggering circuits. We will look at both methods of phase control in this chapter.

7–2 AC Phase Shifting Circuits

The basic SCR phase shift control is shown in Fig. 7–1. The purpose of the ac phase shift circuit is to delay application of triggering voltage at the gate until the ac voltage reaches the phase angle α. Alpha, α, is called the *firing angle*. Current is supplied to the load for the remaining portion of the positive anode voltage. Theta, θ, the portion of

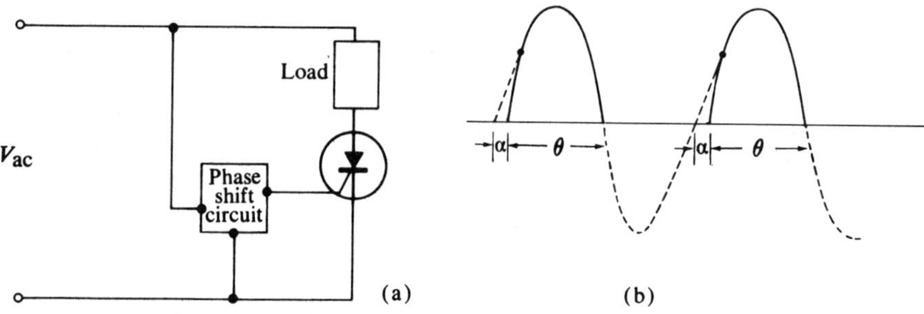

Fig. 7–1 (a) Basic SCR phase shift control, (b) load voltage.

the positive anode cycle when load current flows, is called the conduction angle. Figure 7–2 shows the basic TRIAC phase shift control and the load voltage waveform for a full-wave phase shift control.

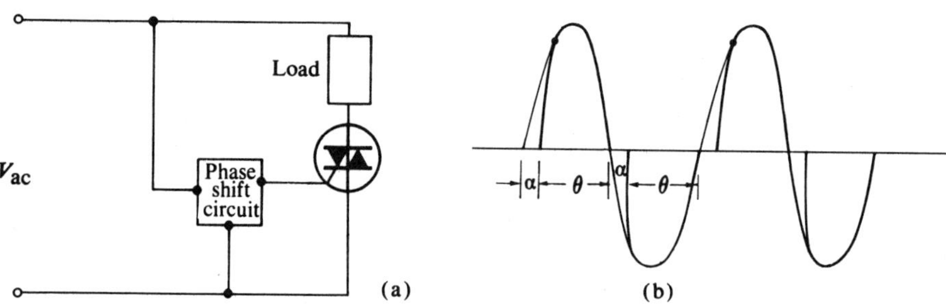

Fig. 7–2 (a) Basic TRIAC phase shift control, (b) load voltage.

Let's take a look at the reactive ac phase shift circuits that are commonly used. Figure 7–3(a) shows the basic *RC* phase shift circuit. Figure 7–3(b) shows the vector diagram of voltage relationships if the output is taken across the capacitor. The input voltage, $V_{1\text{-}3}$, is used as the reference.

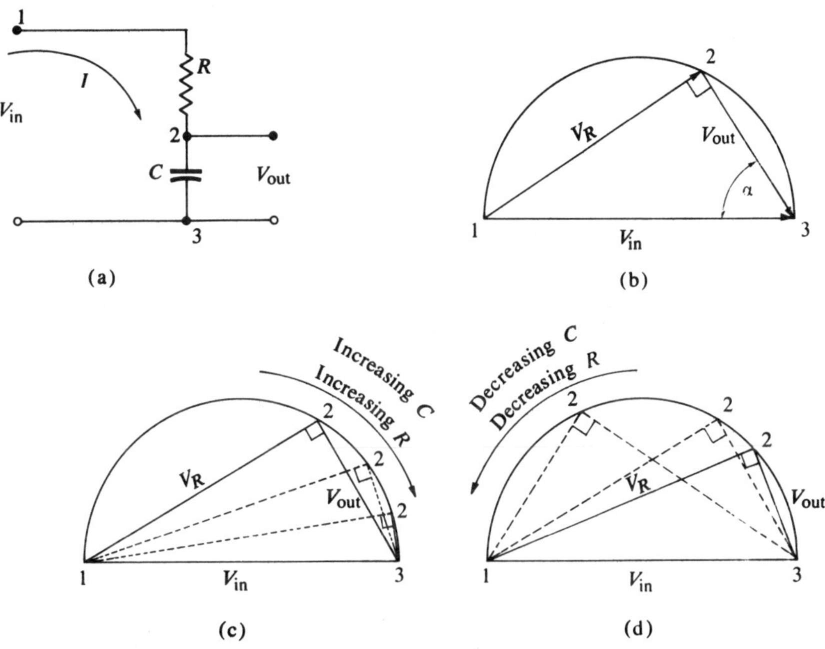

Fig. 7–3 (a) *RC* phase shift circuit, (b) vector diagram for fixed *R* and *C*, (c) vector diagram for increasing *R* or *C*, (d) vector diagram for decreasing *R* or *C*.

The voltage across the resistor, $V_{1\text{-}2}$, will be in phase with the current in the loop, thus leading $V_{1\text{-}3}$. Since $V_{1\text{-}3}$ must be the sum of $V_{1\text{-}2} + V_{2\text{-}3}$, the triangle must be closed, so the vector 2 – 3 is drawn from the tip of $V_{1\text{-}2}$ to the tip of $V_{1\text{-}3}$. You will recall that the current through the capacitor must lead the voltage across the capacitor by 90° in an ac circuit. This accounts for the right angle that must occur at the intersection of $V_{1\text{-}2}$ and $V_{2\text{-}3}$. The phase angle between V_{out} and V_{in} is the delay angle, α. From the triangle of Fig. 7–3(b)

$$\tan \alpha = \frac{V_r}{V_{out}} = \frac{IR}{IX_c} \tag{7.1}$$

where I is the series current in Fig. 7–3(a). Then

$$\tan \alpha = \frac{R}{X_c} = \frac{R}{1/\omega C} = R\omega C \tag{7.2}$$

where R is resistance in ohms
C is capacitance in farads
ω is radian frequency, rad/s

From Eq. (7.2), as R or C is increased, α is increased; as R or C is decreased, α is decreased. These relationships are also shown in the vector triangles of Fig. 7–3(c) and (d). The limits on α are governed by the tan α. As tan α approaches zero, α approaches zero degrees. As tan α approaches α, α approaches 90°. The usual circuit approach is to vary R to control α between the limits of 0° and 90°.

The range of control of α can be extended by using the circuit of Fig. 7–4(a). A bridge circuit is formed by placing two equal resistors across the input. The output is then taken across the bridge. The voltage relationships are shown in Fig. 7–4(b). It can be shown that the locus of

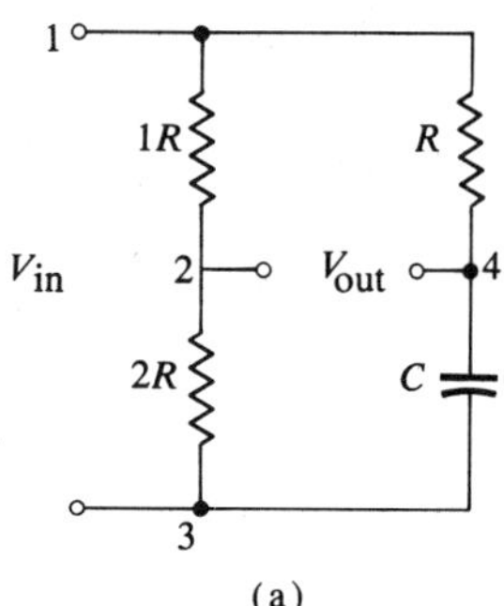

(a)

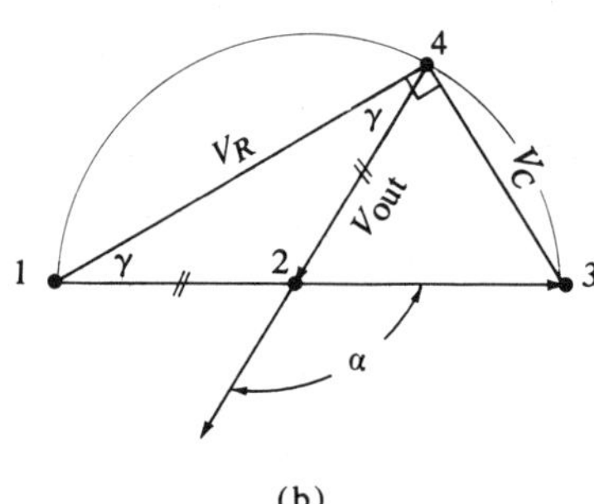

(b)

Fig. 7–4 (a) Extended RC phase shifting circuit, (b) vector triangle showing voltage relationships.

point 4 lies on a semicircle whose diameter is $V_{1\text{-}3}$. The magnitudes of $V_{1\text{-}2}$, $V_{2\text{-}3}$, and $V_{2\text{-}4}$ are all equal. The angle γ is the impedance angle of the RC circuit. Then

$$\tan \gamma = \frac{X_c}{R} = \frac{1}{R\omega C} \tag{7.3}$$

From the triangle 1–2–4

$$\alpha + 2\gamma = 180^\circ \qquad (7.4)$$

$$\alpha = 180^\circ - 2\gamma \qquad (7.5)$$

$$\alpha = 2(90^\circ - \gamma) \qquad (7.6)$$

Using the trigonometric relationship,

$$\tan(90^\circ - \gamma) = \cot\gamma$$

then

$$\tan\alpha/2 = \cot\gamma = R\omega C \qquad (7.7)$$

Again, as R or C increases, α increases; as R or C decreases, α decreases.

The usual design approach is to vary R to control the phase angle α. As R approaches infinity, $\alpha/2$ now approaches 90° or α approaches 180°.

As R approaches zero, α approaches zero degrees. The range of control of α is now 0° to 180°.

Similar phase shifting circuits could be designed using series RL components and taking the output across the coil to get the delay in phase angle.

Let's look at a sample calculation of phase shifting circuits.

Example 7–1. In the circuit of Fig. 7–3(a), $R = 25\text{k}\Omega$, $C = 0.1\ \mu\text{F}$, $V_{in} = 15\sin 377t$. What is the angle α?

From Eq. (7.2), $\tan\alpha = R\omega C$; then

$$\tan\alpha = 25 \times 10^3 \times 377 \times 0.1 \times 10^{-6} \qquad (7.8)$$

$$= 0.945$$

$$\alpha = 43.4^0$$

Example 7–2. In the circuit of Fig. 7–4(a), $R = 25\text{k}\Omega$, $C = 0.1\ \mu\text{F}$, $V_{in} = 15\sin 377\,t$. What is the phase angle α?

From Eq. (7.7) $\tan\frac{\alpha}{2} = R\omega C$

$$\tan\frac{\alpha}{2} = 25 \times 10^3 \times 377 \times 0.1 \times 10^{-6} \qquad (7.9)$$

$$\frac{\alpha}{2} = 43.4^0$$

$$\alpha = 86.8^0$$

The phase angle is exactly twice the phase angle of example 7–1, using the same component values at 60 Hz.

7–3 Design of AC Phase Shift Controls

The design of ac phase shift controls can be stated simply in three steps.

(1) Determine the firing and conduction angles based on the power requirements of the load and the source voltage.

(2) Determine the proper phase shift circuit.

(3) Match the phase shift circuit to the triggering requirements of the thyristor.

Of course, the final test of any design is to build a model and make adjustments in the circuit to meet the specifications.

This is a good point to go back and review the material in Chapter 5 on the design of SCR and TRIAC circuits. You will notice the overlap in the design of phase shift controls and basic thyristor switching circuits.

Let's look at each of the steps in the design of phase shift controls.

(A) Determining the Firing Angle and Conduction Angle Control specifications are generally stated in terms of the average power or rms voltage requirements of the load. The mathematical relationships are:

$$V_{rms} = \left\{ \frac{1}{2\pi} \int_{\alpha}^{\pi} V_{max}^2 \sin^2 \omega t \, d\omega t \right\}^{1/2} \qquad (7.10)$$

$$P_{av} = V_{rms}^2 / R_L \qquad (7.11)$$

In order to determine α, we would have to solve these complicated integral equations. Fortunately, graphs are available from which adequate estimates can be made. Figure 7–5 is a graphical analysis of phase control relationships for half-wave or SCR controls. Figure 7–6 is the similar analysis for full-wave or TRIAC control. Notice that there is very little variation in load power below 20° phase delay and above 160° phase delay. An example will help to clarify the use of the graphical data.

Example 7–3. A 20-Ω resistive load operating from a 230-V, 60-Hz line will be controlled by an SCR phase shift control. Average power to the load will vary between 1100 watts and 750 watts. Determine the range of control of α required.

The maximum average power delivered to the load is calculated first.

$$P_{max} = \frac{V^2_{rms}}{R} = \frac{230^2}{20} = 2650 \text{ W} \qquad (7.12)$$

Then the ratio of P/P_{max} is calculated.

$$\frac{P_{\alpha min}}{P_{max}} = \frac{1100}{2650} = 0.415 \qquad (7.13)$$

$$\frac{P_{\alpha max}}{P_{max}} = \frac{750}{2650} = 0.282 \qquad (7.14)$$

Now α_{min} and α_{max} can be read from Fig. 7–5

$$\alpha_{min} = 55°, \ \alpha_{max} = 85°$$

$$\theta_{max} = 125°, \theta_{min} = 95°$$

Example 7–4. Determine the range of control of α if a TRIAC is used in the design of example 7–3.

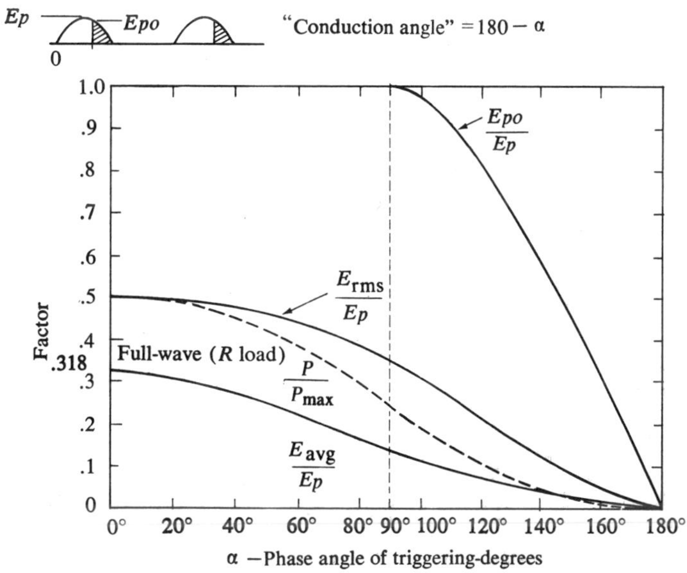

Courtesy of General Electric Semiconductor Dept., Syracuse,,N.Y.

Fig. 7–5 Half-wave phase control analysis.

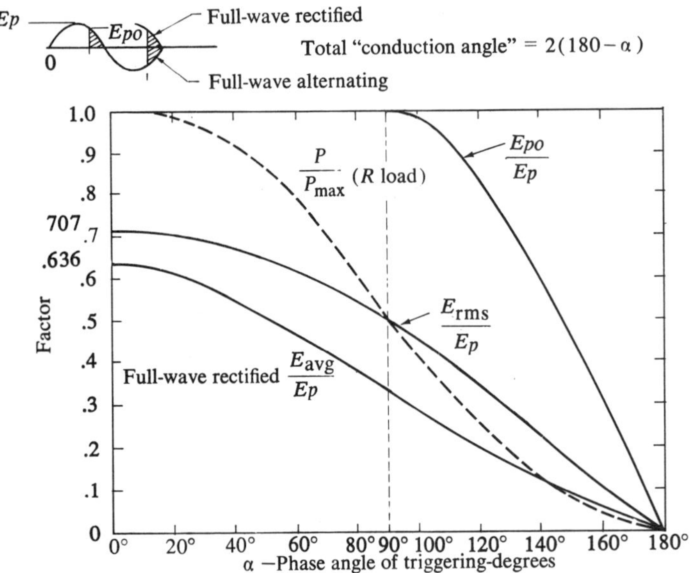

Courtesy of General Electric Semiconductor Dept., Syracuse, N.Y.

Fig. 7–6 Full-wave phase control analysis.

From Eqs. (7.13) and (7.14)

$$\frac{P\alpha_{min}}{P_{max}} = 0.415$$

$$\frac{P\alpha_{max}}{P_{max}} = 0.282$$

Then from Fig. 7–6

$$\alpha_{min} = 95°, \ \theta_{max} = 85°$$

$$\alpha_{max} = 110°, \theta_{min} = 70°$$

This design assumes symmetrical triggering of the TRIAC.

(B) Determining the Proper Phase Shift Circuit The type of phase shift circuit used will depend on the maximum value of α. If α_{max} is less than 90°, a single *RC* phase shift circuit similar to Fig. 7–3(a) would be adequate. As α_{max} approaches and exceeds 90°, a bridged *RC* circuit similar to Fig. 7–4(a) must be used. Let's continue with the design of example 7–3.

Example 7–5. The circuit of Fig. 7–7 is used to control the power delivered to the resistive load. Select the value of R and C for α_{min} and α_{max} of example 7–3. From example 7–3, $\alpha_{min} = 55°$, $\alpha_{max} = 85°$. Since a fixed capacitor is used, its value must be chosen. A 1-μF capacitor will be used. Then from Eq. (7.7)

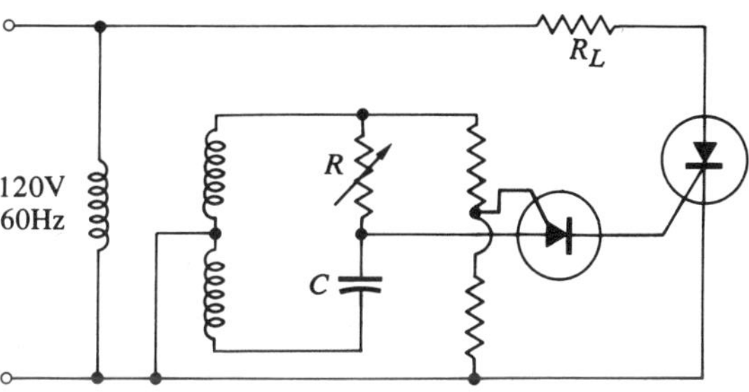

Fig. 7–7 PUT in AC phase shift control.

$$\tan \frac{\alpha}{2} = R\omega C$$

then

$$R_{min} = \frac{\tan \frac{\alpha_{min}}{2}}{\omega C} = \frac{\tan 27.5°}{377 \times 1 \times 10^{-6}}$$

$$R_{min} = 1380\ \Omega$$

$$R_{max} = \frac{\tan \frac{\alpha_{max}}{2}}{\omega C} = \frac{\tan 42.5°}{377 \times 1 \times 10^{-6}}$$

$$R_{max} = 2440\ \Omega$$

In the final circuit you would probably use a fixed 1-kΩ resistor in series with a 2-kΩ potentiometer to get the control required. Notice that we are using the bridged RC circuit even though α_{max} is 85°. Since it is impossible to ever achieve the theoretical 90° phase shift of a single RC circuit, we prefer to be on the safe side and use the extended phase shift circuit.

(C) Matching the Phase Shift Circuit to the Triggering Requirements of the Thyristor The basic phase shift controls of Figs. 7–1(a) and 7–2(a) should be modified to show a triggering device between the phase shift circuit and the power thyristor. The modified SCR control is shown in Fig. 7–8. The triggering device allows us to match the *RC* phase shift circuit to a wide range of thyristors without precision tailoring to the specific requirements of each thyristor. The technique usually employed is to store energy in the capacitor and discharge the capacitor through the triggering device at the proper time. The resulting pulse of energy is usually sufficient to overdrive the gate of the thyristor. The total power dissipation in the gate circuit is still low enough that the gate junction is not damaged.

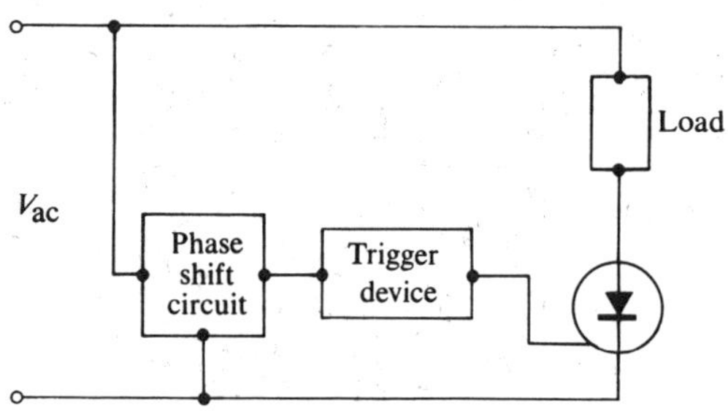

Fig. 7–8 SCR phase shift control with trigger device.

Any semiconductor device with negative resistance characteristics can be used as a triggering device. In Fig. 7–7, the PUT serves this purpose. Other devices often used are neon tubes, UJTs, DIACs, and SCSs. Figure 7–9 shows some basic applications of these trigger devices.

The phase shift circuit that you are most likely to encounter in phase shift controls is the UJT relaxation oscillator. We discussed the design of this triggering circuit in Sec. 5–7–1. Figure 7–10 shows the UJT relaxation oscillator in a TRIAC phase control.

Example 7–6. In Fig. 7–10, R_L = 10 Ω and V_{ac} = 110V, 60 Hz. The load power must be controlled between 600 and 800 watts. Determine the frequency range of the relaxation oscillator.

We must first determine α_{min} and α_{max}.

$$P_{max} = \frac{(110)^2}{10} = 1210\ \text{W}$$

$$\frac{P\alpha_{min}}{P_{max}} = \frac{800}{1210} = 0.66$$

$$\frac{P\alpha_{max}}{P_{max}} = \frac{600}{1210} = 0.495$$

From Fig. 7–6

$$\alpha_{min} = 75^\circ$$

$$\alpha_{max} = 90^\circ$$

The phase angles can be converted to delay time by the relationships

$$\frac{\alpha_{min}}{t_{min}} = \frac{180^\circ}{T/2}, \qquad \frac{\alpha_{max}}{t_{max}} = \frac{180^\circ}{T/2}$$

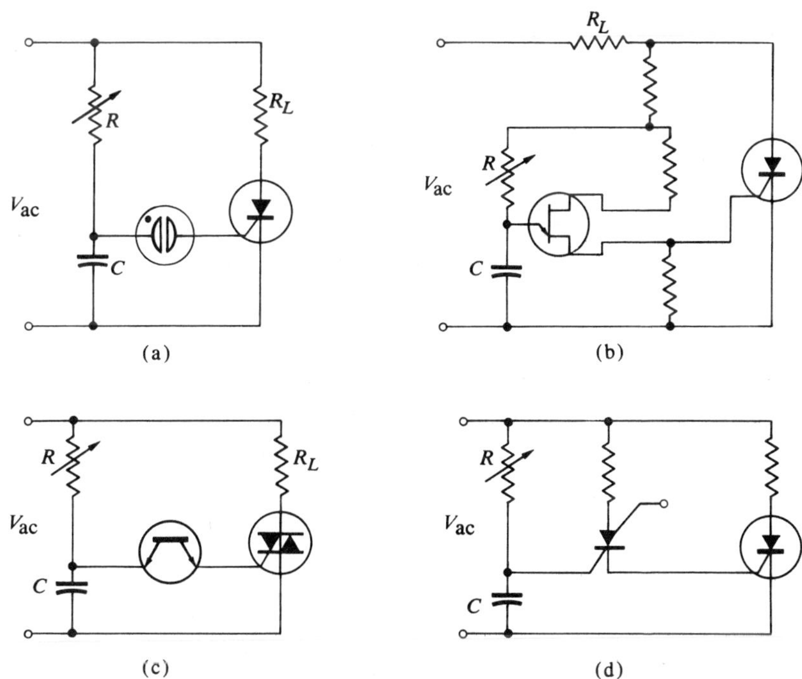

Fig. 7–9 Trigger devices used in AC phase controls: (a) neon tube, (b) UJT, (c) DIAC, (d) SCS.

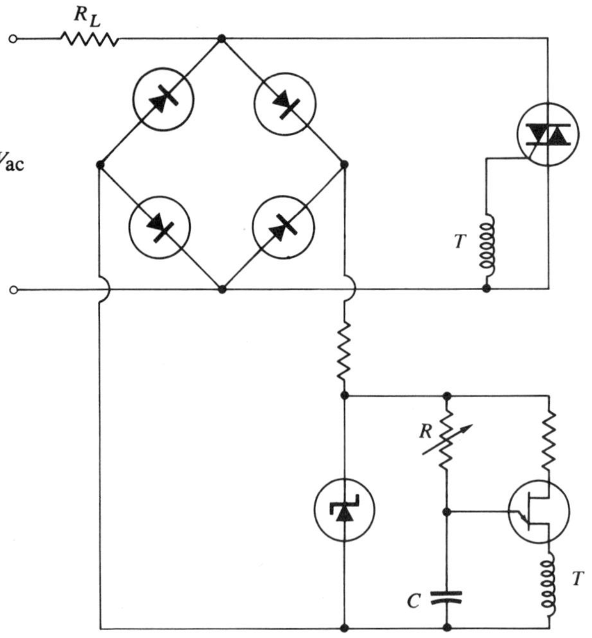

Fig. 7–10 UJT relaxation oscillator in TRIAC phase control.

where $T/2$ is the time for one-half period of the sinusoidal source voltage.

$$\text{At 60 Hz,}\ \frac{T}{2} = 8.3\ \text{ms}$$

$$\text{Then}\ t_{min} = 3.46\ \text{ms}$$

$$t_{max} = 4.15\ \text{ms}$$

Calculations of the frequency range of the oscillator are

$$f_{min} = \frac{1}{t_{max}} = 240\ \text{Hz}$$

$$f_{max} = \frac{1}{t_{min}} = 289\ \text{Hz}$$

The *RC* phase controls that we have discussed so far often require large variations in the charging resistor to get adequate control. This results in low gain and slow, nonlinear response. Gain and linearity can be improved by using the circuit of Fig. 7–11(a). The zener diode clamps the voltage at point 1 at level *D* in Fig. 7–11(b) as long as V_{ac} is

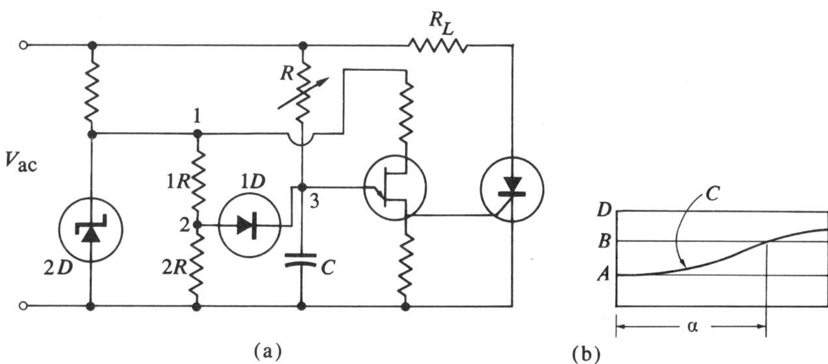

Fig. 7–11 (a) Ramp and pedestal phase control, (b) ramp and pedestal voltage.

greater than V_z. Resistors $1R$ and $2R$ are much smaller than R so that capacitor C charges quickly through $1D$ to the level A. This forms the pedestal for further charging of the capacitor. The capacitor continues to charge through R toward the line voltage. When the voltage at point 3 rises above the voltage at point 2, diode $1D$ becomes reverse-biased and acts like an open circuit. The capacitor charges along the ramp (C), toward the zener voltage level D. The triggering voltage of the UJT,(B), determined by the stand-off ratio, η, is between the pedestal and zener voltages. When the capacitor voltage reaches level B, the UJT fires and triggers the SCR at the phase angle α. The phase angle can be varied by changing the ratio of $1R$ to $2R$ or by adjusting the charging resistor.

7–4 Applications of AC Phase Shift Controls

Let's look closely at two commercial phase shift control circuits employing some of the techniques we have just discussed.

7–4–1 Blower Motor Control

Figure 7–12 shows the ac phase shift control circuit for a blower motor to operate as part of a heating system. The temperature is sensed by thermistor $9R$. The desired temperature is set by adjusting potentiometer $8R$. The blower motor runs continuously with the speed varying between the

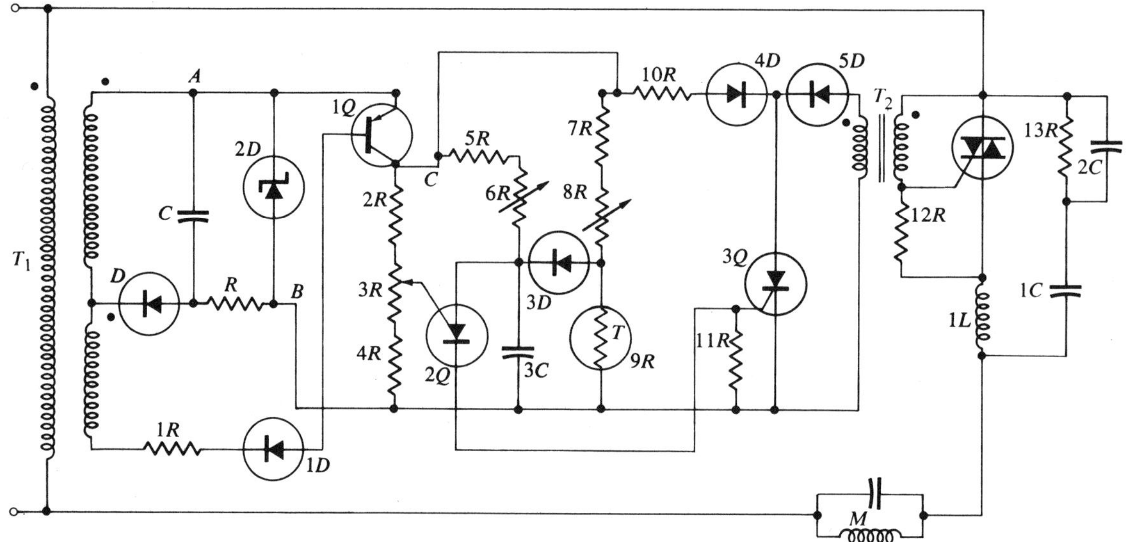

Fig. 7–12 AC phase shift control for blower motor.

maximum and minimum value depending on the difference between the actual temperature and desired temperature. Let's look at the detailed operation of this circuit.

The power control device is the TRIAC 4*Q*. The voltage applied to the motor, *M*, is controlled by the firing angle of TRIAC. The network formed by 1*L*, 1*C*, 2*C*, and 13*R* protects the TRIAC from dv/dt damage and suppresses the *RF* interference due to the switching action (see chapter 5).

The transformer, 1*T*, reduces the line voltage to a safe level to protect the low voltage components of the triggering circuit. Diode, *D*, and capacitor, *C*, form a half-wave rectifier filter to supply the dc voltage for the triggering circuit. Diode 2*D* and Resistor, R, regulate the supply voltage (see chapter 2).

The transistor, 1*Q*, synchronizes the triggering action with the ac line voltage. When the voltage at the upper terminal of 1*T* is positive, the base to emitter of 1*Q* is forward-biased. Base current flows through 1*D* and *R* to drive 1*Q* into saturation. The collector-to-emitter voltage of 1*Q* during saturation is near zero so the voltage between points *C* and *B* is essentially the same as the voltage between points *A* and *B*. When the voltage at the upper terminal of 1*T* is negative, 1*Q* is reverse-biased. 1*Q* is driven to cutoff. The collector-to-emitter impedance of $1Q$ is very high during cutoff. The entire zener voltage is dropped across the collector of $1Q$; the voltage from C to B is now essentially zero. We see that $1Q$ acts as a switch that is closed during the positive half-cycle and open during the negative half-cycle.

The primary source of triggering pulses is the ramp-and-pedestal SCS relaxation oscillator (see Sec. 7–3). At the beginning of each positive half-cycle, $1Q$ applies voltage to the oscillator at point C. Capacitor C charges to the voltage pulse at the gate of SCR, $2Q$. The period of the relaxation oscillator determines the phase angle, α, at which the SCR is triggered. The phase delay can be varied in three ways: by varying $3R$, varying $6R$, or by varying the temperature of the thermistor $9R$. In actual practice $6R$ would be adjusted to control the minimum speed of the motor; $3R$ would be adjusted to control the maximum speed; and a variation of $9R$ would control the motor speed between these limits depending on the actual temperature.

Triggering of the TRIAC, $4Q$, is accomplished by using the reflected impedance property of transformer $2T$. During the positive half cycle before $3Q$ is fired, a high impedance is reflected into the primary of $2T$ at the gate of $4Q$. Gate current flow in the TRIAC is not sufficient to cause triggering. The firing of $3Q$ shortcircuits the secondary of $2T$, reflecting a very low impedance into the primary. Current flow from the line through the primary of $2T$ fires the TRIAC at about the same phase angle as the SCR was fired. At the beginning of the negative half-cycle, $3Q$ turns off and $2T$ returns to the high impedance state. Current flow from the line reverses the flux in $2T$ and eventually saturates the iron core. Again the primary impedance of $2T$ is reduced and gate current flow through the gate of the TRIAC causes triggering. At the end of the negative half-cycle, the TRIAC turns off and triggering control is returned to the relaxation oscillator. With careful selection of the transformer, $2T$, and its core characteristics, nearly symmetrical triggering of the TRIAC can be obtained.

7–4–2 Phase Control for Heating Element

Figure 7–13 is the ac Phase Shift Control for a heating element. Such an element might be used to heat liquid in a tank or to heat an oven. The power control device is the TRIAC $2Q$. The network of $4R$, $1L$, $2C$, and $3C$ protects

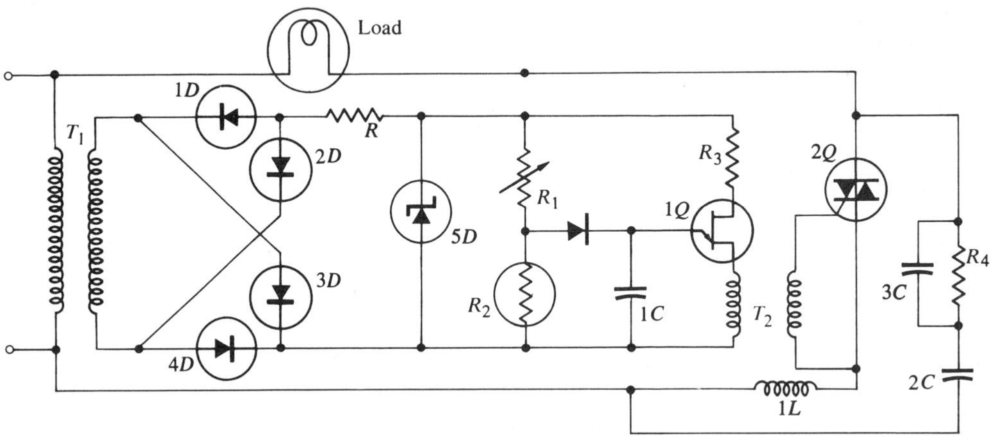

Fig. 7–13 TRIAC phase control for heating element.

the TRIAC against dv/dt damage and suppresses radio frequency noise. Transformer, $1T$, reduces the line voltage to a lower voltage for the triggering circuit. Diodes $1D$, $2D$, $3D$, and $4D$ form a full-wave bridge rectifier to provide positive voltage for the triggering circuit during the positive and negative half cycles of line voltage. Resistor, R, and zener diode, $5D$, regulate the rectified ac voltage to maintain a fixed dc potential for the relaxation oscillator. Triggering pulses are provided by the UJT relaxation oscillator. We have seen this circuit several times so there is no need to detail its operation. The required temperature is set by adjusting potentiometer R_1. If the temperature of the controlled medium rises above the set level, the thermistor, R_2, senses the change and reduces its resistance. R_1 and R_2 form a voltage divider. The capacitor, $1C$, will try to charge to the voltage.

$$V = V_z\left(\frac{R_2}{R_1 + R_2}\right) \tag{7.15}$$

where V_z is the zener regulating voltage.

The period of oscillation is given by the equation

$$T \approx R_1 C_1 \ln\left(\frac{V}{V - V_p}\right) \tag{7.16}$$

where V_p is the peak point voltage of the UJT. Combining Eqs. (7.15) and (7.16)

$$T \approx R_1 C_1 \ln \frac{V_Z}{V_z - V_p\left(1 + \dfrac{R_1}{R_2}\right)} \qquad (7.17)$$

We see that as R_2 varies, the period of oscillation of the relaxation oscillator varies so the angle of phase delay, α, will vary. If $1R$ is fixed at the value corresponding to the desired temperature to be maintained and the temperature falls below desired value, then R_2 increases; R_1/R_2 decreases; T decreases; and the TRIAC is fired earlier allowing more heat to be applied by the element to raise the temperature to the desired level. The relaxation oscillator is reset at the end of each half cycle when the voltage at base 2 of the UJT falls below the voltage on the capacitor, $1C$.

7–5 Digital Phase Shift Controls

We have spent most of our time looking at manual phase shift controls where the output is changed by adjusting a potentiometer or changing a resistor. There are applications where the output must be changed from a remote location in response to a transducer output or other electrical signals. The influence of computers and programmable controllers has dictated that the control signals be digital. The phase shift control must be compatible with digital command signals.

Figure 7–14 shows the basic block diagrams for digital phase shift controls using a digital controller and a digital computer. The operation of both types of controls is essentially the same; the status of the load is sensed and a continuous or analog output is produced; the analog signal is converted to a digital signal and compared with a digital reference or setpoint; a digital command signal is generated proportional to the difference between the reference signal and sensor signal· the command signal is converted to a dc voltage; the dc voltage along with a synchronizing signal is used to control the angle of phase shift of the load voltage. The distinction between a digital computer and a digital controller is sometimes debatable. The computer will generally have these advantages:

(a) more sophisticated mathematical manipulations performed;
(b) more flexibility in stored program capability;
(c) easier access to stored programs;

(d) ability to control more machines or processes simultaneously.

In Fig. 7–14, the only apparent difference between the two types of digital control is that the digital reference signal is internally programmed in the computer. This can be deceptive. The presence of the computer could make a significant difference, especially in closed-loop or servo-controlled systems.

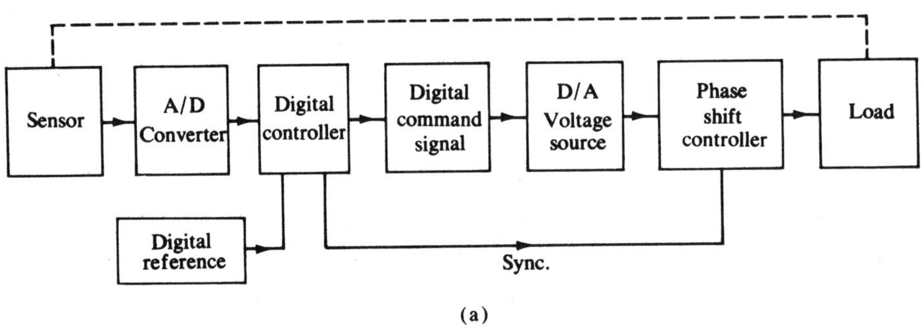

(a)

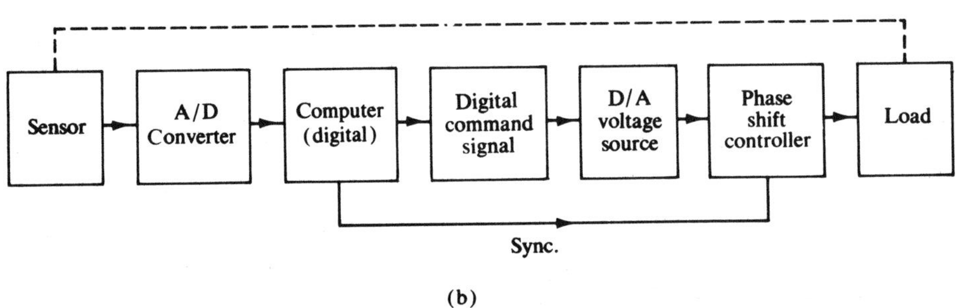

(b)

Fig. 7–14 Digital phase control diagrams: (a) programmable controller, (b) computer.

7–6 Design of Digital Phase Shift Circuits

The major design problem facing us is how to make the phase shift control responsive to the digital command signal. Two methods come to mind:

(1) use of the command signal to control a standard relaxation oscillator;

(2) use of the command signal to provide a delayed pulse to trigger the thyristor.

Let's look at a circuit that uses each method.

The circuit of Fig. 7–15* could be used as the voltage controlled phase shift circuit in the systems of Fig. 7–14. Resistor R and diode $1D$ form the zener regulator that provides the constant voltage for the triggering circuit. Trigger pulses are developed from a ramp-and-pedestal UJT relaxation

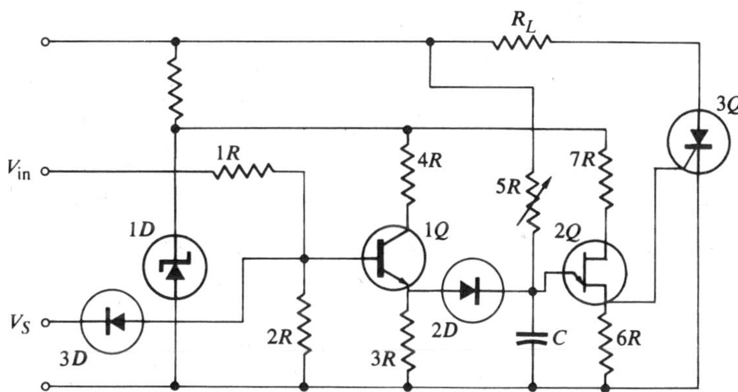

Fig. 7–15 Phase shift circuit using voltage variable relaxation oscillator.

oscillator circuit similar to the circuit of Fig. 7–11. Pedestal voltage level is controlled by transistor $1Q$. Capacitor $1C$ will charge quickly to the voltage drop across emitter resistor $3R$ through $1Q$ and diode $2D$. Ramp voltage across C is taken from the ac line to improve the linearity of power control. The signal from the digital controlled voltage source is applied at V_{in}. The synchronizing signal is applied at V_s. The waveforms of Fig. 7–16 help describe the controlling action of V_s and V_{in}. When V_s is held at zero volts and V_{in} is zero volts the only charge on C will be due to the ramp voltage. This voltage is much less than the peak point voltage required to fire the UJT at the proper phase angle. When the command signal V_{in} is applied at point B in Fig. 7–16, current flows through $1R$ and $3D$ to ground so

*This circuit was described by N. Cheung, University of Hong Kong, in the February 1969 issue of *Electronic Engineering*, Morgan-Grampian (Publishers) Ltd.

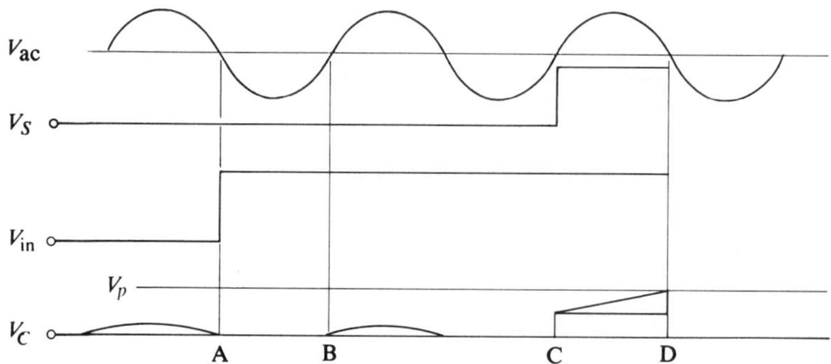

Fig. 7–16 Waveforms of Fig. 7–13.

1Q is clamped in the off state and does not affect the charging of capacitor C. As soon as V_s goes positive, 3D is reverse-biased and isolates V_s from the circuit of 1Q. At the beginning of the next positive half-cycle, point C in Fig. 7–16, 1Q conducts and establishes the pedestal voltage on capacitor C. The addition of the ramp voltage is now sufficient to cause triggering of the UJT at point D, discharging the capacitor through 6R and the gate of the SCR. The angle of phase delay, α, can be varied by varying the value of V_{in}.

The circuit of Fig. 7–17 demonstrates the use of pulse delay triggering to control the angle of phase delay. This circuit might also be used in the systems of Fig. 7–14.

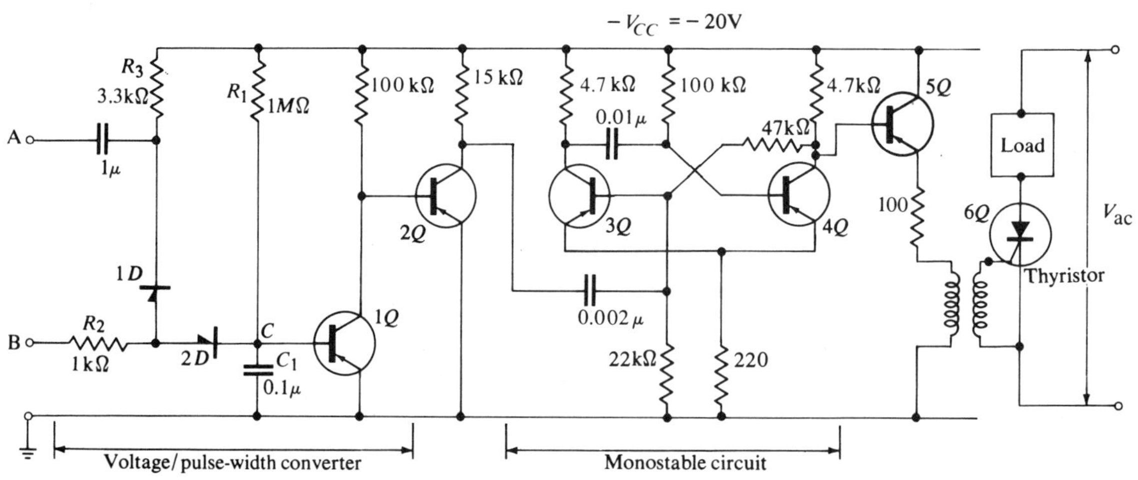

Courtesy of Electronic Engineering

Fig. 7–17 Phase shift circuit using pulse delay triggering.

The essential feature of this circuit is the voltage-to-pulse width converter at the input. The signal from the digital controlled voltage source is applied at input B, the synchronization signal is applied at input A. The control voltage at input B is positive so that when the sync signal is not present diode $1D$ is forward-biased. The voltage at point D in Fig. 7–17 is given by the equation

$$V_D = V_B - \left(\frac{R_2}{R_2 + R_3}\right)(V_B + V_{cc}) \qquad \textbf{(7.18)}$$

Component values are chosen so that V_D is negative. Diode $2D$ is then reverse-biased, isolating the control voltage from the base of $1Q$. Resistor R_1 is chosen so that $1Q$ is saturated.

A positive sync pulse is applied at input A at the beginning of the positive half-cycle of ac line voltage. The voltage at point A rises above the voltage at point D so that $1D$ is now reverse-biased. With $1D$ isolated from the base of $1Q$, capacitor C_1 charges through R_2 and $2D$ toward the voltage at point B forcing $1Q$ into cutoff. The voltage at the collector of $1Q$ drops to $-V_{cc}$. These actions are shown in waveforms a, b, and c of Fig. 7–16.

When the sync pulse is removed, capacitor $1C$ will be charged to the voltage

$$V_{cs} = V_B\left[1 - \exp\left(\frac{-t}{R_2C_1}\right)\right] \qquad \textbf{(7.19)}$$

as shown in Fig. 7–18(b).

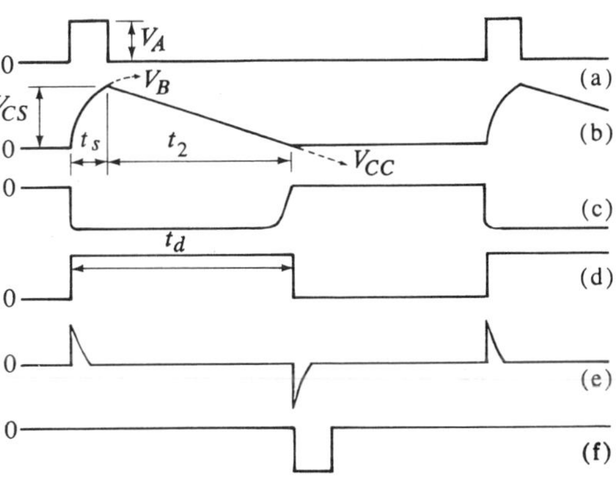

Courtesy of Electronic Engineering

Fig. 7–18 Wave forms for the circuit of Fig. 7–17.

With the voltage at input A removed, diodes $1D$ and $2D$ return to their original states with $1D$ forward-biased and $2D$ reverse-biased. The voltage at input B is again isolated from the base of 1Q. Capacitor C_1 will start to charge toward $-V_{cc}$ by drawing current through R_1. When the voltage of C_1 goes negative, $1Q$ will go into saturation again. The delay time is essentially the time required for C_1 to go negative after the sync pulse is removed. The transient voltages on C_1 are related to t_d by the equation

$$t_d = R_1C_1 \ln\left(1 + \frac{V_{CS}}{V_{CC}}\right) \tag{7.20}$$

where V_{CS} is given by Eq. (7.19).

If the width of the sync pulse, t_s, is held constant, then we see that t_d is varied by varying the amplitude of the voltage at input B.

The output of the pulse width converter is not applied directly to the SCR gate to prevent loading of the converter by the low impedance gate-to-cathode circuit. A multivibrator similar to that of Fig. 6–11 is used. The difference is that $2C$ of Fig. 6–11 is replaced with a resistor so the circuit always returns to a stable state after triggering. This is a *monostable multivibrator*. In the normal state, $4Q$ is off and $3Q$ is in saturation. The output of $1Q$ is amplified at $2Q$, differentiated through the RC network, and used to trigger the multivibrator. Since the multivibrator is triggered on negative going pulses, triggering occurs at the end of time delay t_d. This action is shown in Fig. 7–18(d) and (e). The output of the multivibrator is amplified at $5Q$ and applied to the gate of the SCR through a current transformer. Notice that the negative going output pulse of $5Q$ shown in Fig. 7–18(f) is converted to a positive going current pulse in the SCR gate circuit by the reverse winding of the current transformer.

Let's look at the design of the phase shift circuit of Fig. 7–15 to meet these specifications:

$$\alpha_{\min} = 45°, \ \alpha_{\max} = 110°$$
$$V_Z = 20 \text{ V}$$
$$V_{ac} = 110 \text{ V}$$
$$Q_3 = \text{G.E. C 12B}$$

Specify V_{in}, $1R$, $2R$, $3R$, $4R$, $1Q$, $2Q$, $5R$, and C. We must first select the UJT and transistor to be used. From the G.E. SCR manual, the 2N2646 UJT may be used to trigger the C 12 B SCR with a 20-V dc supply with $6R$ = 27 Ω, $7R$ = 100 Ω, and C = 0.22 μF. The intrinsic standoff ratio of the 2N2646 is 0.65 so the required peak point voltage will be 13 V.

The ramp and pedestal capacitor voltage waveform is shown in Fig. 7–19. We will adjust $5R$ to shape the ramp voltage. The delay angle will be varied by adjusting the pedestal voltage under the influence of the control voltage, V_{in}, at the biasing circuit of $1Q$.

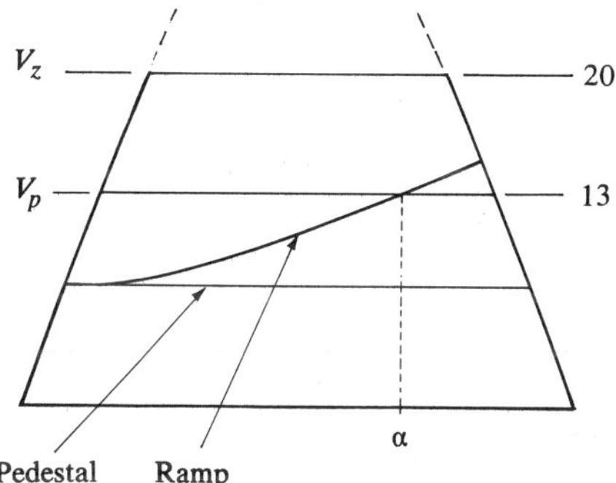

Fig. 7-19 Ramp and pedestal voltage waveform.

It is quite difficult to calculate the ramp voltage for different values of $5R$ and C. The graphical data of Fig. 7–20 can be used for design estimates. Let's say we choose γ = 10. Then at α = 45°, the charge on the capacitor will be 0.01 V_{ac}, at α = 110°. The charge on the capacitor will be 0.06 V_{ac}.

Since V_{ac} is 110 V,

$$V_c = 1.1 \text{ V at } \alpha = 45^\circ$$
$$V_c = 6.6 \text{ V at } \alpha = 110^\circ$$

The total charge on the capacitor must be 13 V at the firing angle so the pedestal voltage must be 11.9 V at 45° and 6.4 V at 110°.

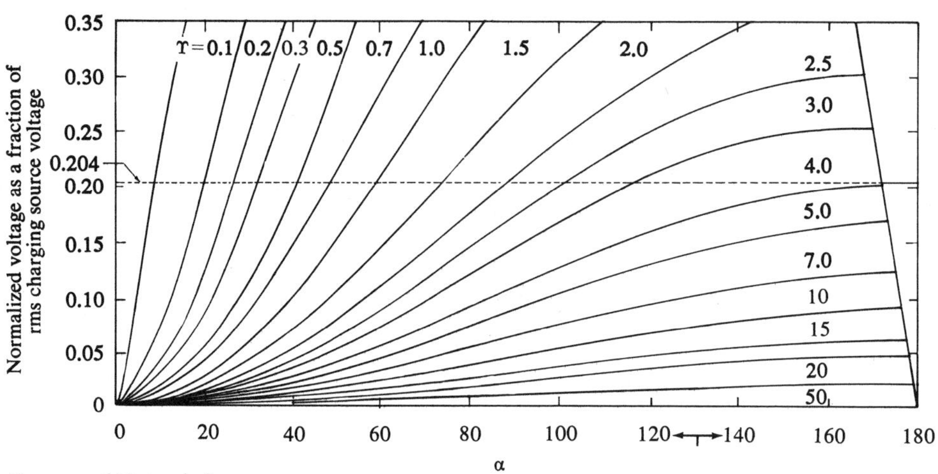

Courtesy of Motorola Inc.

Fig. 7–20 Ramp voltage as a function of delay angle for a capacitor charging from a half-cycle of AC source voltage ($\gamma = 2RCf$).

The value of $5R$ can be calculated from the equation $\gamma = 2RCf$ for

$$\gamma = 10, 5R = \frac{10}{2\ Cf}$$

$$5R = \frac{10}{2 \times 0.22 \times 60} \times 10^6$$

$$5R = 378 \text{ k}\Omega\text{, use } 500 \text{ k}\Omega \text{ pot.}$$

The 2N3014 transistor is chosen to provide the pedestal voltage. The output characteristics are shown in Fig. 7–21. $3R$ and $4R$ are chosen at 500 Ω each. The 1-kΩ load line is drawn on the output characteristics of Fig. 7–21. The operating point will vary between points A and B on the load line. We are now faced with an amplifier design problem. You might want to go back to chapter 3 and review that subject.

At point A, $I_B \approx 0.01$ mA, $V_{CD} = 18$ V.

At point B, $I_B \approx 0.12$ mA, $V_{CD} = 7$ V.

Let $R_B = 10R_E = 5{,}000\ \Omega$.

Then from Eq. (2.70), at point A

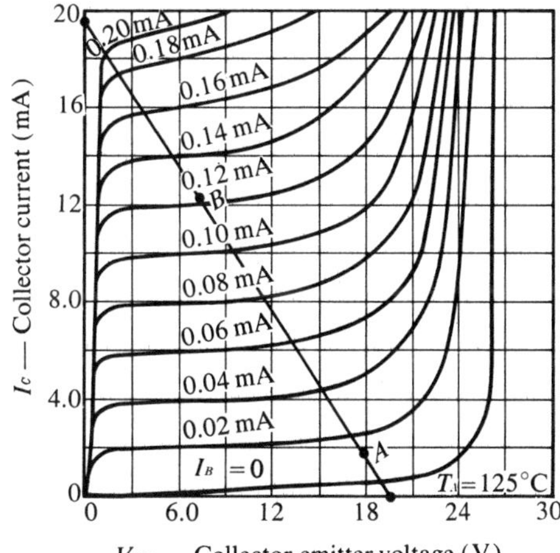

Courtesy of Motorola Inc.

Fig. 7–21 Output characteristics for 2N3014 transistor.

$$V_B = 5.000 \times 0.01 \times 10^{-3} + 0.6 + 500 \times 2 \times 10^{-3} = 1.65 \text{ V} \qquad \textbf{(7.21)}$$

Let $V_{\text{in}} = 6$ V.

Using the equations

$$1R = R_B \frac{V_{cc}}{V_B} \text{ and } 2R = \frac{R_B}{1 - \dfrac{V_B}{V_{cc}}}$$

$$1R = 18 \text{ k}\Omega,\ 2R = 6.8 \text{ k}\Omega$$

At point B,

$$V_B = 5000 \times 0.12 \times 10^{-3} + 0.6 + 500 \times 12 \times 10^{-3} = 7.2 \text{ V}$$

The required value of V_{in} can be calculated from Eq. (2.69)

$$V_{\text{in}} = V_B \frac{(1R + 2R)}{2R} = 26.2 \text{ V}$$

The output of the digital controlled voltage source must vary between 6 and 24 V to cover the required range of phase control. Again, the final step is to build the circuit and make final adjustments in component values to achieve the required performance.

7–7 A Digital Voltage Controlled Phase Shift Circuit

A commercial phase shift control that could be used in the system of Fig. 7–14 is the LPAC-1 SCR power controller. The controller will respond to dc signals from 0 to 5 V. The digital command signal must be converted to a dc voltage by a d/a converter. The dc voltage control signal is applied at terminals 5 and 6 of Fig. 7–22 to adjust the firing angle of the SCR power controller of Fig. 7–23. Let's take a look at the operation of this commercial controller.

7–7–1 The DC Power Supply Section

The dc power supply section of the controller is shown in Fig. 7–22. Rectifier diodes *CR*3, *CR*4, *CR*5, and *CR*6

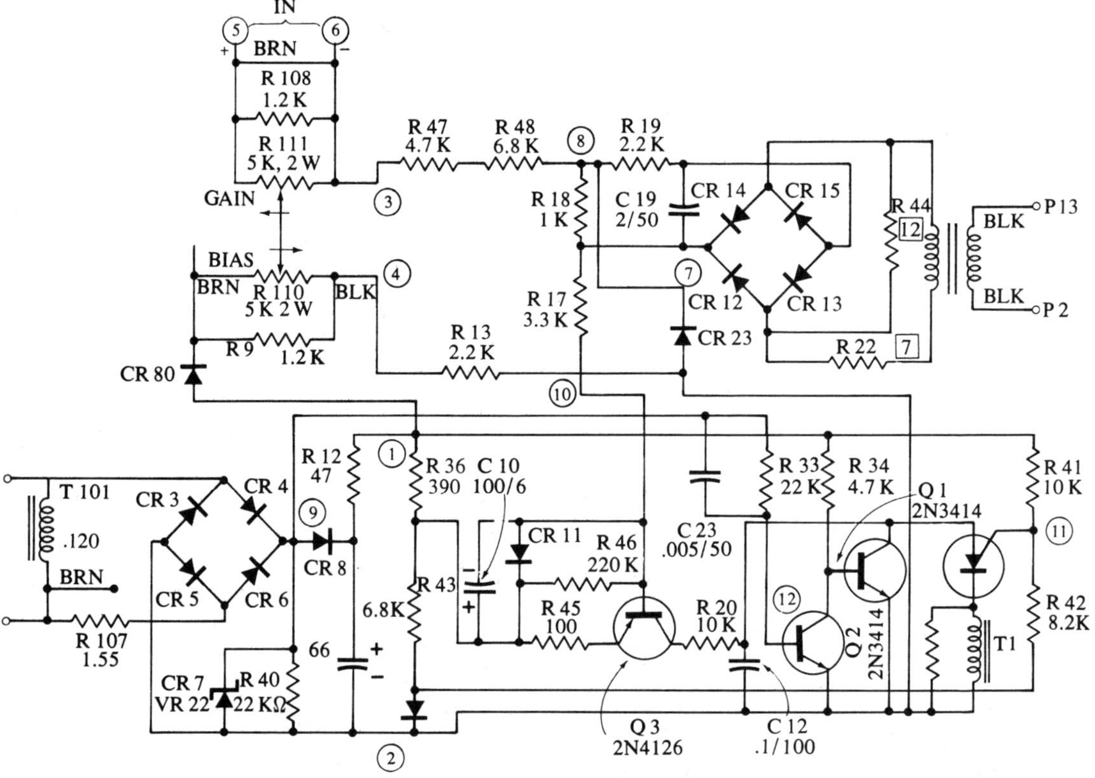

Courtesy of Loyola Industries

Fig. 7–22 Trigger generator of LPAC controller.

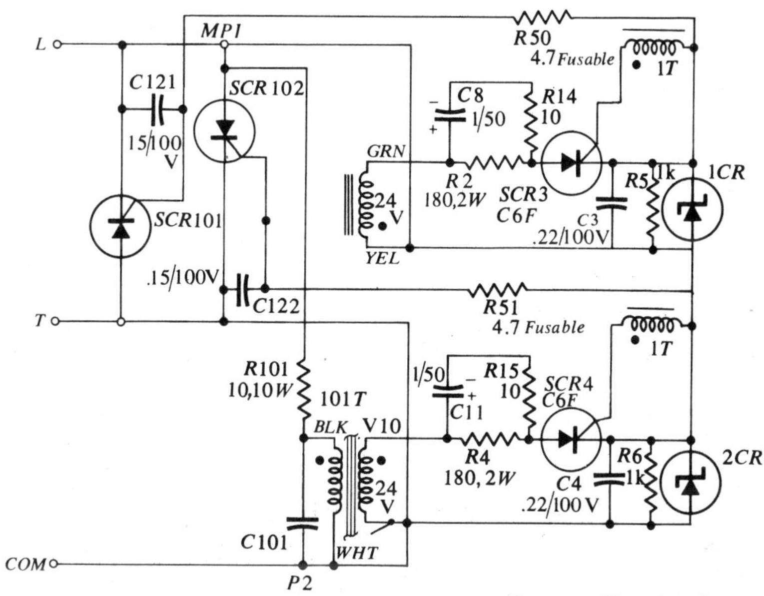

Courtesy of Loyola Industries

Fig. 7–23 SCR power controller of LPAC.

form a full-wave bridge rectifier to invert the negative half-cycle of ac voltage to positive polarity at points 9 to 2. Zener diode 7*CR* regulates the rectified ac voltage at 22 V. Capacitor 6*C* provides the filtering to keep the regulator voltage from dropping to zero at the end of each half-cycle.

7–7–2 The Trigger Control Signal

The trigger circuit accepts a dc command signal and converts it to a time-positioned pulse to control the firing angle of the power SCRs. The trigger control circuit compares the dc command signal with a feedback signal proportional to the load voltage and provides an error signal to the trigger pulse generator. Figure 7–24 shows the simplified trigger control circuit taken from the schematic of Fig. 7–22. The input command signal is connected across the GAIN potentiometer. The gain adjust is used to set the range of power control. The gain pot adjustment depends on the input signal. If the input signal varies from zero to 5 mA, the gain pot is set at maximum and the input voltage will vary from 0 to 5 V. However, if the input signal is 10 to 50 mA, then the gain pot must be adjusted down until 5 V appears at the arm when the input is at 50 mA.

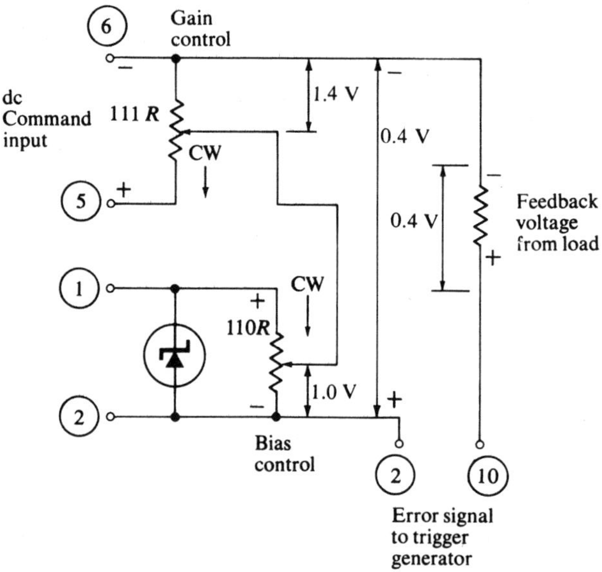

Fig. 7–24 Simplified schematic of trigger control signal circuit showing typical voltages.

The feedback voltage is obtained from the load side of the power controller. A transformer is connected to the load terminals. The secondary voltage of the transformer is rectified by 12*CR*, 13*CR*, 14*CR*, and 15*CR* and stepped down to approximately 5 V at its output when the load is receiving full power. When the command signal gives 5 V at the arm of the gain pot, at 5-V signal will appear at terminals 10 and 6 of Fig. 7–24. If there is a difference between the command signal and the feedback signal, it appears as an error signal and goes to the trigger pulse generator circuit.

Some command signals might not be zero when minimum load power is desired. A very common control signal is 1 to 5 mA, with a 1-mA signal corresponding to no load power and a 5-mA signal corresponding to full load power. The BIAS adjustment allows an adjustable blockout signal to be inserted in series with the command and feedback signals. The zener regulated voltage of the dc power supply is applied across the bias potentiometer at terminals 1 and 2 of Fig. 7–24. The bias pot is set so that the voltage at its arm offsets the minimum input command signal. For a 1- to 5-mA input signal, the bias pot would be set at 1-V output. When the input signal drops below 1 mA, the error voltage goes negative. The trigger pulse generator remains in the "off" condition when the error signal is negative and no trigger pulses are provided to trigger the power SCRs. No output will be realized from the power controller until the command signal is increased to 1 mA. When the voltage at the arm of the gain pot reaches 1.1 V, the controller will produce a very low output. As the input signal is increased, a feedback voltage will appear which will cancel out that portion of the input signal greater than 0.1 V. The feedback voltage will remain within 0.1 to 0.2 V of the gain pot voltage. If the power line voltage tends to vary, the increase or decrease will change the feedback voltage. The change will result in a new error signal. The SCR trigger point will be adjusted to offset the line voltage change and maintain constant load power. Thus, the output is automatically regulated against line voltage variation. Typical voltage levels for this circuit are shown in the diagram of Fig. 7–24.

7–7–3 The Trigger Pulse Generator

The pulse generator for this controller is the PUT relaxation oscillator of Fig. 5–34. If we ignore transistors 1*Q* and 2*Q* for the time being, the simplified schematic is shown in Fig. 7–25. The anode gate voltage is fixed by the voltage divider of 41*R* and 42*R*. When the voltage at the

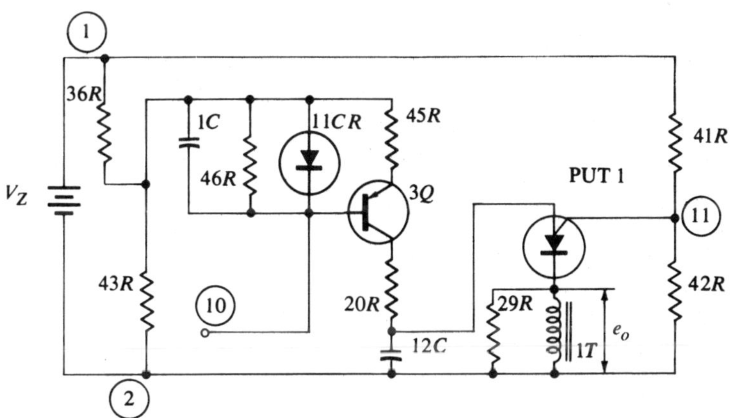

Fig. 7–25 Simplified trigger pulse generator of LPAC controller.

anode becomes more positive than the gate voltage, the PUT conducts and a pulse of current flows through current transformer $1T$ to generate the required trigger pulse. The capacitor charge current flows through transistor $3Q$. The transistor acts like a constant current source; the amount of current flow is controlled by the voltage at the base of $3Q$. This is the trigger control voltage between terminals 10-2 that we discussed in the previous section. As the trigger control voltage increases more current flows through $3Q$, charging capacitor $12C$ faster so that the PUT conducts earlier. Figure 7–26 shows the input and output voltages of the PUT oscillator for increasing trigger control voltage. Only the first output pulse is used for triggering the power control. Capacitor $1C$ is placed across the trigger control input terminals to prevent rapid changes in the trigger output. This provides the "soft-start" action required of most

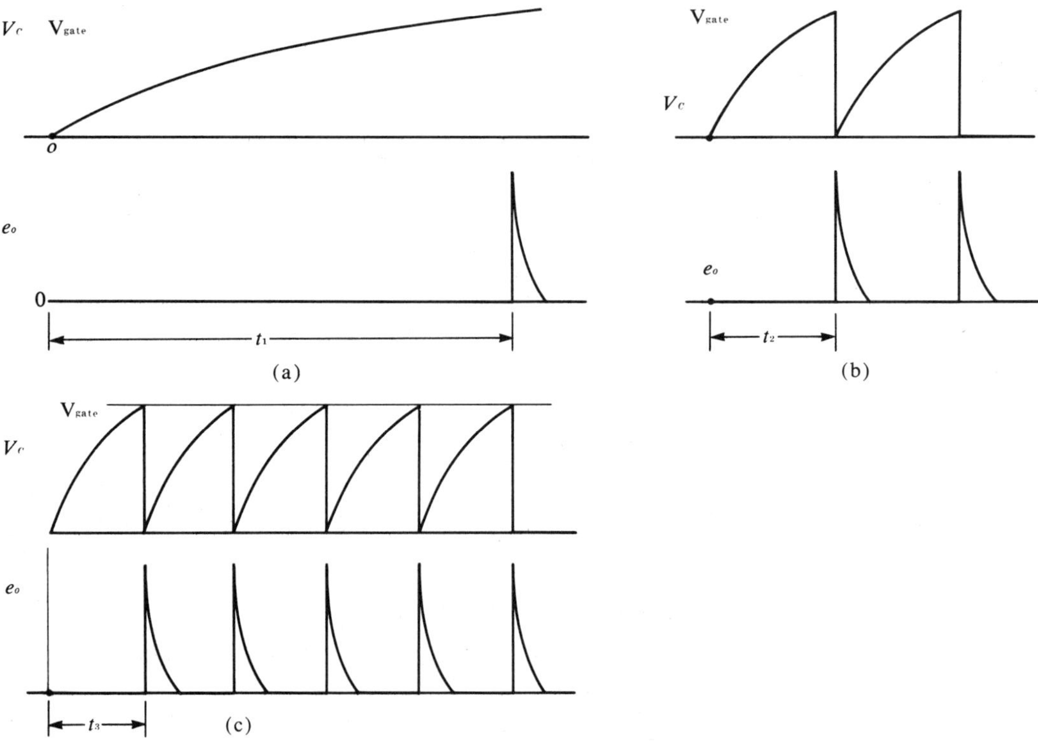

Fig. 7–26 PUT input and output voltages vs time.

controllers. Diode rectifier 11*CR* protects the transistor 3*Q* from excessive reverse bias voltages.

7-7-4 Synchronizing the Trigger Pulse

Output pulses from the trigger generator are independent of the ac line voltage due to the filtering of the rectified voltage. Special circuitry must be provided to insure that the $t = 0$ reference point in Fig. 7-27 coincides with the point at which the ac line voltage goes through 0. Transistors 1*Q* and 2*Q* provide this synchronization in the LPAC

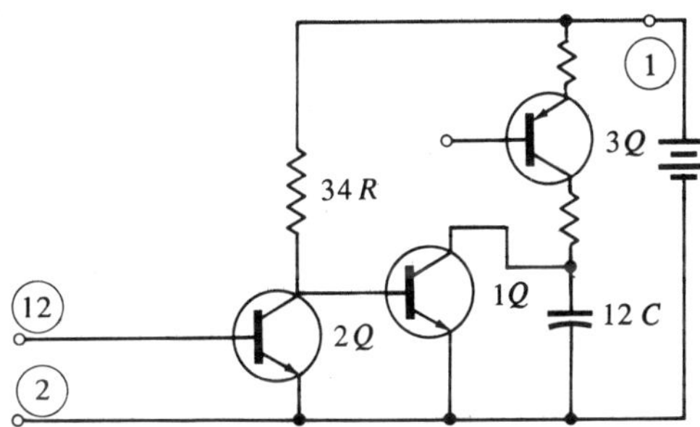

Fig. 7–27 Simplified schematic of synchronizing circuit of Fig. 7–22.

controller. The base-to-emitter voltage of 2*Q* is taken from the unfiltered output of the zener diode regulator so that the level falls to zero at the end of each half-cycle as shown in Fig. 7–28(a). Transistor 2*Q* is saturated while the zener diode is regulating so its collector-to-emitter voltage is near 0 for most of the ac voltage cycle. The base-to-emitter voltage of 1*Q* is clamped at near 0 by the collector of 2*Q* while the zener diode is regulating, as shown in Fig. 7–28(b). Transistor 1*Q* is cut off and has no effect on the charging of capacitor 12*C* for most of the ac cycle. At the end of each half-cycle, the voltage at the base of 2*Q* goes to 0 so 2*Q* is driven to cutoff. Current flow through 34*R* biases 1*Q* into saturation while 2*Q* is cut off; so 1*Q* is driven into saturation. When 1*Q* is saturated, capacitor 12*C* discharges through the collector-to-emitter circuit to ncar 0 as shown in Fig. 7–28(c). At the beginning of the next half-cycle, 2*Q* is saturated again biasing 2*Q* off so the capacitor 12*C* is free to charge through 3*Q* again. The result is that capacitor 12*C* is

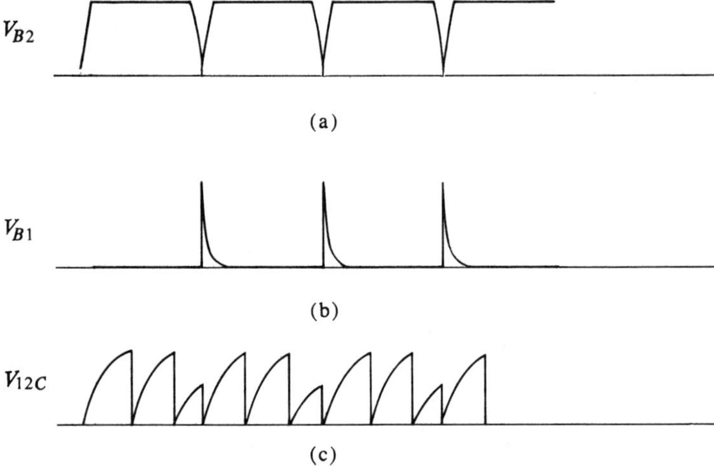

Fig. 7–28 Voltage waveforms in synchronizing circuit.

discharged at the end of each half-cycle so that the time of the first generator output pulse is synchronized with the ac line voltage.

7–7–5 The SCR Power Controller

The power control section of the LPAC controller uses the inverse parallel SCR combination of Fig. 5–2. Each of the high-current output SCRs is triggered by a lower current SCR as shown in Fig. 7–23. Transformer 101 *T* is connected across the ac line at terminals *L* to "common" so the primary voltage of 101 *T* is in phase with the line voltage. The two secondaries of 101 *T* are wound 180° out of phase with each other as indicated by the position of the polarity dots on the circuit diagram. The in-phase secondary of 101 *T* is connected in the anode-to-cathode circuit of SCR 4. A secondary output of the current transformer 1 *T*, in the PUT relaxation oscillator of Fig. 7–22, is connected in the gate-to-cathode circuit of SCR 4. When the line voltage is positive with respect to "common," SCR 4 is forward-biased. The arrival of the trigger pulse through 1 *T* causes the SCR to conduct and deliver current to the gate of SCR 102 as shown in Fig. 7–23. SCR 102 is turned on allowing current to flow between the line at terminal *L* to the load at terminal *T*. When the line voltage is negative with respect to "common," SCR 3 is forward-biased. The secondary of current transformer $1T$ is connected in the gate-to-cathode circuit of SCR 3. Upon arrival of the first trigger pulse from the PUT relaxation oscillator SCR 3 conducts, current flows into the gate of SCR 101 causing it to trigger "on." SCR 101 goes into conduction allowing current to flow from terminal T to L. Figure 7–29 shows the output of the trigger pulse generator, the voltage across SCR 4, and the voltage across terminals L to T for a resistive load.

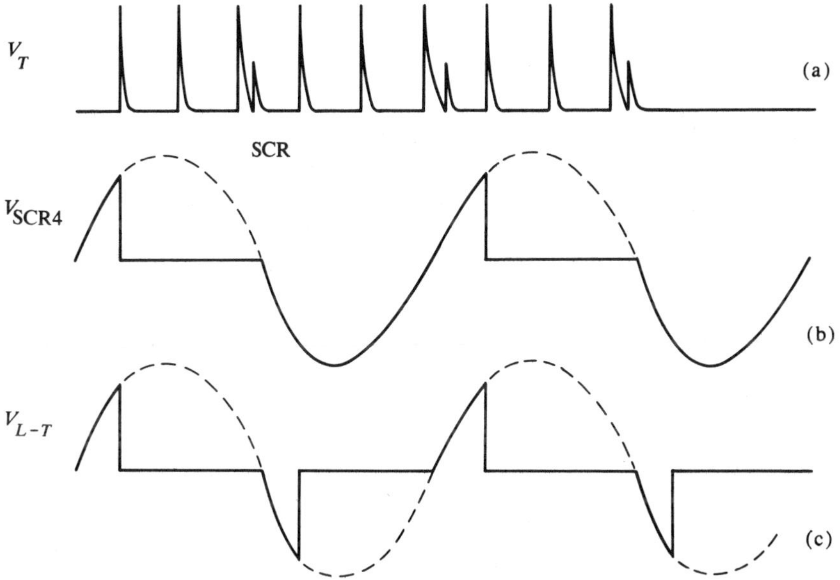

Fig. 7–29 Voltage waveforms in circuit of Fig. 7–23.

Many industrial loads are not purely resistive, but have a positive power factor due to inductance. A welding transformer is a good example of this type of load. With a highly inductive load and a command signal for maximum output, the first pulse from the trigger generator might occur while reactive, or carryover, current is still flowing from the previous half-cycle. We discussed this problem in Sec. 5–2–4. Figure 7–30 shows the relationship between load current and load voltage for an inductive load under full-wave SCR control. The SCR that was conducting on the previous cycle still has current flowing through it because

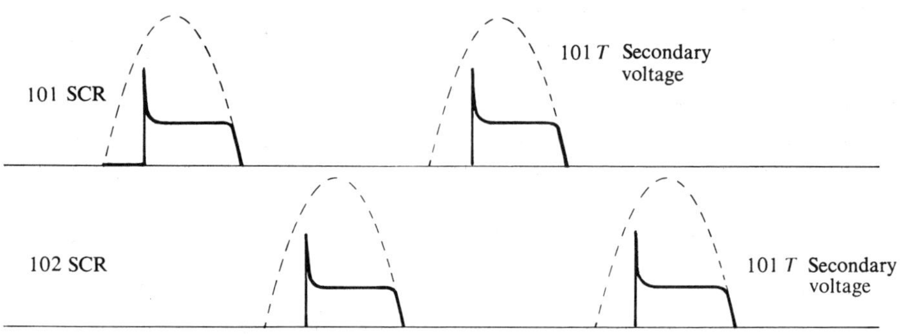

Courtesy of Loyola Industries

Fig. 7–30 Voltage and current waveforms for an inductive load under full wave control.

the inductance of the load is now delivering current back to the power system. Since it is still conducting, the voltage across the SCR is near 1 V. This low voltage is applied across the cathode-to-anode of the SCR receiving the trigger pulse, preventing it from going into conduction. If this single pulse were applied at the gate of the SCR about to turn on, it would have come and gone before the SCR would be allowed to turn on. The usual condition when this occurs is that one power SCR will conduct for a full half-cycle and the other will remain off. The load current will be dc instead of ac. In the case of a transformer, the core will become saturated causing excessive current flow. The excessive current will either blow a fuse, destroy an SCR, or damage the transformer. The low-current SCRs prevent this condition from occurring. SCR 3 and SCR 4 will turn on as soon as the first trigger pulse is applied at the respective gate. Triggering voltage is maintained at the gate of the power SCRs until they are permitted to turn on (Fig. 7–31). The low-current SCRs are automatically turned off at the end of each half cycle as the anode voltage drops below cathode potential. Diodes 1*CR* and 2*CR* protect the gates of SCR 101 and SCR 102 from excessive reverse voltage. Capacitors *C*3, *C*4, *C*121, and *C*122 minimize accidental triggering due to power line transients and noise. The LPAC-1 controller has many of the design features that you are likely to encounter in industrial electronic controls. Understand as thoroughly as you can the techniques used in this controller.

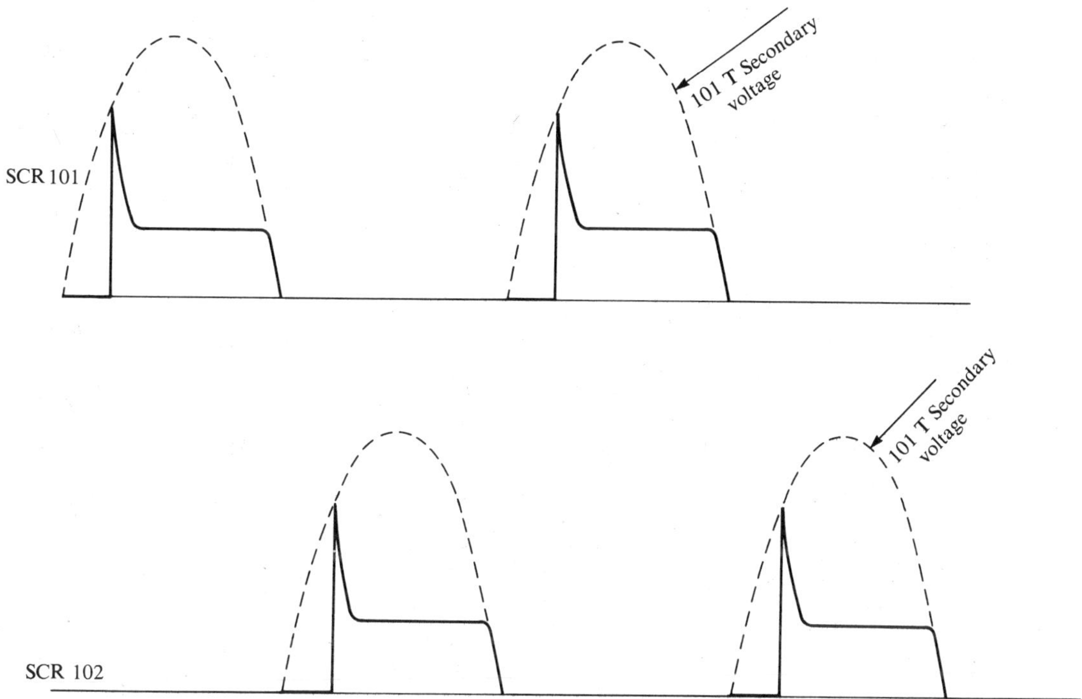

Fig. 7–31 Gate-to-cathode trigger voltage waveforms of SCR 101 and SCR 102.

7–8 Summary

This has been a key chapter in our study of industrial electronics. We have looked at the detailed design of ac phase shift circuits, the selection of SCRs and TRIACs based on power control requirements, the design of analog and digital phase shift controls including the triggering circuits, and some good industrial applications. You will want to return to this chapter and review some of the design techniques for possible application in the future.

EXERCISES

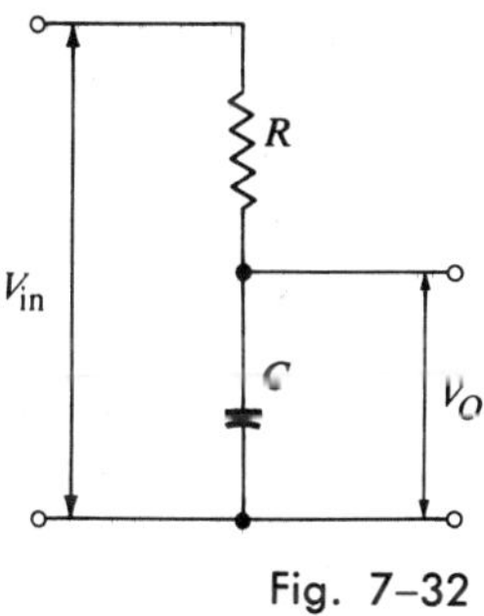

Fig. 7–32

1. In the circuit of Fig. 7–32, $V_{in} = 100 \sin 377\, t$. Calculate α and draw the vector diagram for the following values of R and C.

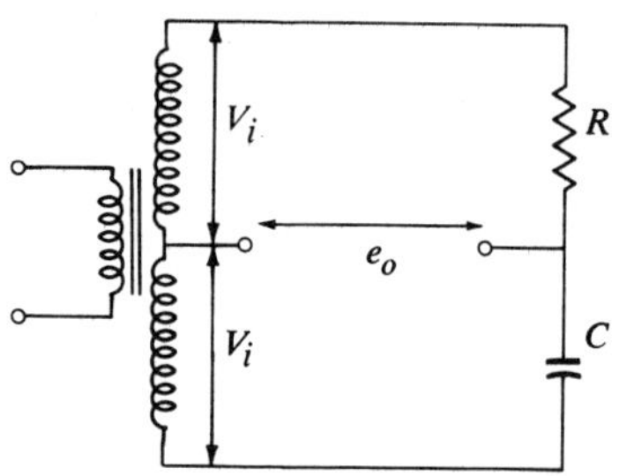

Fig. 7–33

(a) $R = 10\ \text{k}\Omega$, $C = 1\ \mu\text{F}$
(b) $R = 10\ \text{k}\Omega$, $C = 0.1\ \mu\text{F}$
(c) $R = 20\ \text{k}\Omega$, $C = 0.1\ \mu\text{F}$
(d) $R = 25\ \text{k}\Omega$, $C = 0.1\ \mu\text{F}$
(e) $R = 100\ \text{k}\Omega$, $C = 0.1\ \mu\text{F}$
(f) $R = 100\ \text{k}\Omega$, $C = 1\ \mu\text{F}$

2. Determine the peak value of the capacitor voltage for each combination of R and C in problem 1.

3. In the circuit of Fig. 7–33, $V_i = 100 \sin 377\, t$. Calculate α and draw the vector diagram for the values of R and C in problem 1.

4. Determine the peak value of the output voltage for each combination of R and C in problem 1.

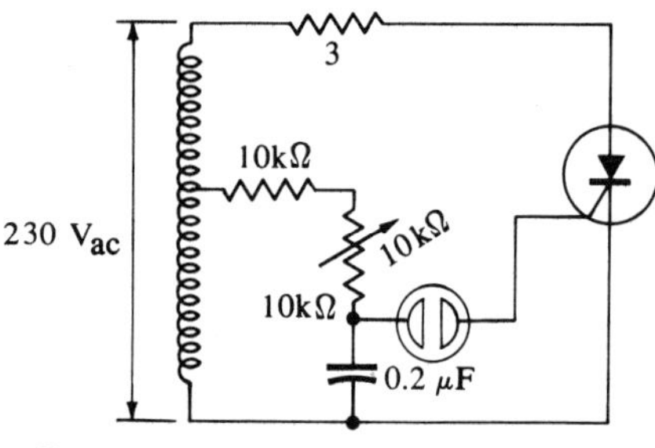

Fig. 7–34

5. The center-tapped transformer of Fig. 7–34 is used to trigger the neon tube which in turn triggers the SCR. The neon tube requires 50 V to trigger.
(a) What is α_{max} and α_{min} for the phase shift network?
(b) What is the minimum and maximum firing angle of the SCR?
(c) What is the maximum and minimum average power that can be delivered to the load?
(d) Specify the SCR current and voltage ratings.

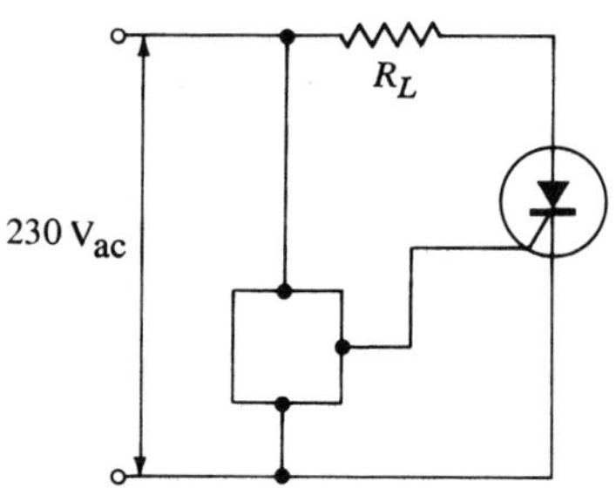

Fig. 7–35

6. The C145D SCR of Appendix B is used in the phase control circuit of Fig. 7–35. If the conduction angle is 60°, what is the maximum allowable rms current?

7. The 20-Ω load of Fig. 7–36 requires a maximum of 1,100 W and a minimum of 750 W average power. Specify:
(a) α_{max} and α_{min}.
(b) θ_{max} and θ_{min}.
(c) I_{av} at α_{max} and at α_{min}.
(d) The voltage and current rating of the SCR.

8. If the SCR of Fig. 7–36 is replaced by a TRIAC and the load requirements are the same, specify
(a) α_{max} and α_{min}.
(b) The voltage and current rating of the TRIAC.

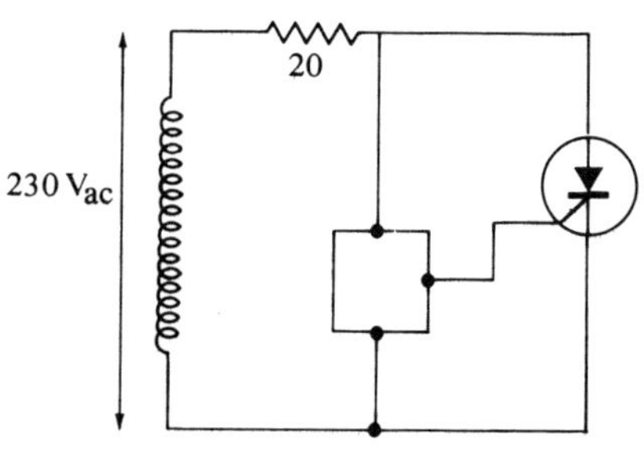

Fig. 7–36

9. The C181D SCR of Appendix B is used in the phase control circuit of Fig. 7–37. The conduction angle varies between 70° and 140°. Determine:
(a) The maximum average load power.
(b) The minimum average load power.
(c) I_{av} at α_{max} and I_{av} at α_{min}.
(d) Maximum allowable case temperature at α_{max} and at α_{min}.

10. If diode $5D$ of Fig. 7–12 is shorted, what effect would you expect this to have on the operation of the blower?

11. The thermistor of Fig. 7–13 replaces the potentiometer $1R$ and $1R$ replaces the thermistor. Explain the operation of this new control circuit.

12. Why are a d/a converter and an a/d converter needed in the block diagram of Fig. 7–14?

13. Increasing the amplitude of V_{in} in the digital phase control of Fig. 7–15 will trigger the SCR sooner. Explain why.

14. What effect would an increase in the width of the sync pulse of the control of Fig. 7–17 have on the firing angle of the SCR?

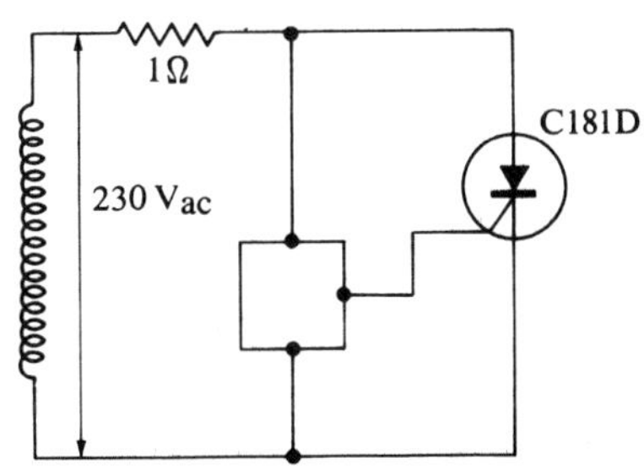

Fig. 7–37

8

Digital Control Concepts

8–1 Introduction

There has been a steady increase in the application of digital devices in industrial electronics since the successful development of the electronic digital computer. Predictions are that this trend will continue. The development of new techniques and advances in solid state technology make digital implementation efficient, reliable, and economically competitive with existing analog equipment in many cases. In this chapter, we will look at some of the concepts that are necessary to the understanding of the design of digital systems in industrial electronics.

8–2 The Binary Number System

The decimal numbering system that we have been taught to use since early childhood uses the ten symbols 0, 1, 2, 3, 4, 5, 6, 7, 8, 9. By the proper placement of these ten symbols, we are able to represent numbers of any size. We are so accustomed to this system that we seldom recognize that it is a specific case of a general method of counting. Let's look at the decimal number 246. It is understood to mean six ones, four tens, and two hundreds. The same number could just as well be written as $2 \times 10^2 + 4 \times 10^1 + 6 \times 10^0$. Notice that the right-most digit tells how many ones to add, the digit to its left tells how many tens to add, the digit to its left tells how many hundreds to add. For numbers larger than 999 we begin to add more digits to the left of the previous ones. In a more general form we are putting digits in the positions

$$\mathcal{A}_{n-1} \times R^{n-1} \cdots \mathcal{A}_3 \times R^3 + \mathcal{A}_2 \times R^2 + \mathcal{A}_1 \times R^1 + \mathcal{A}_0 \times R^0 \quad \textbf{(8.1)}$$

where R is the radix or the number of symbols in the numbering system, n is the position of the last digit counting from right to left, and $\mathcal{A}$ is the value of the digit. The number 246 has the specific values $R = 10$, $\mathcal{A}_0 = 6$, $\mathcal{A}_1 = 4$, $\mathcal{A}_2 = 2$, $n = 3$.

The same position notation is used in numbering systems with less than ten symbols, or with a different radix than ten. Two other common systems for numbering are the octal system where $R = 8$ and the binary system where $R = 2$.

The decimal system is quite convenient for most of our communication needs but quite inconvenient for designing digital equipment.

The binary numbering system is used in the design of digital circuits. The two states required are most natural; for instance, a mechanical switch is either open or closed, a light is either on or off, or an amplifier is in saturation or cutoff. Any two symbols could be used to represent a binary number; however, to avoid introducing new symbols the two lowest order symbols of the decimal system are used, 0 and 1. The number 246_{10} can also be written as 11110110_2. The subscript indicates the radix of the numbering system. Using the position notation of Eq. (8.1) we recognize $1 \times 2^7 + 1 \times 2^6 + 1 \times 2^5 + 1 \times 2^4 + 1 \times 2^2 + 1 \times 2^1$ as 246_{10} when summed in column form as in Table 8–1.

TABLE 8–1

Column Summing for Binary to Digital Conversion

$$
\begin{aligned}
1 \times 2^7 &= 128 \\
1 \times 2^6 &= 64 \\
1 \times 2^5 &= 32 \\
1 \times 2^4 &= 16 \\
0 \times 2^3 &= 0 \\
1 \times 2^2 &= 4 \\
1 \times 2^1 &= 2 \\
0 \times 2^0 &= \underline{\ 0\ } \\
& \quad 246_{10}
\end{aligned}
$$

You will find it necessary at times to convert numbers from the binary to decimal system or from the decimal to binary system. Binary to decimal conversion can be done quite simply with a powers-of-two table using the column addition method of Table 8–1. Decimal to binary conversion can be done by repeated division by two using the carry to generate the binary number. For example, to convert 246_{10} to binary the steps are:

$$
\begin{aligned}
246 \div 2 &= 123 \text{ with carry } 0 \\
123 \div 2 &= 61 \text{ with carry } 1 \\
61 \div 2 &= 30 \text{ with carry } 1 \\
30 \div 2 &= 15 \text{ with carry } 0 \\
15 \div 2 &= 7 \text{ with carry } 1 \\
7 \div 2 &= 3 \text{ with carry } 1 \\
3 \div 2 &= 1 \text{ with carry } 1 \\
1 \div 2 &= 0 \text{ with carry } 1
\end{aligned}
$$

The binary conversion generated by the carry is 11110110_2. The ease with which a binary number can be represented by physical devices is demonstrated in Fig. 8–1

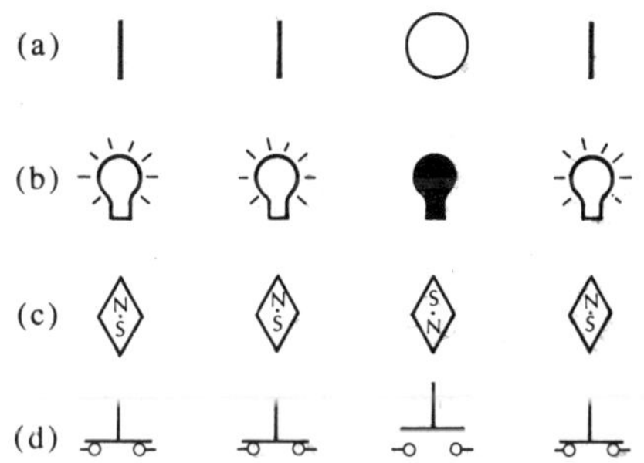

Fig. 8–1 Physical representation of the binary number thirteen using (a) binary digits, (b) lamps, (c) compass needles, (d) mechanical switches.

where the number 1101_2 is represented by electric lamps, compass needles, or mechanical switches.

8–3 Digital Logic

Successful design of digital circuits requires precise, unambiguous definition of the state of each device under all conditions. By definition, only two states are recognized. A switch is either open or closed; there is no in between. Symbology is used to indicate the logical state of each device or combination of devices. Sometimes the states are labeled as 1 or 0, true or false, A or "not A," and high or low. There is still the option of assigning either of the two chosen symbols to either of the two states of the digital device. To avoid confusion, it must be made clear how logic states are assigned to device characteristics.

Let's look at logic state assignments for some simple digital devices. The switch of Fig. 8–2(a) is represented by the letter A when closed and "not A" when open. (The bar above A is read as "not A.") The closed switch is in its "true" state and is usually assigned the binary digit 1. The open switch is in its "false" state and is usually assigned the binary digit 0. The transistor of Fig. 8–2(b) is similarly assigned true and false states. The amplifier is in the true state when its output voltage is +5. The amplifier is in the false state when its output voltage is 0. The true state is represented in this example by the alphabet B or binary digit 1. The false state is represented by $\overline{B}$ or 0. When the more

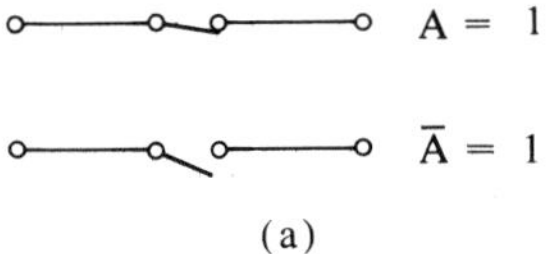

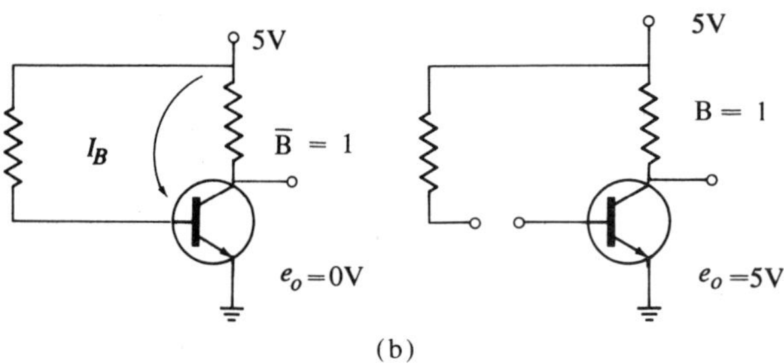

Fig. 8–2 Logic state assignment to digital devices: (a) mechanical switches, (b) transistor amplifiers.

positive of the two voltage outputs is assigned the true state it is called positive logic. If the more negative of the two output voltages were assigned the true state, it would be negative logic. This is purely arbitrary but accepted by most digital designers.

After the logic state assignments are established and agreed upon, we can now start to combine devices to perform logic functions. Consider the parallel combination of switches in Fig. 8-3. Switch $1S$ is represented by A, switch $2S$ is represented by B. The output voltage is represented by C. The true state of the switches is assigned in the

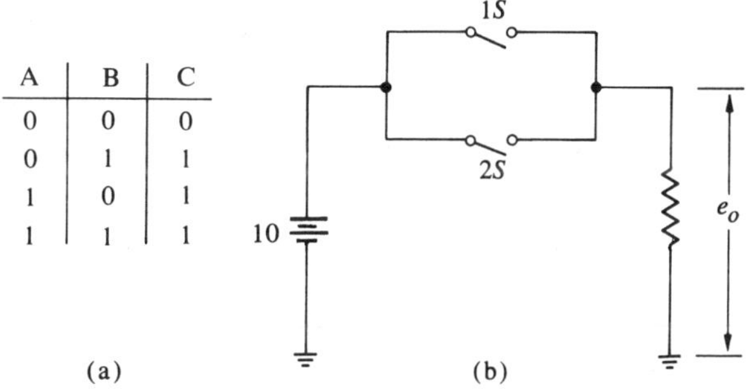

Fig. 8–3 (a) Positive logic truth table, (b) parallel switch combination.

closed position. Using positive logic, the true state of C will be 10 V, $\bar{C}$ will be 0 V. The function of this simple circuit can be stated as C will be true if A is true or B is true or both A and B are true. This is the logic OR function. As the logical function performed by a circuit gets more complex, it is more difficult and clumsy to state in words. A more effective symbolic representation is the *truth table.* The truth table shows all possible combinations of devices in a circuit and the logic output of each combination. Figure 8-3(a) shows the truth table for the simple switching circuit of Fig. 8-3(b). The four logical combinations of switches are:

A open B open	0 0
A open B closed	0 1
A closed B open	1 0
A closed B closed	1 1

The output C is true for the last three combinations in the list, using positive logic.

Figure 8–4(a) shows a series combination of switches. Switch 1*S* is represented by A, switch 2*S* is represented by B. The switches are in the true state when closed. The output voltage is represented by C. Positive logic is used. The function performed by this circuit can be stated as *the output C is true only when A and B are both true*. The truth table of Fig. 8–4(b) says the same thing in terms of binary digits. This is the logical AND function. Notice that in making the truth table there are 2^n possible combinations of the digital switches, where *n* is the number of switches. In order to avoid missing any of the possible combinations, entries in the input columns of the truth tables (A and B) are the binary representation of the first four decimal numbers: 0, 1, 2, and 3. If there were three switches, the entries would be 0, 1, 2, 3, 4, 5, 6, and 7. This technique will reduce the chance of making errors in listing input combinations in truth tables.

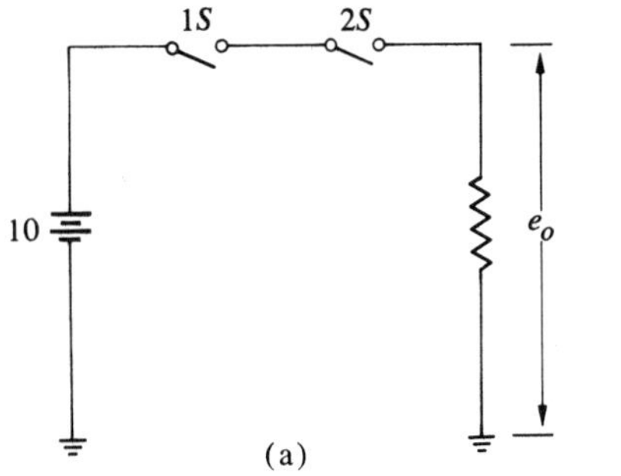

A	B	C
0	0	0
0	1	0
1	0	0
1	1	1

(b)

Fig. 8–4 (a) Series switch combination. (b) Positive logic truth table,

8–4 Electronic Logic Gates

Electronic circuits designed to perform specific functions on a number of digital inputs are called *gates*. Every industrial electronics designer or technician should be familiar with the characteristics of the fundamental logic gates. We will look at the terminal characteristics and simplified circuits of these gates.

The development of integrated circuit technology has had tremendous impact on the manufacture of logic gates. The circuit schematics are much more complicated than the simplified circuits presented here but the terminal characteristics are very similar. A key to successful application of IC (integrated circuit) logic gates to the design of industrial electronics systems is the truth table. The truth table is to IC logic gates what the characteristic curves are to discrete transistors. Both provide information about the terminal characteristics of the device.

8–4–1 The AND Gate

Let's look at the diode logic (DL) circuit of Fig. 8–5(a), designed to perform the logical AND function. The two possible values of e_1, e_2, and e_3 are +6 V and 0 V. Using positive logic assignment, +6 V represents logic "1," 0 V represents logic "0." The two possible values of e_0 are +6 V and 0.6 V. Again using positive logic assignment, +6 V represents logic "1" and 0.6 V represents logic "0." There are eight possible combinations of the three input voltages to consider. Consider first the case in which all three input voltages are at 0 V or logic 0. Diodes 1*D*, 2*D*, and 3*D* are all forward-biased so the output voltage e_0 is at logic 0. This condition is the first row of the truth table of Fig. 8–5(b). When e_1 and e_2 are at 0 V but e_3 goes to +6 V, 3*D* is reverse-biased, 1*D* and 2*D* are still conducting.

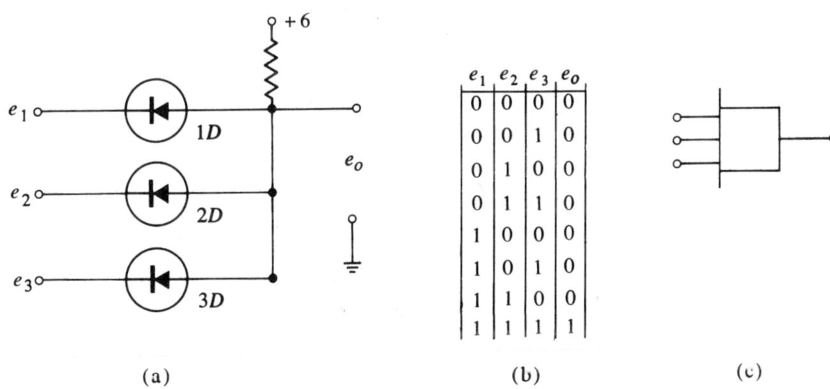

e_1	e_2	e_3	e_o
0	0	0	0
0	0	1	0
0	1	0	0
0	1	1	0
1	0	0	0
1	0	1	0
1	1	0	0
1	1	1	1

Fig. 8–5 (a) DL positive logic AND gate, (b) truth table, (c) NEMA symbol.

The output voltage is still clamped at 0.6 V or logic 0. This is the condition of row two of the truth table. We could go through each row of the truth table and prove that if any or all of the inputs are at logic zero, the output will be at logic zero. The output will be at logic "1" only if all inputs are at logic 1. This is the positive logic AND gate. There are several sets of symbols used to identify logic gates including MIL STD—806B, ASA, and NEMA. We will use the NEMA symbology, more appropriate for industrial applications. Figure 8–5(c) is the NEMA symbol for the AND gate. Figure 8–6 is the manufacturer's schematic for a DUAL 3 input AND gate of integrated circuit construction. It is common to package more than one gate on a single IC chip to reduce costs. Notice that the input diodes are replaced by three emitter terminal transistors that perform the same function. This is the T^2L (Transistor-Transistor Logic) AND gate.

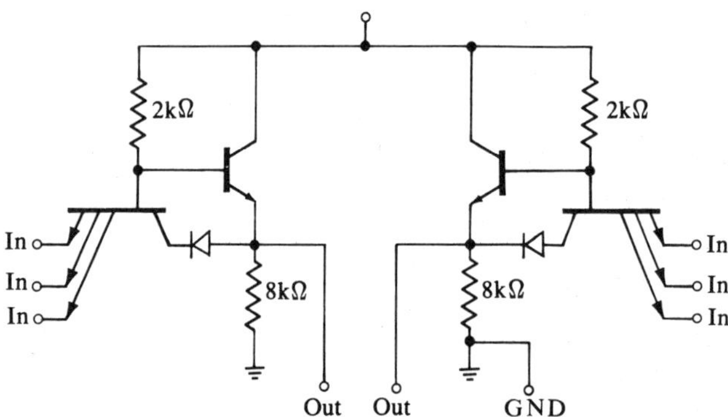

Courtesy of Signetics Corporation

Fig. 8–6 Dual 3-input AND gate.

8–4–2 The OR Gate

Figure 8–7(a) is the diode logic circuit to perform the OR function. The two possible values of the input voltages e_1, e_2, and e_3 are −6 V and 0 V. Using the positive logic assignment, 0 V represents logic 1, −6 V represents logic 0. The two possible values of the output voltage are −0.6 V and −6 V. Using positive logic assignment, −0.6 V repre-

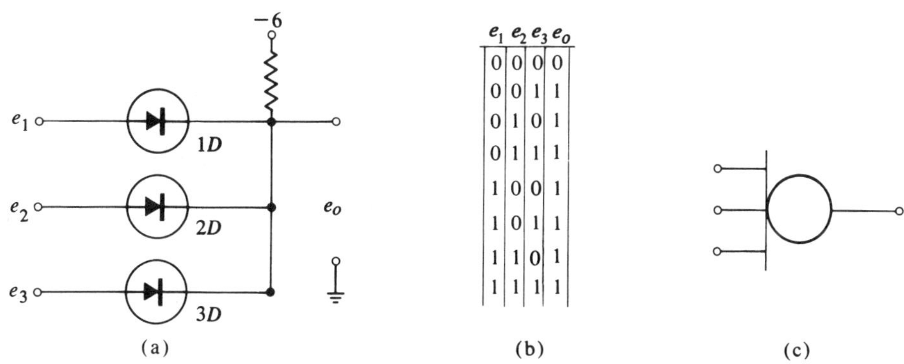

e_1	e_2	e_3	e_o
0	0	0	0
0	0	1	1
0	1	0	1
0	1	1	1
1	0	0	1
1	0	1	1
1	1	0	1
1	1	1	1

Fig. 8–7 (a) DL positive logic OR gate, (b) truth table, (c) NEMA symbol.

sents logic 1, −6 V represents logic 0. There are 2^n possible combinations of inputs to consider where n is the number of inputs. In this case, $2^n = 8$ possible combinations. When all three inputs are at logic 0 or −6 V, $1D$, $2D$, and $3D$ are reverse-biased so that e_0 is at −6 V or logic 0. This is the condition of the first line of the truth table of Fig. 8–7(b). If e_1 and e_2 are at logic 0 and e_3 goes to ground potential, then diode $3D$ becomes forward-biased and conducts. The output voltage will be clamped at the diode forward conducting voltage of −0.6 V, or logic 1. This is the condition of the second line of the truth table. We could go through the entire truth table and show that if any or all of the inputs are at logic 1, then the output will be at logic 1. This is the positive logic OR gate. The NEMA symbol for the OR gate is shown in Fig. 8–7(c). Figure 8–8 shows a commercial transistorized IC 3-input OR gate. The diodes of Fig. 8–7(a) are replaced by transistors. Transistor amplifiers are used in the output circuit to maintain the output level and improve switching.

8–4–3 The NOT Gate

The simplest logic function to implement is the NOT or INVERTER. In simple language, "The logic state of the output is the opposite of the logic state of the input." Figure 8–9 shows how a transistor amplifier performs the logic NOT function. The two possible values of e_{in} are 0 V and −6 V. Using positive logic assignment, 0 V

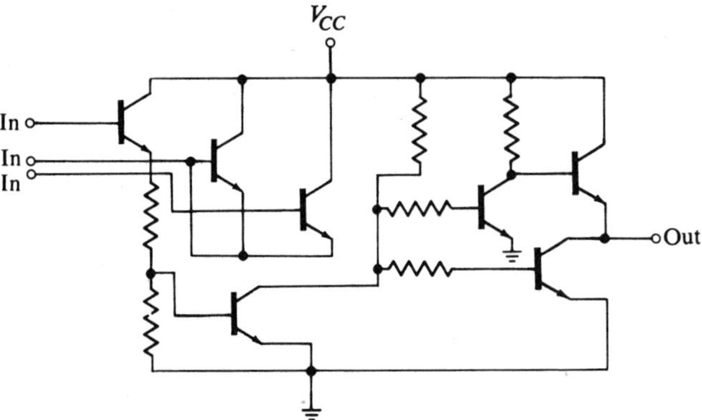

Courtesy of Signetics Corporation

Fig. 8–8 3-input OR gate.

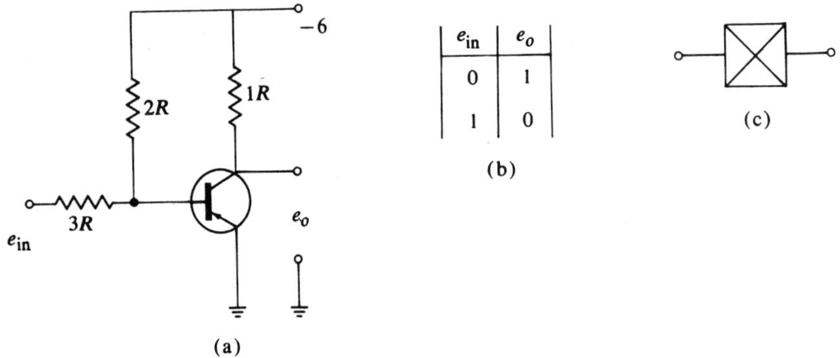

e_{in}	e_o
0	1
1	0

Fig. 8–9 (a) Transistor inverter, (b) truth table, (c) NEMA symbol.

represents logic 1, −6 V represents logic 0. The two possible output voltage levels are −0.2 and −6 V. Using positive logic assignment, −0.2 V represents logic 1, −6 V represents logic 0. When e_{in} is at 0 V, current flow is shunted around the base-to-emitter junction through $3R$ and the transistor is cut off. The output is at −6 V or logic 0 and the input is at 0 V or logic 0. When the input is at −6 V, current flow through the base-to-emitter circuit forces the transistor into saturation; e_0 is at −0.2 V. The input is now at logic 0 while the output is at logic 1. This is the logic NOT gate or INVERTER. The truth table is shown in Fig. 8–9(b). The NEMA symbol is shown in Fig. 8–9(c).

8–4–4 The NAND and NOR Gates

The AND, OR, and NOT gates that we have already looked at are the basic building block from which many other digital gating functions can be derived. Two combinations that are so common as to require special attention are the NAND and NOR gates. The NAND gate is a combination of AND and NOT gates, or NOT-AND. Figure 8–10(a) shows the combination of a DL AND gate and a transistor INVERTER to perform the positive logic NAND function. The truth table is shown in Fig. 8–10(b). The NEMA symbol is shown in Fig. 8–10(c). Figure

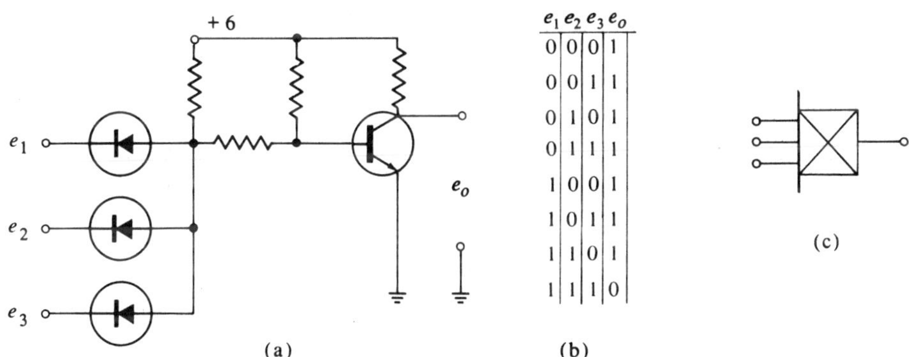

e_1	e_2	e_3	e_o
0	0	0	1
0	0	1	1
0	1	0	1
0	1	1	1
1	0	0	1
1	0	1	1
1	1	0	1
1	1	1	0

Fig. 8–10 (a) Diode-transistor NAND gate, (b) truth table, (c) NEMA symbol.

8–11(a) shows the combination of the DL OR gate and transistor inverter to perform the positive logic NOR function. The truth table is shown in Fig. 8–11(b). The NEMA symbol is shown in Fig. 8–11(c). Figure 8–12 shows the manufacturer's schematic for a Quad-2-input positive logic NAND gate integrated circuit. Four NAND gates are manufactured on the same chip to reduce costs. The logic equations for "1" outputs at terminals 3, 6, 8, and 11 are:

$$3 = \overline{1 \cdot 2} \qquad 6 = \overline{4 \cdot 5}$$

$$8 = \overline{9 \cdot 10} \qquad 11 = \overline{12 \cdot 13}$$

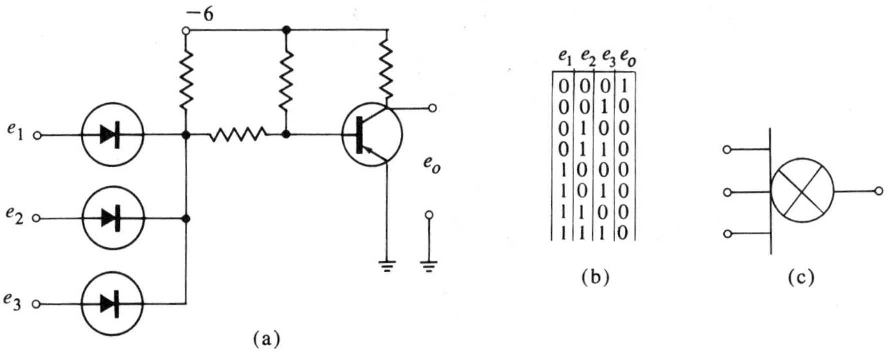

e_1	e_2	e_3	e_o
0	0	0	1
0	0	1	0
0	1	0	0
0	1	1	0
1	0	0	0
1	0	1	0
1	1	0	0
1	1	1	0

Fig. 8–11 (a) Diode-transistor NOR gate, (b) truth table, (c) NEMA symbol.

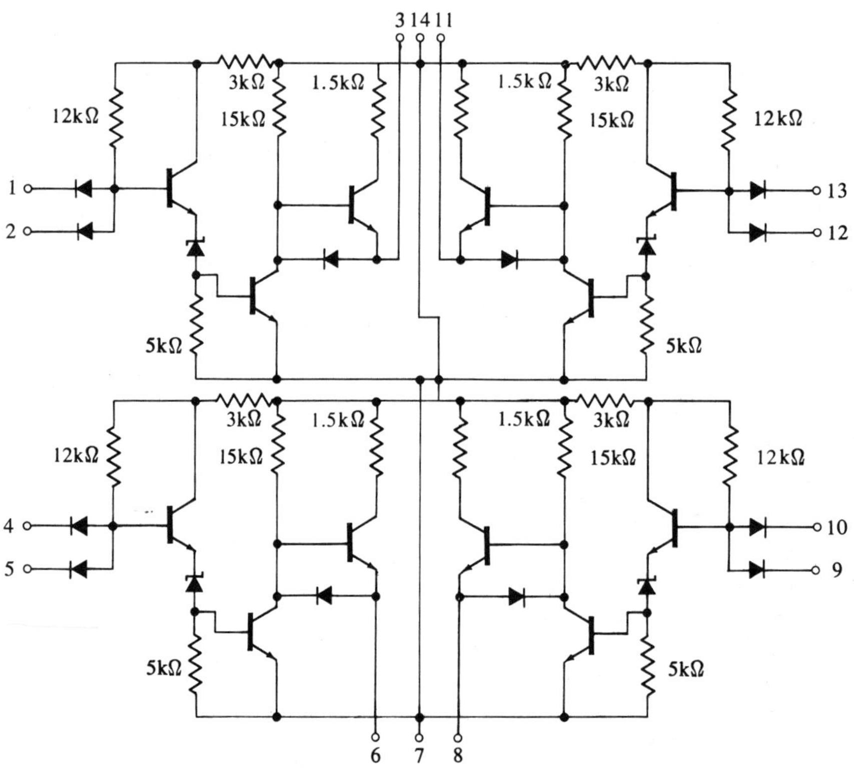

Courtesy of Motorola Inc.

Fig. 8–12 Quad 2-input NAND gate.

Figure 8–13 shows the circuit of a commercial 7-input positive logic NOR gate.

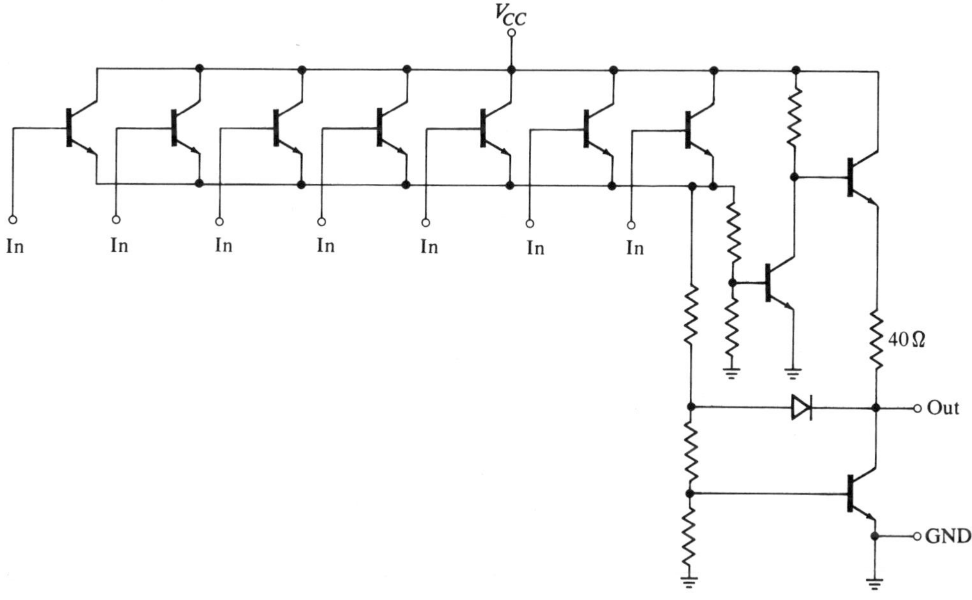

Fig. 8–13 7-input transistor NOR gate.

8–5 Boolean Algebra

Boolean algebra is the algebra of two-state variables. It is ideally suited to the design and analysis of digital circuits based on binary or two-state logic. By using Boolean algebra, we will be able to write mathematical equations to describe logical functions. Complex logical functions can sometimes be described more compactly by Boolean algebraic equations than with truth tables or word statements. The laws of Boolean algebra can be used to simplify the mathematical equations describing logical functions, thus reducing the number of gates required for implementation.

The letters of the alphabet are used to represent variables in a Boolean algebraic equation. The variable can only have two values, 1 or 0. The true state of the variable is represented by a letter, say A. The false state of the variable is represented by placing a bar above the letter, $\overline{A}$. The usual convention is to assign the numeric value 1 to the true state and 0 to the false state. For the variable A,

$$A = 1 = \text{true}$$
$$A = 0 = \text{false}$$

We will follow this convention throughout these discussions.

The symbols used in Boolean algebra are:

$$A \text{ and } B \rightarrow A \cdot B$$
$$A \text{ or } B \rightarrow A + B$$
$$\text{not } A \rightarrow \overline{A}$$
$$A \text{ equals } B \rightarrow A = B$$

Let's look at a word statement, its truth table, and the mathematical equation describing a logical operation.

Example 8–1. The output of a logical circuit is controlled by four input variables, A, B, C, and D. The output, E, will be true if only B is true; if only B, C, and D are true; if only A and D are true; if only A and C are true; or if only A, B, and C are true. Construct the truth table and write the equation describing the operation of the circuit.

There are 16 possible combinations of the input variables (2^n, where n is the number of variables) so the truth table has 16 rows. The truth table of Table 8–2 is constructed by listing the binary numbers in sequence from 0000_2 to 1111_2 and the corresponding numerical value of the output E.

TABLE 8–2
Truth Table for Word Statement of Example 8–1

A	B	C	D	E
0	0	0	0	0
0	0	0	1	0
0	0	1	0	0
0	0	1	1	0
0	1	0	0	1
0	1	0	1	0
0	1	1	0	0
0	1	1	1	1
1	0	0	0	0
1	0	0	1	1
1	0	1	0	1
1	0	1	1	0
1	1	0	0	0
1	1	0	1	0
1	1	1	0	1
1	1	1	1	0

The algebraic equation describing the operation of the word statement is formed by summing all possible combinations of inputs for which the output variable is true.

$$E = \bar{A}B\bar{C}\bar{D} + \bar{A}BCD + A\bar{B}\bar{C}D + A\bar{B}C\bar{D} + ABC\bar{D}. \quad \textbf{(8.2)}$$

Since the only possible numerical values of E are 1 or 0, input combinations not included in Eq. (8.2) are automatically equal to 0.

We observed earlier that implementation of logical functions could be simplified by manipulation of the Boolean algebraic equation. The basic laws of Boolean algebra that must be followed in manipulation of algebraic equations are shown in Table 8–3. Some of the identities that are useful in simplification of algebraic equations are shown in Table 8–4.

TABLE 8–3

Basic Laws of Boolean Algebra

(1)	$A + B = B + A$ $A \cdot B = B \cdot A$	commutative
(2)	$A + B + C = A + (B + C) = (A + B) + C$ $A \cdot B \cdot C = A \cdot (B \cdot C) = (A \cdot B) \cdot C$	associative
(3)	$A \cdot (B + C) = A \cdot B + A \cdot C$ $A + (B \cdot C) = (A + B) \cdot (A + C)$	distributive

TABLE 8–4

Basic Identities of Boolean Algebra

(1)	$A + A = A$	(10)	$A + AB = A$
(2)	$A \cdot A = A$	(11)	$AB + A\bar{B} = A$
(3)	$A \cdot \bar{A} = 0$	(12)	$A + B = \overline{\bar{A}\bar{B}}$
(4)	$A + 1 = 1$	(13)	$AB = \overline{\bar{A} + \bar{B}}$
(5)	$A \cdot 1 = A$	(14)	$A(\bar{A} + B) = AB$
(6)	$A + 0 = A$	(15)	$A + \bar{A}B = A + B$
(7)	$A \cdot 0 = 0$	(16)	$\bar{A} + AB = \bar{A} + B$
(8)	$A + \bar{A} = 1$	(17)	$\bar{A} + A\bar{B} = \bar{A} + \bar{B}$
(9)	$\bar{\bar{A}} = A$	(18)	$(A + B)(A + \bar{B}) = A$
(19)	$AC + AB + B\bar{C} = AC + B\bar{C}$		
(20)	$(A + B)(B + C)(\bar{A} + C) = (A + B)(\bar{A} + C)$		

Let's look at an example of simplification.

Example 8–2. Simplify the equation

$$L = ABC + AB\bar{C} + ABD$$

AB can be factored from the first two terms,

$$L = AB(C + \overline{C}) + ABD$$

Using identity 8 of Table 8–4

$$L = AB + ABD$$

Again AB can be factored,

$$L = AB(D + 1)$$

using identity 4. L = AB is the simplified equation.

Two of the identities of Table 8–4 that are important to note are 12 and 13. These two identities are statements of DeMorgan's Theorem. They are useful in eliminating AND functions or OR functions from algebraic equations. Any algebraic equation can then be implemented using NOR gates and inverters or NAND gates and inverters.

Figure 8–14 shows equivalent logic diagrams according to DeMorgan's theorem. The simple rules for "DeMorganizing" are:

(1) Negate the algebraic expression.
(2) Change all OR operations to AND operations or change all AND operations to OR operations.
(3) Negate each term in the expression.
(4) On complex expressions, work from the inside out.

An example will demonstrate these rules.

Example 8–3. Use NAND gates and inverters to implement the equation

$$L = \overline{A\overline{B}C + (A + C)(B + \overline{C})} \qquad \textbf{(8.3)}$$

and draw the logic diagram.

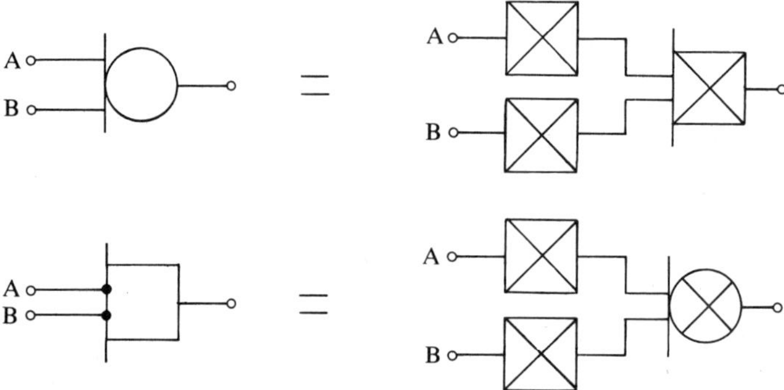

Fig. 8–14 Equivalency of DeMorgan's theorem.

Each of the OR operations must be eliminated by DeMorganizing. Starting with the expression A + C and applying the rules

$$A + C \rightarrow \overline{A + C} \rightarrow \overline{AC} \rightarrow \overline{\overline{A}\,\overline{C}}$$

$$B + \overline{C} \rightarrow \overline{B + \overline{C}} \rightarrow \overline{B\overline{C}} \rightarrow \overline{\overline{B}C}$$

Then

$$\overline{A\overline{B}C + (\overline{\overline{A}\,\overline{C}})(\overline{\overline{B}C})} \rightarrow \overline{\overline{A\overline{B}C + (\overline{\overline{A}\,\overline{C}})(\overline{\overline{B}C})}} \rightarrow$$

$$A\overline{B}C + (\overline{\overline{A}\,\overline{C}})(\overline{\overline{B}C}) \rightarrow (A\overline{B}C)(\overline{\overline{A}\,\overline{C}})(\overline{\overline{B}C}) \rightarrow \overline{\overline{(A\overline{B}C)}\,\overline{(\overline{\overline{A}\,\overline{C}})(\overline{\overline{B}C})}}$$

The NAND-NOT diagram is shown in Fig. 8–15.

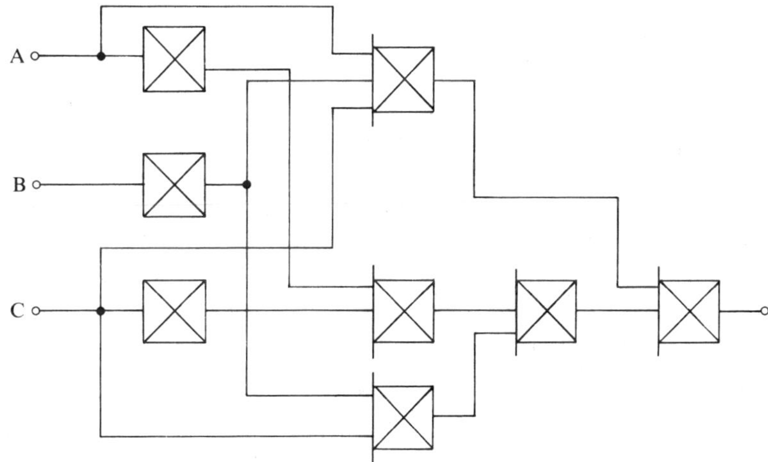

Fig. 8–15 Logic diagram to implement Eq. (8.3).

The Boolean algebra we have just discussed is very useful in the design of digital control systems. It provides us with the mathematical tools to express logical operations and reduce the complexity and cost of the circuitry required for implementation.

8–6 Electronic Logic Devices

The gates that we discussed in Sec. 8–4 are the basic building blocks of digital control systems, but there are several other electronic logic devices that are so common as to deserve our attention. We will look at the truth table and characteristics of some of the more commonly used logic devices.

8–6–1 The Flip-Flop

The digital gates that we have looked at so far require an input in order to produce the specified output. If the inputs are removed the output condition will change. Most electronic digital controls require that outputs remain fixed even if the input conditions change. Devices with "memory" are designed to hold the output condition while the input conditions change. In Sec. 6–4–2 we discussed the design of the bistable multivibrator as a two-state memory device. That same basic design, with modifications, is used to make digital integrated circuit bistable multivibrators called flip-flops for applications in digital controls. Digital flip-flops are designed by combining basic logic gates. There are several families with similar but different triggering characteristics.

The *R-S* Flip-Flop is frequently used in digital control circuits. It has two inputs, SET and RESET. The logic diagram of the *R-S* flip-flop is shown in Fig. 8–16(a) and the commonly used industrial symbol is shown in Fig. 8–16(b).

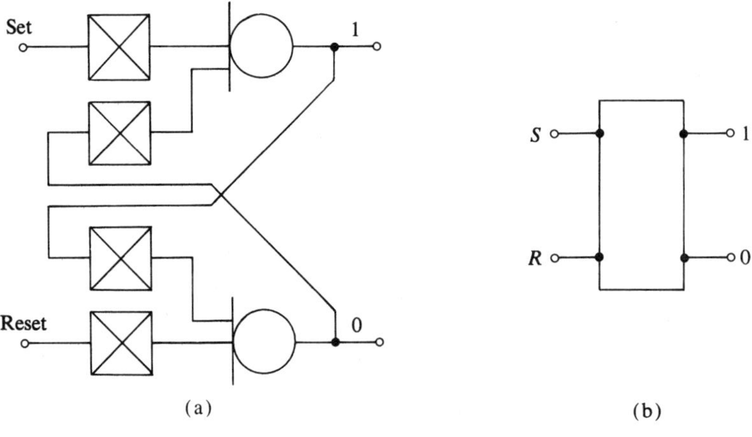

Fig. 8–16 (a) *R-S* flip-flop diagram, (b) *R-S* flip-flop symbol.

The truth table for digital devices with memory is more complicated than the tables we have used so far. The truth table must allow for all possible states of the device before inputs are applied, all possible input combinations, and all possible outputs after the new inputs are applied. The truth table is shown in Table 8–5. Notice that simultaneous 0 inputs should not be applied at the *R-S* terminals.

TABLE 8–5

Truth Table for *R-S* Flip-Flop

Previous Outputs		Inputs Applied		Final Outputs	
1 Output	**0 Output**	**S**	**R**	**1 Output**	**0 Output**
1	0	0	0	undefined	
1	0	0	1	1	0
1	0	1	0	0	1
1	0	1	1	1	0
0	1	0	0	undefined	
0	1	0	1	1	0
0	1	1	0	0	1
0	1	1	1	0	1

There are many control applications where actions are based on a real timeclock output. Even if inputs to the flip-flops are changed, we would like the output to remain in its previous state until the clock output occurs. A third terminal is sometimes added to the *R-S* flip-flop for this purpose. The symbol for a clocked *R-S* flip-flop is shown in Fig. 8–17. Figure 8–18 shows the modification of the *R-S* flip-flop to produce the clocked *R-S* flip-flop. NAND gates are added at the input to isolate the *R-S* terminals until the clock pulse occurs. Simultaneous 1 inputs are not allowed at the clocked input terminals.

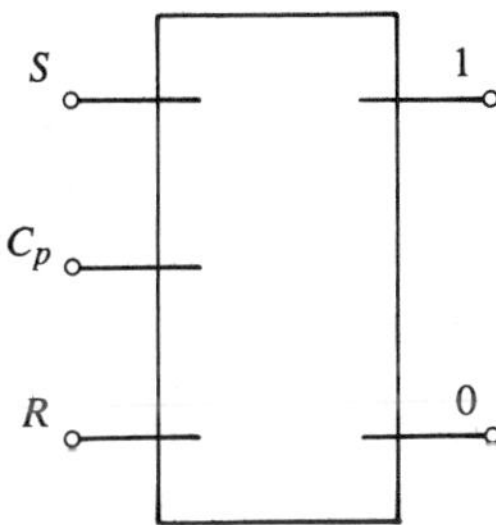

Fig. 8–17 Symbol for *R-S* flip-flop with clocked trigger.

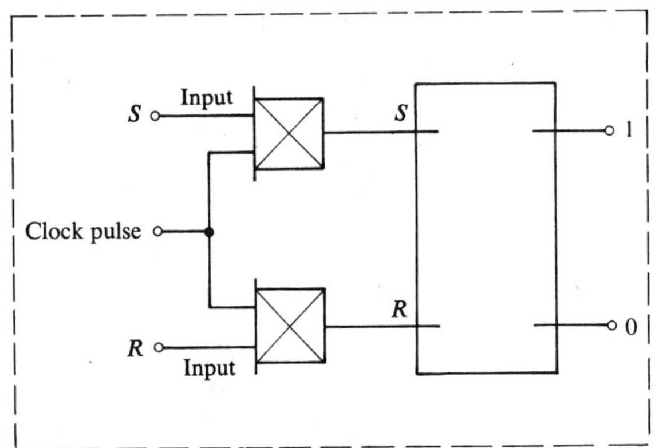

Fig. 8–18 Modification of basic *R-S* flip-flop to produce clocked *R-S* flip-flop.

Another flip-flop that has become quite popular since the development of IC technology is the *J-K* flip-flop. One of the desirable characteristics of this device is that it has no undefined output states. If both input signals are true the output is negated; if both input signals are false the output remains in the previous state. Figure 8–19 shows the logic diagram for the *J-K* flip-flop. A clock signal is required to produce output changes. The truth table is shown in Table 8–6. Figure 8–20 is the manufacturer's schematic for an IC *J-K* flip-flop. The input terminals are S and C. The

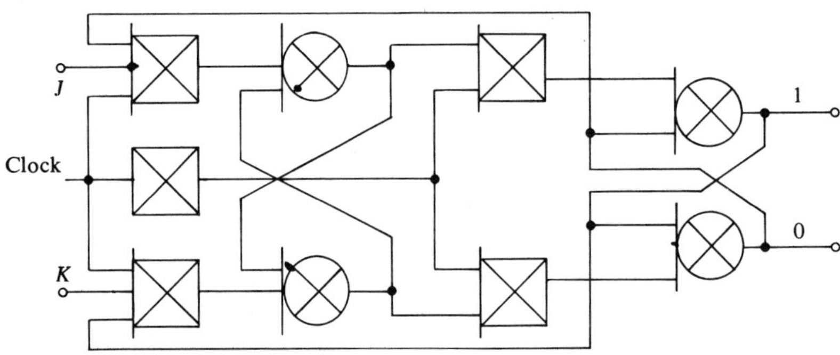

Fig. 8–19 Logic diagram of *J-K* flip-flop.

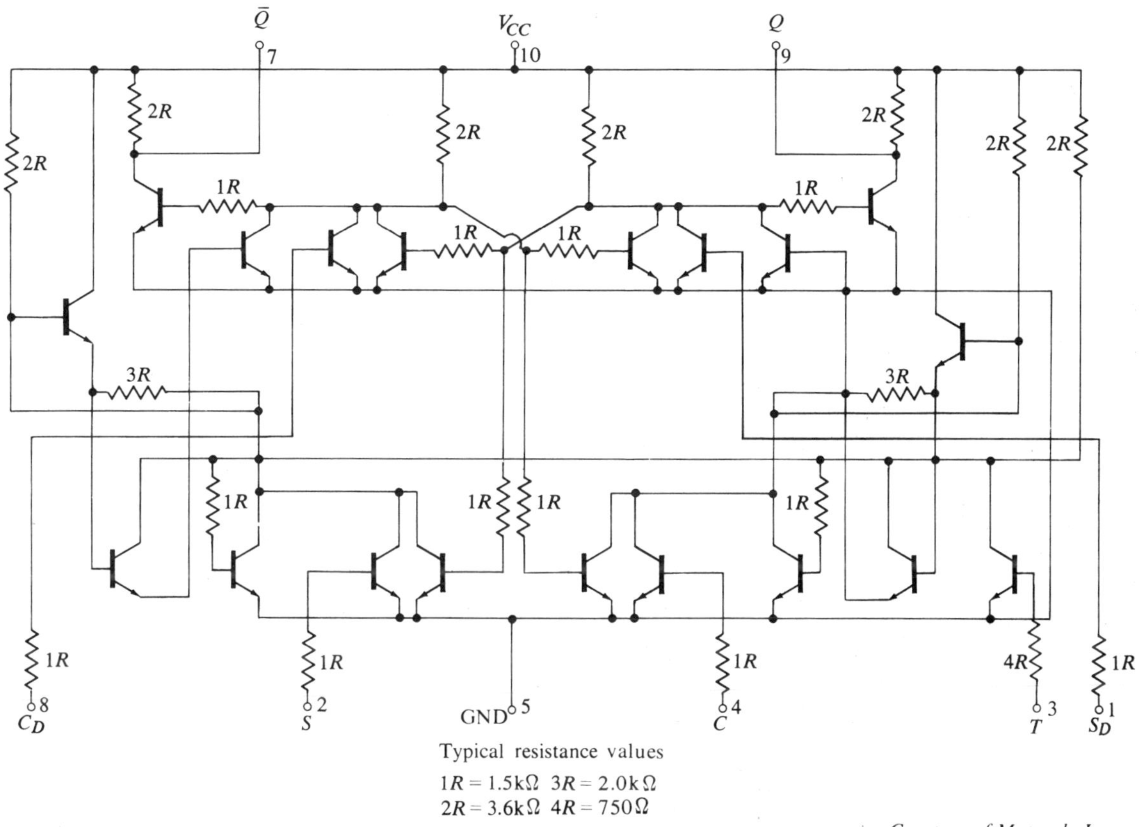

Courtesy of Motorola Inc.

Fig. 8–20 Commercial IC *J-K* flip-flop.

clock input is T. The outputs are Q and $\overline{Q}$. The S_D and C_D terminals are for direct setting of the output in either state. These terminal inputs override the normal input signals.

TABLE 8–6

Truth Table for *J-K* Flip-Flop

Previous States		Input Signals		Final States	
1	**0**	**J**	**K**	**1**	**0**
0	1	0	0	0	1
0	1	0	1	0	1
0	1	1	0	1	0
0	1	1	1	1	0
1	0	0	0	1	0
1	0	0	1	0	1
1	0	1	0	1	0
1	0	1	1	0	1

8-6-2 Counters

We emphasized in Sec. 6-4-2 the application of digital counters for controlling time delay in industrial electronics systems. Counters find many other applications, such as counting the number of parts produced by a machine during a period of time, determining the frequency of a radio signal by counting the number of cycles that occur in a second, or determining when a given number of parts have been placed in a carton for packaging. The simplified diagram of a counter was shown in Fig. 6-21 using flip-flops. Let's look at the way *J-K* flip-flops can be used to design a counter.

A common technique in counter design is to use *J-K* flip-flops of Fig. 8-19 with the *J* and *K* inputs held at logic 1. The output of the flip-flop is inverted each time the clock pulse is applied. Figure 8-21 shows how four *J-K* flip-flops are cascaded to make a divide by sixteen ripple counter. A pulse is applied at the clear terminal to set all of the 1 outputs at logic 0. Positive pulses are applied at the clock

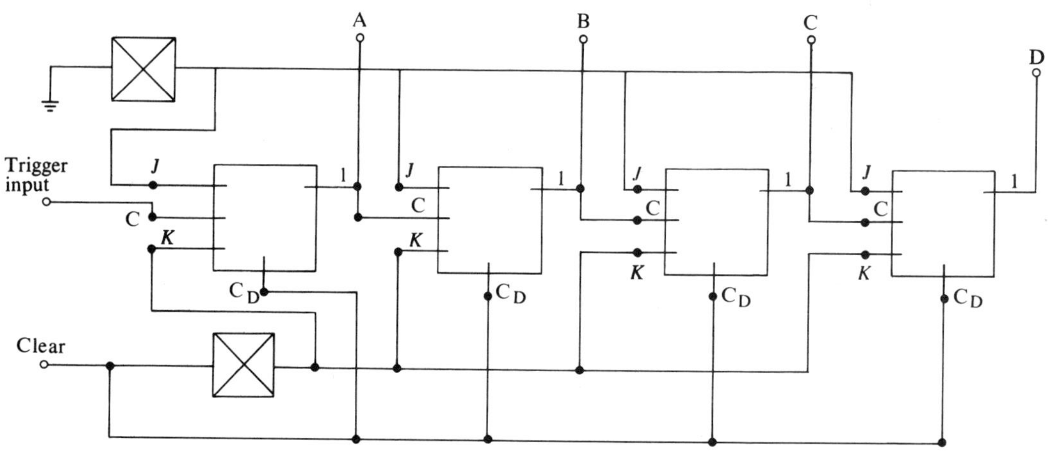

Fig. 8-21 Binary ripple counter.

terminal of the first flip-flop. The output of the flip-flop changes when the clock pulse goes back to logic 0. Figure 8-22 shows how the flip-flop outputs change as a train of pulses is applied at the clock terminal of the first flip-flop.

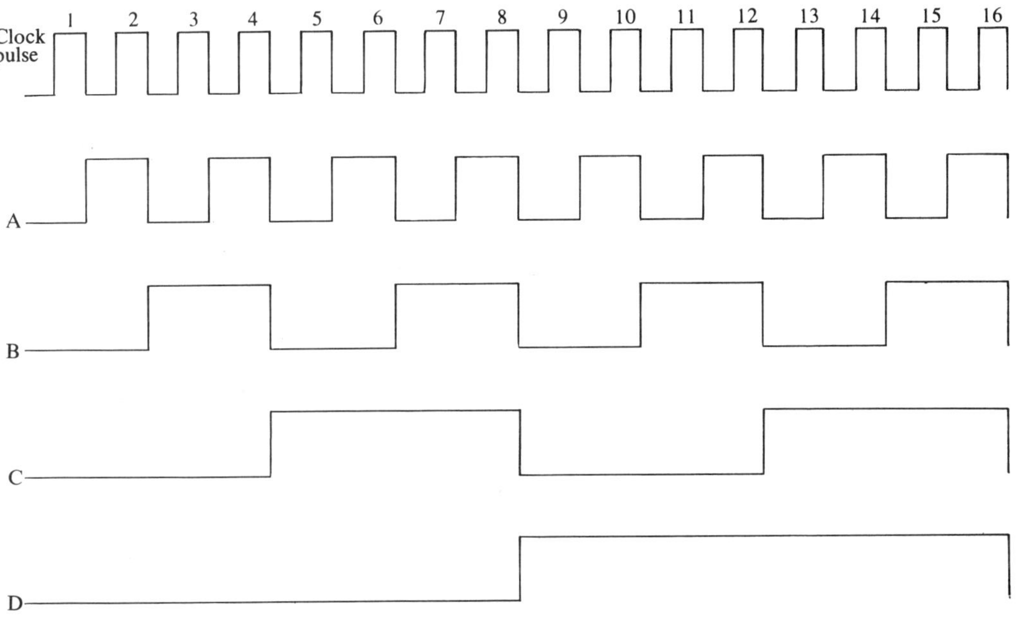

Fig. 8–22 Pulse diagram of the ripple counter of Fig. 8–21.

A complete pulse output at the fourth flip-flop requires sixteen input pulses at the first flip-flop. The count can be terminated at any point by applying a pulse at the clear terminal.

The decimal count stored in the counter can be determined by looking at the state of the flip-flops. The least significant digit is stored at A, the most significant digit is stored at D. On the count of six, D = 0, C = 1, B = 1, A = 0, or binary 0110.

D	C	B	A
0	1	1	0

8–6–3 Decoders

The binary number system is quite convenient for the hardware implementation of digital functions, but the machine operator, the housewife, the supervisor, the inspector, and even the engineer would prefer to read decimal numbers. Indeed, the decimal system is more efficient for displaying the results of a measurement or the frequency of a radio transmitter. A "decoder" is used to convert the

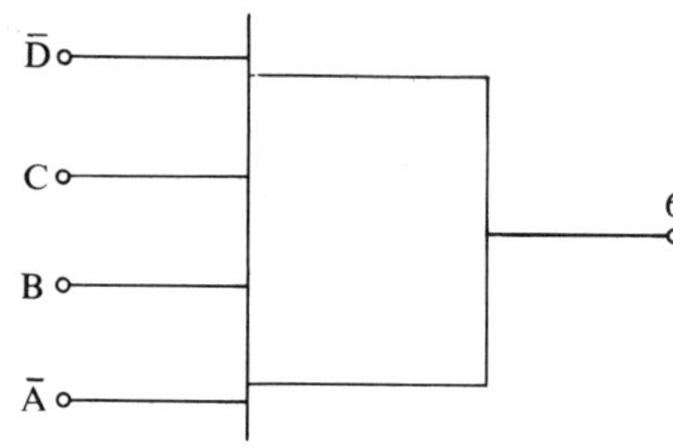

Fig. 8–23 Binary to decimal six decoder.

output of a binary counter such as that of Fig. 8–21 to a decimal number. The decimal number might be displayed for visual indication, recorded, or used to control other operations.

We could use AND gates to decode any decimal digit given the output signals of the binary counter. As an example, on the count of six, logic "1" outputs occur at $\bar{D}$, C, B, and $\bar{A}$. Figure 8–23 shows the binary to decimal six decoder. The complete decoder is shown in Fig. 8–24.

8–7 Implementing Logic Functions

The ultimate desire of the designer of digital circuits for industrial electronics is to specify the hardware necessary to perform the functions required by the customer. The concepts of this chapter can be used to progress from a word statement of the requirements of the customer to the logic design of the circuitry. The designer must then decide what hardware is best suited to the particular logic design. Factors such as cost, environmental conditions, compatibility, and availability of a specific line of hardware must be considered. An entire book could be written on this subject. Such a book would be obsolete before it was printed due to technological advances and the introduction of new products. You must rely on trade literature for most of the information required.

There are some things that we can do to aid in the hardware selection, such as reducing the number of gates required, minimizing the number of inputs, and reducing the loading requirements. We will look at some very general examples of implementing logic functions in this section.

8–7–1 Converting the Word Statement

The prospective user usually comes to the designer with a problem in the form of a word statement. The designer must convert the word statement to a logic function before it can be implemented. The procedure is to make logic assignments to the customer inputs and outputs, formulate the truth table, and write the Boolean equation. From the Boolean equation we can proceed to specify the hardware. Let's look at examples of converting the word statement to a Boolean equation.

Example 8–4. A rotary assembly machine has three stations. Two of the three stations are manned by different employees during the assembly operation. The third station is a spare. Industrial safety requires that a lock-out circuit be installed to allow the machine to rotate if, and only if, both employees actuate two palm buttons indicating that the operation is complete. Write the Boolean equation to perform this logic function.

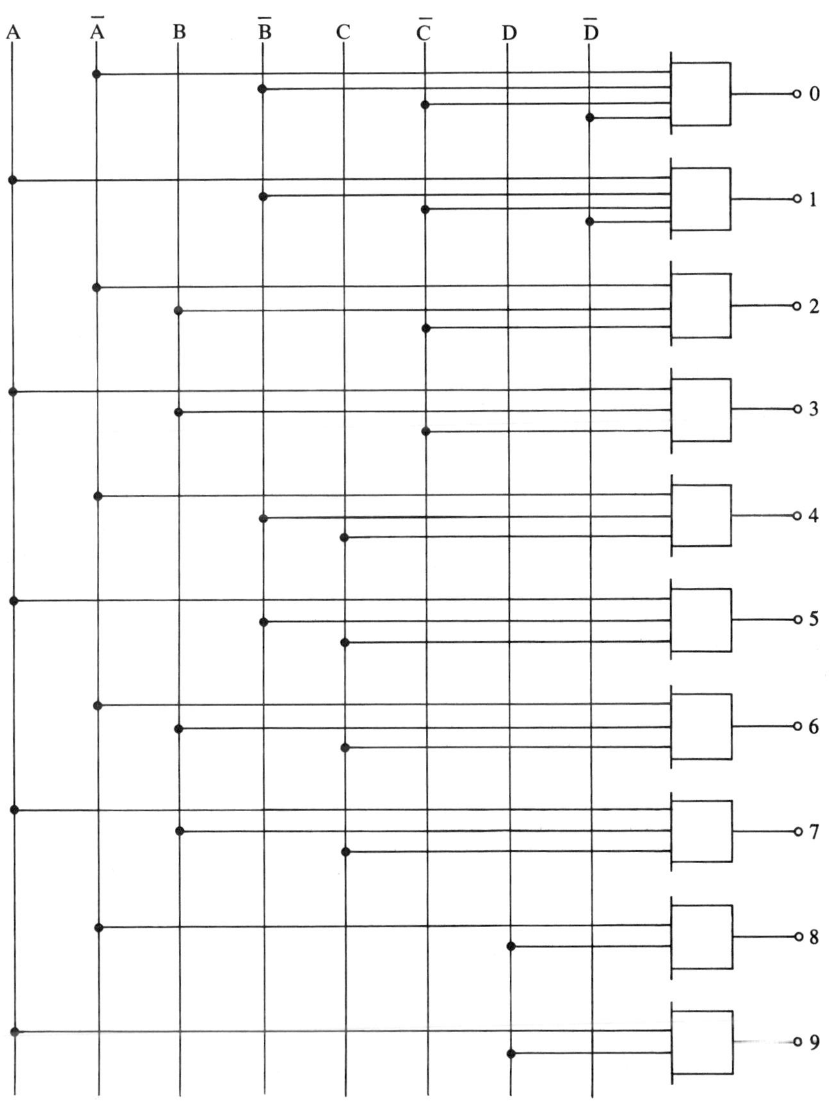

Fig. 8–24 Binary to decimal decoder using AND gates.

First, alphabets are assigned to the inputs and outputs. Figure 8–25 shows that the palm buttons at station 1 are called A and B, palm buttons at station 2 are called C and D, palm buttons at station 3 are called E and F. The output signal to rotate the machine is called K. Logic assignments will be A = 1 when switch is closed, L = 1 when rotation is signaled. All other operator switches will be similarly assigned.

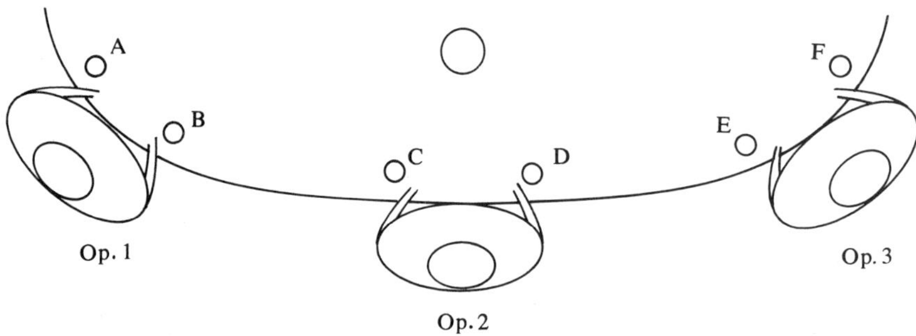

Fig. 8–25 Assignment of inputs for example 8–4.

There are six inputs to be considered. The complete truth table would have 64 rows. We can reduce the number of inputs by first performing the AND operation at each station and using the output from each station as the input to the circuit. Figure 8–26 shows the new inputs: G, H, and J. The truth table of Table 8–7 shows the three combinations of inputs that result in a true output.

The Boolean equation is:

$$L = \bar{G}HJ + G\bar{H}J + GH\bar{J} \qquad \textbf{(8.4)}$$

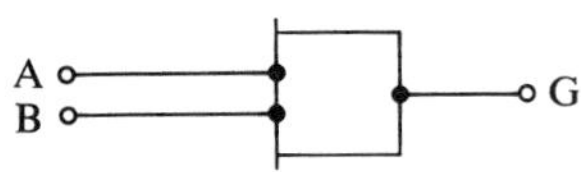

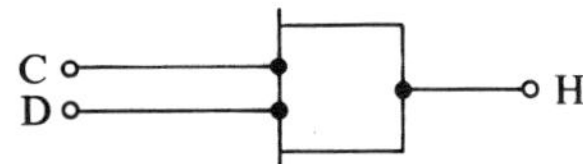

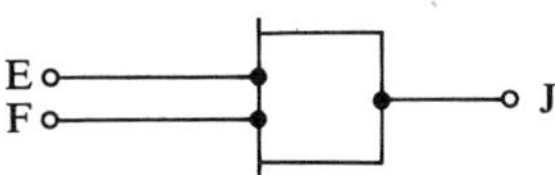

Fig. 8–26 Reduced inputs of example 8–4.

TABLE 8–7
Truth Table of Example 8–4

G	H	J	L
0	0	0	0
0	0	1	0
0	1	0	0
0	1	1	1
1	0	0	0
1	0	1	1
1	1	0	1
1	1	1	0

Example 8–5. A 4-bit binary counter having both true and false outputs available is used to sort parts made on two different machines. The parts from machine #1 are assigned the numbers 0 to 5. Parts from machine #2 are assigned the numbers 6 to 12. Write the Boolean equation of the logic circuit to perform the sorting.

The outputs of the counter are labeled A, B, C, and D as shown in Fig. 8–27. The outputs of machine #1 are represented by E. The outputs of machine #2 are represented by F. The truth table is shown in Table 8–8.

TABLE 8–8
Truth Table of Example 8–5

Inputs				Outputs	
A	**B**	**C**	**D**	**E**	**F**
0	0	0	0	1	0
0	0	0	1	1	0
0	0	1	0	1	0
0	0	1	1	1	0
0	1	0	0	1	0
0	1	0	1	1	0
0	1	1	0	0	1
0	1	1	1	0	1
1	0	0	0	0	1
1	0	0	1	0	1
1	0	1	0	0	1
1	0	1	1	0	1
1	1	0	0	0	0
1	1	0	1	0	0
1	1	1	0	0	0
1	1	1	1	0	0

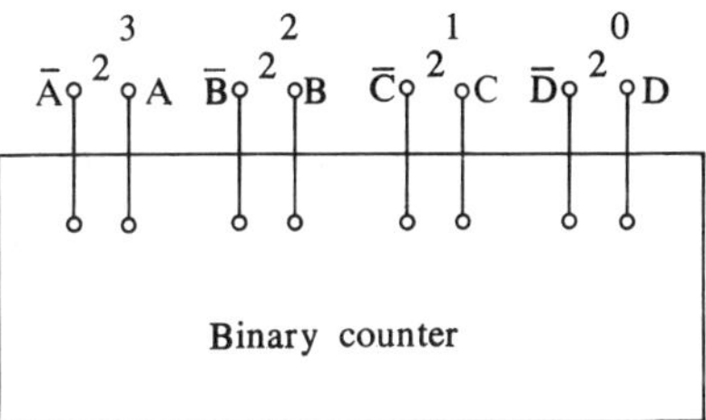

Fig. 8–27 Binary counter output assignments of example 8–5.

The Boolean equations are:

$$E = \overline{A}\,\overline{B}\,\overline{C}\,\overline{D} + \overline{A}\,\overline{B}\,\overline{C}\,D + \overline{A}\,\overline{B}\,C\,\overline{D} + \overline{A}\,\overline{B}\,C\,D + \overline{A}\,B\,\overline{C}\,\overline{D} + \overline{A}\,B\,\overline{C}\,D \quad (8.5)$$

$$F = \overline{A}\,B\,C\,\overline{D} + \overline{A}\,B\,C\,D + A\,\overline{B}\,\overline{C}\,\overline{D} + A\,\overline{B}\,\overline{C}\,D + A\,\overline{B}\,C\,\overline{D} + A\,\overline{B}\,C\,D \quad (8.6)$$

8–7–2 Minimizing Gates and Inputs

The cost, speed, and complexity of logic circuits can sometimes be reduced by minimizing the number of gates required and the number of inputs to each gate. There are several mathematical techniques that can be applied to minimization including mapping and Quine-McCluskey methods. We will look at a simple Boolean equation that does not warrant these sophisticated techniques.

Let's look at the Boolean equation (8.5).

$$E = \overline{A}\ \overline{B}\ \overline{C}\ \overline{D} + \overline{A}\ \overline{B}\ \overline{C}\ D + \overline{A}\ \overline{B}\ C\ \overline{D} + \overline{A}\ \overline{B}\ C\ D + \overline{A}\ B\ \overline{C}\ \overline{D} + \overline{A}\ B\ \overline{C}\ D \quad (8.7)$$

It would require 6 AND gates with 4 inputs each and 1 6-input OR gate to implement the function.

By simple factoring, we can reduce Eq. (8.5) to

$$E = \overline{A}\ \overline{B}\ \overline{C}\ (\overline{D} + D) + \overline{A}\ \overline{B}\ C\ (\overline{D} + D) + \overline{A}\ B\ \overline{C}\ (\overline{D} + D)$$

but

$$\overline{D} + D = 1 \quad (8.8)$$

so

$$E = \overline{A}\ \overline{B}\ \overline{C} + \overline{A}\ \overline{B}\ C + \overline{A}\ B\ \overline{C} + \overline{A}\ \overline{B}\ (\overline{C} + C) + \overline{A}\ B\ \overline{C} \quad (8.9)$$

but

$$\overline{C} + C = 1,$$

then

$$E = \overline{A}\ \overline{B} + \overline{A}\ B\ \overline{C} = \overline{A}\ (\overline{B} + B\ \overline{C}) \quad (8.10)$$

from Eq. (16) of Table 8–4,

$$\overline{B} + B\ \overline{C} = (\overline{B} + \overline{C}) \quad (8.11)$$

so

$$E = \overline{A}\ (\overline{B} + \overline{C}) \quad (8.12)$$

Implementing Eq. (8.12) requires a 2-input AND gate and a 2-input OR gate. There is no question about the reduction in cost and complexity of the hardware by simple Boolean algebra manipulation.

8–7–3 Loading of Electronic Logic Gates

Electronic logic gates are made of active semiconductor devices. Excessive loading will affect the output level just as in amplifier design. We discussed the loading effects on amplifiers in chapter 3 under the subject of multiple input and output switching circuits. Let's look at the same effects in the context of logic gates. Consider the positive logic RTL NAND gate of Fig. 8–28. A typical manufacturer's specification might be

$$\text{logic 1; } e_0 \geq 5 \text{ V}$$
$$\text{logic 0; } e_0 \leq 0.8 \text{ V}$$

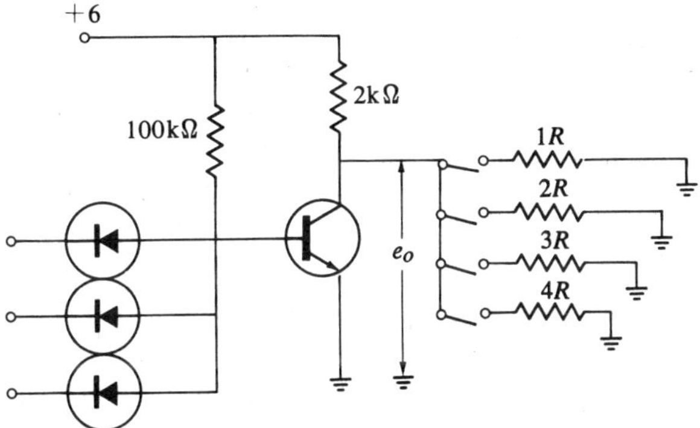

Fig. 8–28 RTL NAND GATE (positive logic) with multiple loading.

These voltage "thresholds" must be maintained at all times if proper logic interpretation is to be guaranteed. The NAND gate will be used to drive up to 4 loads represented by $1R$, $2R$, $3R$, and $4R$. If the loads are switched in one at a time while the gate is in the logic 1 state, we can see how loading could result in faulty logic levels. Let $1R = 2R = 3R = 4R = 40\text{k}\Omega$ and the dynamic output impedance of the transistor be infinite. Then Fig. 8–29 shows the equivalent output circuits. When the first load is applied, the output voltage will be

$$e_0 = \frac{40\text{k}\Omega}{2\text{k}\Omega + 40\text{k}\Omega} 6 = 5.7 \text{ V} \tag{8.13}$$

The output is above the 5-V threshold. The addition of the second load drops the output voltage to 5.45 V. Addition of the third load drops the output voltage to 5.22 V. The addition of the fourth load reduces the output voltage to 5.00 V. Since 5.00 V is the lower limit for logic 1 output, no more loads can be driven by the gate. The gate is said to be at its maximum load driving capability.

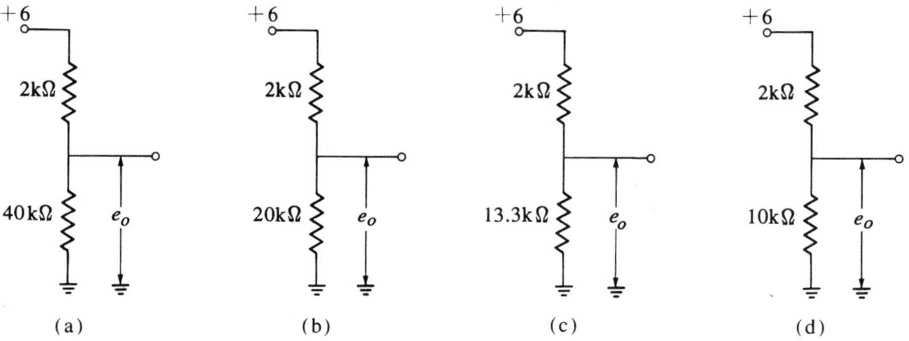

Fig. 8–29 Loading of NAND gate at logic 1: (a) 1*R*; (b) 1*R* and 2*R*; (c) 1*R*, 2*R*, and 3*R*; (d) 1*R*, 2*R*, 3*R*, and 4*R*.

The manufacturer will specify the maximum number of loads that can be driven by a gate. Since there is usually a variety of loads that can be driven by each gate, the "unit load" is defined and used to specify driving capacity. For the NAND gate of Fig. 8–28 a 40-kΩ load drawing 0.125 mA is a *unit load*. The NAND gate then has 4 unit load capacity. A similar rating is applied to the input of each gate. A unit load input specifies the amount of current a gate must be able to sink while in the logic 0 state without exceeding the threshold voltage. Manufacturers will sometimes indicate the number of unit loads that can be driven by a gate and its input loading factor on the logic diagram such as in Fig. 8–30.

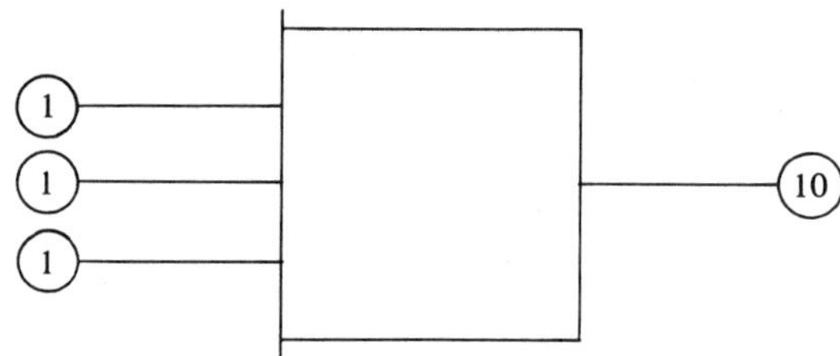

Fig. 8–30 Input and output loading factors indicated on logic diagram.

The driving capacity of logic gates can be increased by replacing the 2-kΩ "pull-up" resistor of Fig. 8–28 by another transistor. This is known as *active pull-up.*

The designer must keep track of the number of unit loads at each gate during the process of minimization.

8–8 Summary

The basic digital concepts that we discussed in this chapter will aid in your understanding of electronic circuits and systems. They will apply to your future experiences in industrial electronics, communication electronics, computer electronics, and other phases of the electronics world. We introduced binary numbers, logic state assignments, truth tables, logic gates, Boolean algebra, simple logic elements, and implementing logic functions. Even if you have had some of these concepts before, I hope we have related them to industrial electronics.

EXERCISES

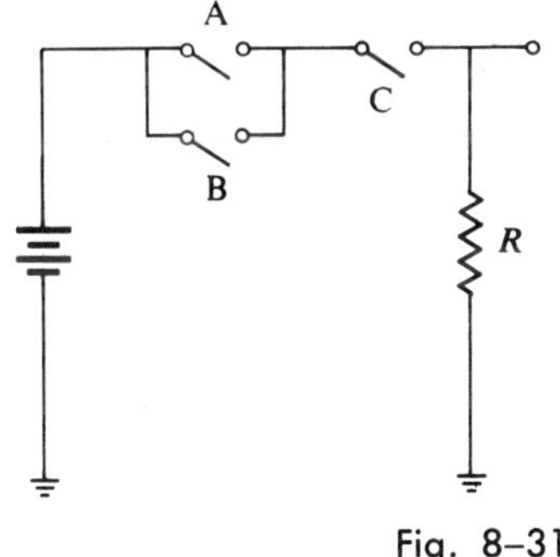

Fig. 8–31

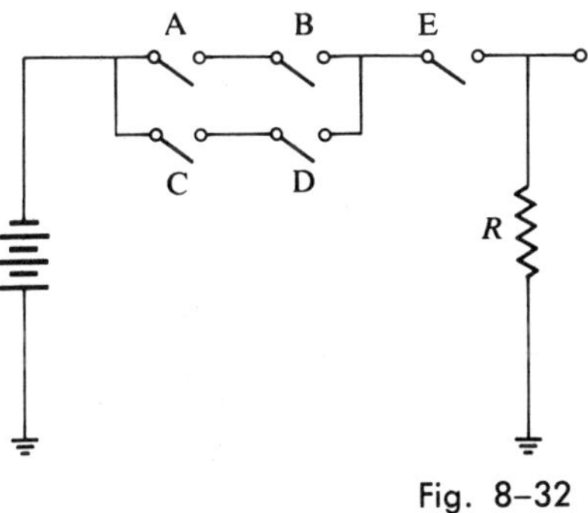

Fig. 8–32

1. Convert the decimal number 86 to binary.
2. Convert the decimal number 227 to binary.
3. Convert the binary number 100110101. to decimal.
4. Convert the binary number 1001111010. to decimal.
5. Explain the meaning of "positive logic." Explain the meaning of "negative logic."
6. Make state assignments to the switches of Fig. 8–31 and make the truth table describing the operation of the circuit.
7. Make state assignments to the switches of Fig. 8–32 and construct the truth table describing the operation of the circuit.
8. Draw the logic diagram represented by the truth table of Table 8–9.
9. Draw the logic diagram represented by the truth table of Table 8–10.
10. Write the Boolean algebraic equation for the truth table of Table 8–9.
11. Write the Boolean algebraic equation for the truth table of Table 8–10.

TABLE 8–9

Inputs			Outputs	
A	B	C	D	E
0	0	0	0	1
0	0	1	1	0
0	1	0	1	0
0	1	1	0	1
1	0	0	0	0
1	0	1	0	0
1	1	0	1	1
1	1	1	1	0

TABLE 8–10

Inputs			Outputs	
A	B	C	D	E
0	0	0	0	1
0	0	1	0	0
0	1	0	0	0
0	1	1	0	0
1	0	0	1	1
1	0	1	1	0
1	1	0	1	0
1	1	1	0	0

12. How many rows will there be in a truth table representing a digital circuit with 8 input variables?

13. Use the rules of Boolean algebra from Table 8–4 to simplify the equation

$$L = \bar{A}\ \bar{B}\ C\ D + \bar{A}\ B\ C\ D + A\ B\ \bar{C}\ \bar{D} + A\ B\ \bar{C}\ D + A\ B\ C\ D$$

14. Draw the logic diagram to implement the Boolean equation of problem 13 before and after simplification.

15. Use DeMorgan's theorem to implement the equation $L = (A + B)\ (C + D) + E$ using only NAND gates.

16. Use DeMorgan's theorem to implement the equation $L = [A.B + C.D]\ E$ using only OR gates with inverters.

17. Draw the logic diagram of problem 15 before and after the application of DeMorgan's theorem.

18. Draw the logic diagram of problem 16 before and after the application of DeMorgan's theorem.

19. How many gates would be required to implement the logic function

$$L = [(AB + C.D)\ E]\ [.\bar{A}\ (C + \bar{D}) + F]$$

20. What is meant by a unit load in the language of digital logic hardware?

9

Digital Sequence Control

9–1 Introduction

Most industrial processes require the completion of several operations to produce the required output. Manufacturing, machining, assembling, packaging, finishing, or transporting of products requires the precise coordination of several tasks for an efficient and economical system to function. Electronic controls have long provided the necessary coordination and monitoring of industrial operations. The controls can be divided into two general categories: sequential and combinatorial. Sequential controls are required for processes that demand that certain operations be performed in a specific order. Combinatorial processes require that certain operations be performed without regard to the order in which they are performed.

Consider the operations of Figs. 9–1 and 9–2. Both operations are a part of a bottling process. In the filling and capping operation, the essential tasks are:

(a) Fill bottle.
(b) Press on cap.

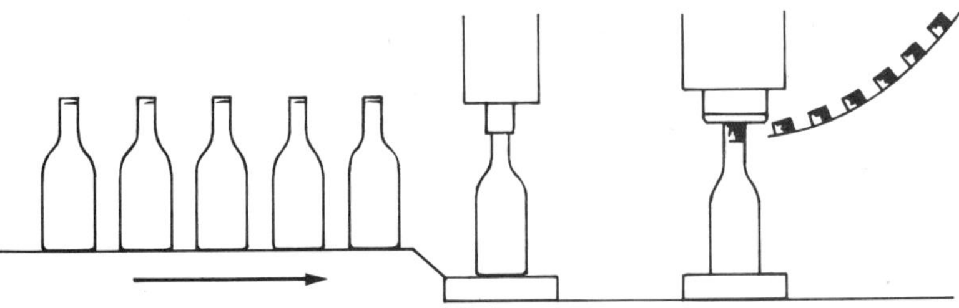

Fig. 9–1 Bottle filling and capping process.

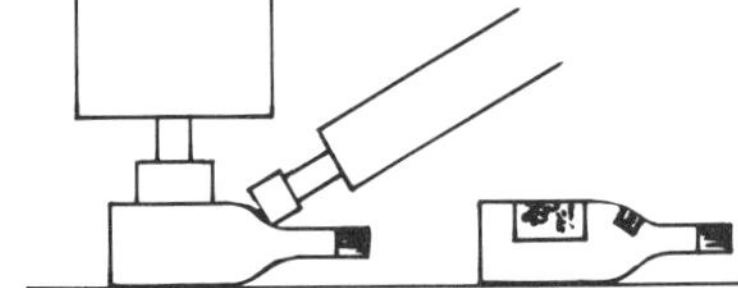

Fig. 9–2 Bottle labeling process.

These tasks must be performed in the proper order. Surely we couldn't fill the bottle after the cap was pressed on. This operation requires sequential control. In Fig. 9–2, two labels are placed at different locations on the bottle. The essential tasks are:

(a) Place label #1 on bottle.
(b) Place label #2 on bottle.

The order in which the tasks are performed does not matter. This operation requires only combinatorial control. The processes we looked at in Examples 8 and 9 of chapter 8 were both combinatorial. The majority of industrial control problems are sequential. Indeed, processes that are not inherently sequential are performed in a sequential manner because work analysis has shown that there is some economic advantage to a specific order of operations.

In this chapter, we will look at some of the techniques and hardware for designing digital sequential controls.

9–2 The Relay Ladder Diagram

Earlier sequence controls used multicontact electromechanical relays in switching circuits. Logic functions were performed by state assignment of open or closed relay contacts. The closed contact was usually the true or 1 state.

The open contact was usually the false or 0 state. Figure 9–3 shows some elementary relay contact connections and the equivalent logic functions. Let's look at the interpretation of the symbols. They are very basic to understanding "relay logic." In Fig. 9–3(a) we will get continuity between points 1 and 2 if relay coils A and B and C are energized. The symbology of Fig. 9–3(b) says "there will be continuity between points 1 and 2 if relay coil A is not energized and relay coils B and C are energized." Figure 9–3(c) says there will be continuity between points 1 and 2 if either relay coil A or relay coil B is energized. Figure 9–3(d) says there will be continuity between points 1 and 2 if relay coil A is energized or relay coil B is not energized.

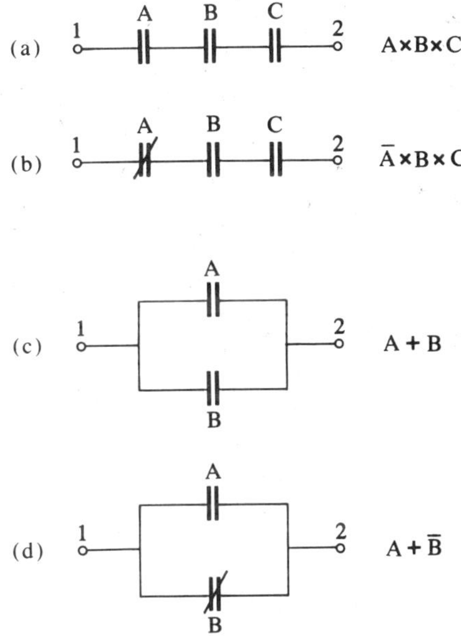

Fig. 9–3 Relay contact logic functions.

The entire operating sequence can be determined from the circuit diagram showing the interconnection of all the relay coils, relay contacts, and actuating devices. The diagram is usually drawn in ladder form as shown in Fig. 9–4. The control transformer secondary terminals are extended vertically (after fusing) to form the boundary lines for the diagram. Relay coils, contacts, and actuating devices are

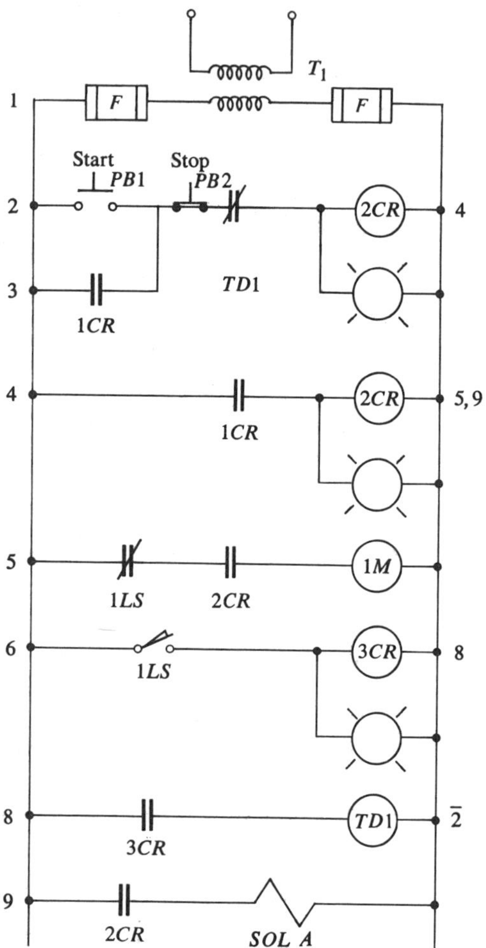

Fig. 9–4 Relay ladder diagram.

then shown connected to the transformer output. Relay coils are shown in the order in which they are energized during the normal sequence of operations. It is customary to show the start push button on the top line of the diagram. Relay contacts and actuation devices are placed on the left side of the diagram. Relay coils and the indicator lamps are placed on the right side of the diagram. Each line is numbered for easy referencing. Line numbers are placed on the left side of the vertical line. The numbers next to each relay coil indicate the lines on which the relay contacts are used. We can list the operation sequence by following the ladder diagram line by line:

Line 2—Start push button is actuated; 1*CR* is energized if 3*CR* is unenergized and the stop push button is not actuated.

Line 3—Contact 1*CR* closes, sealing in 1*CR* even if the start push button is released.

Line 4—Contact 1*CR* closes, energizing relay coil 2*CR* if limit switch 1 is not actuated.

Line 5—Contact 2*CR* closes, energizing motor control relay 1*M* if limit switch 1 is not actuated.

Line 9—Contact 2*CR* closes, energizing solenoid A which might control an air valve or hydraulic valve.

Line 6—Limit switch 1 is actuated, energizing relay coil 3*CR*.

Line 5—Contact 1*LS* opens, de-energizing relay coil 1*M*, shutting off the motor.

Line 8—Contact 3*CR* closes, energizing the time delay relay coil.

Line 2—After the time delay, contact 1*TD* opens, de-energizing relay coil 1*CR*.

Line 4—Relay contact 1*CR* opens, de-energizing relay coil 2*CR*.

Line 9—Contact 2*CR* opens, de-energizing solenoid coil A to end the sequence.

9–3 Design of Sequencing Systems

Sequential systems require that a specific order of operations must be performed if the predicted outcome is to be realized. The simple techniques used to design combinatorial systems are no longer adequate. We must go to more complicated and sophisticated "synthesis" techniques. However, *sequential circuit synthesis* would be a subject for an entire book itself. We do not pretend to explore it thoroughly in a single section of one chapter. What we do intend to do is demonstrate a couple of techniques that will work for simple physical systems. As the systems become more complicated, we would be involved in massive state diagrams, transition tables, and minimization techniques that are beyond the scope of this text.

9–3–1 The "Commonsense" Approach

Most requests for electronic control systems come to us in the form of word statements, specifications, or manufacturing process statements. The first step is to convert the problem to the language of digital logic. Without any formal techniques in mind, simple sequential systems can be designed by using the commonsense or intuitive approach. Let me show you what I mean.

We would like to design the digital control for an assembly machine fixture. The fixture will start in the center position. When the start push button is actuated the fixture must move left to a preset position, move up to a preset position, hesitate for 10 seconds, move right to a preset position, move down to the original level, and back to center. Figure 9–5 shows the diagram of fixture motion. We might start by assigning alphabets to the mechanical devices.

Start push button	=	A
Stop push button	=	B
Left limit switch	=	$1LS$
Up limit switch	=	$2LS$
Time delay	=	$1TD$
Right limit switch	=	$3LS$
Down limit switch	=	$4LS$
Center limit switch	=	$5LS$
Left motor control relay	=	$1M$
Up motor control relay	=	$2M$
Right motor control relay	=	$3M$
Down motor control relay	=	$4M$

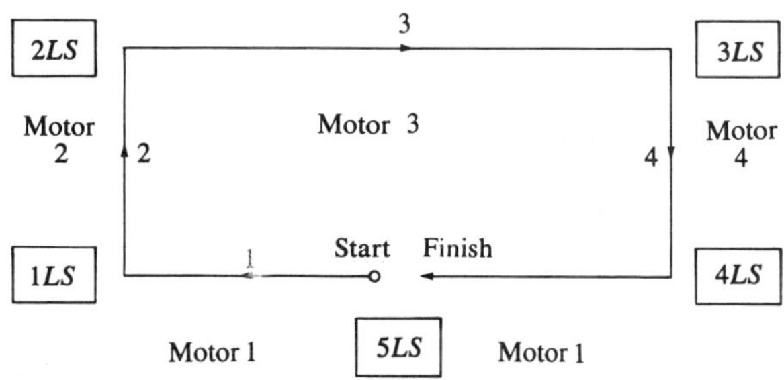

Fig. 9–5 Motion diagram of fixture.

We might then build the ladder diagram as we go through a normal machine cycle. Follow the development of the ladder diagram in Fig. 9–6.

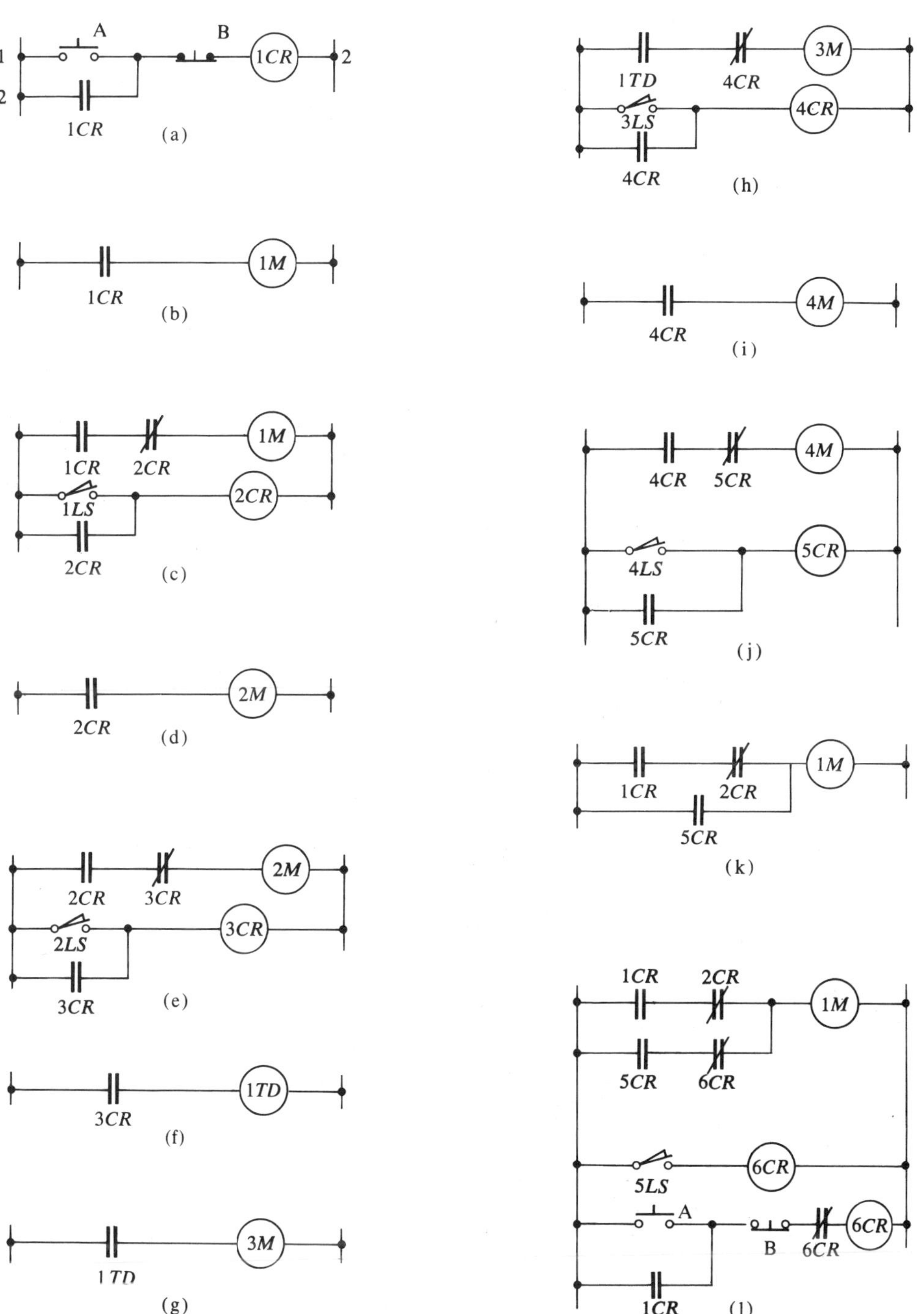

Fig. 9–6 Development of the relay ladder diagram.

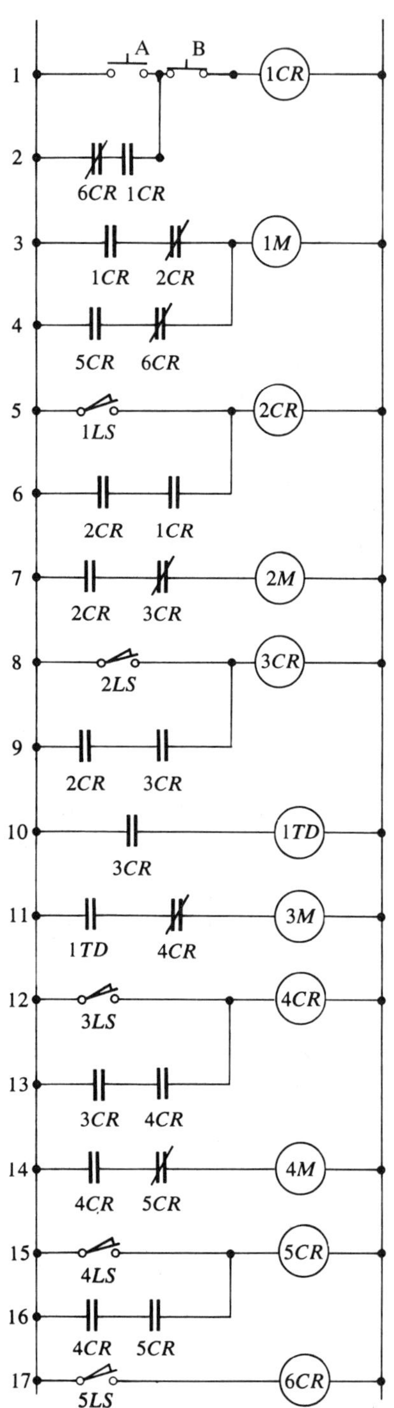

Fig. 9–7 Complete relay ladder diagram of Fig. 9–6.

(a) Push start button
(b) Move fixture left
(c) Stop at left limit switch
(d) Move fixture up
(e) Stop at upper limit switch
(f) Initiate time delay
(g) Move fixture right
(h) Stop at right limit
(i) Move fixture down
(j) Stop at lower limit
(k) Move fixture left
(l) Stop at center

The fixture has completed one cycle, but the ladder diagram is not complete. What would happen if the start button were actuated again? The cycle would not repeat because all control relays are still *sealed in*. We must go one step further to be sure the circuit is *reset* at the end of the cycle. Placing another contact in the seal-in circuit will reset all relays. The complete ladder diagram is shown in Fig. 9–7. Notice the start button must be held until the fixture moves off center. From the ladder diagram, we can now write the Boolean equations for actuation of each of the relay coils. They are:

$$
\begin{aligned}
1CR &= [A + \overline{6CR} \cdot 1CR] \cdot \mathrm{B} \\
1M &= [1CR \cdot \overline{2CR}] + [5CR \cdot \overline{6CR}] \\
2CR &= 1LS + [2CR \cdot 1CR] \\
2M &= 2CR \cdot \overline{3CR} \\
3CR &= 2LS + [2CR \cdot 3CR] \\
1TD &= 3CR \\
3M &= 1TD \cdot \overline{4CR} \\
4CR &= 3LS + [3CR + 4CR] \\
4M &= 4CR \cdot \overline{5CR} \\
5CR &= 4LS + [4CR \cdot 5CR] \\
6CR &= 5LS
\end{aligned}
$$

We should now look at some of the properties of this design:

(1) Relays appear on both sides of the equation.
(2) Only one relay is actuated at each step in the sequence.
(3) Sequence must start at step one.

Some designers get around the first problem by assigning lowercase letters to the coil energizing signal and uppercase letters to the relay contacts. Whenever a relay term appears on both sides of the same equation, such as $5CR = 4LS + [4CS \cdot 5CR]$, there is some feedback or memory in the circuit.

One of the limitations of this technique is that only one relay should be actuated at each step. All relays require a finite time to actuate the contacts. This time is usually on the order of 10 milliseconds, but will vary considerably. If more than one relay is actuated at a given point in the sequence, then there is the problem of which will be the fastest. Undesired conditions sometimes occur where there are "races" between relays energized at the same time. It is usually difficult to analyze all the possible outcomes of "races" in multiple actuating circuits unless a more systematic synthesis technique is available.

Having a well defined starting point is almost a must if the commonsense approach is to be effective. Problems multiply rapidly when multiple starting conditions are allowed and, again, a more systematic synthesis technique is required.

9–3–2 The Sequence Chart

A technique commonly used by industrial electronic designers of digital sequence systems is to make up a sequence chart listing the logical state of each device in the system. Table 9–1 is the sequence chart for the fixture operation of

TABLE 9–1
Sequence Chart for Fixture Operation

		Inputs							Agents	Outputs			
Phase	**Action**	**A**	**B**	**1LS**	**2LS**	**3LS**	**4LS**	**5LS**	**1TD**	**1M**	**2M**	**3M**	**4M**
	Start position	0	1	0	0	0	0	1	0	0	0	0	0
1	Push start button	1	1	0	0	0	0	1	0	0	0	0	0
2	Move fixture left	1	1	0	0	0	0	0	0	1	0	0	0
3	Stop at left limit	0	1	1	0	0	0	0	0	0	0	0	0
4	Move fixture up	0	1	0	0	0	0	0	0	0	1	0	0
5	Stop at upper limit	0	1	0	1	0	0	0	0	0	0	0	0
6	Initiate time delay	0	1	0	1	0	0	0	1	0	0	0	0
7	Move fixture right	0	1	0	0	0	0	0	0	0	0	1	0
8	Stop at right limit	0	1	0	0	1	0	0	0	0	0	0	0
9	Move fixture down	0	1	0	0	0	0	0	0	0	0	0	1
10	Stop at lower limit	0	1	0	0	0	1	0	0	0	0	0	0
11	Move fixture left	0	1	0	0	0	0	0	0	1	0	0	0
12	Stop at center	0	1	0	0	0	0	1	0	0	0	0	0

Sec. 9–3–1. The system devices are divided into three categories:

- inputs—those devices that provide the external stimulus to the system; typical inputs are push buttons, limit switches, pressure switches, etc.
- outputs—those devices that control the loads affected by the system; typical outputs are relays, amplifiers, solenoids, etc.
- agents—those devices that operate within the control system to provide feedback, memory, or timing; typical devices are relays, flip-flops, time delays, clocks, etc.

Each step of the sequential operation is listed and the resulting action is recorded. Inputs are traditionally listed first, agents second, and then outputs. The initial logical state of each device is listed as the first operation.

Input devices, by definition, do not require electronic energization. Agents and outputs must be energized at the proper time if the sequence is to progress in the desired manner. The next step, then, is to list the conditions for energizing agents and outputs. Energization must occur on the phase prior to actuation. For instance, at phase 2 of Table 9–1 motor control relay $1M$ must be actuated. The conditions of phase 1 must be used to energize relay $1M$. The Boolean equation is:

$$1M = A \cdot B \cdot 5LS \tag{9.1}$$

The Boolean equations for the system are determined by logical summing of all conditions, such as Eq. (9.1), that demand energizing each output or agent. The list of equations from Table 9–1 is:

$$1M = A \cdot B \cdot 5LS + B \cdot 4LS \tag{9.2}$$

$$2M = B \cdot 1LS \tag{9.3}$$

$$3M = 1TD \cdot 2LS \cdot B \tag{9.4}$$

$$4M = B \cdot 3LS \tag{9.5}$$

$$1TD = B \cdot 2LS \tag{9.6}$$

Figure 9–8 shows the NEMA symbolic diagrams for Eqs. (9.2) to (9.6).

But wait, look at the physical devices involved. The electronic logic gates of Fig. 9–8 require that the inputs be maintained if the output must be maintained. The limit switches

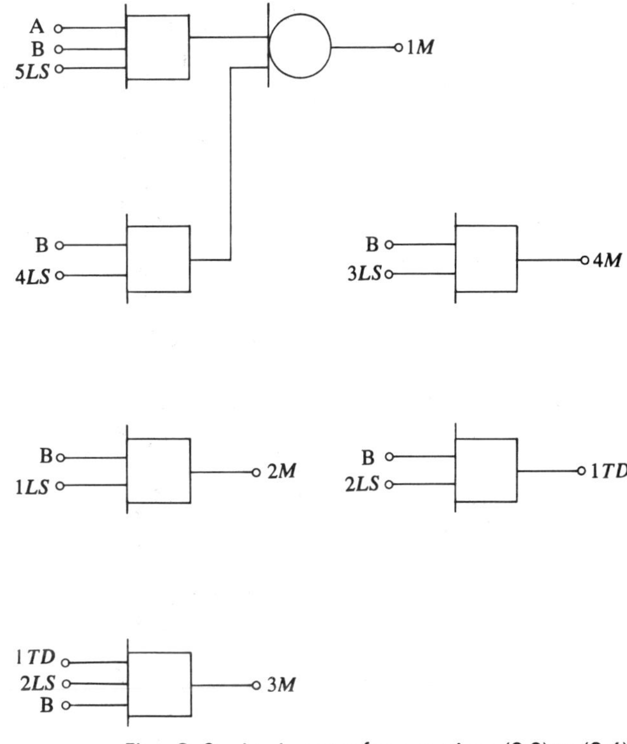

Fig. 9–8 Logic gates for equations (9.2) to (9.6).

are only actuated while the fixture actually engages them. During travel between limits the limit switch inputs will be lost. We must provide memory elements at each output. The *R-S* flip-flop of Sec. 8–6–1 will be adequate. Figure 9–9 shows the NEMA logic diagram including the input devices. This circuit will perform the same sequence as the relay logic circuit of Fig. 9–7.

9–4 Application of Solid State Logic to Sequencing Systems

We looked at the digital logic diagram in Fig. 9–9 to replace the relay ladder diagram of Fig. 9–7. Indeed, that is what has occurred in the evolution of sequential controls. Solid state digital logic has replaced many of the conventional relay control panels. The advantage of higher reliability, lower cost, smaller size, higher speed, increased

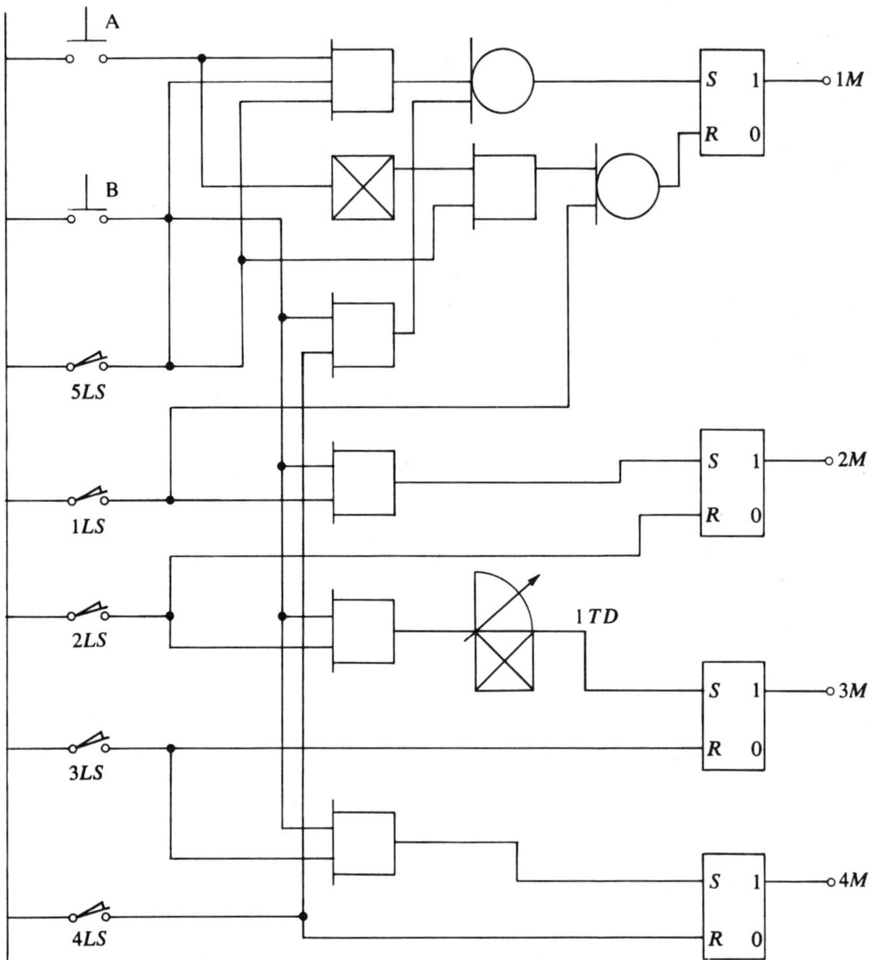

Fig. 9–9 NEMA logic diagram for operation of Table 9–1.

flexibility, and compatibility with computerized control has made solid state logic the obvious choice when compared with electromechanical relays.

Solid state logic components can be classified in four categories: input interfacing, logic gates, output interfacing, and accessory components. The design process involves converting your control requirements to Boolean equations, choosing from the proper components to perform the logical decision making, and selecting the proper interfacing devices to match your system or process to the solid state circuitry. We will look at the implementation of the control of Fig. 9–9 using solid state logic components of two major manufacturers—General Electric and Digital Equipment Company.

9–4–1 General Electric Transistorized Static Control

The G. E. line of solid state logic elements are transistorized, phenolic-base, plug-in modules (Fig. 9–10). The five basic logic functions are: AND, OR, NOT, MEMORY, and DELAY. Figure 9–11 shows the NEMA symbols and manufacturer's identification of the basic modules. The AND and OR functions are the same as we discussed in chapter 8. The NOT function is another name for the inverter. The retentive memory is a reset-set (*R-S*) flip-flop. The output, C, will go to the logical 1 state whenever a logic

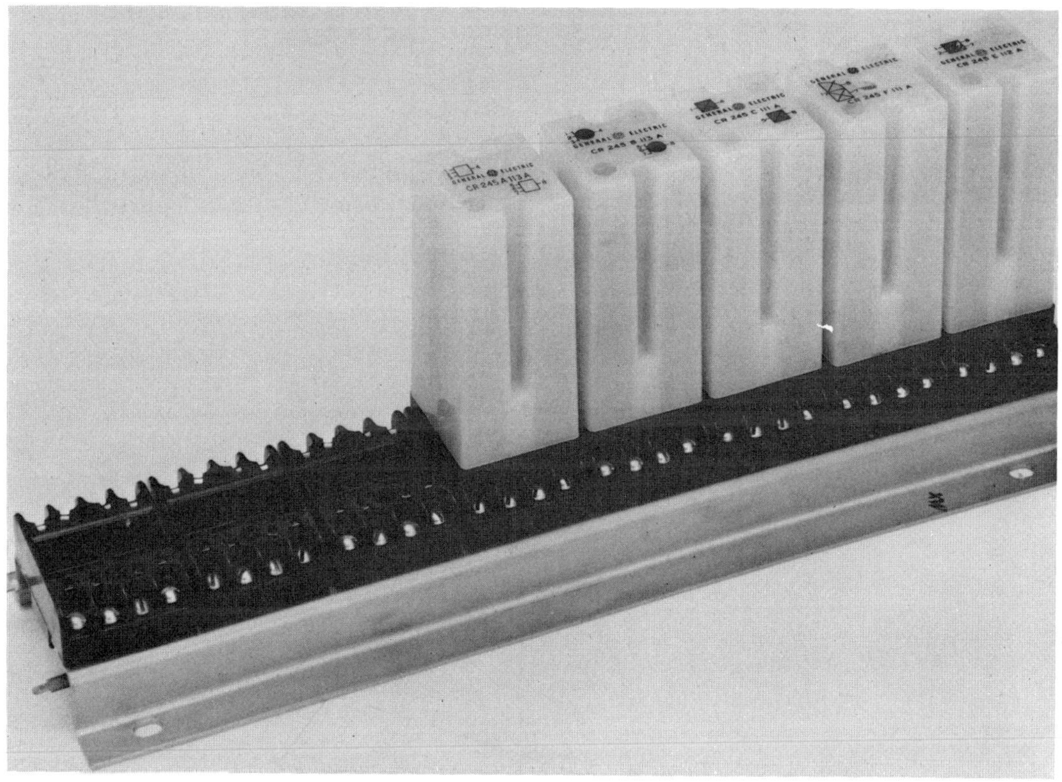

Courtesy of General Electric Company

Fig. 9–10 G.E. static control elements.

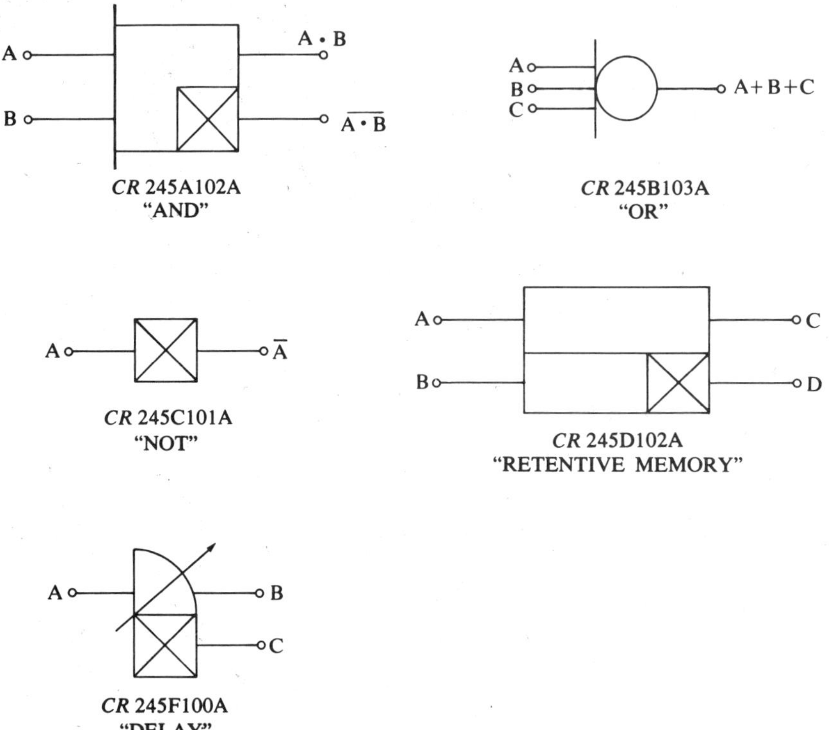

Fig. 9–11 G.E. static control basic functions.

1 input is applied at terminal A. The output will remain in the 1 state until a reset pulse is applied at terminal B. Logic 1 inputs are not allowed at both A and B at the same time. The D output is the complement of C, or $D = \overline{C}$. The electrical voltage levels used are 0 V for logic 1 and −4 V for logic 0. This is positive logic as we defined it in chapter 8. The time delay module is a basic UJT relaxation oscillator with gated inputs and outputs. Figure 9–12 is a simplified schematic of the time delay module. The output of the relaxation oscillator is at logic 1 when the input, A, is at logic 0. Then output B is at logic 0 and output C is at logic 1. When a 1 input is applied at A, the relaxation oscillator capacitor charges toward the emitter triggering voltage of the UJT. The negative-going voltage applied at the input of NOT-AND gate #1 causes the flip-flop outputs to change. B goes to logic 1 and C goes to logic 0. The delay module is automatically reset when the 1 input at A is

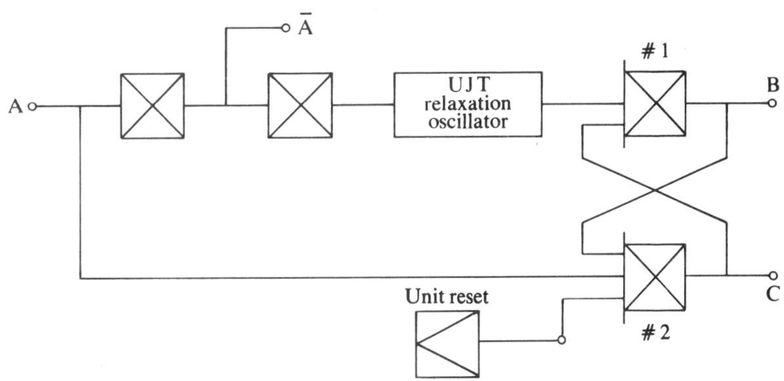

Courtesy of General Electric Company

Fig. 9–12 Simplified schematic of digital time delay module.

removed. The unit RESET is an accessory module used to delay the input to flip-flops when power is first applied.

The basic input interfacing modules are 125-V dc original input, 24-V dc original input, and 115-V ac original input. The original inputs act as signal conditioners between the actuating devices and the solid state logic circuitry. Figure 9–13 shows commercial 125-V input conditioners.

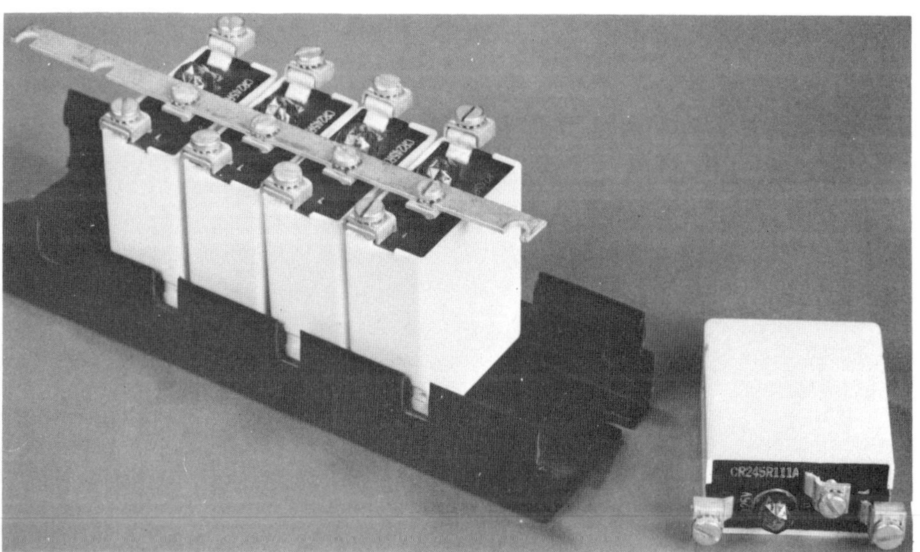

Courtesy of General Electric Company

Fig. 9–13 Commercial input source voltage conditioners.

The function of the original input modules is to convert the input voltage to logic level, suppress transients, and reduce contact bounce noise. Figure 9–14 shows the NEMA symbols for ac and dc inputs. The output of logic modules must be amplified to provide sufficient drive for most loads such as relays, solenoids, and motor starters. Output amplifiers are used to boost the output of logic modules. A wide variety of ac and dc amplifiers is available within the load current range of 1 to 10 A. Figure 9–15 shows a typical amplifier load configuration.

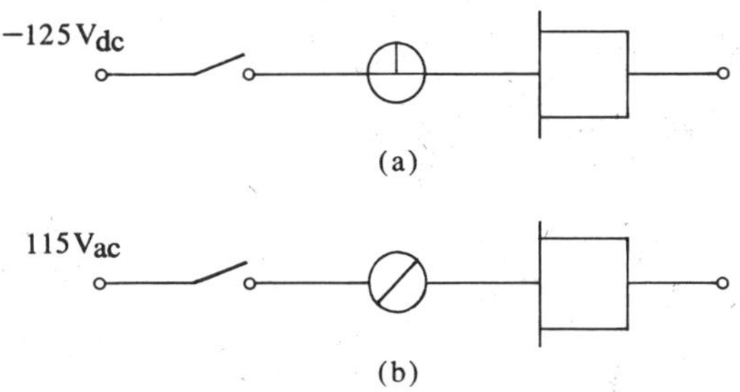

Fig. 9–14 Original inputs in digital logic circuits.

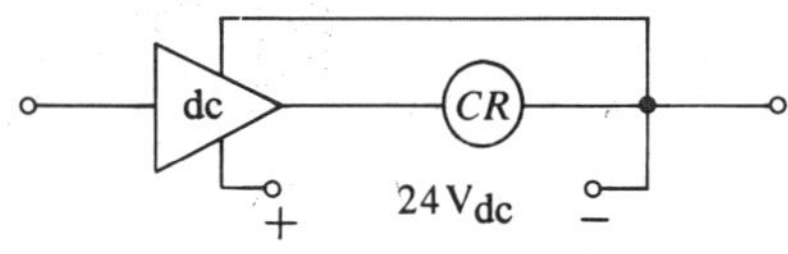

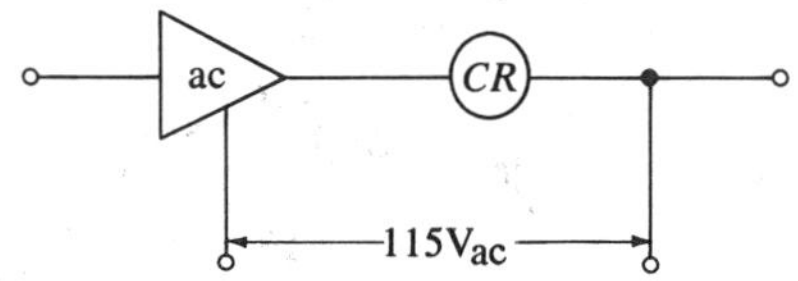

Fig. 9–15 AC and DC output amplifiers driving a control relay.

Let's look at the logic diagram to implement the sequential control of Fig. 9–7 using these basic logic modules. The Boolean equations are:

$$1M = A \cdot B \cdot 5LS + B \cdot 4LS \tag{9.7}$$

$$2M = B \cdot 1LS \tag{9.8}$$

$$3M = 1TD \cdot 2LS \cdot B \tag{9.9}$$

$$4M = B \cdot 3LS \tag{9.10}$$

$$1TD = B \cdot 2LS \tag{9.11}$$

Figure 9–16 shows the logic diagram. Notice that the original input signal conditioners are placed between the mechanical switches and the logic inputs. The signal conditioners should be placed as close as possible to the physical logic input with minimum lead length. Retentive memory elements are used at each limit switch to retain the logic signal after the fixture moves away from the switch position. AC amplifiers are used to match the logic elements with the motor control relays. The unit reset module establishes the proper state of the memories and time delay at turn on. One of the problems with memory elements in sequential circuits is that they must be reset when the memorized information is no longer necessary. The Boolean equations for actuation of the motors do not reflect the reset conditions. The sequence chart of Table 9–1 can be used to determine when memory elements must be reset but there is no simple general procedure.

After the basic logic diagram is drawn and checked, we start looking for ways to eliminate gates to reduce the cost and complexity of control.

9–4–2 Digital Equipment Company K Series Logic

Another commercial line of transistorized digital logic elements designed for industrial applications is the D.E.C. K-series logic. The logic levels are 0 V at logic 0 and +5 V at logic 1. The basic logic elements are a multiple input combination AND-OR gate, input signal converters, output signal converters, an *R-S* flip-flop, a time delay, and a power source module. Figure 9–17 shows the NEMA logic symbols and manufacturer's designation. The dotted lines on the gates indicate that other inputs may be connected through the use of a *gate expander*. This provides a versatile and powerful logic module. Figure 9–18 shows some of the expansions possible using the basic AND-OR

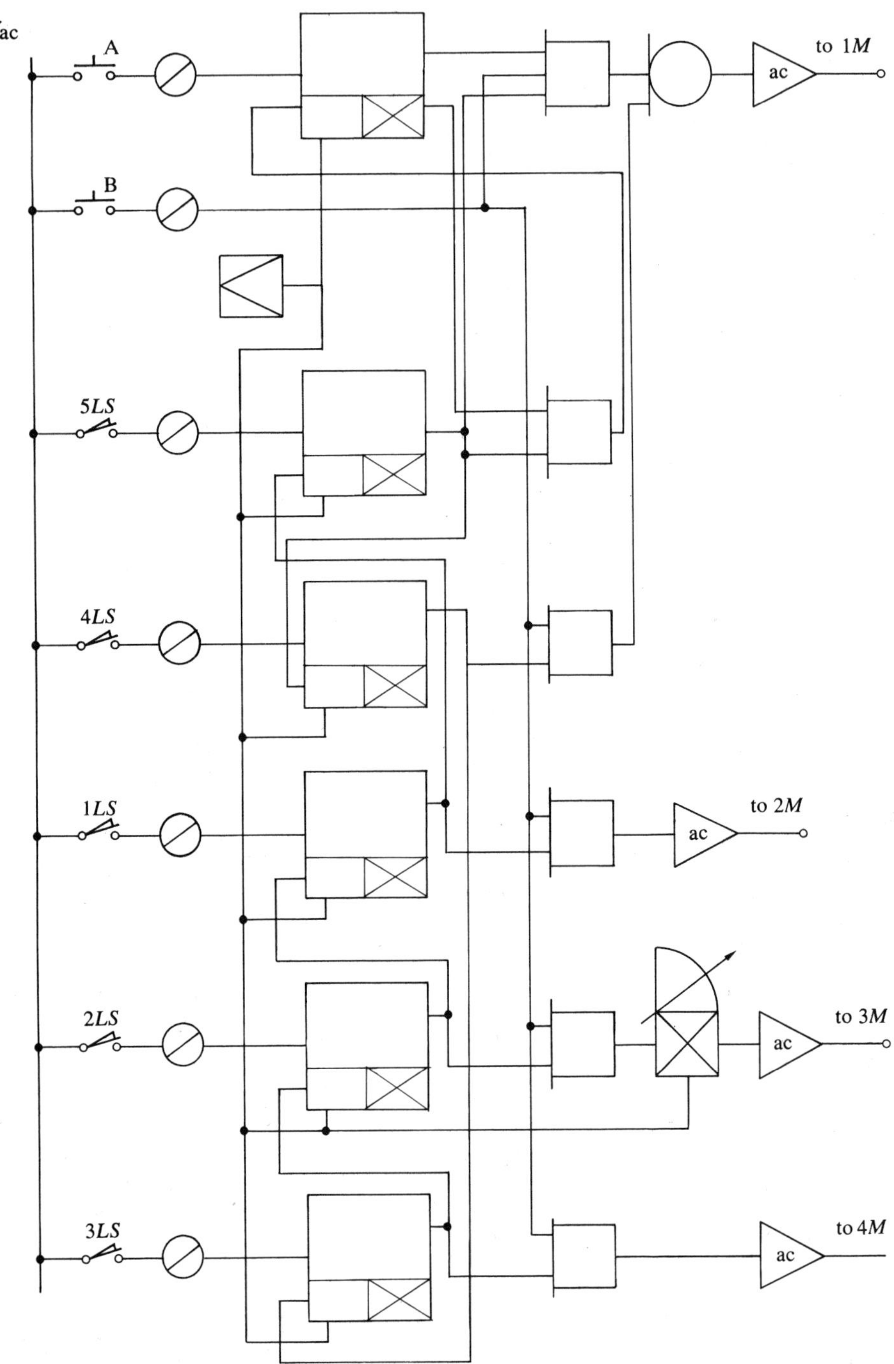

Fig. 9–16 Basic static control elements used to implement control of Fig. 9–7.

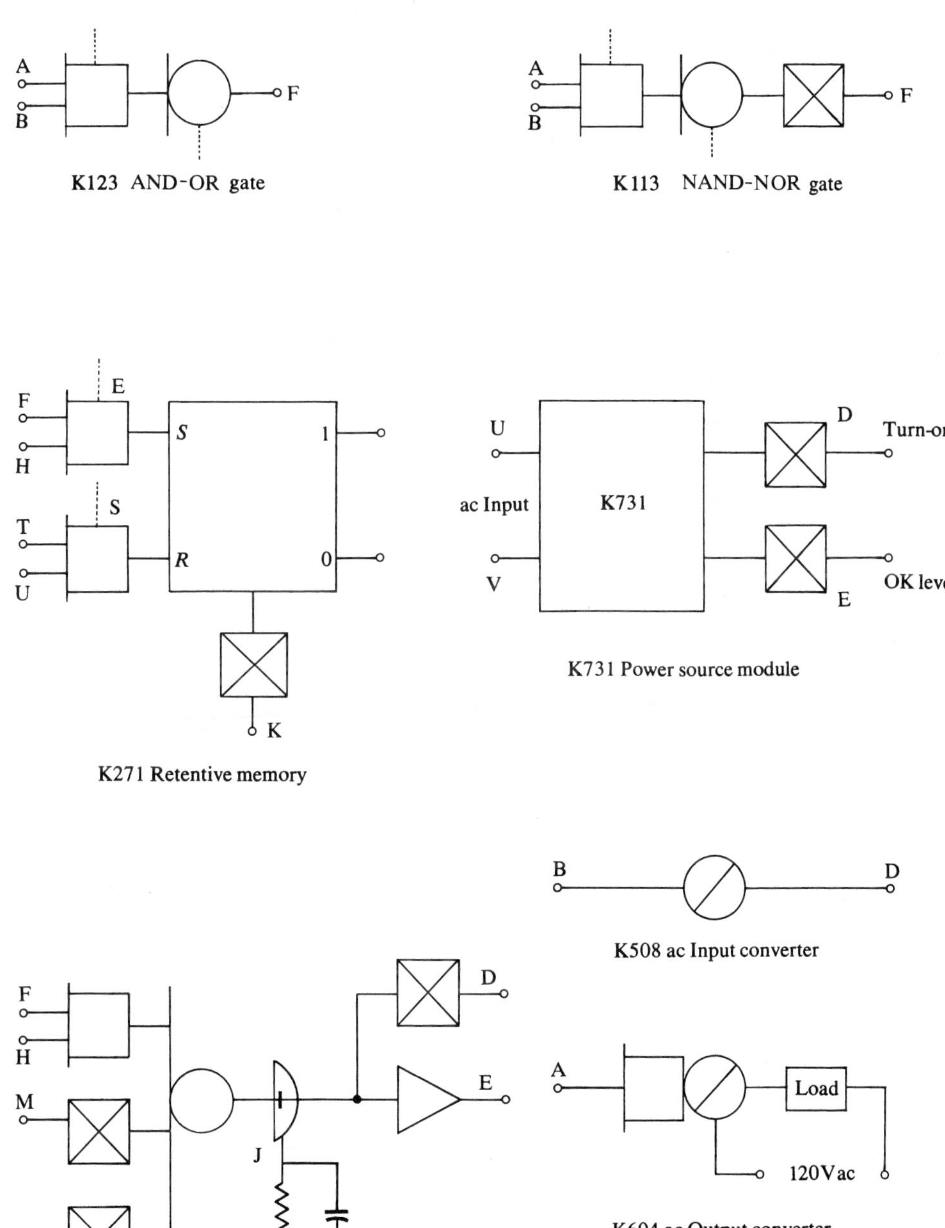

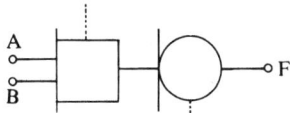

Fig. 9–17 Basic logic modules of D.E.C. K series.

gate. The retentive memory also has multiple inputs with expandable terminal connections. Pin K of the memory module is held low by output E of the power source module K731 to retain the proper output state in case of temporary power failure. The D.E.C. logic modules can be used to implement the control of Fig. 9–7. The NEMA symbolic logic diagram for the control is shown in Fig. 9–19.

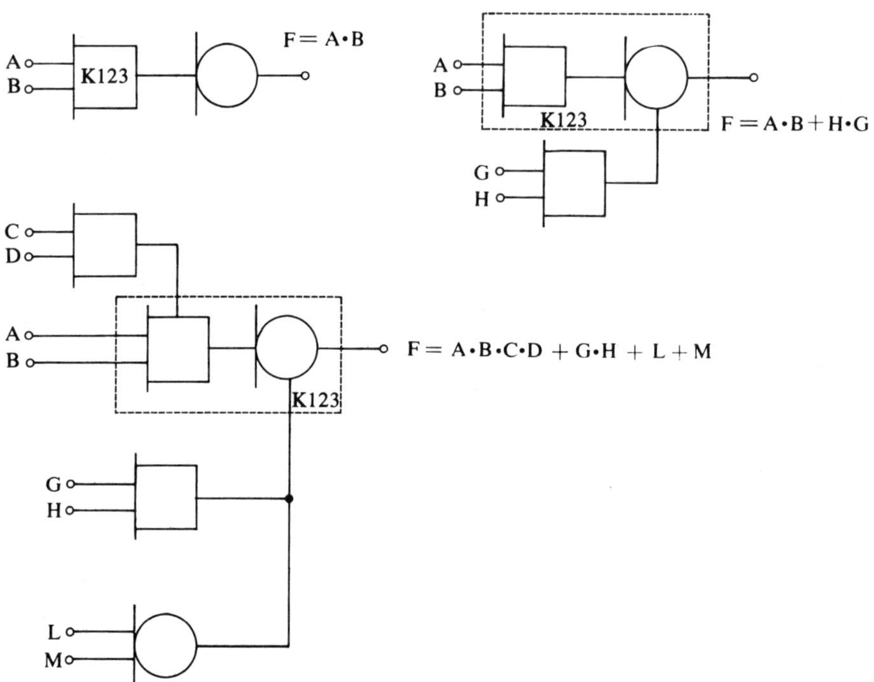

Fig. 9–18 Expansion of the basic AND-OR logic function.

9–5 The Programmable Controller

Digital controls made of solid state logic elements are designed and built for a specific operation or process. While the construction is simpler than an equivalent relay panel, the cost of making changes in the circuit is still quite high. The programmable controller is an assembly of solid state logic elements packaged so that the interconnections between elements can be conveniently changed by a stored program. The design of most programmable controllers is very similar to that of a small computer but there are some distinct differences. The programmable controller is designed to make logical decisions and provide output signals. It usually has no arithmetic calculating capability. The programming language of the controller is very simple, unsophisticated, and oriented toward the industrial designer. Much of the terminology is taken from the jargon of relay control designers. The interfacing of the programmable controller is designed to match the typical machine control components such as relays, motor starters, push buttons, limit switches, and solenoids. The programmable controller bridges the gap between solid state logic modules and computers.

There are several controllers on the market but the basic organization is essentially the same. We will look at the characteristics of the Modicon 084. It will provide considerable insight into the organization of programmable logic controllers.

9–5–1 The Modicon 084 Controller

The Modicon 084 controller has four major sections: the central processor, memory, registers, and signal conditioning equipment. Figure 9–20(a) shows the controller mounted on a vertical rack. Figure 9–20(b) shows how the controller is wired into the panel. The central processor is the brain of the controller. It executes the stored programs designed to perform control functions. The memory is a 16-bit, 1024-word read/write magnetic core structure. It contains the control instructions for the central processor and the stored program of the machine or process. The registers provide the interfacing between the central processor/memory section and the machine actuators. The signal conditioners convert the input signals to logic level and provide the isolation between the low voltage logic circuits and the electrical environment of the machine or process. Figure 9–21 is a block diagram of the controller organization.

9–5–2 Designing Sequential Controls With the Programmable Controller

Designing control logic with the 084 controller is similar to designing conventional relay control logic. The ladder

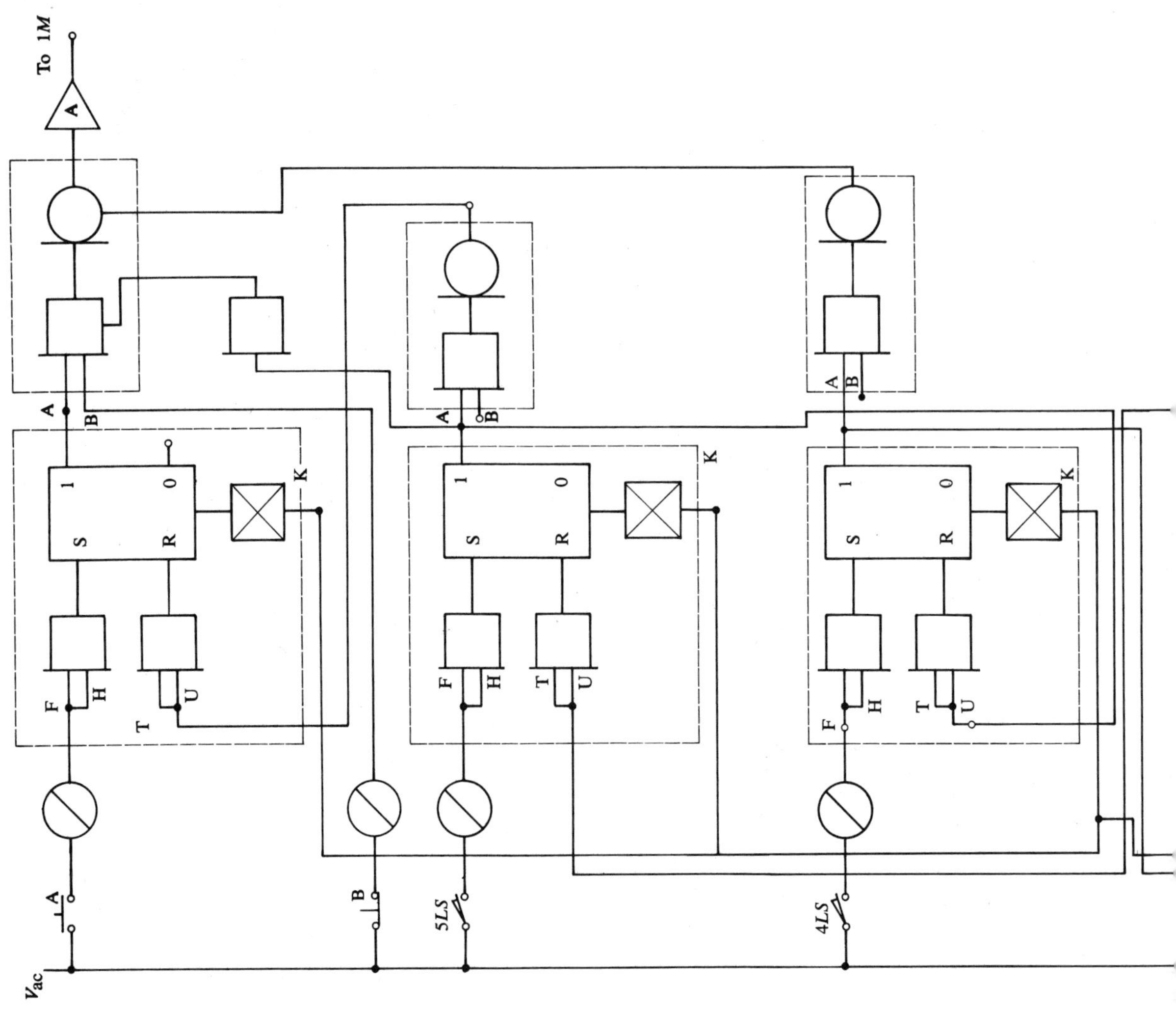
To 1M
A
B
1
0
S
R
K
F
H
T
U
5LS
4LS
V_{ac}

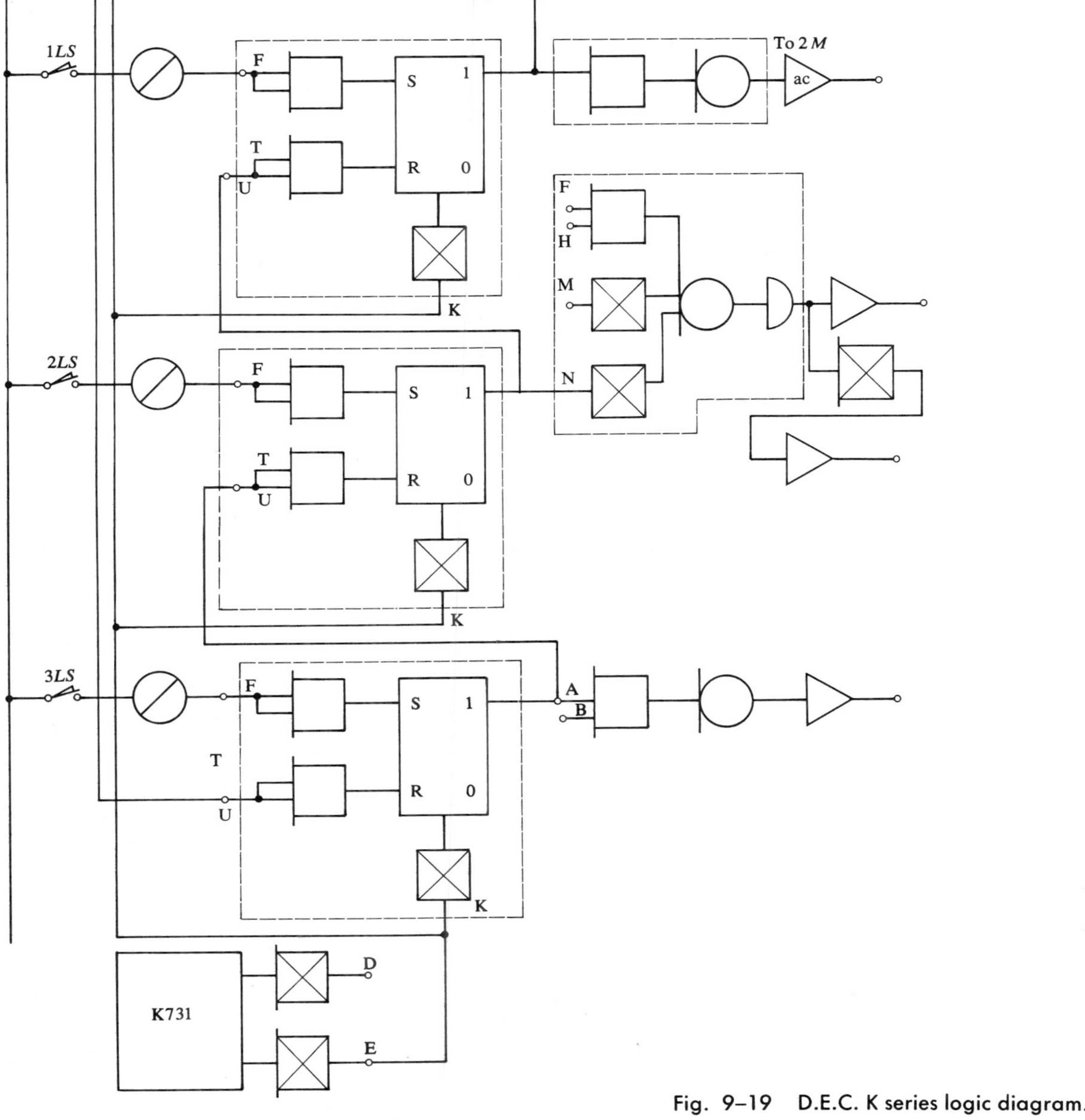

Fig. 9–19 D.E.C. K series logic diagram.

Courtesy of Modicon Corporation

Fig. 9–20 (a) The Modicon 084, a programmable controller,

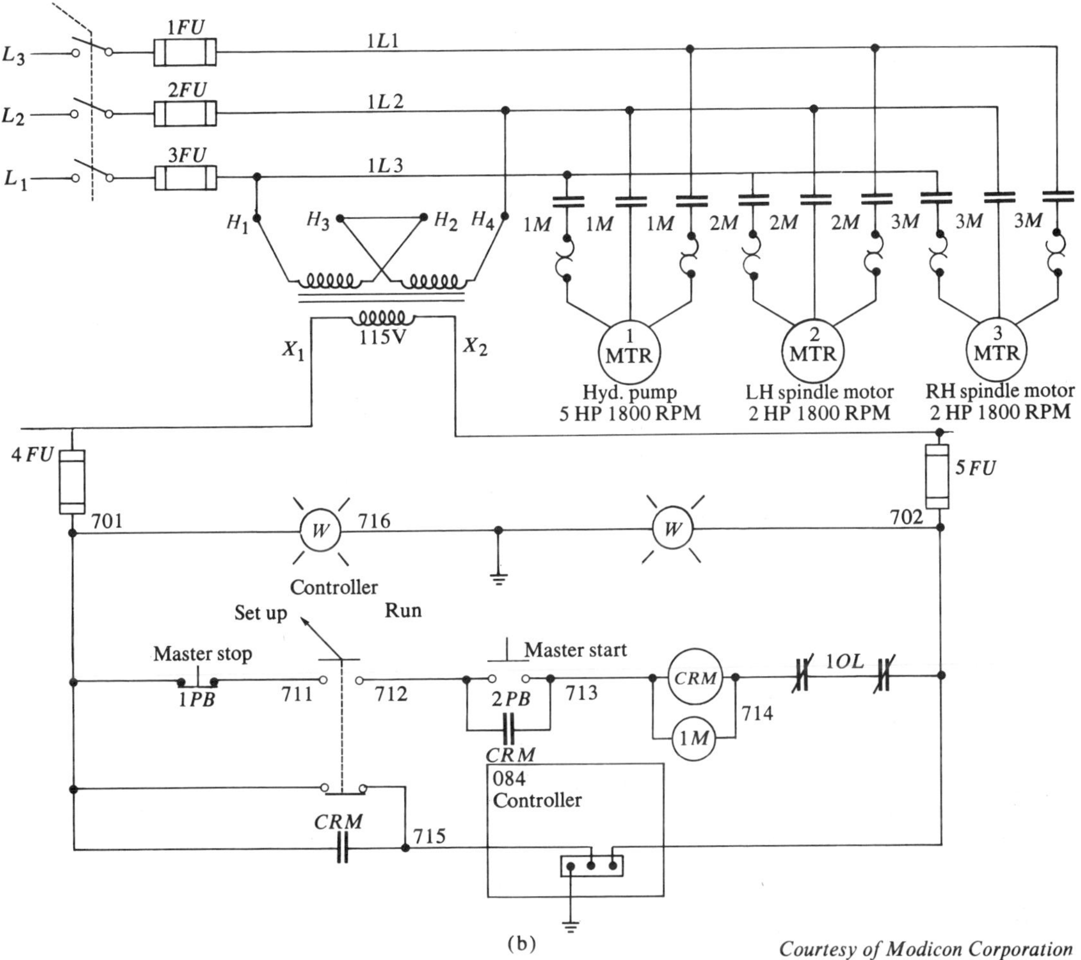

(b)

Courtesy of Modicon Corporation

Fig. 9–20 (continued): (b) Modicon controller wiring.

diagram is first drawn using the procedures of Sec. 9–2. The same format of line numbering and contact identification is used to program the controller. Figure 9–22 shows a typical ladder diagram prepared for programming the 084 controller. Relay contacts are placed in four columns, A, B, C, and D. The control relay coils are in the normal position, on the right side of the ladder diagram. One difference between the ladder diagram and the 084 format is that the line number is the same as the relay coil number. In Fig. 9–22, we have shown lines 1, 2, and 8.

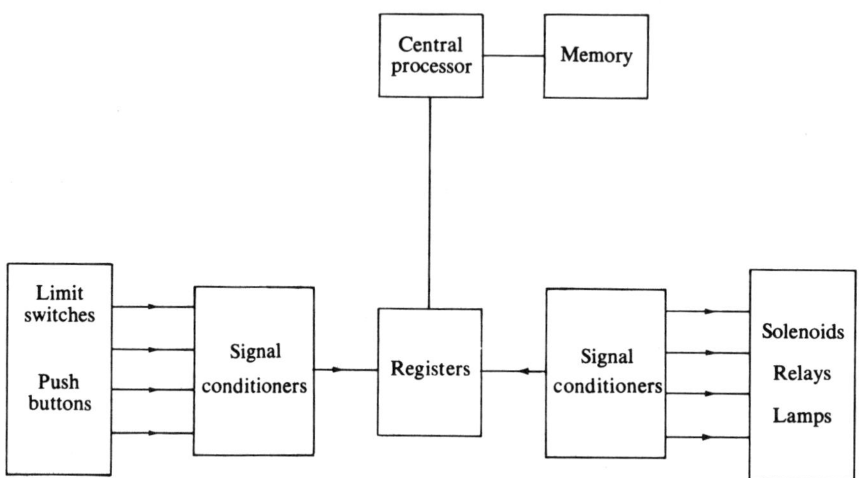

Fig. 9–21 Block diagram of the programmable controller.

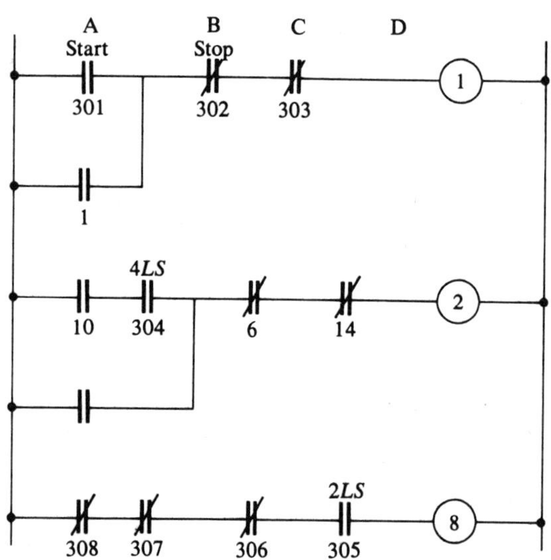

Fig. 9–22 Programming format for controller.

Four types of contacts are allowed:

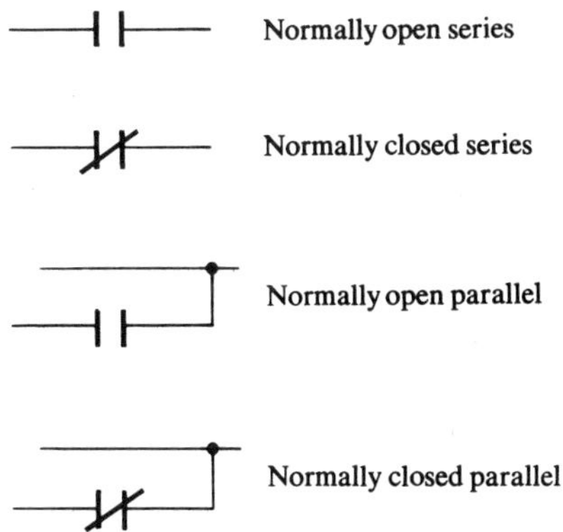

Parallel contacts automatically connect back to the left side of the ladder diagram. The control relays are numbered from 1 to 255. Inputs such as limit switches, push buttons, and pressure switches are numbered from 301 to 555.

When power is turned on, the controller starts at line one and compares the actual state of contacts in each line with the logic diagram. The control relays are energized or de-energized according to the programmed sequence. The controller runs through all 255 relay lines in 40 ms. At the end of a complete search of all 255 logic lines, the controller starts at line one again and repeats the cycle. For example, if the controller found contacts 301, 302, 303 all closed, output 1 would be energized.

9–5–3 Programming the 084 Controller

The 084 controller is programmed through the model PC-45 Programming Panel of Fig. 9–23. The panel is a portable unit that plugs into the controller allowing the designer to enter the program for a specific process or machine into the memory section of the controller. The program is entered into memory one line at a time. The steps we would go through to enter a line of the ladder diagram using the portable programming panel are:

(1) Set line number switches (Item A, Fig. 9–23); the line to be programmed is selected.
(2) Select line type (Item B, Fig. 9–23). The choices of line type are logic, timing, or counting.

(3) Select element push button (Item C, Fig. 9–23). The column in which the element is located is selected, either A, B, C, or D.

(4) Select element type (Item D, Fig. 9–23). The contact type is specified as N.C. series, N.O. series, N.O. parallel, or N.C. parallel.

(5) Select panel mode (Item J, Fig. 9–23). The panel will operate in the DATA mode for changing the program and MONITOR mode for troubleshooting.

(6) Display the reference number (Item F, Fig. 9–23). The reference number is the input or line number that controls the contact in the selected line and column.

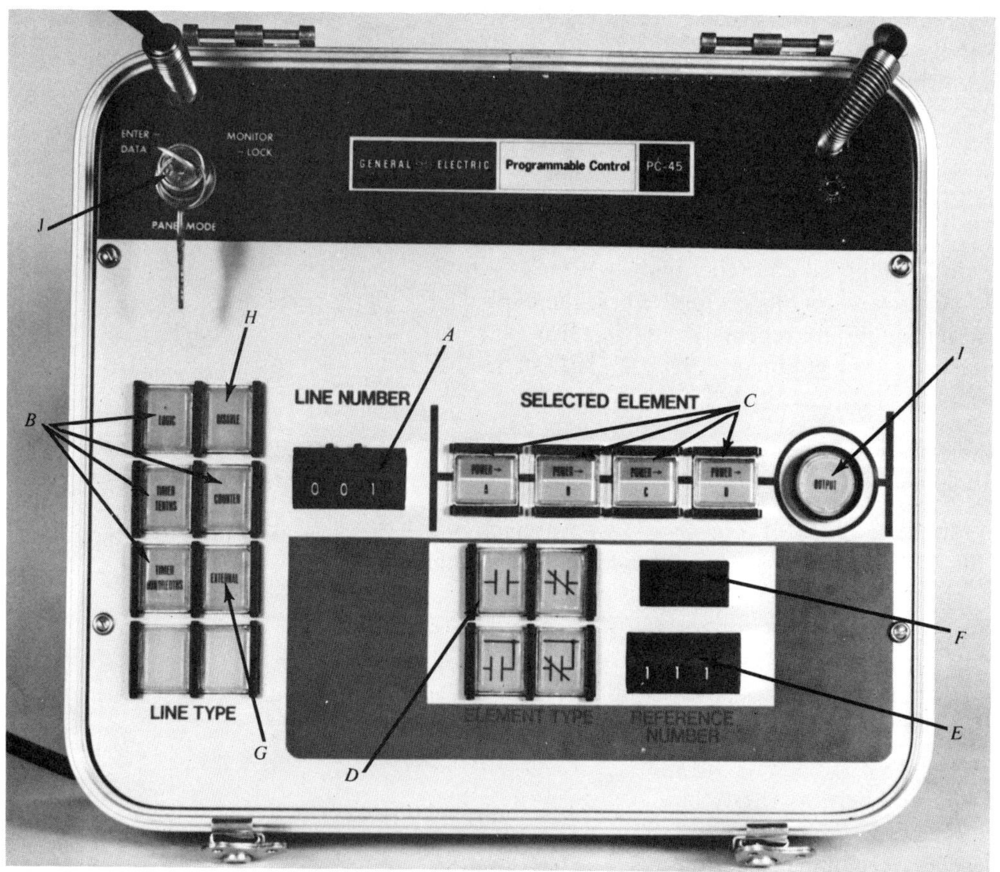

Courtesy of Modicon Corporation

Fig. 9–23 PC-45 control panel.

Let's look at the entry of one line of a control sequence shown in Fig. 9–24:

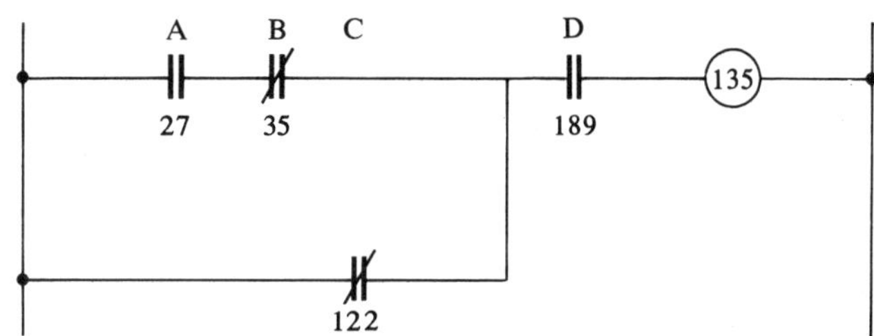

Courtesy of Modicon Corporation

Fig. 9–24 Logic line for entry into 084 controller.

(a) Set line number switches to 135
(b) Press logic push button
(c) Press "A" element push button
(d) Set REFERENCE NUMBER switches to 27
(e) Press normally open ELEMENT TYPE push button
(f) Press "B" element push button
(g) Set REFERENCE NUMBER switches to 35
(h) Press normally closed ELEMENT TYPE push button
(i) Press "C" element push button
(j) Set REFERENCE NUMBER to 122
(k) Press parallel normally closed ELEMENT TYPE push button
(l) Press "D" element push button
(m) Set REFERENCE NUMBER switches to 189
(n) Press normally open ELEMENT TYPE push button

Line 135 is now entered. We would proceed to the next line and repeat the entry process. After the program is entered, input and output devices are connected to the signal conditioning terminals and the controller is ready to run. Fig. 9–25 shows a typical connection diagram of input and output devices to the 084 signal conditioning cards. The programming panel can be used to troubleshoot the control before the machine is connected. Some programmable controllers use a compatible computer to simulate the machine actuation for trial runs and troubleshooting.

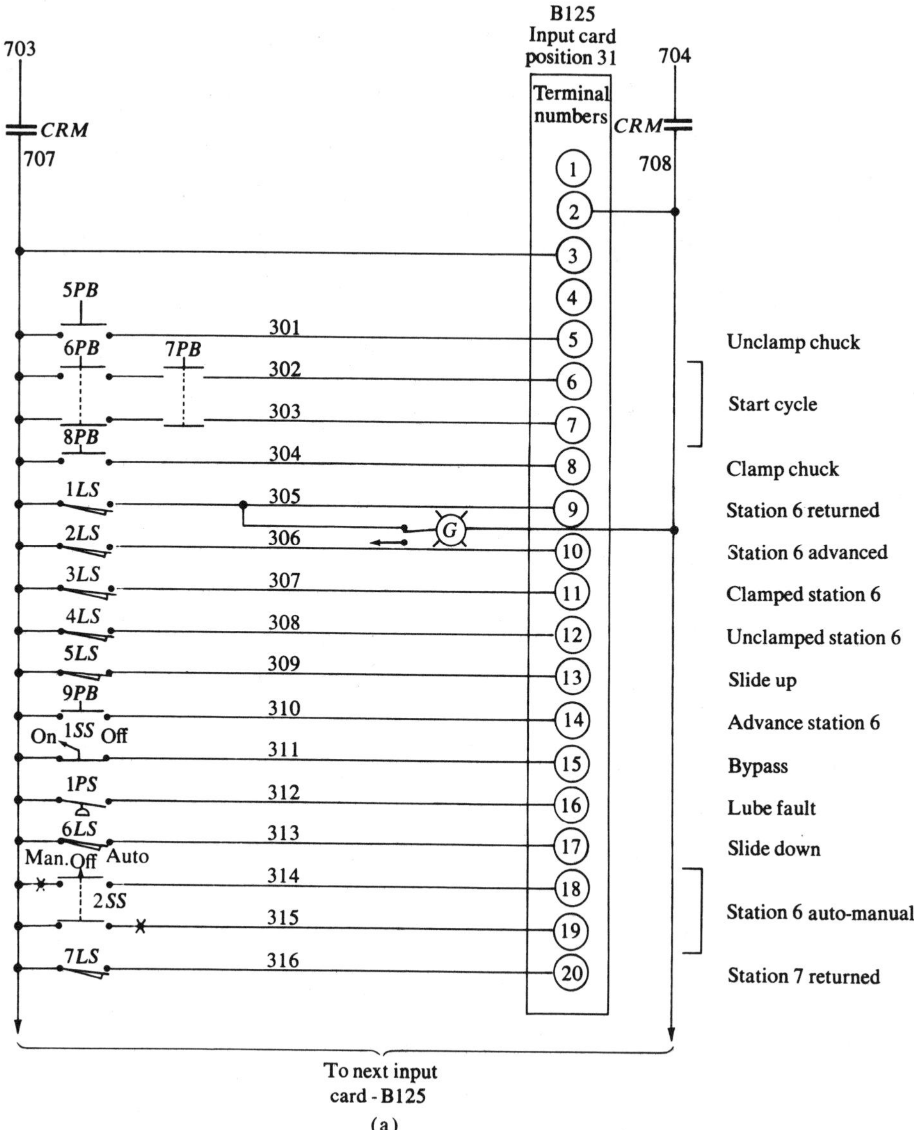

(a)

Courtesy of Modicon Corporation

Fig. 9–25 (a) Typical connection diagram of input devices using the 084 controller,

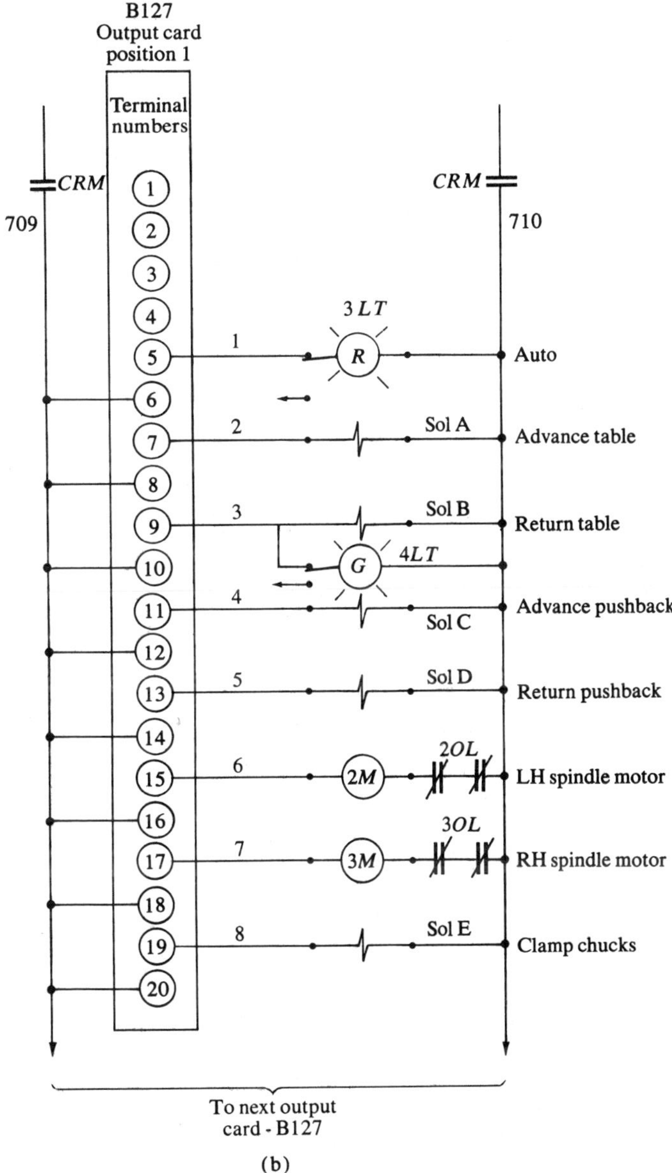

(b)

Courtesy of Modicon Corporation

Fig. 9–25 (continued): (b) typical connection diagram of output devices using the 084 controller.

9–6 Summary

The digital concepts of the previous chapter were applied to industrial electronics circuits and systems in this chapter. The relay ladder diagram was demonstrated as a tool for implementing digital functions. Simple methods for designing sequential systems were discussed mainly to provide the conceptual foundation on which more formal design discipline can be developed. Two commercial lines of digital hardware were introduced as well as a programmable controller. You will want to refer to professional journals from time to time to keep abreast of current hardware developments. The examples we have used will help establish a reference.

EXERCISES

1. What is the important difference between sequential controls and combinational controls?

2. Write the Boolean algebraic equation and draw the logic diagram for the relay contact diagrams of Fig. 9–26.

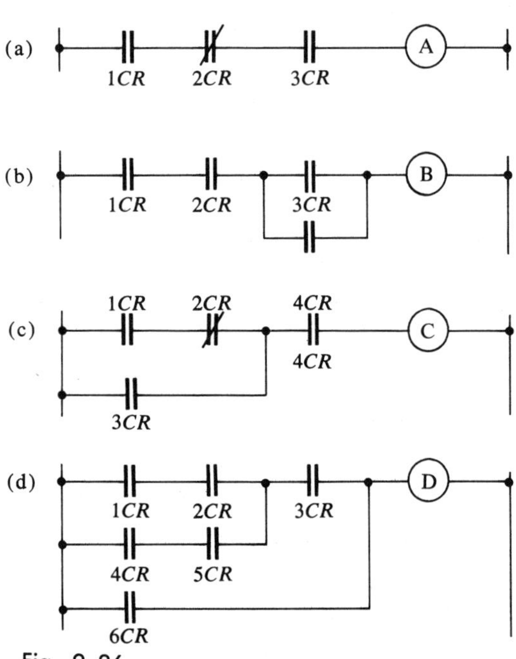

Fig. 9–26

3. The package of Fig. 9–27 is powered by the motor M. When the start button is pushed, the package should travel to the position of limit switch #1 and stop. The relay ladder diagram of the control circuit is shown in Fig. 9–27.
   (a) Analyze the ladder diagram at each step of the operation.
   (b) List each step of the sequence.
   (c) Draw the logic diagram to replace the relay control with an electronic digital control.

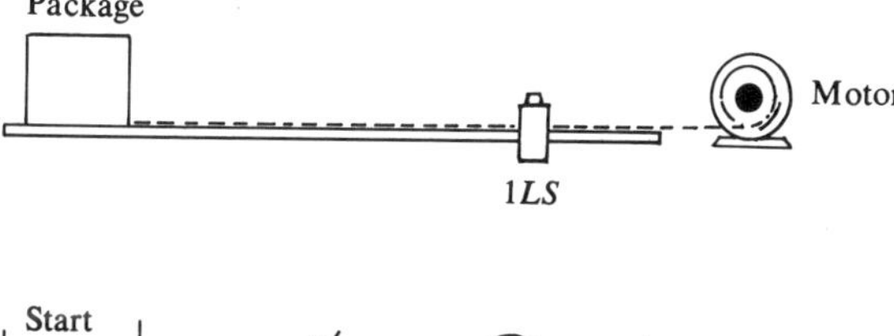

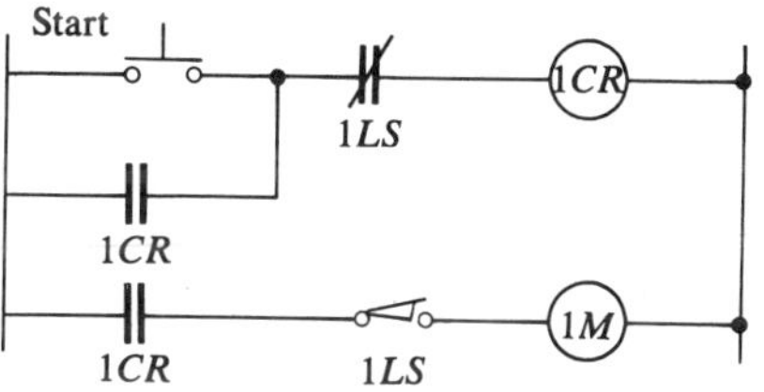

Fig. 9–27

4. An electric lamp must be operated from either of two remote locations. It is desired to be able to turn the lamp on or off from either location. Assuming that two multicontact mechanical switches are used:
   (a) Construct the truth table describing the control circuit to be used.
   (b) Write the Boolean equation from the truth table for the lamp "on" conditions.
   (c) Draw the connection diagram showing all switch contacts used.

5. Design the electronic control circuit to replace the mechanical control of problem 4. Use the G.E. logic components.

6. Design the electronic control circuit to replace the mechanical control of problem 4 using the K-series logic components.

7. An electric lamp must be operated from either of three remote locations. It is desired to be able to turn the lamp on or off from either location. Using multicontact mechanical switches:

(a) Construct the truth table describing the control circuit to be used.
(b) Write the Boolean equation from the truth table for the lamp "on" conditions.
(c) Draw the connection diagram showing all switch contacts used.

8. Design the electronic control circuit to replace the mechanical control of problem 7 using the G.E. logic components.

9. Design the electronic control circuit to replace the mechanical control of problem 7 using the K-series logic components.

10. Two different types of petroleum are stored in tank #1 and tank #2, and an additive in tank #3. When tank #1 solenoid valve is actuated (valve open), tank #2 solenoid must be unactuated (valve closed). Additive from tank #3 must be mixed with petroleum from either tank #1 or tank #2 but not with both at the same time. An electronic control circuit for this operation is shown in Fig. 9–28. Make up the sequence chart for the operation of drawing petroleum from tanks #1 and #2.

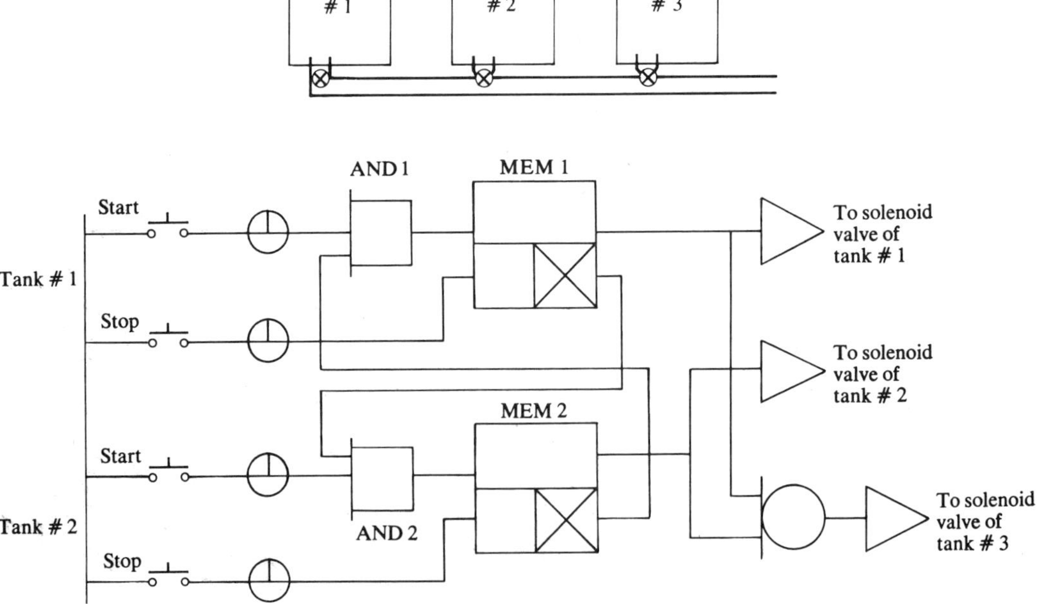

Courtesy of General Electric Company

Fig. 9–28

11. The boy and girl of Fig. 9–29 are both heavy enough to keep their platforms from actuating the limit switches. If either leaves his platform, the spring will raise the platform and actuate the limit switch. A digital circuit is required to turn on lamp A if the boy leaves his platform first, turn on lamp B if the girl leaves her platform first, and turn on both lamps at the end of two minutes. If the boy leaves first, then the girl must be allowed to leave without her lamp turning on.

12. How is a programmable controller different from a mini computer?

13. Why is a signal conditioner required between mechanical actuators and electronic gates?

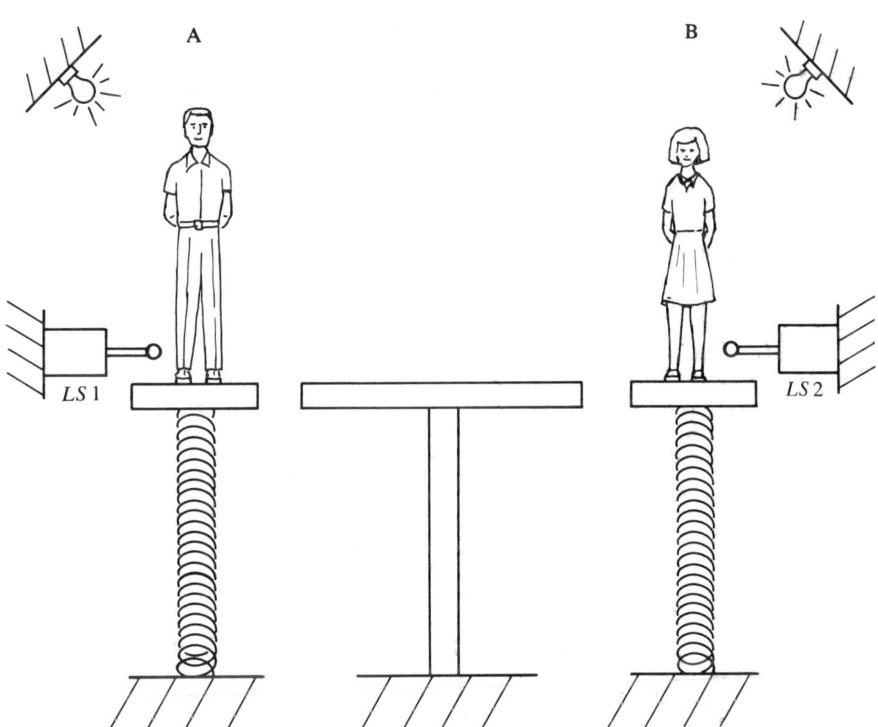

Fig. 9–29

10

Electronic Control of Motors

10–1 Introduction

Many times the industrial electronics engineer is called upon to design or select an electronic control for the motor that provides the drive for a machine or device. Quite often the customer knows what the drive system requirements are but is unsure what kind of motor or electronic control is needed. The electronics engineer sometimes feels that someone else should select the motor and allow him to provide the electronic control, but it usually doesn't work out that way. The electronics designer often bears the responsibility for selecting the motor and providing the control.

Some basic questions must be answered before much progress is made, such as:

(a) What horsepower is required?
(b) What speed or speed control must be provided?
(c) What torque is required and how does it vary with speed?

(d) What are the control requirements in terms of speed variation, sequencing, direction of rotation, and torque?

Given the answers to these questions, we must then match the customer's requirements with the proper type and size motor and electronic control. In this chapter we will review briefly the types of motors available and their characteristics as well as the design of electronic controls.

10–2 Motors and Characteristics

Successful application of electric motors depends on matching the performance characteristics of the motor with the load requirements of the machine or device to be driven. Let's look at some of the types of motors available and typical performance characteristics.

Electric motors are classified first according to the horsepower rating. The three general classes are: miniature or subfractional horsepower, fractional horsepower, and integral horsepower. Miniature motors are usually rated in millihorsepower or in ounces of torque output; fractional horsepower motors are rated at less than one horsepower such as 1/6, 1/4, 1/2, or 3/4 horsepower; integral horsepower motors have rated outputs of one horsepower or more. Large integral horsepower motors are rated as high as 10,000 horsepower. The horsepower is a unit of measure of rate of work. One horsepower is equivalent to lifting 33,000 pounds to a height of one foot in one minute. The equation for calculating horsepower is

$$\text{hp} = \frac{\text{torque} \times \text{speed}}{5250} = \frac{T \times N}{5250} \tag{10.1}$$

where T = torque in foot-pounds; N = speed in revolutions per minute; 5250 = constant of conversion.

10–2–1 The Split-Phase Induction Motor

The split-phase induction motor operates on single-phase ac line voltage. It is usually rated at fractional horsepower output. The split-phase motor requires an auxiliary stator winding to start the motor action when operating from single-phase line voltage. Figure 10–1(a) shows the winding connections. When the motor reaches 75 to 80% of its final speed the centrifugal switch opens and the motor operates on its main stator winding under continuous duty.

Figure 10–1(b) shows the speed-torque characteristic curves for a typical split-phase induction motor. It has relatively good or high starting torque. Speed is fairly constant over the operating region dropping off about 4 to 6% as the load

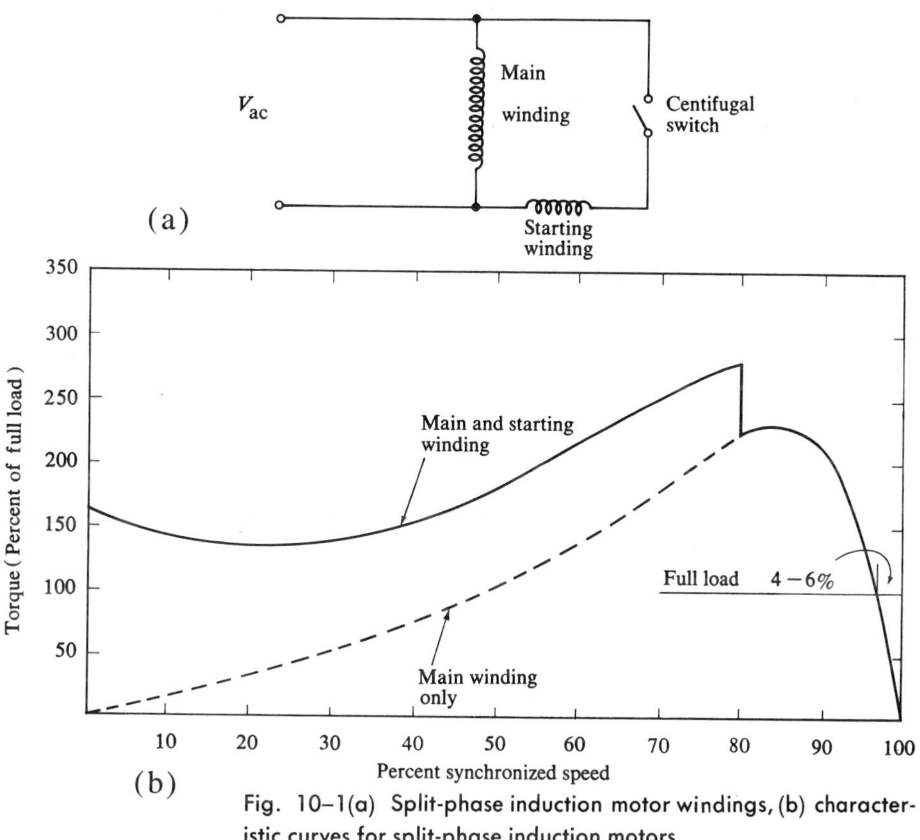

Fig. 10–1(a) Split-phase induction motor windings, (b) characteristic curves for split-phase induction motors.

increases. Torque is normalized to the rated full load external torque that the motor can deliver. Figure 10–2 shows the definitions of torque specifications that motor manufacturers normally provide. Figure 10–3 shows the characteristic curves for a commercial split-phase induction motor.

10–2–2 Capacitor Motors

The performance of the split-phase ac motor can be improved by using a capacitor in the auxiliary winding. There are three types of capacitor motors: the capacitor-start

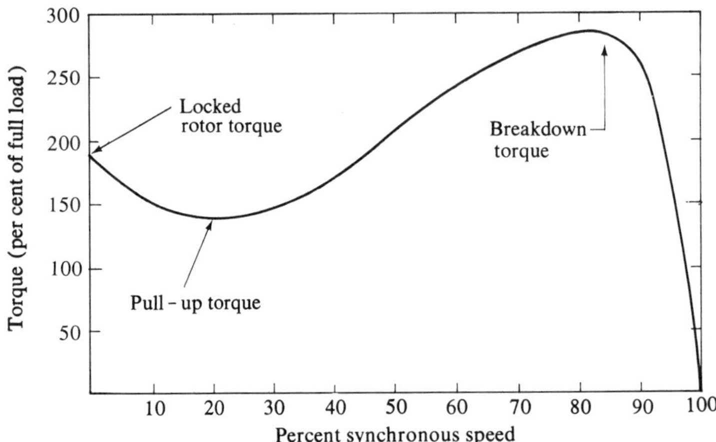

Fig. 10–2 Speed torque definitions for induction motors.

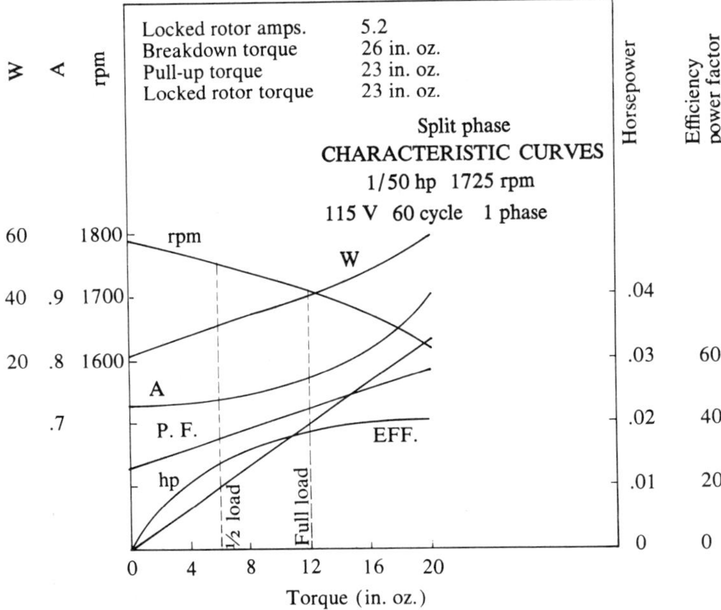

Courtesy of Bodine Electric Company

Fig. 10–3 Characteristic curves for fractional horsepower split-phase motor.

motor, which uses a capacitor in the auxiliary winding only during starting to increase the starting torque; the permanent-split capacitor motor, which leaves the capacitor in the auxiliary winding during starting and running; and the two-value capacitor motor, which uses a different capacitor in the auxiliary winding during starting than during running. In the permanent-split capacitor and two-value capacitor motors, the starting capacitor increases the starting torque while the running capacitor increases the maximum torque and efficiency allowing the motor to run at a cooler temperature. This type motor is used in fractional and low-integral horsepower applications. Capacitor motors are used in business machines, fans, blowers, desk calculators, and other applications where low starting torque and frequent operation are required.

10–2–3 The Shaded Pole Motor

The simplest, lowest cost single-phase ac motor is the shaded pole motor. The horsepower rating is usually in the lower fractional and subfractional range. The starting torque, running torque, and efficiency are low. The shaded pole motor sometimes requires ventilating air to keep it cool. Figure 10–4 shows the characteristic curves of a commercial shaded pole motor. The 1/200 horsepower rating is common for this type motor. The speed is relatively constant but not as good as the split-phase motor. The dominant characteristic of the shaded pole motor is its low cost.

10–2–4 The Universal Motor

The universal motor is designed to operate on ac or dc voltage. The speed can be varied by controlling the applied voltage. The universal motor usually operates at very high speeds with good efficiency. The speed will vary considerably with external loading. This motor finds widespread use in many consumer appliances such as sewing machines, vacuum cleaners, mixers, blenders, garden tools, hand drills, and floor polishers.

10–2–5 The Polyphase Induction Motor

Fractional horsepower ac motors are also designed to operate on two-phase or three-phase line voltages. These motors are generally very efficient with high starting and running torque. Speed is relatively constant from no load to full load. Industrial machine tools, air compressors, and pumps are often driven by three-phase induction motors.

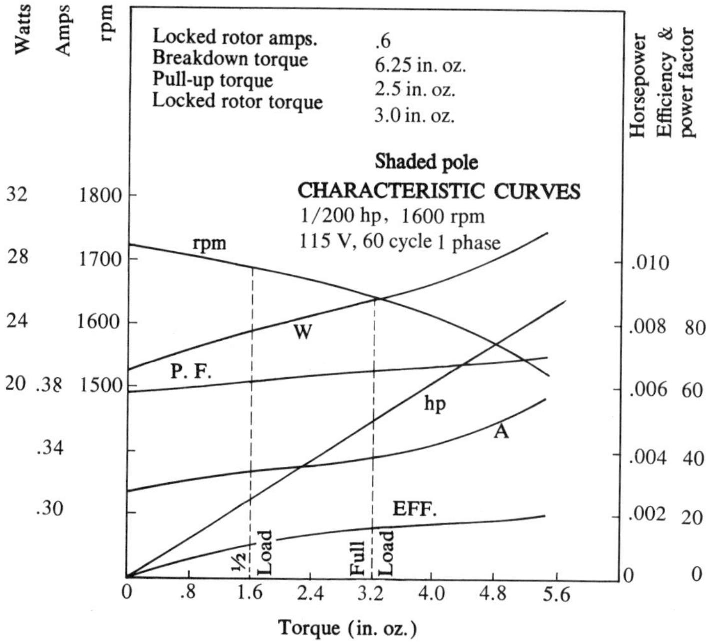

Courtesy of Bodine Electric Company

Fig. 10-4 Characteristic curves of commercial shaded pole motor.

10–2–6 The Synchronous Motor

The synchronous motor is used in fractional as well as larger integral horsepower motors. Three-phase ac voltage is used although single-phase synchronous motors are available. Split-phase, capacitor-start, and shaded pole single-phase synchronous motors have characteristics similar to induction motors of the same type. The speed of a synchronous motor is determined by the frequency of the line voltage. The speed remains constant regardless of the load applied. Three-phase synchronous motors are used to drive heavy industrial equipment requiring hundreds of horsepower.

The efficiency of the integral synchronous motor is higher than that of a comparable induction motor. One of the disadvantages of this motor is its low starting torque. The starting torque of many synchronous motors is less than 75% of rated full load torque. Synchronous motors are generally not used for heavy start-stop operation; they are ideal for continuous processing mills, compressors, or motor-generator sets.

10–2–7 The Series-Wound DC Motor

The series-wound dc motor has the field coil in series with the armature. The dc voltage is applied across the series combination as shown in Fig. 10–5(a). When we say dc voltage, we aren't talking about the same filtered, regulated voltage used to operate the electronic circuitry. The dc voltage to operate motors and other heavy industrial equipment is usually a rectified ac voltage (see Sec. 12–2). If low ripple is required, three-phase ac voltage is rectified to provide the dc excitation. The series-wound dc motor has the

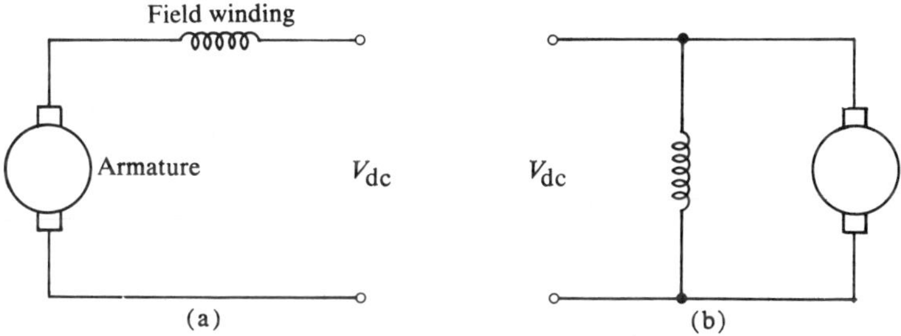

Fig. 10–5(a) Series-wound DC motor, (b) shunt-wound DC motor.

advantage of very high starting torque. Figure 10–6 shows typical characteristic curves for a series-wound dc motor. Notice the torque is nearly 300% of full load at low speeds. As the motor speed increases, the torque and horsepower output decrease. This is not a constant speed motor. The series-wound dc motor is ideal for applications where various torques and speeds are required, such as cranes, hoists, lifts, trolley cars, and railway cars. The ability to start slow with heavy loads, operate fast with light loads, brake quickly, and reverse makes the series-wound dc motor very popular.

10–2–8 The Shunt-Wound DC Motor

The shunt-wound dc motor has the field coil in parallel with the armature. The dc voltage is applied across both

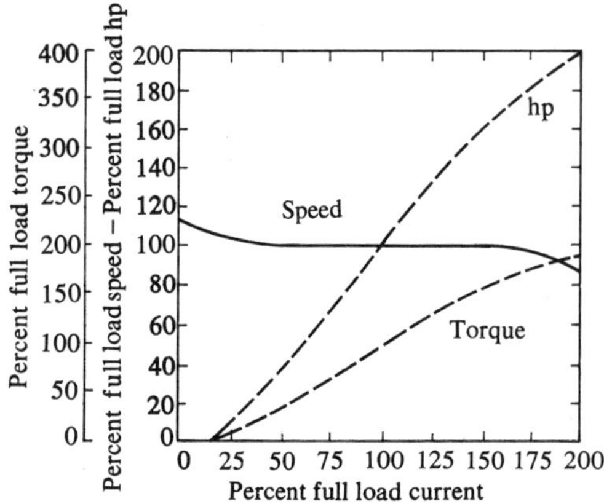

Source: *American Electrician's Handbook*, by Croft, Carr, and Watt. Copyright 1970, McGraw-Hill Book Company. Used with permission of McGraw-Hill.

Fig. 10–6 Typical characteristics for a shunt-wound DC motor.

elements at the same time as shown in Fig. 10–5(b). Sometimes the field is supplied from an entirely different dc voltage source. This has no effect on the motor operation but allows for control of field current or armature voltage. Unlike the series-wound motor, the speed of the shunt-wound dc motor remains relatively constant over the range of loads as shown in Fig. 10–6. Sometimes a combination series-wound, shunt-wound motor is used to get speed-torque characteristics between those of Fig. 10–6 and Fig. 10–7. Such a motor is called a dc compound motor. The shunt-wound dc motor has the advantage of controlling the speed by varying the field current or armature voltage. A wide range of speed control is possible—a 20-to-1 ratio between maximum and minimum speed is not uncommon. The shunt motor is most suitable for applications requiring a wide range of operating speeds with easy reversibility and good braking. Typical applications are rolling mills, strip welders, printing presses, elevators, and machine tools. If constant speed is required for a range of loading, the dc shunt motor might also be used.

10–2–9 The Digital Stepper Motor

The stepper motor is a synchronous motor designed to operate with a pulsed input voltage rather than a continuous ac voltage. The motor is designed to move the output shaft

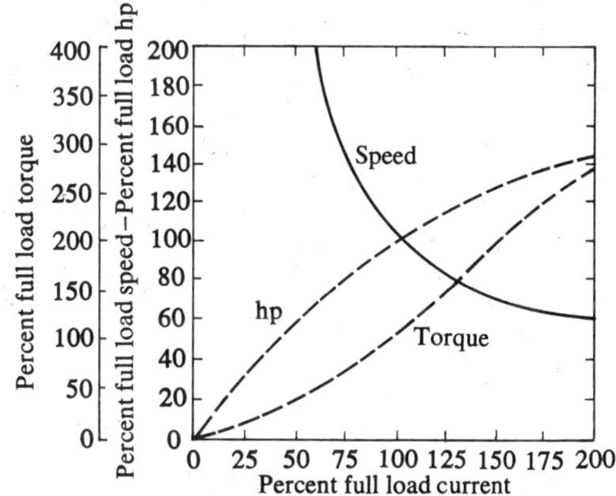

Source: *American Electrician's Handbook*, by Croft, Carr, and Watt. Copyright 1970, McGraw-Hill Book Company. Used with permission of McGraw-Hill.

Fig. 10–7 Typical characteristics for a series-wound DC motor.

a fixed number of degrees each time the proper pulsed voltage is applied. The polarity and sequence in which pulses must be applied to the motor windings are determined by the specific manufacturer's design. Figure 10–8 shows the connection diagram for a typical two-phase dc stepper motor. The sequence of pulses is demonstrated by the use

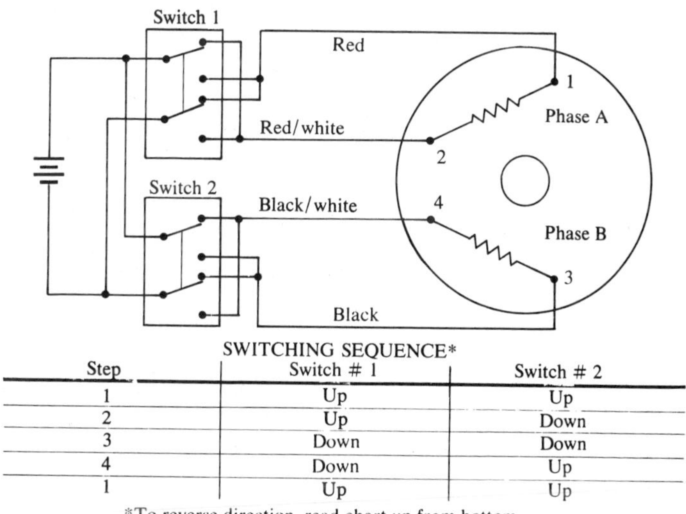

SWITCHING SEQUENCE*

Step	Switch # 1	Switch # 2
1	Up	Up
2	Up	Down
3	Down	Down
4	Down	Up
1	Up	Up

*To reverse direction, read chart up from bottom

Courtesy of Superior Electric Company

Fig. 10–8 Wiring diagram for typical DC stepper motor.

of mechanical switches at the input terminals. The sequence table is shown directly beneath the connection diagram. The actual switching is done electronically. We will look at the electronic switching circuitry later in this chapter. Stepper motors are available with control up to 200 steps per revolution, or more. The motor characteristics are specified in terms of output torque vs steps per second. Figure 10–9 is typical of the characteristics of a dc stepper motor. Outputs are generally in the fractional horsepower range. Since the stepper motor operates on pulses, the maximum speed is limited by the impedance of the windings and the pulse source. Particular care must be paid to impedance matching if high speed is to be attained.

The stepper motor is used in digital control systems where command signals are from a computer, programmer, tape reader, or digital logic circuit. Numerical control machines are an ideal application. Other applications are remote control of potentiometers, plotter and recorder drives, camera focusing, and machine tool carriage feeds.

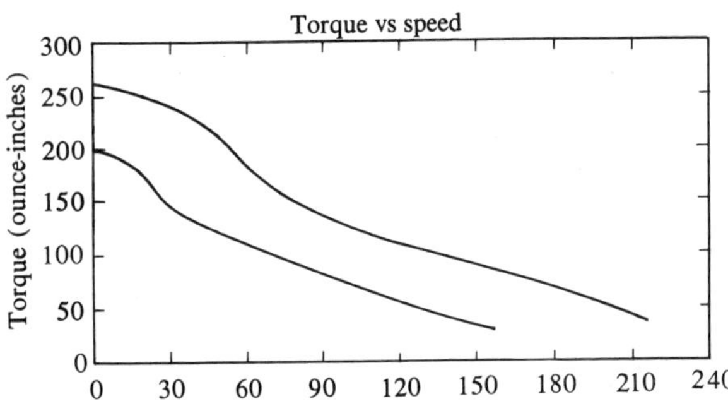

Courtesy of Superior Electric Company

Fig. 10–9 Characteristic curves for stepper motor of Fig. 10–8.

10–3 Control of Fractional Horsepower AC Motors

Most induction and synchronous motors are not designed for voltage-sensitive speed control. As we saw from the characteristic curves earlier, the speed of most ac motors is dependent on the frequency of the line voltage. Small fractional horsepower motors and universal motors are adaptable to the phase control techniques of chapter 7 to vary the operating speed in certain applications. Let's look at some of these techniques used in industry.

10–3–1 A High Torque Speed Control

One of the characteristics of the universal motor is that the speed decreases as the external load is increased. The circuit of Fig. 10–10 uses current feedbacks to maintain the speed of a universal motor constant over the entire load range. Diodes 1*D* and 2*D* along with SCRs 1*Q* and 2*Q* form the full-wave bridge rectifier (see Sec. 12–2) to supply the armature current for the motor. The armature current also flows through feedback resistor R_f. Let's assume the ac voltage is applied at terminal 1 at the moment that the sinusoidal voltage is going through zero with positive slope. SCRs 1*Q* and 2*Q* are both "off." Diodes 1*D*, 2*D*, 3*D*, and 4*D* form a full-wave bridge rectifier to supply the dc voltage for the triggering circuit. Resistor 1*R* and zener diode 7*D* form the voltage regulator (see Sec. 3–2). Resistors 1*R* and 3*R*, capacitor C_1, and UJT 3*Q* make up the basic UJT relaxation oscillator (see Fig. 5–38). Capacitor C_1 charges through 4*D*, R_1, R_2, 2*D*, and the motor armature circuit.

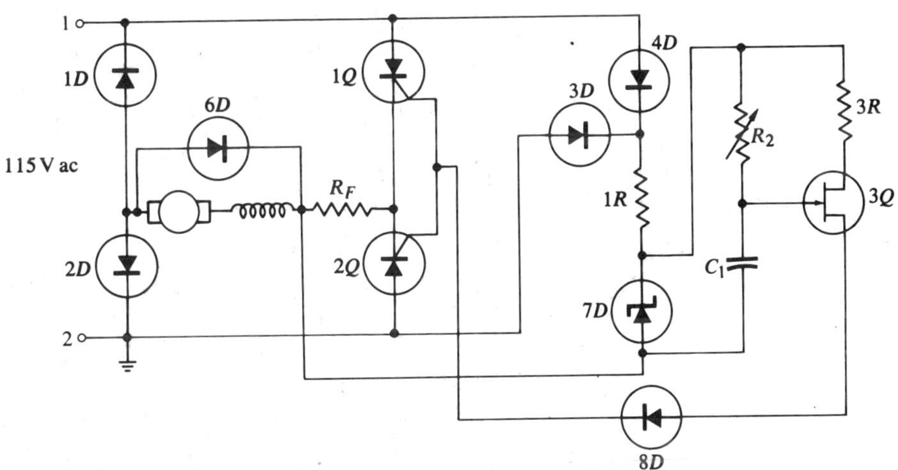

Courtesy of Motorola Inc.

Fig. 10–10 High torque motor speed control.

When the capacitor charges to the UJT firing voltage, $3Q$ conducts. Current flows through $8D$ to trigger SCR $1Q$. When $1Q$ conducts, the voltage supplied to the trigger circuit is not high enough to keep zener diode $7D$ regulating. Capacitor $1C$ will charge to the value of the voltage drop across R_f and remain fixed for the remainder of the half-cycle of line voltage. At the end of the half-cycle, $1Q$ will turn off when its anode current falls below the holding current required to maintain conduction. The ac voltage at terminal 2 will now go positive with respect to terminal 1. Current flow through $3D$, R_1, and R_2 will charge C_1 toward the UJT firing voltage; but since C_1 is already charged to the voltage across R_f it reaches the UJT trigger voltage sooner. The firing angle of $2Q$ will be advanced by the initial charge on C_1. The equation relating these variables is:

$$t_\alpha = R_2 C_1 \ln \left[\frac{V_z}{V_z - \eta V_z + I_f R_f} \right] \qquad \textbf{(10.2)}$$

where

t_α = time of firing the SCR
V_z = zener regulating voltage
η = stand-off ratio of UJT
I_f = current in armature circuit

As the load increases, the current I_f increases. From Eq. (10.2), the SCRs are fired earlier applying more voltage to the armature to maintain the motor speed constant. As the load decreases, the SCRs are fired later, reducing the voltage applied to the armature again maintaining constant motor speed. In this circuit, the firing angle of the SCRs is adjusted according to the amount of current flowing in the motor armature circuit.

10–3–2 TRIAC Control of a Universal Motor

Many applications of universal motors do not require the speed regulation provided in the control of Fig. 10–10. A simple, open-loop full-wave control can be made quite economically using a TRIAC. Figure 10–11 shows the circuit using an RC phase shift circuit and a DIAC. We looked at DIAC characteristics in Fig. 2–5. They are presented

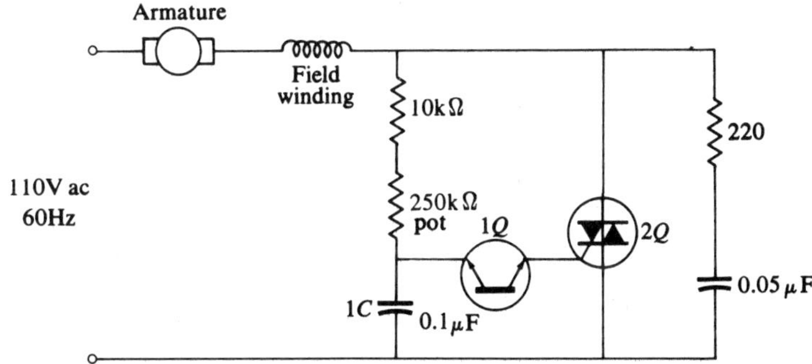

Fig. 10–11 TRIAC control of a universal motor.

again in Fig. 10–12 as a reminder. The voltage across the capacitor lags behind the line voltage by the angle α as shown in Fig. 10–13. From Eq. (7.2), $\tan \alpha = R\omega C$.

The breakover voltage of the DIAC is between 28 and 35 V. When the capacitor voltage reaches the DIAC breakover voltage, the DIAC conducts, delivering a pulse of current to the gate of the TRIAC. The TRIAC conducts, applying voltage to the motor armature. At the end of each half-cycle the TRIAC cuts off due to the anode current falling below the holding current level. Since the universal motor is an inductive load, anode current will actually flow

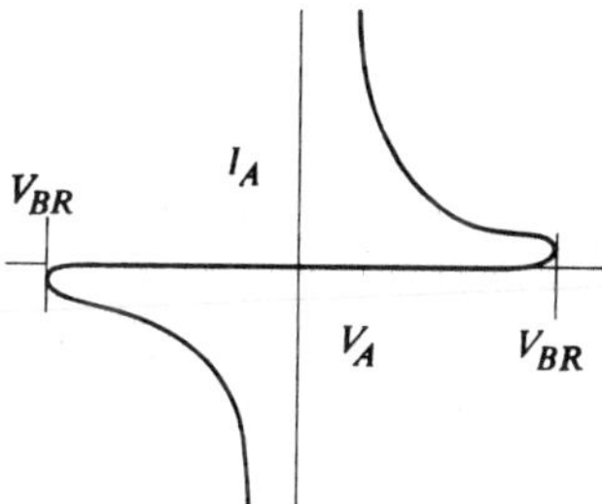

Fig. 10–12 DIAC characteristic curve.

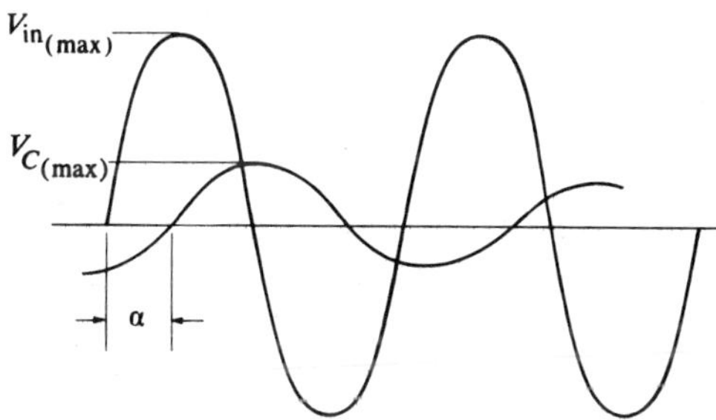

Fig. 10–13 Phase lag of capacitor voltage in Fig. 10–11.

until the fields are completely collapsed. The current flow through the motor armature and the voltage across the armature-field combination are shown in Fig. 10–14. Let's look at a design example of this control.

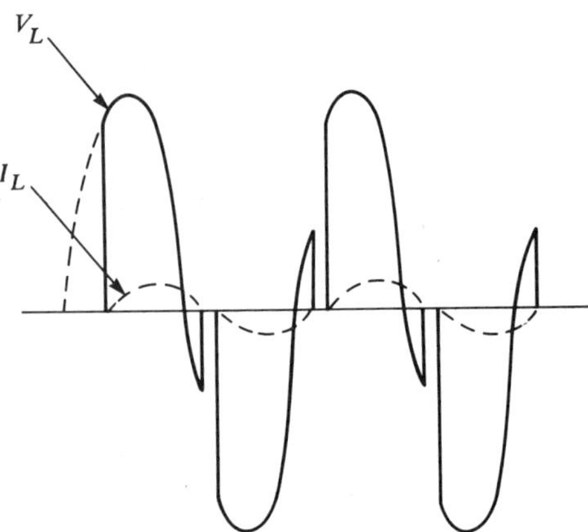

Fig. 10–14 Voltage and current waveforms in the circuit of Fig. 10–11.

Example 10–1. In the circuit of Fig. 10–11, the breakover voltage of the DIAC is 30 V. Determine the maximum and minimum firing angles of the TRIAC.

The minimum firing angle will occur when the pot is set at zero. From Eq. (7.2)

$$\tan \alpha_{\min} = \omega RC = 377 \times 10^4 \times 0.1 \times 10^{-6} = 0.377$$

$$\alpha_{\min} = 20.5°$$

The DIAC will not fire until the capacitor voltage reaches 30 V. In order to determine the time at which $V_c(t) = 30$ V we must solve the equation:

$$30 = V_{C\max} \sin(\omega t - \alpha) \qquad \textbf{(10.3)}$$

where

$V_{C\max}$ = peak value of the voltage across the capacitor.

ωt = the angle at which the TRIAC is triggered including the additional delay required for the capacitor voltage to reach the DIAC triggering voltage.

From Fig. 7–3(b)

$$\cos \alpha_{\min} = \frac{V_{C\max}}{V_{\text{in max}}} \qquad (10.4)$$

$$V_{C\max} = V_{\text{in max}} \cos \alpha_{\min} = 110\sqrt{2} \cos 20.5^\circ \qquad (10.5)$$

$$V_{C\max} = 145 \text{ V}$$

Then from Eq. (10.3)

$$\sin(\omega t - \alpha) = \frac{30}{145} = 0.207 \qquad (10.6)$$

$$\omega t = 12^\circ + \alpha$$

The minimum TRIAC firing angle is:

$$\alpha_{\min} + 12^\circ = 20.5^\circ + 12^\circ = 32.5^\circ$$

The maximum TRIAC firing angle occurs when $V_{C\max} = 30$ V. From Eq. (10.5)

$$V_{C\max} = 30 = \sqrt{2}\,110 \cos \alpha_{\max} \qquad (10.7)$$

or

$$\cos \alpha_{\max} = \frac{30}{155} = 0.194$$

$$\alpha_{\max} = 78.8^\circ$$

The maximum TRIAC firing angle is

$$\alpha_{\max} + 90^\circ = 78.8^\circ + 90^\circ = 168.8^\circ$$

From Eq. (7.2)

$$\tan \alpha_{\max} = \omega RC = 377 \times R \times 10^{-7} = 5.05 \qquad (10.8)$$

$$R = \frac{5.05}{377} = 10^7 = 134 \text{ k}\Omega \qquad (10.9)$$

If the pot resistance is increased beyond 124 kΩ, the TRIAC will not conduct.

The range of control of α between 32.5° and 168.8° is quite sufficient for this type of control circuit.

10–4 Adjustable Speed DC Motor Controls

The workhorse of the industry as far as adjustable speed motors is concerned is the dc shunt-wound motor. Speed is conveniently controlled by varying the armature voltage or field current. This can be seen by examining the equations for the shunt motor. The total voltage in the armature circuit is

$$V_A = V_{emf} + I_A R_A \tag{10.10}$$

where I_A = armature current
R_A = the resistance of the armature windings
V_{emf} = the counterelectromotive force generated by the motor

The counter-emf generated by the motor is

$$V_{emf} = K_1 N \phi \tag{10.11}$$

where N = the motor speed in r/min
ϕ = the strength of the field
K_1 = a conversion constant

Voltage is applied to the field windings to establish the value of ϕ before the motor is started. When the armature voltage is applied, the motor speed increases but so does the counter-emf. The counter-emf will increase until the armature current is just sufficient to overcome the inertia and losses in the motor. As the external load is applied, the armature current will increase to provide the torque necessary to match the load according to the equation

$$T_2 = K \phi I_A \tag{10.12}$$

The counter-emf must decrease to maintain the balance in Eq. (10.10). The motor speed decreases proportionately. Equations (10.10) and (10.12) can be solved simultaneously for speed.

$$N = \frac{V_A - I_A R_A}{K_1 \phi} \tag{10.13}$$

The motor speed can be controlled by varying V_A or ϕ. The usual technique is to vary armature voltage for speed control up to the rated speed and to vary the field to control speed above the rated speed of the motor. Figure 10–15 shows the ranges for armature control and weak field control.

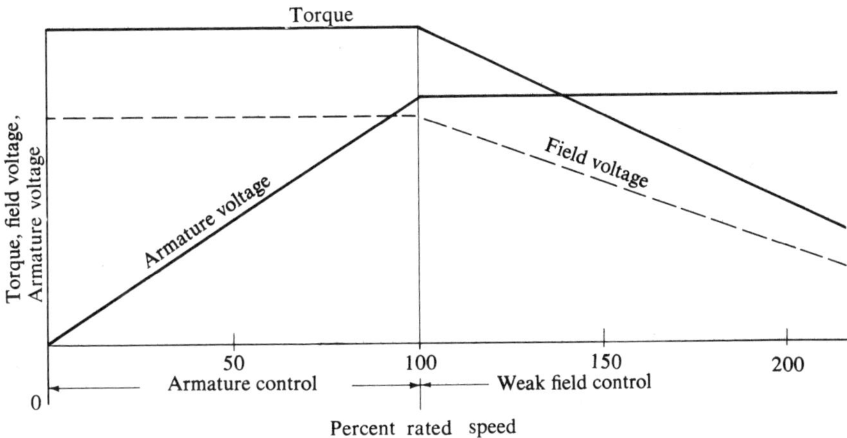

Fig. 10–15 DC shunt motor control.

The basic elements of most electronic dc motor controls are a reference signal to set the desired speed, a control device to vary the armature or field voltage, and a feedback mechanism to compare the motor speed with the reference setting. The simplest, most economical design is the SCR control of Fig. 10–16. The counter-emf generated by the

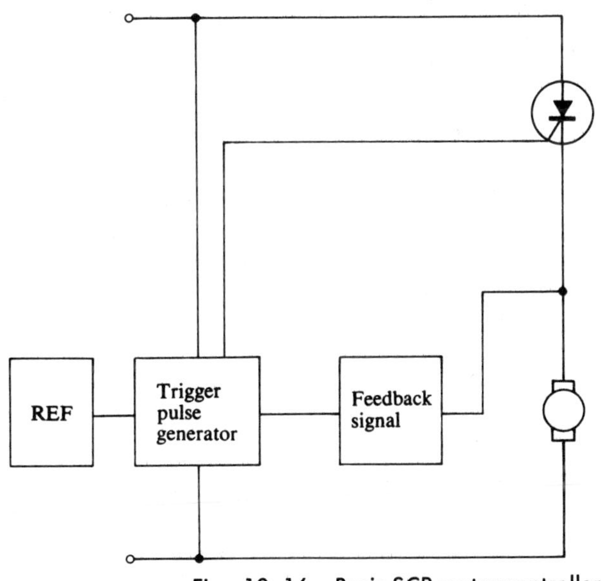

Fig. 10–16 Basic SCR motor controller.

motor or the armature current serves as the feedback mechanism. The SCR trigger timing is dependent on the difference between the reference setting and the feedback signal. The single-phase SCR control does not offer the smooth, precise control required for some applications but it is quite common for the control of fractional horsepower motors. Larger motors use three-phase rectified control to achieve smooth, precise speed control under heavy loading. The basic principles of single-phase control can be extended to polyphase controls. Let's look at some commercial SCR single-phase dc shunt motor controls to see how solid state devices are used to solve the problem of speed control.

10–4–1 A Speed Control for Small Fractional Horsepower Motors

Small dc shunt motors are often used in applications where adjustable speed control is required. Figure 10–17 shows the SCR control for a 1/50 horsepower dc shunt

Courtesy of Bodine Electric Company

Fig. 10–17 Typical DC fractional horsepower motor control.

motor. The schematic for the same control is shown in Fig. 10–18. Let's look at some of the features of this design. The ac line voltage is applied to the control at terminals 1 and 2. The bipolar ac voltage is converted to dc by the full-wave bridge rectifier of diodes 2-*CR*, 3-*CR*, 4-*CR*, and 5-*CR*.

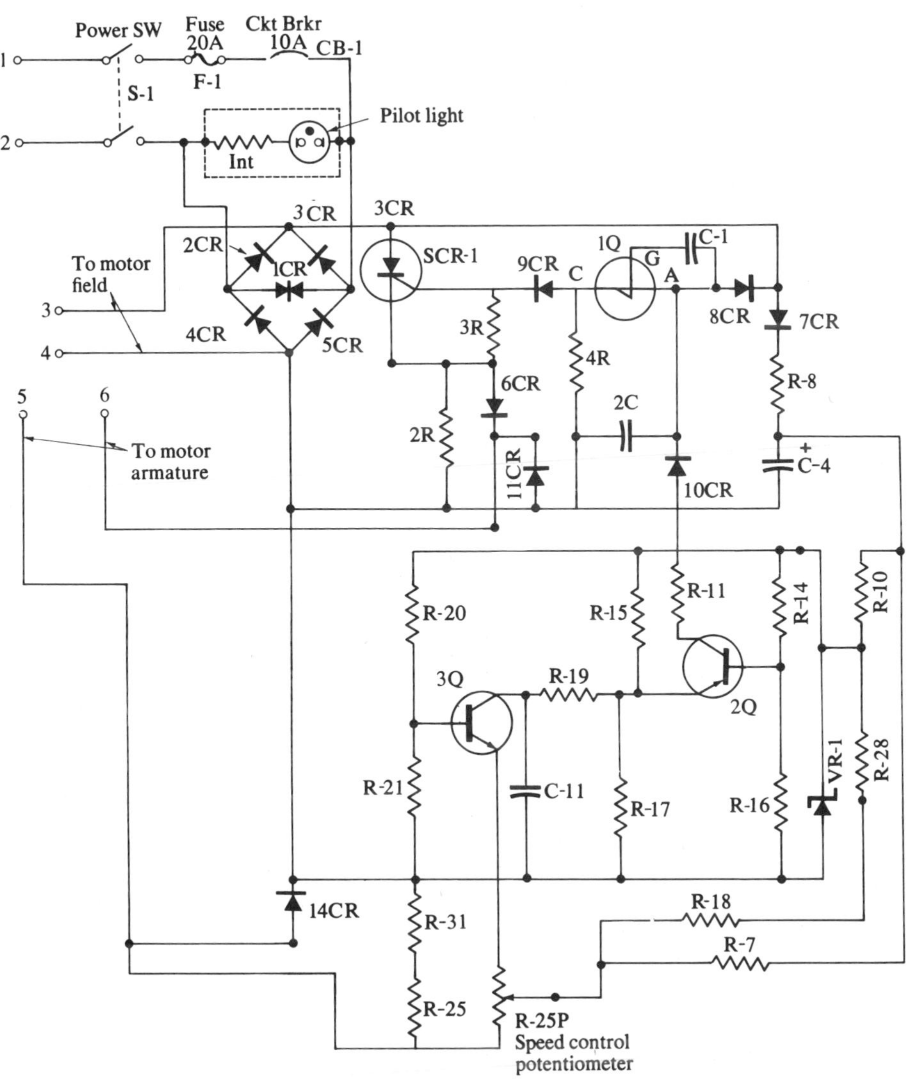

Circuitry part of Bodine Electric Company patented circuit #3475672.

Fig. 10–18 Simplified schematic of controller of Fig. 10–17.

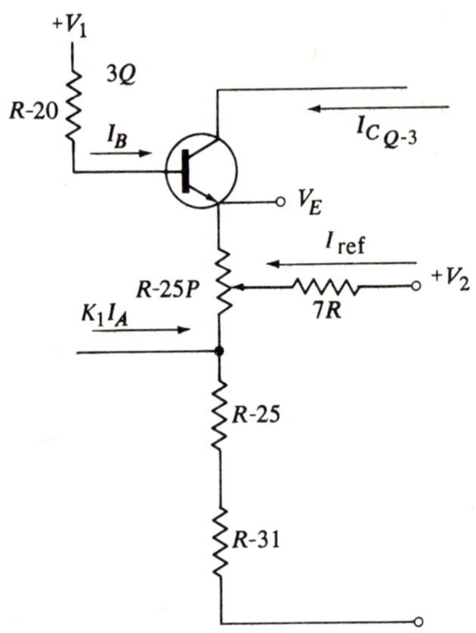

Fig. 10–19 Difference circuit of Fig. 10–18.

The full-wave rectified voltage is applied directly to the field windings at terminals 3 and 4. The armature current flows through SCR 1, diode 6-*CR*, and 14-*CR*. The rest of the circuitry provides the proper triggering pulses to maintain the motor speed constant. The reference signal to set the desired motor speed is entered at pot *R*-25P. Capacitor *C*-4b filters the full-wave rectified input voltage to provide a stable reference supply. The feedback is provided by armature current flow through pot *R*-25. The emitter voltage of 3*Q* varies with motor speed and the setting of pot *R*-25P. Figure 10–19 shows the circuitry surrounding 3*Q*. From Fig. 10–19

$$\begin{aligned} V_E = {} & R\text{-}25\text{P}\,(I_E + K_2 I_{\text{ref}}) \\ & + R\text{-}25\,(K_1 I_A + I_{\text{ref}} + I_E) \\ & + R\text{-}31\,(K_1 I_A + I_{\text{ref}} + I_E) \end{aligned} \tag{10.14}$$

Where *R*-25P, *R*-25, and *R*-31 represent the value of each resistance in ohms. K_1 is the portion of the armature current flowing into *R*-25, and K_2 is the setting of *R*-25P pot wiper.

Varying the emitter voltage of 3-*Q* causes the collector current to vary, so, in effect, the collector current of 3-*Q* is controlled by the speed of the motor and the setting of the reference speed pot. The triggering pulse for SCR-1 comes from the silicon switch relaxation oscillator (see Sec. 5–4), made of 1*Q*, 4*R*, and 2*C*. The current to charge 2*C* flows through 15*R*, 2*Q*, 11*R*, and 10*CR*. When the voltage on 2*C* reaches the breakover voltage of the silicon switch, 1*Q* conducts and 2*C* discharges through 4*R*, 3*R*, 2*R*, and the gate of the SCR. The SCR conducts, applying voltage to the motor armature to maintain a fixed speed. The amount of current that flows to capacitor 2*C* is controlled by the collector current of transistor 3*Q*. Figure 10–20 shows the capacitor charging circuit. Increasing $I_{CQ\text{-}3}$ decreases the current flow through 2*Q* to charge the capacitor. The time to charge the capacitor increases so the frequency of the relaxation oscillator is reduced. Let's put these actions together and see how speed control is affected. Assume that the speed control pot is set and the motor is running at a constant speed. Now the mechanical load on the motor shaft is suddenly increased. The motor speed decreases and the armature current increases to support the additional load. From Fig. 10–19, we see that increasing the armature current increases

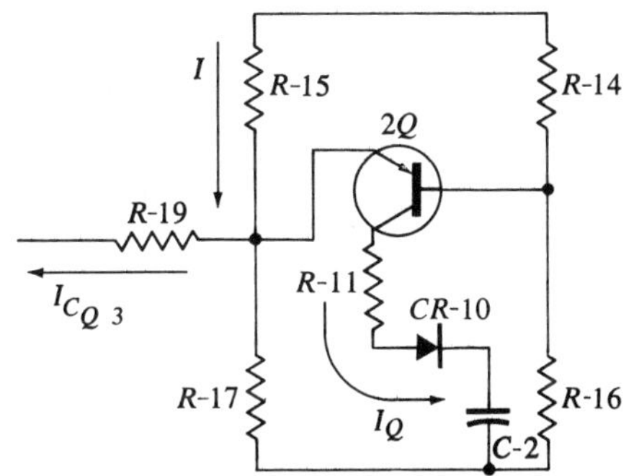

Fig. 10–20 Capacitor charging circuit of Fig. 10–18.

the emitter voltage of $3Q$. Increasing the emitter voltage of the NPN transistor decreases the forward bias at the base-to-emitter junction so $I_{CQ\text{-}3}$ is decreased. From Fig. 10–20, we see that decreasing $I_{CQ\text{-}3}$ allows the capacitor charging current to increase. Capacitor $2C$ charges faster, the SCR trigger pulse occurs earlier, and more voltage is applied to the motor armature to increase the motor speed to its original value. Now let's assume that we wish to slow the motor down. The speed control pot R-25P should be moved downward in Fig. 10–18. From Fig. 10-19, we see that the emitter voltage of $3Q$ will be reduced. Reducing the emitter voltage of the NPN transistor increases the base-to-emitter forward bias. The collector current will increase. From Fig. 10–20 we see that increasing $I_{CQ\text{-}3}$ causes the capacitor charging current to decrease. The capacitor takes longer to charge to the breakdown voltage of the silicon switch. The SCR triggering pulse arrives later and the voltage applied to the motor armature is decreased causing the motor to slow down. Figure 10–21 shows this SCR control in an electropneumatic industrial system.

10–4–2 A 3-hp DC Shunt Motor Control

The MinPak Drive control of Fig. 10–22 and Fig. 10–23 can be used to control dc shunt motors rated up to 3 horsepower. Integrated circuit operational amplifiers are used to control the timing of the SCR trigger pulses. Let's look at some of the techniques used in this circuit. Figure 10–24 shows the schematic for the control with its operator controls. The ac line voltage is applied at the L_1 and L_2 terminals. Diodes $1D$ and $2D$ along with SCRs $1Q$ and $2Q$

Courtesy of Bodine Electric Company

Fig. 10–21 Industrial application of SCR motor control.

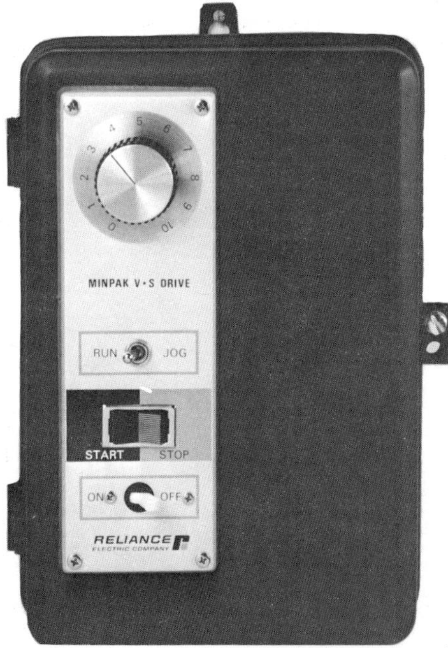

Courtesy of Reliance Electric Company

Fig. 10–22 Commercial dc shunt motor control.

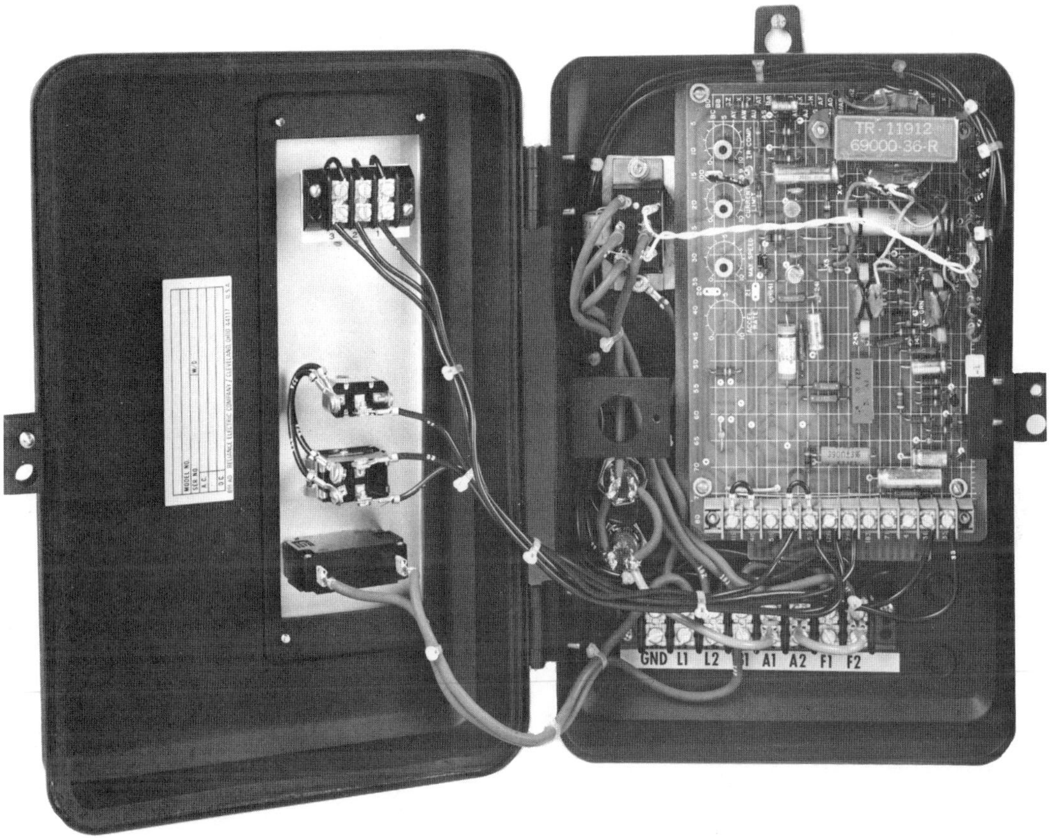

Courtesy of Reliance Electric Company

Fig. 10–23 Open enclosure view of the control of Fig. 10–22.

form a full-wave bridge rectifier to supply the dc voltage to the motor armature. The speed of the motor is adjusted by phase control of the SCRs. The SCR trigger pulses are applied at terminals 1*G* and 2*G*. Diode 4*D* is a thyrector (see Sec. 1–2). The thyrector protects the control from high voltage transients that may occur on the line. The motor field coil uses diode 2*D* as a half-wave rectifier. Another full-wave bridge rectifier is used to energize the starting coil, *CR*. Diodes 5*D*, 6*D*, 7*D*, and 8*D* provide the rectification of the 56-V output of transformer 1*T*. When the start button is pressed, current flow through the rectifier energizes the relay coil. The normally open contacts across the starting switch are closed, "sealing-in" the start circuit. A normally

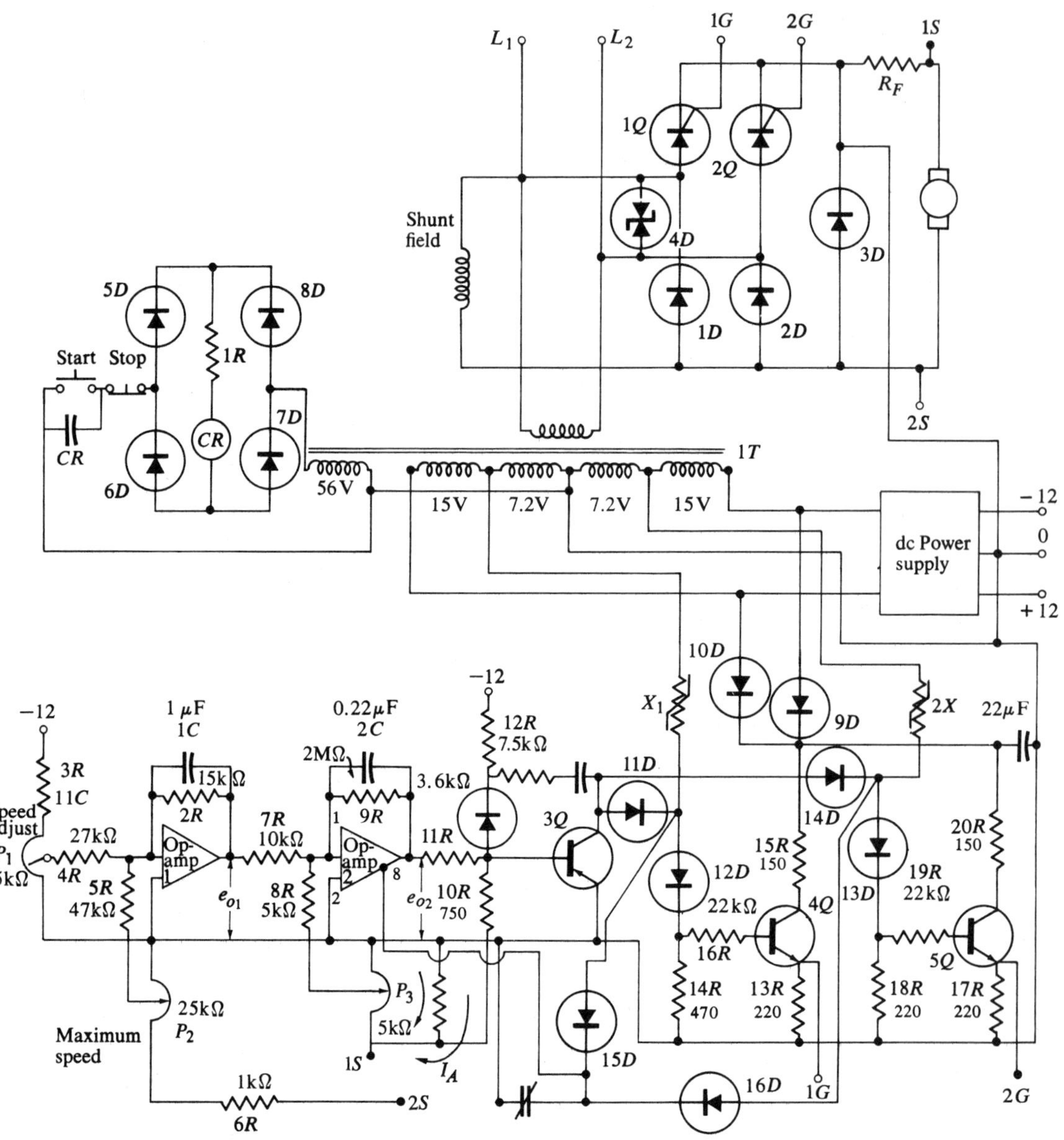

Courtesy of Reliance Electric Company

Fig. 10–24 Simplified SCR motor control.

closed contact on the start relay keeps transistors 4Q, 5Q, and op amp 2 turned off when the circuit is not energized. Transistors 4Q and 5Q provide the triggering pulses for the SCRs as we shall see later. When the start relay is energized, the normally closed contact opens; 4Q and 5Q are now free to provide the SCR trigger pulses at the proper time. The reference voltage for the desired motor speed is adjusted by the setting of pot $P1$. The maximum speed is adjusted by the setting of pot $P2$. Both voltages are applied at the 1 input of operational amplifier #1. Figure 10–25 shows the function of the operational amplifier in this circuit. (See Sec. 12–6–1 for a more complete discussion of the op-amp.) Then

$$e_{o_1} = -K_1 E_{\text{ref}} - K_1 E_{\max} \quad \textbf{(10.15)}$$

where E_{ref} is the output of P_1

$E_{\max}$ is the output of $P2$

$E_{\max}$ is dependent on the counter-emf of the motor armature and the IR drop of the armature circuit. At op-amp #2, the output of op-amp #1 is summed with the output of pot $P3$. The voltage across $P3$ is dependent on armature current flow. Then

$$e_{o_2} = -e_{o_1} - K_4 E_F \quad \textbf{(10.16)}$$

where E_f is the output of Pot $P3$.

Combining Eqs. (10.15) and (10.16)

$$e_{o_2} = K_3 K_1 E_{\text{ref}} + K_3 K_2 (I_A R_A + C_{\text{emf}}) - K_4 I_A \quad \textbf{(10.17)}$$

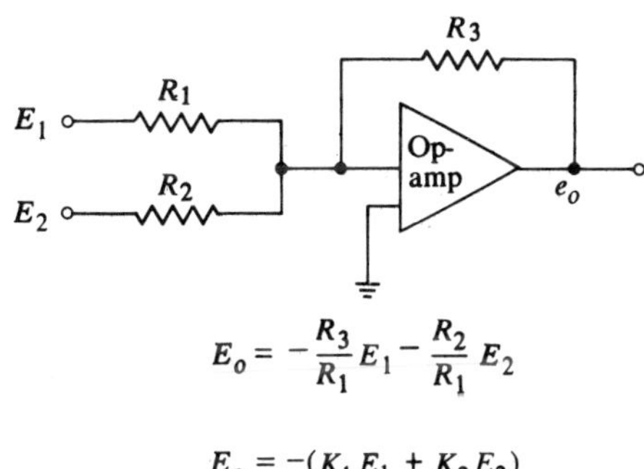

$$E_o = -\frac{R_3}{R_1} E_1 - \frac{R_2}{R_1} E_2$$

$$E_o = -(K_1 E_1 + K_2 E_2)$$

Fig. 10–25 Operational amplifier as a summer.

By increasing *P*3, the speed decrease due to motor $I_A R_A$ drop can be eliminated. The output of op-amp #2 is used to control the bias of transistor 3*Q*. Reactors *X*1 and *X*2 along with 3*Q* form a magnetic amplifier triggering circuit. If we take 3*Q* and 4*Q* out of the total circuit it will be easier to see how the magnetic action produces the triggering signal at the proper time. If we look at the circuit of Fig. 10–26 when the voltage at point A is positive, 4*Q* is forward-biased by current flow through *X*1, 12*D*, and 16*R*. The voltage at terminal 1*G* will be positive due to the emitter current flow of 4*Q*. The saturable reactor, *X*1, will become saturated due

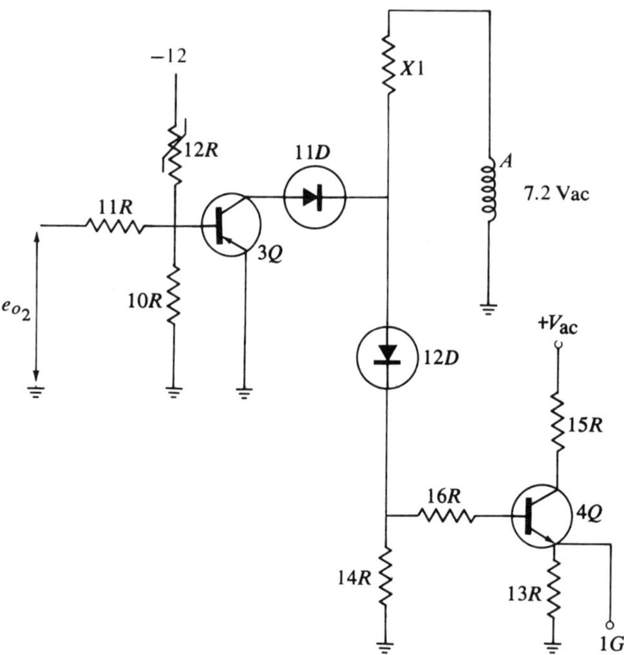

Fig. 10–26 Magnetic amplifier triggering circuit of Fig. 10–24.

to the unidirectional current flow. The operating point of the reactor will be at point 1 of Fig. 10–27. At the end of the positive half-cycle, 4*Q* will be reverse-biased so the voltage at terminal 1*G* drops to zero. Current flows through 3*Q*

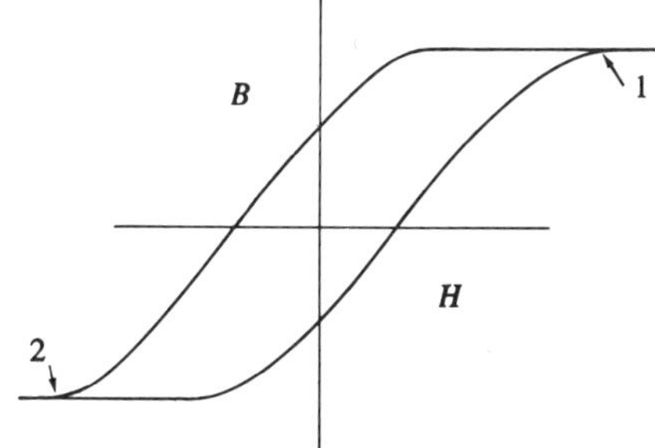

Fig. 10–27 Magnetization curves for typical saturable reactor.

and 11D to reverse the magnetization of the saturable reactor to point 2 of Fig. 10–27. The amount of time required to reverse the magnetization of the reactor will depend on the value of e_{o2} at the base of 3Q. The more negative e_{o2}, the faster the magnetization will be reversed. At the beginning of the next positive half-cycle, 4Q will try to turn on again, but it must wait until the magnetization of the saturable reactor is reversed to point 1 of Fig. 10–27. The time required to reverse the magnetization will be approximately the same on both half-cycles and will be determined by the value of e_{o2}. So we see that the time lag of the positive voltage at terminal 1G is controlled by the value of e_{o2}. The action of $X2$, 5Q, and the voltage at terminal 2G is analogous. The time lag of the voltages at terminals 1G and 2G determine the firing angle of the SCRs. Varying the firing angle of the SCRs causes the voltage applied to the armature to vary and controls the speed of the motor. Let's put these actions together and see how the motor speed is adjusted.

Assume that the motor is running at a constant speed and suddenly the mechanical load is increased. The motor will slow down and the armature current will increase to support the increased load. The increase in armature current will cause E_f in Eq. (10.17) to increase so e_{o2} becomes more negative. Making e_{o2} more negative decreases the time required to reverse the magnetization of $X1$ and $X2$ so that the positive voltage appears at terminals 1G and 2G earlier each half-cycle. The SCRs will fire earlier applying more voltage to the armature of the motor. The motor speed will increase toward the original value. What if we would like to decrease the speed of the motor? Speed adjust pot $P1$ should be rotated counterclockwise, decreasing the magnitude of E_{ref} in Eq. (10.14). Since E_{ref} is taken from the -12-V supply, e_{o_2} will become more positive. Making e_{o_2} more positive will increase the time required to reverse the magnetization of $1X$ and $2X$. The result will be later firing of the SCRs and a decrease in the voltage at the motor armatude of E_{ref} in Eq. (10.14). Since E_{ref} is taken from the -12-V supply, e_{o2} will become more positive. Making e_{o2} more positive will increase the time required to reverse the magnetization of $1X$ and $2X$. The result will be later firing of the SCRs and a decrease in the voltage at the motor armature the motor will slow down.

The two dc shunt motor controls that we have just examined employ many of the techniques that you are likely to encounter in the field today.

10–5 Control of Digital Motors

The digital motors that we discussed in Sec. 10–2 require the proper sequence and amplitude of pulses at the motor windings to advance the motor shaft properly. The command signals are usually a train of low energy pulses. The motor must advance the output shaft one step for each pulse of the command signal. The motor controller must translate each pulse of the command signal into the proper motor excitation. Figure 10–28 shows a block diagram of the system. The motor controller is usually a digital logic circuit. Figure 10–29(a) shows a simple digital control circuit to provide the proper pulse sequence to drive the motor forward. The truth table is shown in Fig. 10–29(b) with the

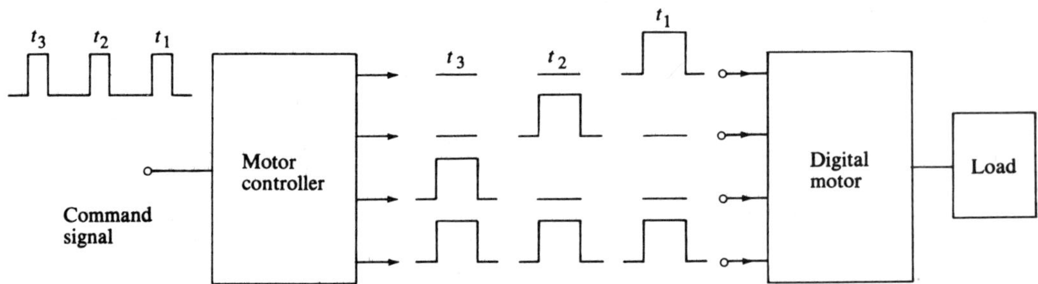

Fig. 10–28 Digital motor drive system.

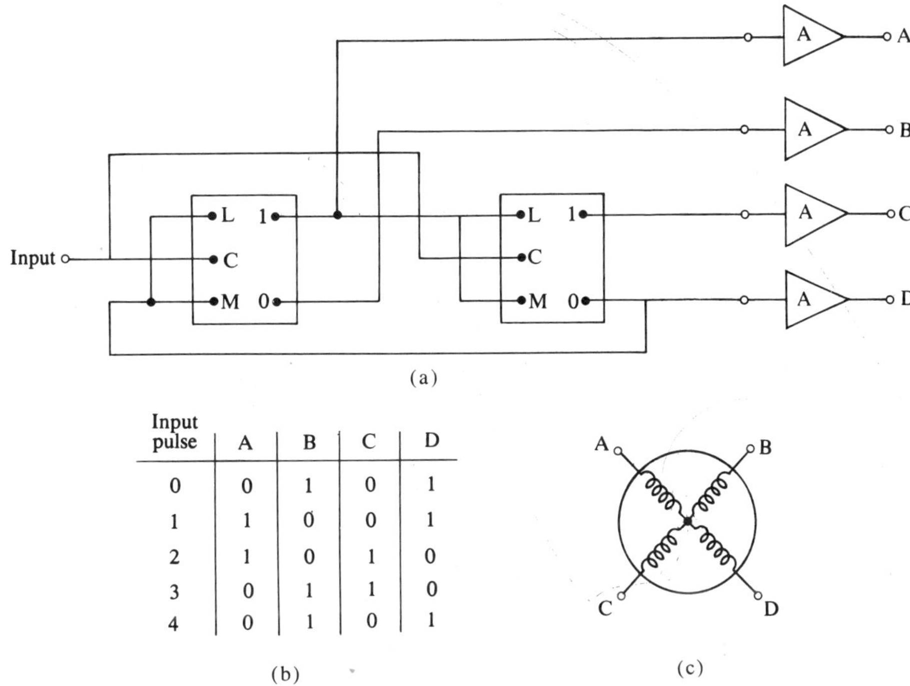

Input pulse	A	B	C	D
0	0	1	0	1
1	1	0	0	1
2	1	0	1	0
3	0	1	1	0
4	0	1	0	1

(b)

Fig. 10–29 (a) Controller circuit, (b) truth table, (c) motor connection.

motor connections shown in Fig. 10–29(c). The "D" flip-flop is used to provide the digital gating. The L and M inputs to the flip-flop are tied together. When the clock input is applied, the output 1 terminal will take on the state of the L-M terminals. Amplifiers at the flip-flop outputs provide the drive for the motor. Figure 10–30 shows how the controller can be revised to operate the same motor in reverse rotation. Figure 10–31 shows the use of additional gates to allow the selection of forward or reverse rotation by applying the proper pulse at the F or R terminal. A variable voltage oscillator is added at the input so that the speed of the motor is programmable by varying the amplitude of the command signal.

10–6 Battery-Operated DC Motor Controls

Many battery-operated vehicles require variable speed control. Golf carts, forklift trucks, wheelchairs, and, more

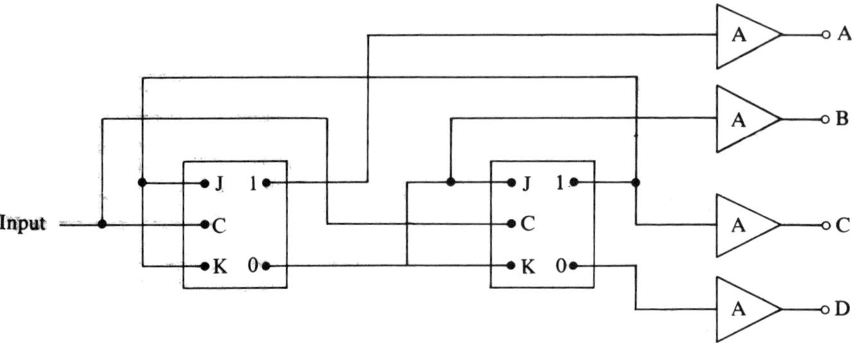

Fig. 10–30 Revision of Fig. 10–29(a) for reverse drive of the motor.

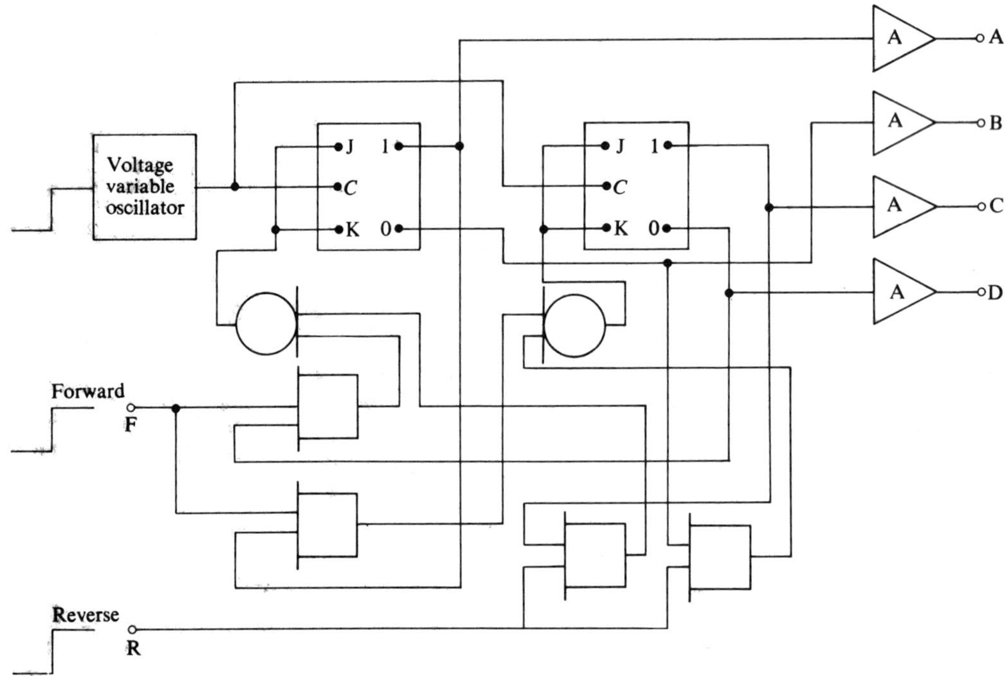

Fig. 10–31 Adjustable speed controller for motor of Fig. 10–27(c).

recently, the electric automobile are some of the vehicles that use electronic speed controls. Most vehicles use the series-wound dc motor because of its high starting torque. The motor speed is controlled by varying the average voltage at the motor armature-field terminals. There are two ways to vary the average voltage. Both involve *chopping* of the dc voltage to reduce the average voltage applied to the motor. Figure 10–32 shows the application of constant pulse width, pulse frequency modulation. Figure 10–33 shows constant pulse frequency, pulse width modulation. In

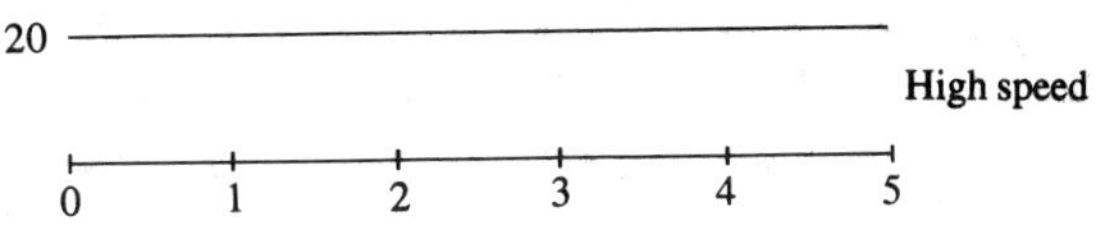

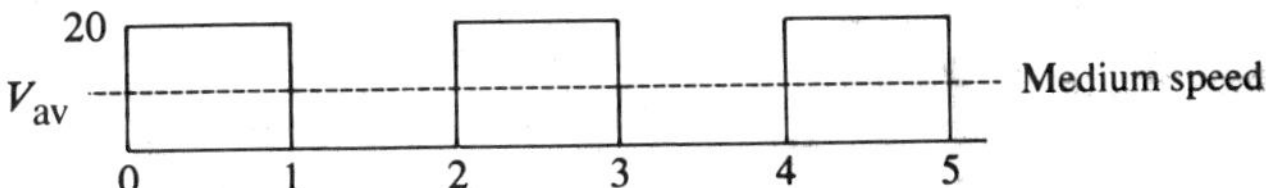

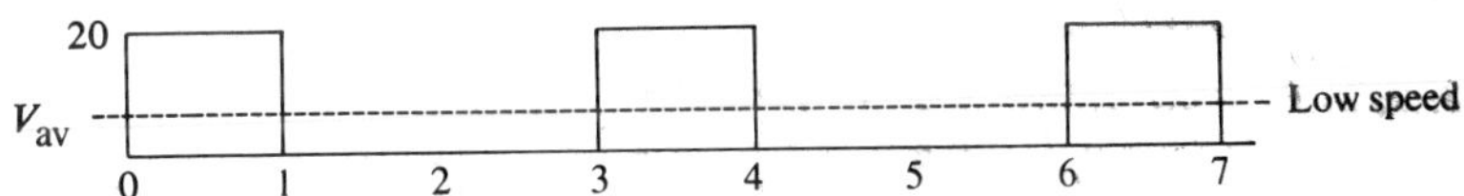

Fig. 10–32 Pulse frequency modulation for DC motor speed control.

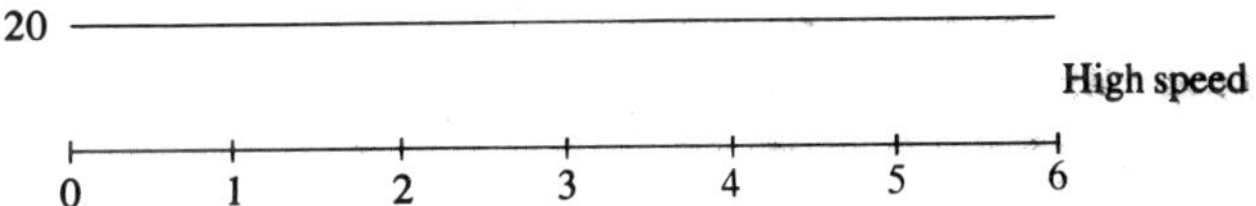

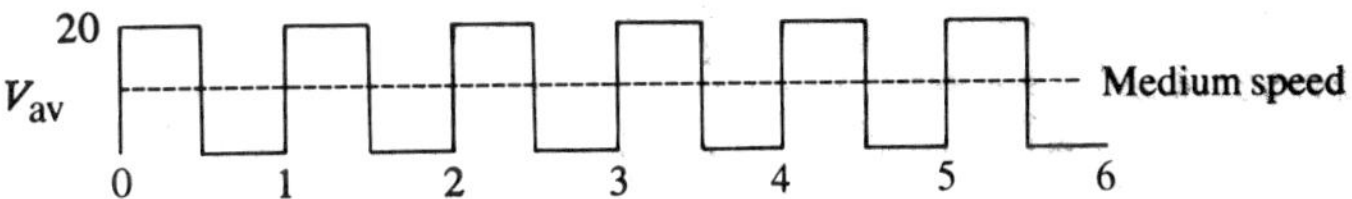

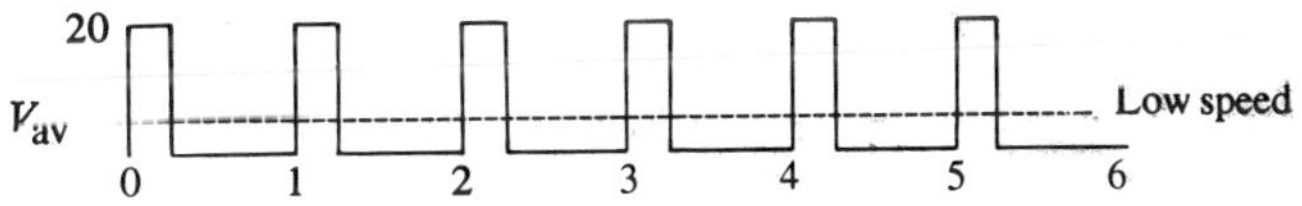

Fig. 10–33 Pulse width modulation for DC motor speed control.

either case, the motor speed is reduced by reducing the average voltage applied to the motor. Figure 10–34 shows the block diagram of a battery-operated drive system. The SCR is used to perform the electronic switching. Remember, from chapter 5, the SCR in a dc circuit does not turn off automatically. Some form of turn-off on *commutation* circuit must be provided. The two basic methods of commutation are *active commutation*, where another semiconductor device is used to switch the SCR off, or *reactive commutation*, where a reactive oscillator-type circuit is used to switch the SCR off. We will look at control designs using both techniques.

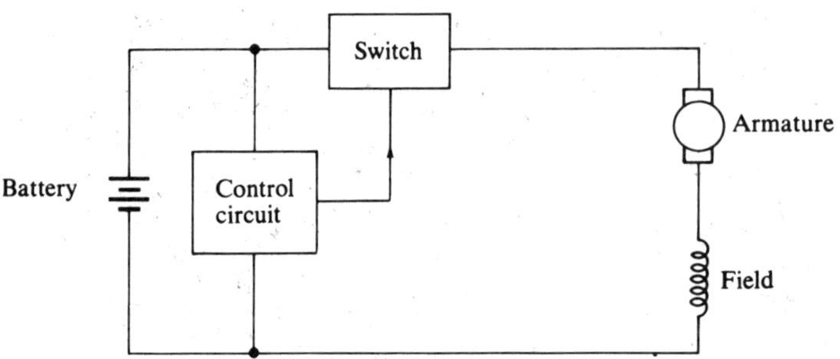

Fig. 10–34 Block diagram of a battery-operated drive.

10–6–1 Speed Control of a Fractional Horsepower Battery-Operated Vehicle

An electronic control was developed at the Western Regional Hospital in Glasgow for a power-operated carriage.* The 100-pound carriage operates between zero and 3 m/ph using a 1/10 horsepower series motor. Efficiency of greater than 90% is achieved. The schematic of the control circuit is shown in Fig. 10–35. Let's look at the operation of this circuit. Resistor 11*R* and zener diode 5*Z* form the voltage regulator for the turn-on relaxation oscillator. The

*This circuit was developed by R. J. Eadie, Western Regional Hospital Board Physics Dept., Glasgow, Scotland, and described in the Jan. 1968 issue of *Electronic Engineering*.

turn-on frequency is controlled by adjusting pot RV_3. This indicates that constant pulse frequency, pulse width modulation is used since the SCR triggering pulses from transformer $1T$ will occur at a fixed rate. When the SCR conducts, voltage is applied at the motor terminals and capacitor $1C$ charges through $1R$. Freewheeling diode $3MR$ keeps current flowing through the motor when the SCR is not conducting. On the other side of the diagram, resistor $2R$ and zener diode $1Z$ form a voltage regulator for the turn-off relaxation oscillator. The speed of the motor is controlled by varying the emitter resistor $2RV$ in the transistor amplifier circuit of $2VT$. The output of the amplifier is used to charge the capacitor $2C$ in the UJT oscillator. Increasing the resistance $1RV$ will cause the capacitor to charge slower, allowing the SCR to conduct longer and increasing the speed of the motor. The output pulse from the UJT oscillator is passed through the buffer amplifier of $3VT$. When $3VT$ conducts, the driver transistor $4VT$ conducts, forcing the power transistor $1VT$ into saturation. The switching "on" of $1VT$ applies a positive 12-V pulse at capacitor $1C$. Since the capacitor voltage cannot

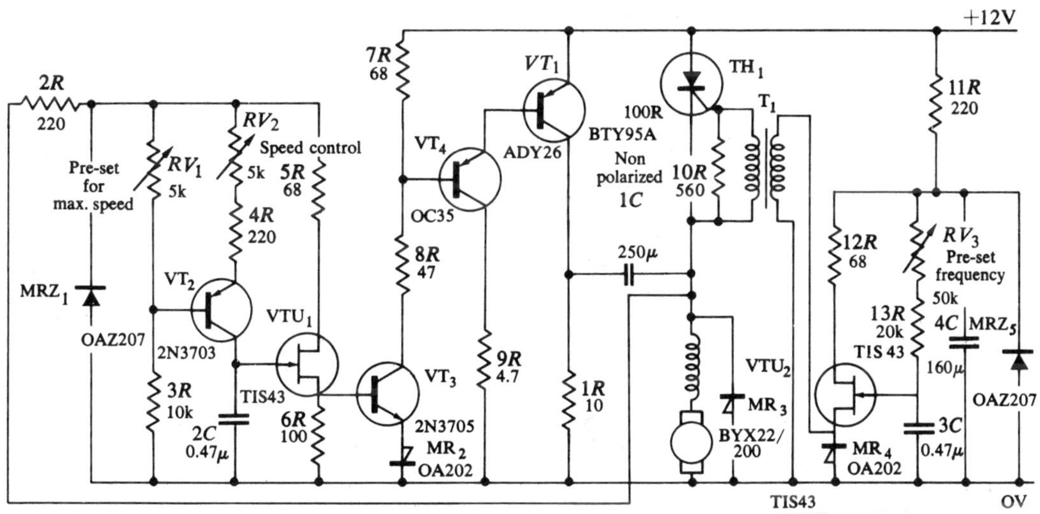

Courtesy of Electronic Engineering

Fig. 10–35 Electronic control for a fractional horsepower battery-operated vehicle.

change instantly, the 12-V pulse is transmitted to the cathode of the SCR. Raising the cathode potential of the SCR forces the anode current below the holding current level and the SCR shuts off. The SCR will remain off until the next triggering pulse from the turn-on relaxation oscillator is applied at 1*T*; then the cycle will repeat.

10–6–2 Control of Higher Horsepower Battery-Operated Vehicles

Higher horsepower battery-operated vehicles such as industrial forklift trucks and proposed electric automobiles for consumer production avoid the use of active turn-off techniques due to the power loss in the turn-off circuit. Various modifications of the Jones circuit of Fig. 5–18 are often used. A simplified schematic of a battery-operated vehicle control using a Jones turn-off circuit is shown in Fig. 10–36. The direction switch should be set in forward

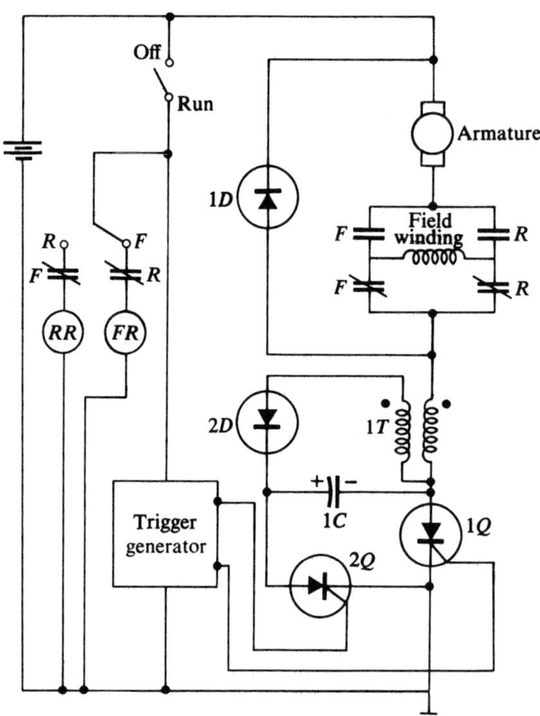

Fig. 10–36 Simplified schematic of Jones circuit in battery-operated vehicle control.

or reverse before the master control switch is placed in the "run" position. When the master switch is closed, the forward relay or reverse relay contacts in the field circuit are actuated, directing the current flow in the field windings. The trigger pulse is applied to the high-current SCR, 1*Q*, first. The increasing current flow through the primary of transformer 1*T* generates a positive voltage in the circuit of 2*D* and 1*C*. Capacitor 1*C* is charged with the polarity shown in Fig. 10–36. A trigger pulse is applied to the commutating SCR 2*Q*. When 2*Q* conducts, the positive side of the charged capacitor, 1*C*, goes to ground potential. Since the capacitor voltage cannot change instantly, the anode of SCR 1*Q* drops below ground. SCR 1*Q* is cut off to end the cycle. The motor speed is adjusted by controlling the timing of the triggering pulse to 1*Q*. The interval between the firing of 1*Q* and the firing of 2*Q* is fixed so that the pulse width remains constant. Figure 10–37 shows a trigger generator that could be used for this application. When the

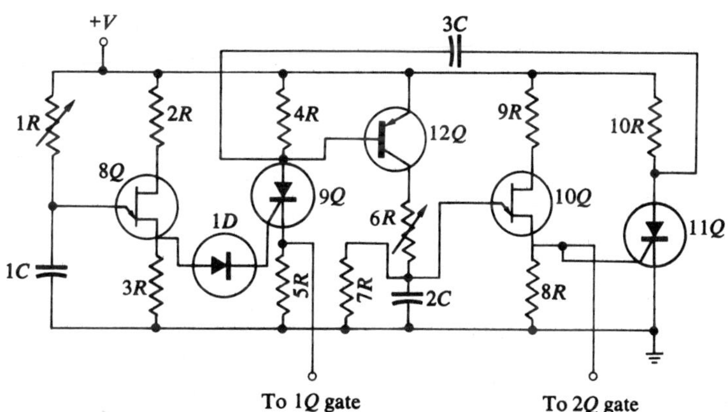

Fig. 10–37 Trigger generator for battery-operated vehicle control of Fig. 10–36.

master control switch is closed, voltage is applied to the trigger generator circuit. Capacitor 1*C* charges through 1*R* toward the emitter breakover voltage of UJT 8*Q*. Capacitor 2*C* cannot charge because SCR 9*Q* is cut off. When 8*Q* conducts, a portion of the output current flows into the gate of SCR 9*Q*, causing 9*Q* to turn on. A part of the current flow through 9*Q* goes into the gate of 1*Q* to start the current flow from battery to motor. The turning on of 9*Q* allows current to flow through 4*R* into the base of transistor 12*Q*. Collector current flow through 12*Q* charges capacitor 2*C* toward the emitter breakover voltage of UJT 10*Q*. When 10*Q* conducts, a portion of the output current goes into the gate of 2*Q* to shut off current flow from battery to motor. A part of the output current from 10*Q* flows into the gate of 11*Q*. When 11*Q* conducts, capacitor 3*C* charges by drawing current through 4*R*. The additional current flow through 4*R* causes the anode potential of 9*Q* to drop below ground potential. SCR 9*Q* turns off, shutting off transistor 12*Q*. Capacitor 2*C* discharges through resistor 7*R* and the cycle begins all over.

Most vehicles incorporate other design features such as acceleration control, braking control, battery-saving techniques, and safety interlocks. The circuit of Fig. 10–36 contains only the basic elements of adjustable speed control.

10–7 Summary

The control of motors is one of the major tasks of industrial electronics. The industrial electronics designer must be knowledgeable about the characteristics of motors as well as the requirements of the load if he is to achieve a successful design. For this reason, we started this chapter with a brief discussion of motors and their characteristics. We then proceeded to demonstrate some of the design techniques by examples and commercial applications. The circuitry might have seemed complicated at times but you must learn to identify, design, and analyze parts of the total system in much the way we have done in this chapter.

EXERCISES

1. A load requires 14 inch-ounces at 3300 r/min. What is the horsepower required?
2. What is the main advantage of the series-wound dc motor over the shunt-wound dc motor?
3. Is there a motor that operates on ac or dc line voltage? If so, what is it called?

4. If the feedback resistor, R_f, of Fig. 10–10 is increased but all other conditions remain the same, will the motor speed increase or decrease? Explain why.

5. In the control circuit of Fig. 10–11, the control potentiometer is changed to 150 kΩ.
   (a) Determine α_{min} and α_{max}.
   (b) Using the same DIAC specifications, determine the maximum and minimum TRIAC firing angle.

6. What are the three possible ways to control the speed of a dc shunt motor?

7. What are the two methods used in modern electronic motor controls to adjust the speed of dc shunt motors?

8. What is the function of diode 11*CR* in the control of Fig. 10–18?

9. In the control of Fig. 10–18, if the ratio $R14/R_{16}$ is increased, what will be the effect on the speed of the motor?

10. What is diode 4*D* of Fig. 10–24 and what is its function?

11. Why is it not necessary to control the conduction of diodes 1*D* and 2*D* in Fig. 10–24?

12. What would be the effect of increasing 10*R* in the trigger circuit of Fig. 10–26 if everything else remained the same?

13. Redesign the digital control circuit of Fig. 10–29(a) for the following pulse sequence:

INPUT PULSE	A	B	C	D
0	1	1	0	0
1	0	1	1	0
2	0	0	1	1
3	1	0	0	1
4	1	1	0	0

14. Explain the operation of relays *RR* and *FR* in the drive control schematic of Fig. 10–36.

15. What would be the effect of disconnecting capacitor 3*C* in Fig. 10–37?

11

Control of Large Currents

11–1 Introduction

Many industrial systems utilize low-impedance, high-current loads. Typical systems are resistance welders, furnace heaters, plating tank temperature controls, and high intensity lamps. A transformer is usually required to match the load to the power distribution system. SCR phase controls are often used in the transformer primary to control the average power delivered to the load. These high-current phase control systems encounter problems significant enough for our special consideration.

The main problems may be categorized as:

(a) Form factor
(b) Power line fluctuations
(c) Triggering variance
(d) Voltage and current transients

We will look at a commercial resistance welder as an example of high-current phase control systems, but first let's discuss the nature of the problems.

(a) Form factor—The form factor is defined as the ratio of rms current to average current. Reducing the conduction angle of an SCR in a transformer circuit as shown in Fig. 11-1 increases the form factor. Figure 11-2 shows the form factor increase at $\alpha = 0°, 45°$, and $90°$. Figure 11-3 shows how the form factor varies over the usual range of phase

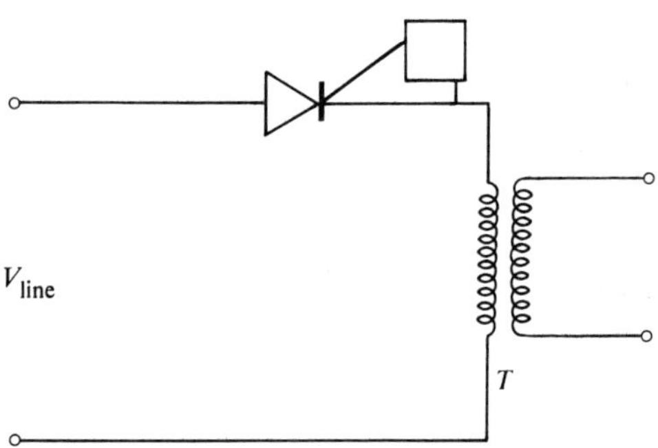

Fig. 11-1 Single SCR in primary of transformer.

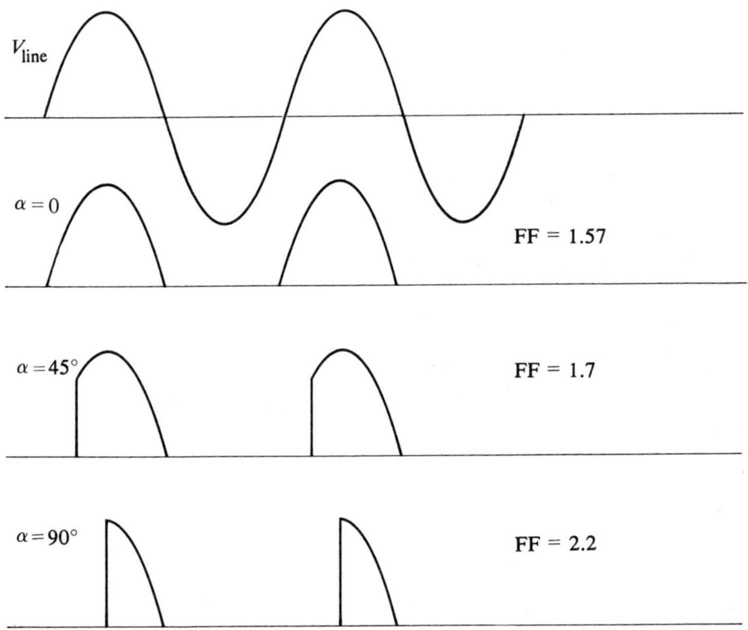

Fig. 11-2 Variation of form factor with phase control of SCR.

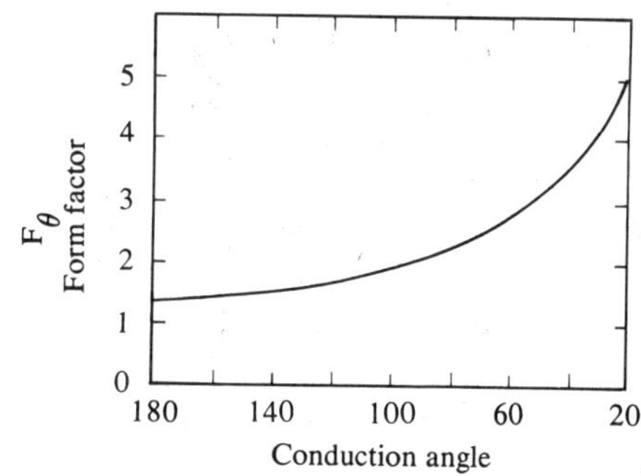

Fig. 11–3 Form factor vs conduction angle for a resistive load.

control. Remember from chapter 5, Sec. 2, that the maximum allowable average current of an SCR decreases as the form factor increases. Increased form factor also results in additional transformer losses and poorer utilization of the power distribution system. As shown in Fig. 11–2, the form factor can be limited to 2.2 if the phase control is limited to a maximum of 90° delay. Additional control can be achieved by using transformer taps to control the peak ac voltage.

(b) Power line fluctuations—Besides the inefficient utilization of the power distribution system, phase control of high currents causes fluctuations in the line voltage. Twenty to thirty percent line drops are sometimes experienced. The solid state control circuitry must be designed to tolerate the major line fluctuations. Other systems and equipment in the vicinity of the high-current systems are also affected by the fluctuation in voltage. A generally hostile electrical environment is created.

(c) Triggering variance—Two SCRs are usually operated in back-to-back inverse parallel connection on the primary side of the current transformer as in Fig. 11–4. The inductance of the transformer causes current to flow beyond the normal commutation point. The amount of extra conduction depends on the impedance of the transformer and the peak current. Figure 11–5(b) shows that the additional conduction, θ', causes a voltage transient.

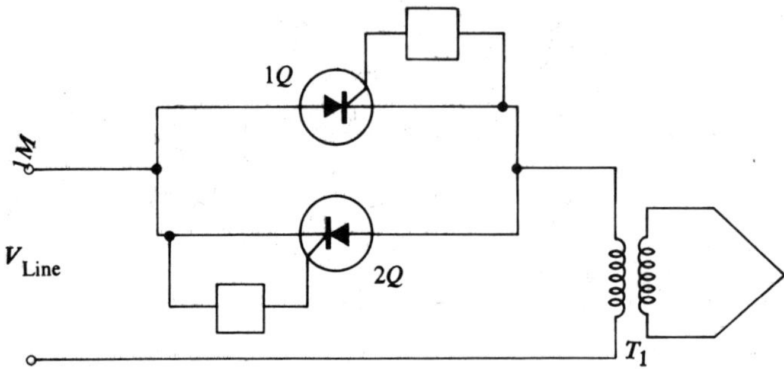

Fig. 11–4 Inverse parallel connection of SCRs in welding transformer primary.

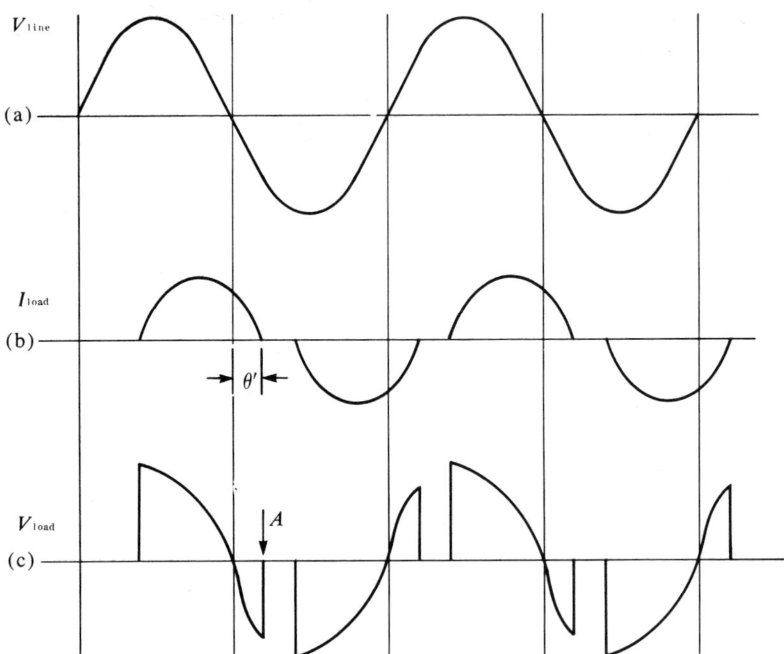

Fig. 11–5 Current and voltage waveform in circuit of Fig. 11–4.

(d) Voltage and current transients—The sudden increase in voltage across the SCRs of Fig. 11–4, at point A, might exceed the dv/dt rating of the devices if there is not sufficient capacitance at the anode-to-cathode terminals. The high di/dt that occurs while switching the SCRs on might exceed the maximum *di/dt* rating of the devices if there is not sufficient inductance in the load circuit. The same types of precautions that we discussed in chapter 5 are usually taken to protect the SCRs.

11–2 A Solid State Resistance Welder

The resistance welder is a good example of a machine that requires electronic control of large currents. All of the problems we discussed in Sec. 11–1 must be considered in the design of solid state controls for resistance welders if high quality welds are to be achieved. Let's look closely at a commercial design of the controls for a resistance welder.

11–2–1 The Welding Sequence

Resistance welding is a sequential process requiring precise control and timing of each phase of the operation. The four phases are squeeze, weld, hold, and off. Two pieces of metal to be welded are placed between the copper welding tips and the control circuit is initiated. A brief description of the four phases of the sequence that follows is:

SQUEEZE—The welding tips are forced together under pressure. Squeeze time allows the tip pressure to stabilize.

WELD—Current flow through the transformer secondary and welding tips fuses the two pieces of metal together.

HOLD—Pressure is applied to the welding tips after the welding current stops to allow the weld to crystallize.

OFF—The welding tips are removed from the work pieces. The squeeze cycle cannot be initiated again until the off time has expired.

11–2–2 The Robotron Solid State Resistance Welder

Let's look at a commercial solid state welder. Figure 11–6 is a photograph of the Robotron model 3060 UJT Resistance Welder. This welder has a total capability of 120 cycles of squeeze, 60 cycles of weld, 60 cycles of hold, and 61 cycles of off. The block diagram of Fig. 11–7 will hclp us get an overview of the general operation of the control. We will then take a detailed look at each portion of the machine cycle.

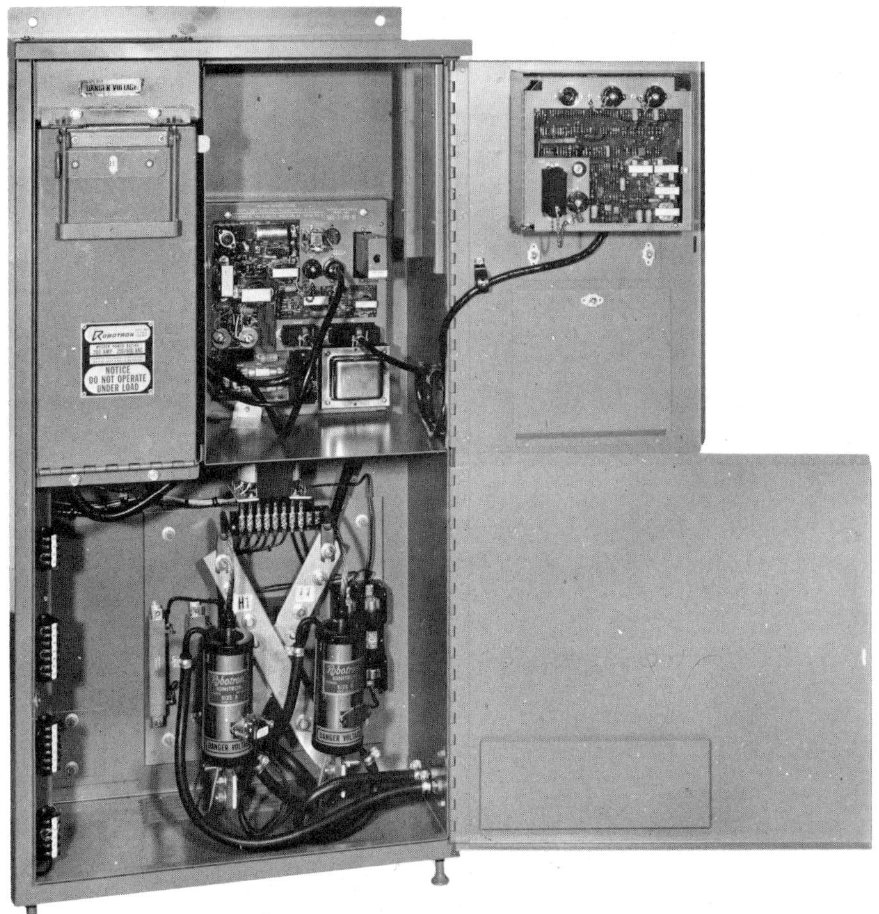

Courtesy of Robotron Corporation

Fig. 11–6 A commercial solid state resistance welder.

After the work pieces are placed in position, the weld control is activated by closing the initiation switch in the upper left-hand corner of Fig. 11–7. SCR 1Q is energized by the closing of the initiation switch. Current flow through 1Q energizes transformer 103T. The secondary voltage of 103T turns off transistor 2Q and turns on 3Q. When 2Q stops conducting, transistor 4Q conducts, applying voltage to the primary of transformer 104T. Another secondary winding of 103T causes transistor 5Q to conduct. The squeeze timing circuit and UJT timer 13Q are unclamped when 2Q

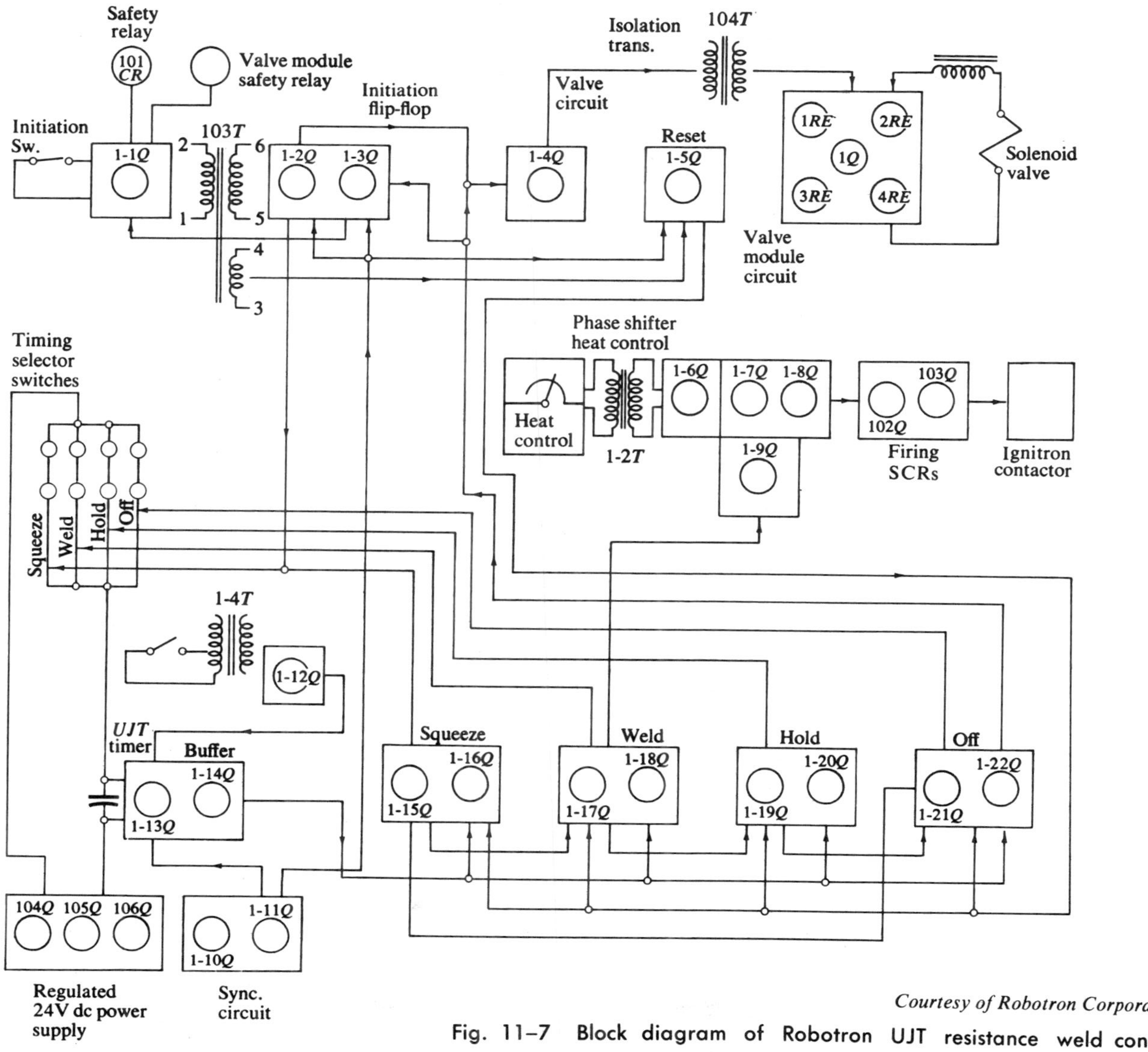

Courtesy of Robotron Corporation

Fig. 11–7 Block diagram of Robotron UJT resistance weld control.

stops conducting, allowing the squeeze phase to begin. Closure of the initiation switch also energizes relay 101 *CR*, which closes the safety contacts in the ignitor circuit preparing for weld current flow. When 2*Q* stops conducting, transistor 4*Q* conducts, energizing the primary of transformer 104*T*. Voltage at the secondary of 104*T* triggers SCR 101*Q*, energizing relay coil *SVCR*, closing contacts in the solenoid valve coil circuit. The welding tips are closed, applying pressure to the work piece. Contacts on *SVCR* in the initiation circuit are closed, sealing in transistor 1*Q*. The UJT timing capacitor charges toward the required trigger voltage at a rate determined by the setting of the squeeze time selector switch. At the end of the squeeze time, an output pulse from the UJT timer occurs in coincidence with the sync pulse from 11*Q* at the emitter of transistor 14*Q*. The output of 14*Q* switches transistor 15*Q* on and 16*Q* off, ending the squeeze phase.

The conduction of transistor 15*Q* causes a negative pulse to be applied to the base of 17*Q*, turning it off; 18*Q* is turned on by the flip-flop action of 17*Q*. The turning off of 17*Q* unclamps capacitor 17*C* in the PUT relaxation oscillator timer circuit. The output pulse of PUT 9*Q* triggers SCR 8*Q*. SCR 7*Q* conduction is controlled by a phase shift circuit and fires at the same point on each positive half-cycle. Conduction of both 7*Q* and 8*Q* completes the path of current flow through transformers 105*T* and 106*T* to provide the trigger pulses for SCRs 102*Q* and 103*Q*. The conduction of 102*Q* and 103*Q* in the ignitor circuits of the ignitron tubes allows welding current to flow through the two pieces of metal.

Capacitor 17*C* in the UJT timer circuit now charges toward the required triggering voltage at a rate determined by the setting of the weld time selector switches. At the end of weld time, the trigger pulse from 14*Q* turns 18*Q* off and 17*Q* on. The turning on of 17*Q* forces SCR 8*Q* to stop conducting and the welding current is interrupted. This is the end of the weld phase.

The turning on of 17*Q* at the end of the weld phase causes a negative pulse to be applied to the base of 19*Q*, turning it off and turning 20*Q* on. Capacitor 17*C* in the UJT timer circuit now charges toward the required triggering voltage at a rate determined by the setting of the hold selector switches. At the end of the hold time, the trigger pulse from 14*Q* turns 20*Q* off and 19*Q* on. The turning on of 19*Q* provides a negative pulse to turn 21*Q* off, causing

22*Q* to turn on. The conduction of 22*Q* turns off SCR 4*Q*, which interrupts current flow through transformer 104*T*. The trigger pulse at 101*Q* is lost, 101*Q* turns off, opening the relay contacts in the solenoid valve coil energizing circuit. The weld tip pressure is removed and the tips retract. This is the end of the hold phase.

The turning off of SCR 101*Q* causes the "seal in" contacts to open; but if the initiation switch is held closed, transistor 1*Q* will continue to conduct even though 22*Q* is turned on. The circuit resets after the off phase and the machine cycle repeats.

Let's look at the details of this sequential operation as an example of how the solid state electronic circuits we have been studying can be applied to one of the most difficult industrial designs.

11–2–3 The Initiation Circuit

The initiation circuit is shown in Fig. 11–8. Closure of the initiation switch completes the circuit to the 102*T* transformer primary winding and connects the cathode of SCR 1*Q* to the secondary of transformer 101*T*. The voltage at the anode of 1*Q* is 180° out of phase with the voltage applied to 102*T*. The secondary of 102*T* is applied to the gate-to-cathode of SCR 1*Q*. The inductive kick voltage of 102*T* is used to trigger 1*Q*. This guarantees that the SCR will start conducting at the beginning of the positive anode voltage cycle as shown in Fig. 11–9 and will not trigger on line transients or noise. The 101*CR* relay is energized by current flow through 102 *RE* when its anode voltage is positive. The contacts of 101*CR* in the ignitor circuit are normally open to prevent the accidental firing of the high-current ignitron tubes. The energizing of 101*CR* causes these contacts to close, preparing the ignitrons for firing later in the sequence. Transformer 103*T*, in the anode circuit of 1*Q*, is energized when 1*Q* conducts. The secondary voltage of 103*T* is used in the initiation flip-flop circuit of 2*Q*–3*Q* to start the sequence.

11–2–4 The Sequence Start Circuit

The sequence start circuit is shown in Fig. 11–10. Transistors 2*Q* and 3*Q* make up a bistable flip-flop. Transistor 2*Q* is normally biased on with 3*Q* normally biased off. The ac voltage across the output of 103*T*–*SA* in series with the dc supply voltage is applied to the base of 2*Q* through 4*RE*,

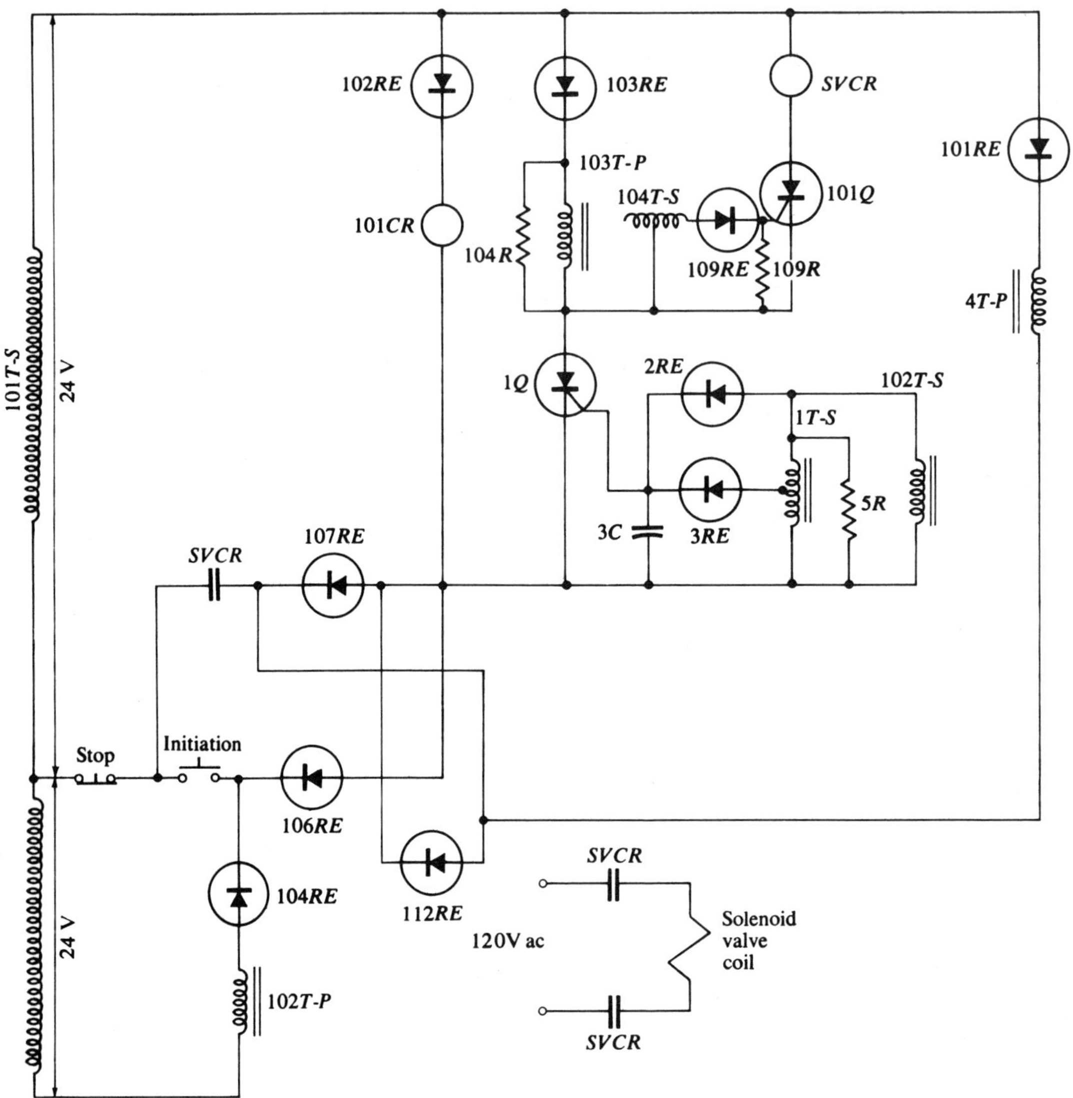

Fig. 11–8 Initiation circuit of MOD 3060 welder.

11*R*, and 12*RE*. This voltage is not quite sufficient to switch the flip-flop. A synchronizing pulse is required in coincidence with the output of 103*T–SA* to begin the sequence. The sync pulses are developed from the differentiated output of a Schmitt trigger. (See section 6–3–1 for a review of the

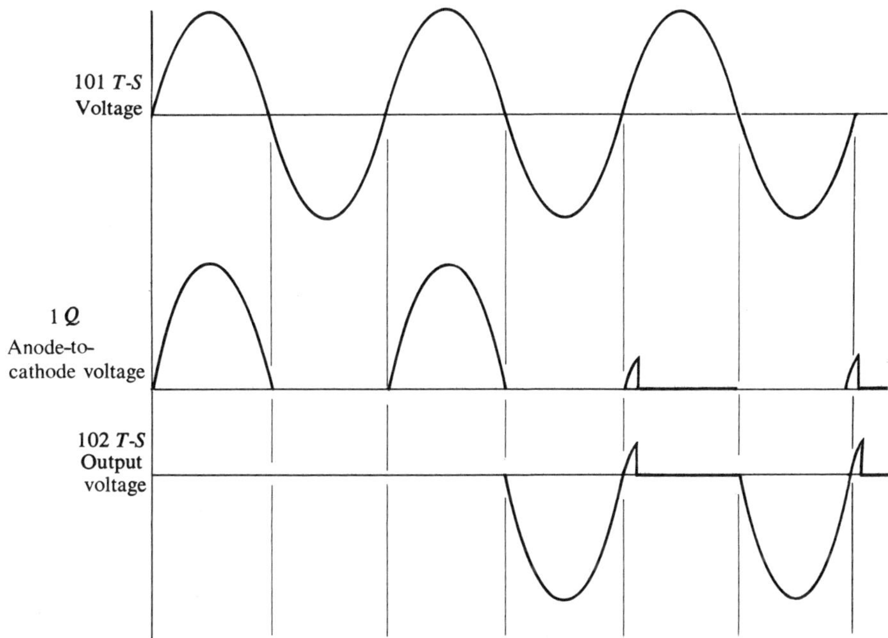

Fig. 11–9 Trigger voltage waveforms at initiation.

Schmitt trigger.) The Schmitt trigger circuit is shown in Fig. 11–11. Transistors 10*Q* and 11*Q* are used in a regenerative switching amplifier. The input voltage is taken from the secondary of 101*T*, but 180° out of phase with the line voltage as shown in Fig. 11–12(a). The square wave output of the Schmitt trigger is shown in Fig. 11–12(b). The square wave output is differentiated by capacitor 6*C* and applied to the base of 2*Q* through 12*RE*. The negative going pulses used to synchronize the start of the sequence occur at the beginning of the positive ac line voltage as shown in Fig. 11–12(c). Transistor 2*Q* is switched off at the beginning of the positive half-cycle of ac line voltage after the initiation switch is closed. The turning off of 2*Q* turns 3*Q* on, allowing current to flow through the primary winding of transformer 1*T*. The secondary winding of 1*T* is in the gate-to-cathode circuit of SCR 1*Q*. The positive voltage of 1*T–S* keeps 1*Q* conducting throughout the sequence. When conduction of 2*Q* ceases, the gate of 4*Q* is no longer clamped

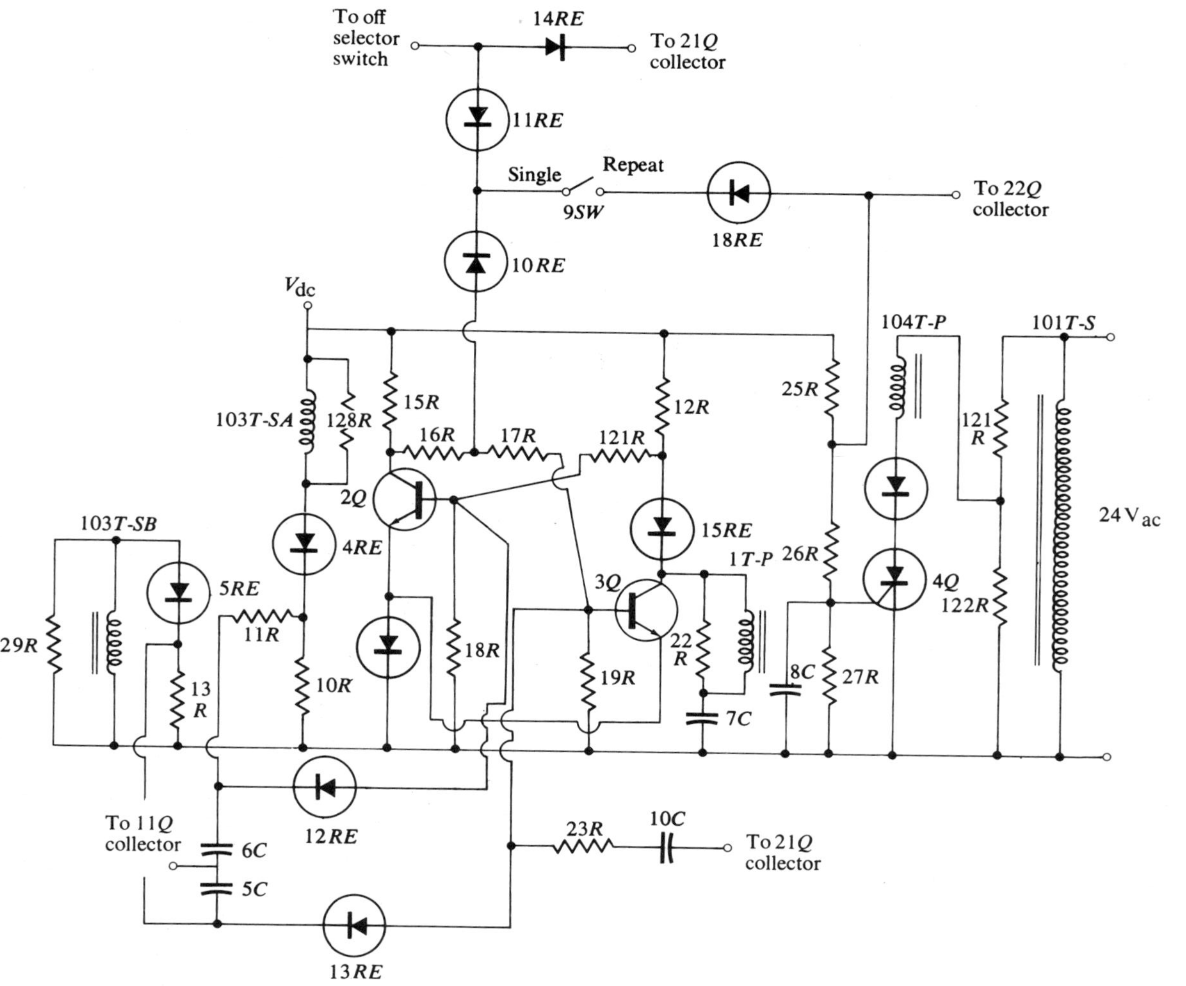

Courtesy of Robotron Corporation

Fig. 11–10 Sequence start circuit of MOD 3060 solid state welder.

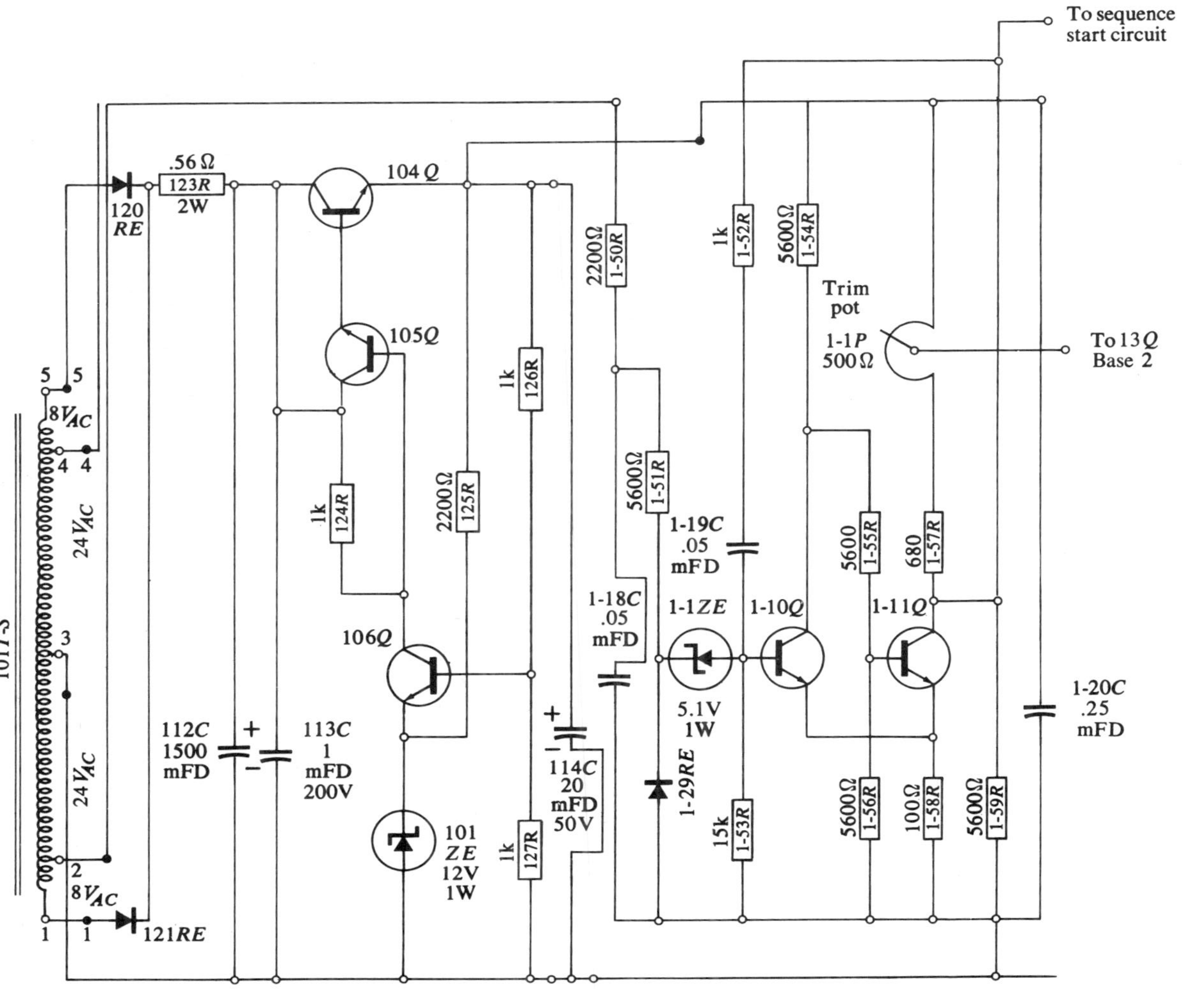

Courtesy of Robotron Corporation

Fig. 11–11 DC power supply and synchronizing circuits.

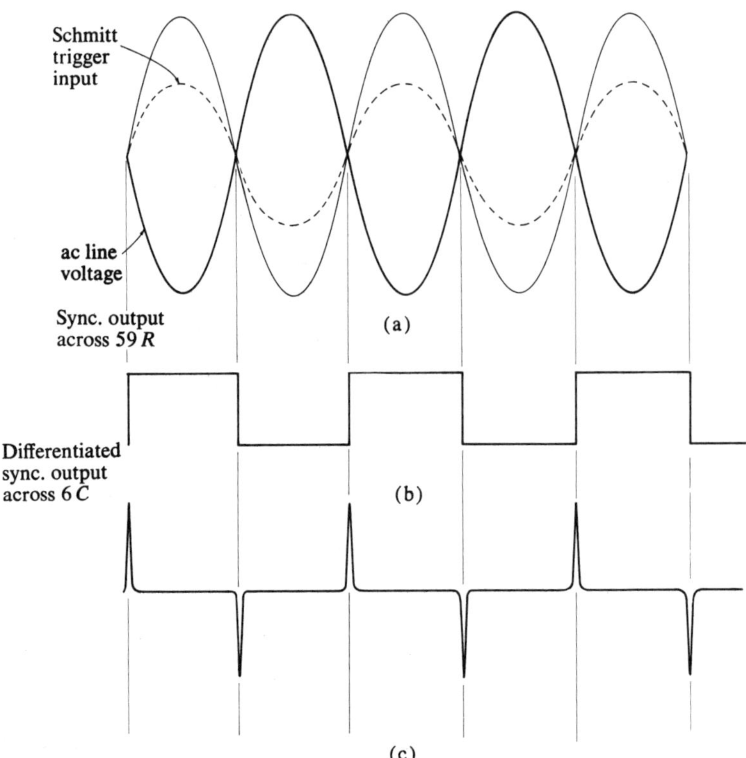

Fig. 11–12 (a) AC line voltage, (b) Schmitt trigger output, (c) differentiated Schmitt trigger output.

at the saturation voltage of 2*Q* by current flow through 17*RE*. The gate voltage at 4*Q* rises due to current flow through 25*R*, 26*R*, and 27*R*. SCR 4*Q* is switched on, energizing the primary winding of transformer 104*T*. The secondary of 104*T* is in the gate-to-cathode circuit of SCR 101*Q*. The positive voltage of 104*T*–*S* forces current through diode 109*RE* to trigger 101*Q*. Solenoid valve relay *SVCR* is energized by current flow through 101*Q*. Contacts in the solenoid valve coil power circuit are closed, activating the solenoid valve and applying pressure to the welding tips. This is the beginning of the squeeze phase of the machine cycle. A pair of contacts in the "seal-in" circuit are also closed by the solenoid valve relay, sealing the initiation circuit so that the initiation switch can be released if desired.

The output of 103*T–SB* applies a positive voltage at the base of 3*Q* through 5*RE* and 14*R* to prevent switching of 3*Q* when the negative-going sync pulse is applied. The sum of the two opposing signals must be positive to keep 3*Q* conducting during the entire sequence.

11–2–5 Timing the Squeeze Phase

The heart of this control circuit is the UJT timer of Fig. 11–13. Prior to initiation of the sequence, resistor 60*R* is connected to the negative terminal of the dc supply by conduction through 7*RE* and 2*Q*. The voltage divider formed by 61*R* and 60*R* across the dc supply limits the charging voltage on capacitor 21*C* to a very low level. Capacitor 21*C* is across the emitter to base 1 junction of the UJT but the voltage never reaches the level required to trigger the UJT. Prior to initiation of the sequence, current also flows through 1–*S*, 2–*S*, 66*R*, 39*RE*, and 2*Q* further clamping the voltage across 21*C* to a very low value. The UJT cannot fire while 2*Q* is conducting. When 2*Q* stops conducting, capacitor 21*C* charges through the squeeze timing resistors on the selector switches 1–*S* and 2–*S*. When the voltage on 21*C* reaches the required triggering voltage the UJT is ready to conduct. Resistors 62*R*, 63*R*, and diode 31*RE* limit the maximum charge on 21*C* to a level slightly below the triggering voltage required by the standoff ratio of the UJT. Triggering of the UJT is accomplished by the negative-going sync pulses applied at base 2 from the Schmitt trigger. The square wave output of the Schmitt trigger is differentiated through 22*C* so that the negative-going sync pulses occur at the beginning of the positive half-cycle of ac line voltage [see Fig. 11–12(c)]. Transistor 12*Q* is normally biased at saturation by current flow through 73*R* and 42*RE*. So long as this conduction exists, base 2 of the UJT is clamped to the positive side of the dc supply and the negative going sync pulses cannot cause triggering. One more condition must be met before the squeeze phase can end; that is, 12*Q* must be turned off. Transistor 12*Q* is turned off by the inductive kick of transformer 4*T–S*. The primary of 4*T* is connected to 101*S* through diode 101*RE*. The positive "kick" voltage applied at the base of 12*Q* at the beginning of the positive half-cycle forces this transistor off long enough for the sync pulse to cause the UJT to trigger. Capacitor 21*C* discharges through the emitter to base 1 junction causing a positive voltage pulse across 71*R* and

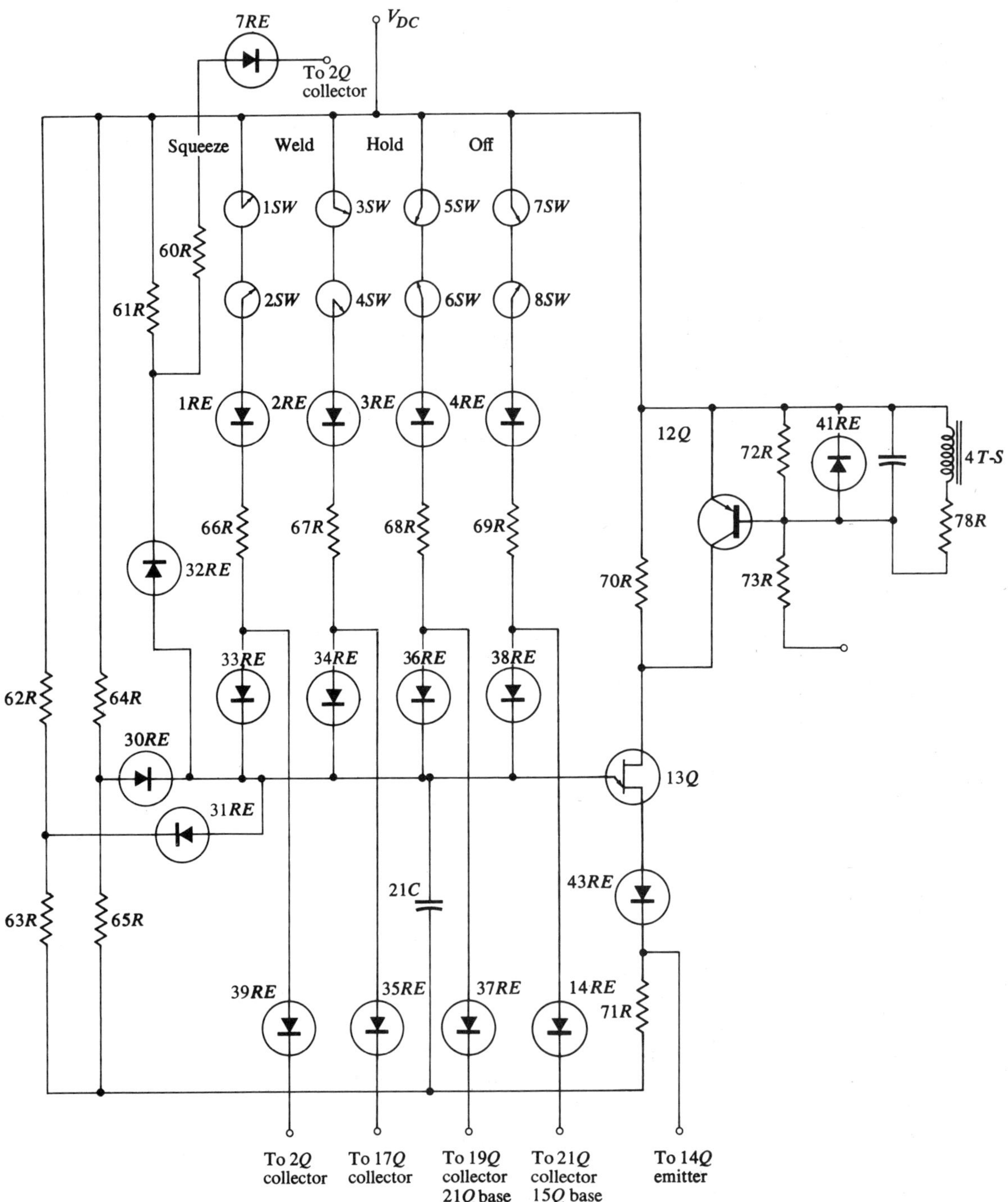

Fig. 11–13 UJT timer circuit.

75*R*. The output of the UJT cannot provide sufficient output to drive the flip-flops of the sequence control circuit. A buffer amplifier is required to boost the drive capability. Transistor 14*Q* is used to buffer the loading on the UJT. Before the arrival of the UJT output pulse, 14*Q* is biased on. The sudden increase in voltage across 75*R* raises the emitter of 14*Q* above the level of the base voltage and 14*Q* is turned off. The rise in voltage at the collector of 14*Q* is transferred through 45*RE* to the emitters of 16*Q*, 18*Q*, 20*Q*, and 22*Q* of Fig. 11–13, causing whichever one is conducting to turn off. At the beginning of the sequence only 16*Q* is conducting, so it is turned off. Transistors 15*Q* and 16*Q* are in a bistable flip-flop circuit so when 16*Q* turns off, 15*Q* turns on. Current flow through 1*S*, 2*S*, 66*R*, 40*RE*, and 15*Q* prevents capacitor 21*C* from recharging through the squeeze circuit. Resistors 64*R*, 65*R*, and diode 30*RE* form the charging platform for the ramp and pedestal action that we discussed in chapter 7, section 6.

11–2–6 Timing the Weld Phase

Sequencing from one phase to the next is accomplished by using four bistable flip-flops as shown in Fig. 11–14. Conduction of 15*Q* causes a negative-going pulse to be applied at the base of 17*Q* through capacitor 28*C*. The base voltage of 17*Q* is driven below the emitter voltage, causing 17*Q* to turn off and 18*Q* to turn on. Before 17*Q* was turned off, current flow through 49*R*, 28*RE*, and 17*Q* kept the voltage across capacitor 17*C* clamped at a very low level so that the PUT timer was inactive. With 17*Q* off, 17*C* is free to charge to the required anode voltage to trigger PUT 9*Q*, providing the delayed firing pulse for SCR 8*Q*. We will look at how this pulse is used to control the flow of weld current later. At the end of the present weld time, UJT 13*Q* will conduct, 14*Q* will be forced off by the positive going pulse at its emitter. The collector voltage of 14*Q* will rise, turning off 18*Q*. The turning off of 18*Q* will force 17*Q* on again, limiting the voltage on capacitor 17*C*. The PUT timer output is no longer available at the gate of 8*Q*. SCR 8*Q* is not triggered, preventing conduction of the high current ignitron tubes. This is the end of the weld phase.

11–2–7 Timing the Hold Phase

The turn-on of 17*Q* at the end of the weld phase sends a negative-going pulse through 29*C* to the base of 19*Q*. The

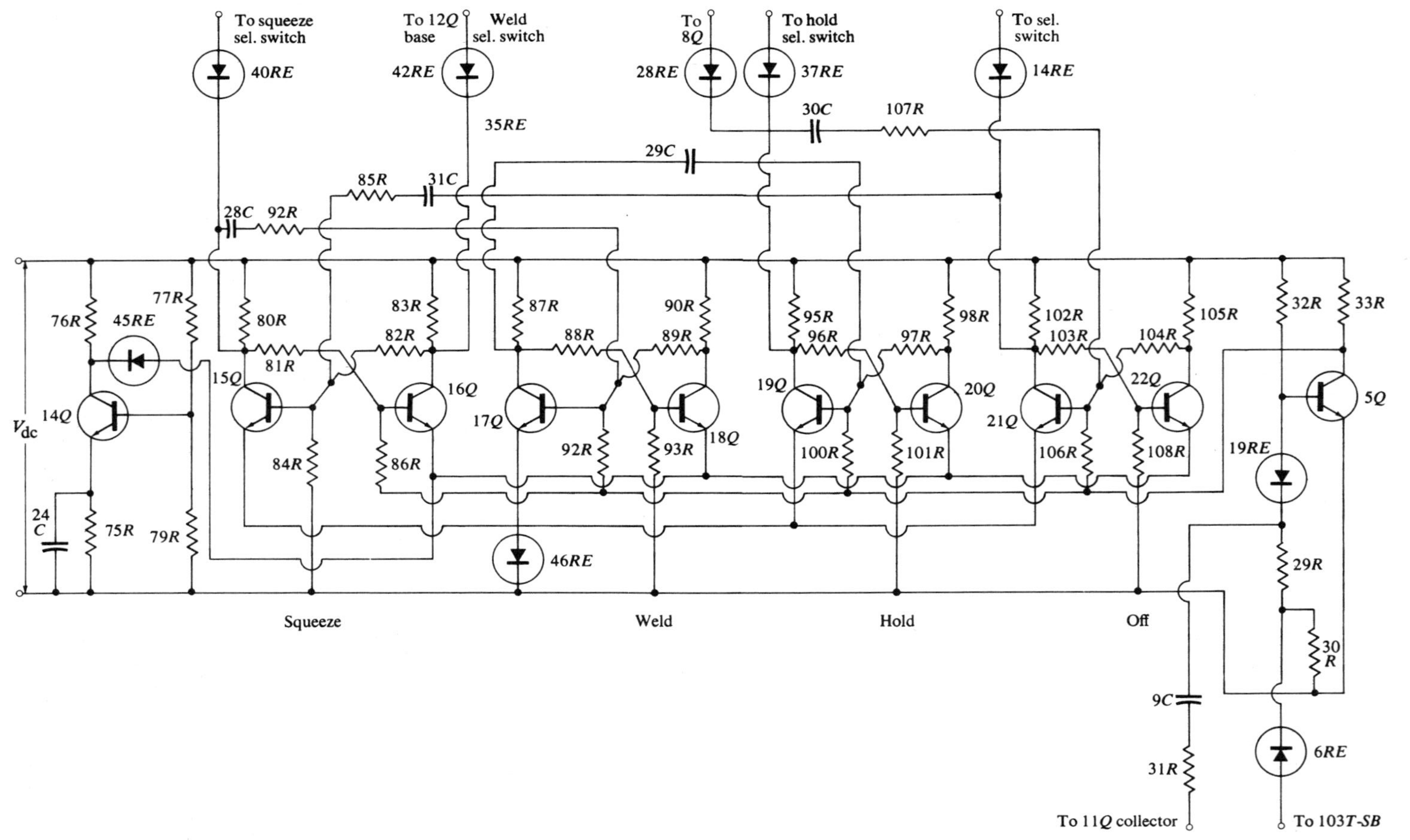

Courtesy of Robotron Corporation

Fig. 11–14 Sequence control circuit for MOD 3060 solid state resistance welder.

voltage at 19*Q* is driven below the emitter voltage and 19*Q* is turned off. Transistor 20*Q* is switched on by the flip-flop action of 19*Q*. The voltage rise at the collector of 19*Q* prevents current flow through 37*RE*. Capacitor 21*C* is now free to charge through the hold timing circuit of 5*S*, 6*S*, 68*R*, and 36*RE*. When the voltage on 21*C* reaches the required triggering voltage, UJT 13*Q* is fired, discharging 21*C* through 43*RE*, 71*R*, 44*RE*, and 75*R*. The emitter of 14*Q* is raised above the voltage level at the base long enough to turn 14*Q* off. The increase in voltage at the collector of 14*Q* is transmitted through 45*R* to the emitter of 20*Q*. Transistor 20*Q* will be turned off, causing 19*Q* to switch on. Conduction of 19*Q* caused a negative-going pulse to be applied at the base of 21*Q* through 30*C*, turning 21*Q* off. Transistor 22*Q* will be turned on by the flip-flop action of 21*Q*. When 22*Q* is turned on, current flow through 25*R* will be diverted from the gate of SCR 4*Q* through 18*RE* and 22*Q*. SCR 4*Q* will cause conduction at the end of the positive half-cycle of its anode voltage. Since there is no current flow through 104*T–P* when 4*Q* is not conducting, the voltage across 104*T–S* in the gate circuit of 101*Q* will be lost. SCR 101*Q* will stop conducting at the end of its positive half-cycle of anode voltage. The solenoid valve relay coil is no longer energized; contacts in the solenoid valve power circuit are opened, releasing the pressure on the weld tips. This is the end of the hold phase of the machine cycle.

11–2–8 Timing of the Off Phase

If switch 9*S* of Fig. 11–10 is in the repeat position, the machine will remain at rest or in the off phase for up to 61 cycles and then repeat the machine sequence as long as the initiation switch is closed. The amount of off time is controlled by setting switches 7*S* and 8*S* at the proper precalibrated position. When 21*Q* turns off at the end of the hold phase, current flow through 7*S*, 8*S*, 69*R*, 14*RE*, and 21*Q* is discontinued. Capacitor 21*C* then charges through 7*S*, 8*S*, 69*R*, and 38*RE* at a rate determined by the selection of resistors in 7*S* and 8*S*. When the voltage on capacitor 21*C* reaches the required triggering voltage, UJT 13*Q* conducts, discharging 21*C* through 43*RE*, 71*R*, 44*RE*, and 75*R*. The positive voltage pulse at the emitter of 14*Q* turns 14*Q* off long enough for the rise in voltage at its collector to turn off 22*Q*. When 22*Q* turns off, 21*Q* will turn on again. The drop in voltage at the collector of 21*Q* sends a negative-going pulse to the base of 3*Q* through 10*C* and 23*R*, causing 3*Q* to turn off. The positive voltage at the gate of 1*Q* will not be present however; 102*T* will supply the gate triggering signal to 1*Q* as long as the initiation switch is closed. Transistor 2*Q* will be turned on by the flip-flop action of 3*Q*. Current flow through 1*S*, 2*S*, 1*RE*, 66*R*, and 2*Q* will clamp the voltage on 21*C* to a very low level. The UJT will not fire and the timer remains at rest. Since the initiation switch is held closed, 2*Q* will be driven negative by the next synchronizing pulse applied at its base through 6*C*. This is the end of the off phase. The machine cycle will start over again. If the initiation switch is opened before the end of the off phase, when 21*Q* goes into conduction again, applying a negative pulse to the base of 3*Q* to turn it off, transformer 1*T* will be de-energized. Since the initiation switch is open, 102*T*–*S* can no longer keep 1*Q* conducting. When 1*Q* stops conducting, the voltage at 103*T*–*SB* is lost so the next negative-going sync pulse will be applied at the base of 5*Q* (Fig. 11–14) through 31*R* and 9*C*. Transistor 5*Q* will stop conducting. The rising voltage at the collector of 5*Q* will be applied to the bases of 16*Q*, 17*Q*, 19*Q*, and 21*Q*, resetting the sequence control flip-flop to the starting position. The machine will remain at rest awaiting the closing of the initiation switch.

11–2–9 The Heat Control Circuit

The amount of heat developed during the welding phase is controlled by phase control of the firing of the ignitron tubes. The extended *RC* phase shift control circuit of Fig. 7–4(a) is used to provide the variable amount of phase shift for delayed triggering. The center-tapped secondary winding of 101*T*, capacitor 110*C*, 1*P*, and 2*P* are used in the phase shift circuit as shown in Fig. 11–15. The phase-shifted output voltage is developed across the primary of transformer 2*T*. The amount of phase lag is determined by the setting of pots 1*P* and 2*P*. If both pots are set at maximum resistance, the phase lag will be maximum as shown in Fig. 7–4(b). The secondary windings of 2*T* are in the full-wave rectifier circuit of 20*RE*, 21*RE*, 12*C*, and 13*C*. The inverted output of the rectifier is applied at the base of transistor 6*Q* through the voltage divider action of 34*R* and 37*R*. Transistor 6*Q* is normally biased on when there is no output from the rectifier. Application of the rectified output of the phase shift circuit causes 6*Q* to conduct only during the short time

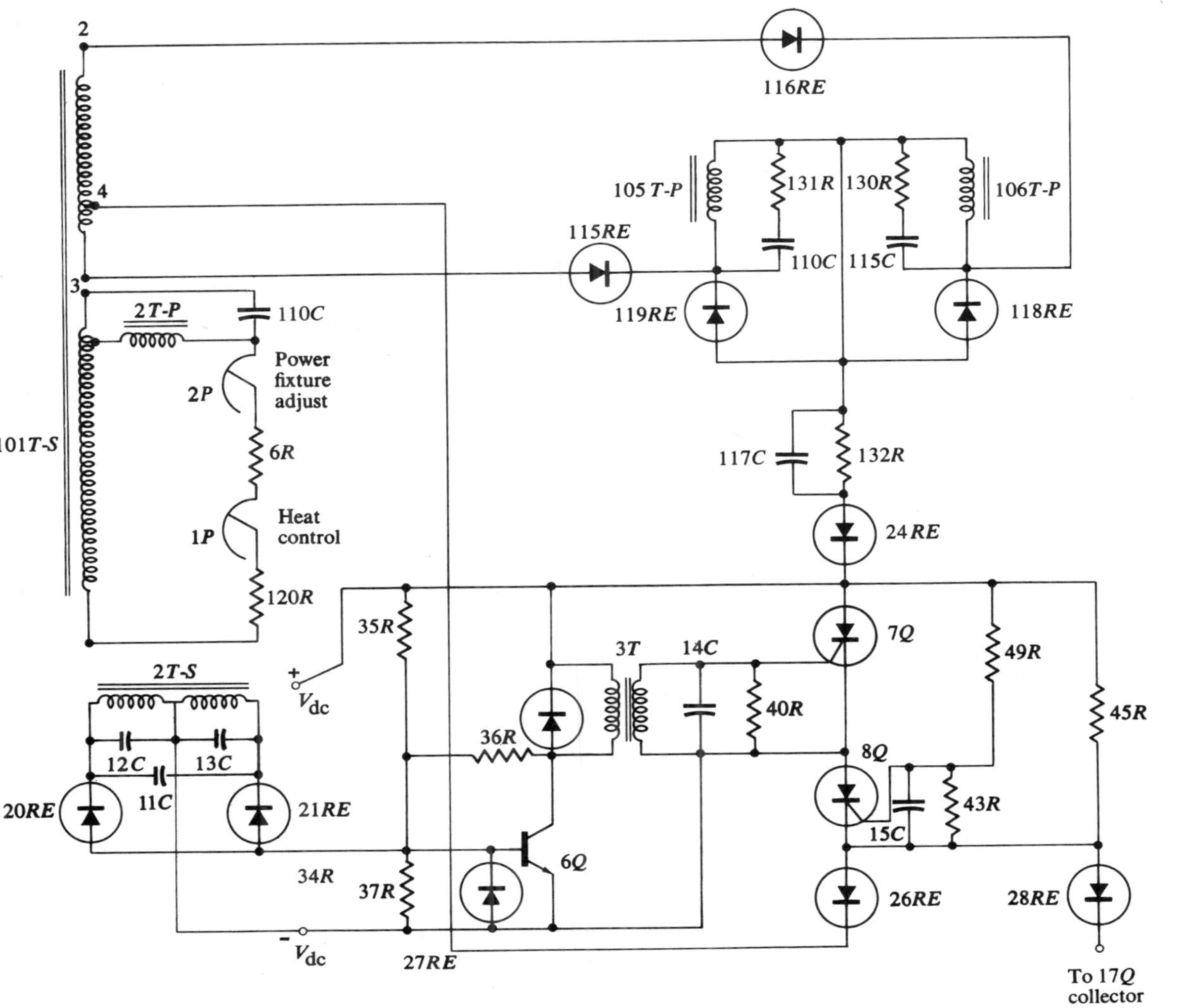

Courtesy of Robotron Corporation

Fig. 11–15 - Heat control circuit of MOD 3060 welder.

when the rectifier voltage is near zero. Transformer 3*T* is energized during the short time that 6*Q* conducts. Positive voltage pulses at the secondary of 3*T* are used to trigger SCR 7*Q*. SCR 7*Q* receives a positive triggering pulse every half-cycle while the machine is operating but can only conduct when SCR 8*Q* is conducting. At the beginning of the machine sequence, the gate of 8*Q* is clamped by diode 28*RE* conducting through 45*R* and 17*Q*. When 17*Q* is switched off at the beginning of the weld cycle, the gate of 8*Q* is driven positive by current flow through 45*R* and 43*R*. SCR 8*Q* is turned on when 7*Q* is conducting by positive voltage from terminal 2 of 101*T–S*. Current flow through 116*RE*, 132*R*, and 24*RE* energizes 106*T–P*, providing the positive gate pulse to trigger SCR 103*Q* in the weld current circuit of Fig. 11–16. During the other half-cycle of line voltage, 8*Q* is turned on by positive voltage at terminal 3 of 101*T–S*. Current flow through 115*RE*, 132*R*, and 24*RE* energizes 105*T–P*, providing the positive gate pulse to trigger SCR 102*Q* in the weld circuit of Fig. 11–16. During the weld phase, SCR 8*Q* determines the number of cycles of line voltage that the ignitron tubes conduct by a control signal from 17*Q*, but SCR 7*Q* determines the exact point at which the ignitron tubes fire during each half-cycle. At the end of the weld phase, 18*Q* is turned off and 17*Q* is turned on, clamping the gate of 8*Q* very close to the voltage level of its cathode. Since 8*Q* can no longer conduct, 105*T* and 106*T* are not energized and current flow through the ignitron tubes is blocked, ending the flow of welding current.

11–2–10 The Ignitron Weld Current Circuit

The ignitron weld current circuit shown in Fig. 11–16 consists of two SCRs and two ignitron tubes. The ignitron is a gas tube capable of conducting thousands of amperes of current under low-duty-cycle operations such as welding. Figure 11–17 is a simplified view of a typical ignitron. The cathode is made of a pool of liquid mercury at the bottom of a large glass tube. When a positive voltage is applied at the ignitor terminals, a high-intensity electric field is developed between the ignitor tip and the surface of the mercury pool. Electrons are pulled from the surface and form a cloud of charges. If a positive voltage is applied at the anode at the same time, the electrons will be accelerated from the cathode to the anode, ionizing the gas in the tube and causing conduction. You can see the similarity between

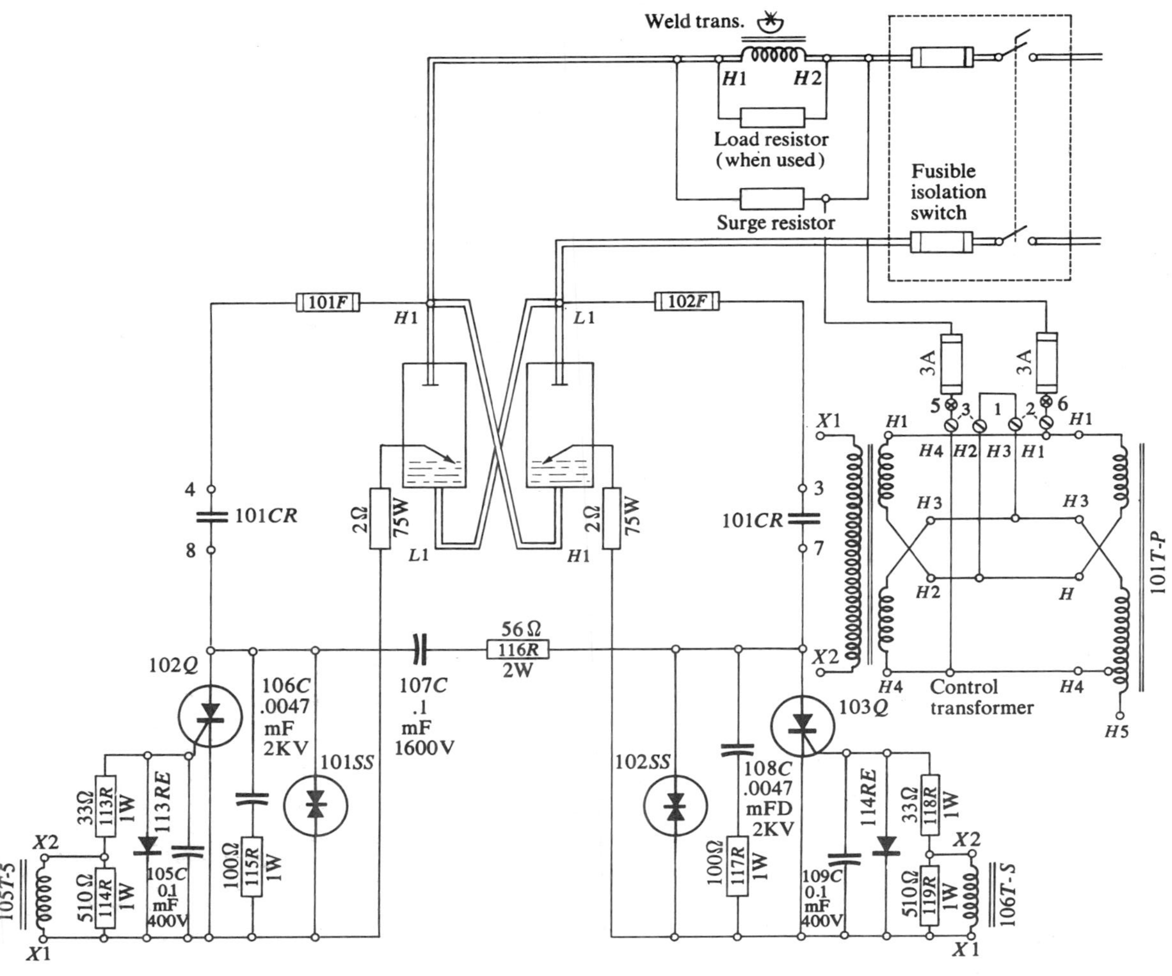

Courtesy of Robotron Corporation

Fig. 11–16 Ignitron weld current circuit.

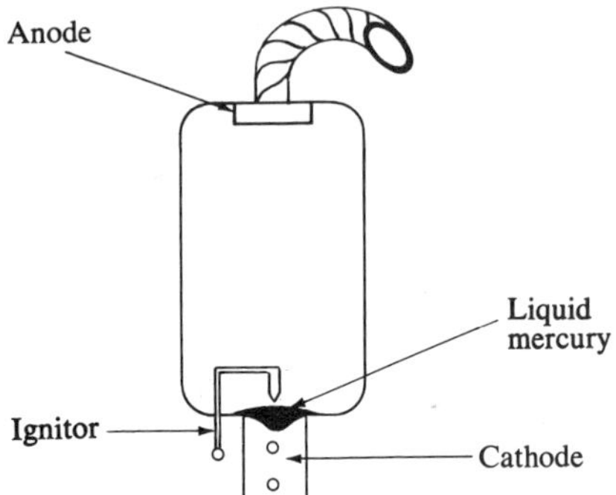

Fig. 11–17 Simplified view of an ignitron tube.

the SCR and the ignitron. SCRs are replacing ignitron tubes in some applications, but there are still many in use in industrial applications such as welding. The mercury pool in the ignitron tube is not contained so the tube must be operated in the vertical position. The ignitron tube is usually kept cool by circulating water through a metal jacket enclosing the tube itself.

When positive pulses occur across 105*T–S* and 106*T–S*, the SCRs conduct, applying a positive voltage at the ignitors of the tubes. Notice the ignitrons are connected in an inverse parallel or back-to-back configuration just like SCRs. They are sometimes referred to as a "contactor." The safety contacts are actuated by 101*CR* in the initiation circuit to prevent accidental firing of the ignitron tubes. Thyrectors 101*SS* and 102*SS* (see section 1–2c) across the SCR anode-to-cathode protect the SCRs from damage by high voltage transients. Resistor 115*R* and capacitor 106*C* prevent high-frequency transients from affecting SCR 102*Q*. 117*R* and 108*C* provide the same protection for 103*Q*. The fusible disconnect switch shown in Fig. 11–16 is almost an industry standard requirement on heavy equipment of this type. It is on the upper left-hand side of the enclosure of Fig. 11–6. The main control transformer, 101*T*, and most of the other components are mounted on the power panel shown in Fig. 11–18.

11–3 The SCR Contactor

The ignitron tubes used in the contactor circuit of Fig. 11–14 are being replaced in most applications by SCRs.

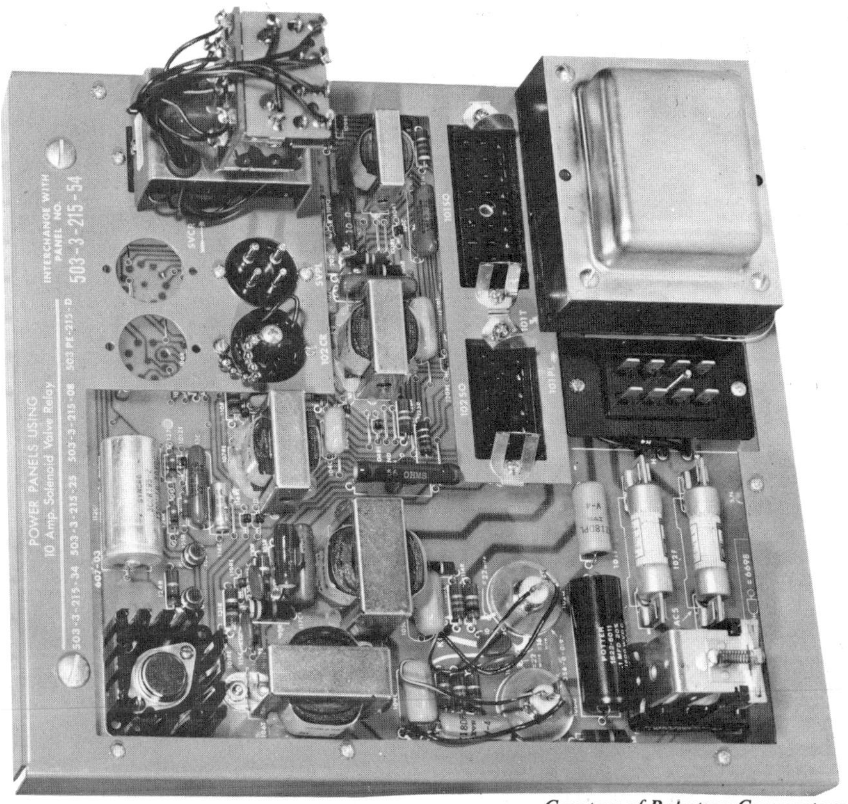

Courtesy of Robotron Corporation

Fig. 11–18 Power panel of the MOD 3060 welder.

Two high-current SCRs are connected in an inverse parallel configuration to make a full-wave contactor. Figure 11–19 shows an SCR contactor used in the high-current circuit of a Robotron Resistance Welder. The simplified schematic is shown in Fig. 11–20. Control pulses to trigger the SCR contactor are supplied by the timing and sequencing circuit. A fusible isolation switch is used in most industrial circuits. This switch is on the upper right-hand door of the NEMA enclosure of Fig. 11–19. Resistor $1R$ and capacitor $1C$ limit dv/dt to protect the SCRs. Resistor $2R$ dissipates the surge of the load transformer during the phase-controlled switching. Fuse F_3 provides fault protection for the SCR pair. You will notice that there is no large finned heat sink to cool the SCRs. It is more common to use a circulating water system to cool the SCRs. The large water hoses are visible in the lower chamber of the enclosure of Fig. 11–19.

Fig. 11–19 SCR contactor in Robotron resistance welder.

11–4 Summary

We have devoted most of this chapter to a commercial solid state resistance welder. The weld control circuitry contains a great variety of applications of the electronics that we have been discussing in earlier chapters. The UJT, SCRs, transistors, *RC* phase shift circuits, multivibrators, the Schmitt trigger, and time delays are used in this complex circuit. Some of the techniques used in weld control circuits can be applied to other designs that you might encounter. The noisy electrical environment in which the welder operates demands extra design precautions to prevent false triggering due to electrical transients. Keep this in mind. Refer to this chapter when you have similar requirements.

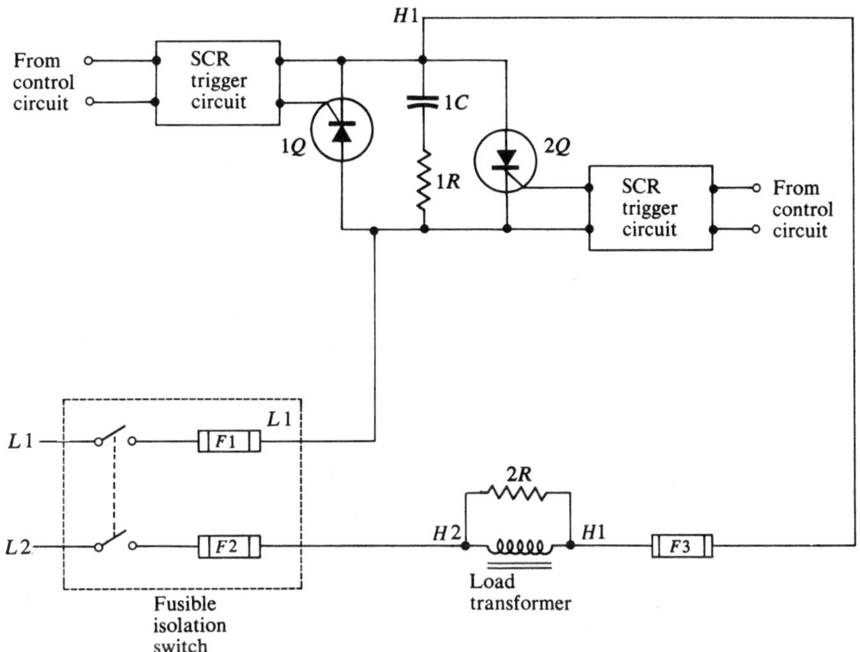

Fig. 11–20 SCR contactor circuit.

EXERCISES

1. Explain why a tapped transformer allows for reduction in the form factor of a high-current phase control system.

2. The firing angle of the SCR of Fig. 11–1 is 80°.
   (a) What is the form factor?
   (b) If the average current is 800 A, what is the rms current?

3. What is the application of the Schmitt trigger of Fig. 11–11?

4. The UJT circuit from Fig. 11–13 is shown in Fig. 11–21 for a typical setting of one of the selector switches. Determine:
   (a) The pedestal level
   (b) The UJT trigger voltage if $\eta = 0.75$
   (c) The magnitude of the synchronization pulse at base 2 to trigger the UJT

5. Follow the weld control circuit operation if $9S$ is in the "single" position.

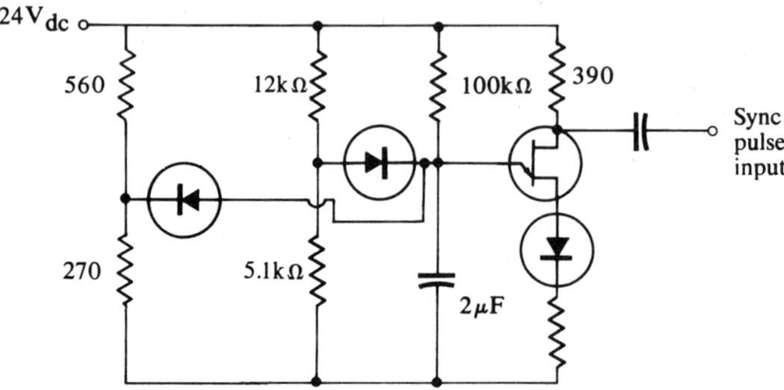

Fig. 11–21

6. In the phase shift circuit of Fig. 11–15, 110*C* is 0.5 μF, 1*P* is a 10-kΩ potentiometer, 2*P* is a 3.5-kΩ potentiometer, and 120*R* is 330 Ω. The transformer is 48 V, center-tapped.
(a) What is α_{max}?
(b) What is α_{min}?
(c) What is the maximum and minimum output voltage of the *RC* phase shift circuit?

7. Ignitron tubes must always be operated in the vertical position with the anode terminal up. Why?

8. A thermal circuit interrupt is usually mounted on the jacket of an ignitron tube to avoid operation above the maximum allowable temperature. Where would you place the contacts in the circuit of Fig. 11–16?

12

DC Power Supplies

12–1 Introduction

Many devices in industry such as welders, motors, and solid state digital controls require dc power sources. Since the power distribution system is usually single-phase or three-phase ac, a *power supply* is required to convert the ac voltage to dc voltage. The power supply might be as crude as a simple half-wave diode rectifier or as sophisticated as a highly regulated, temperature-stabilized laboratory supply. The power supply might be built onto the same chassis as the electronic circuitry or it might be a separate chassis. In some cases, such as the high-current dc welder, the dc power package constitutes a major portion of the machine.

The dc power supply circuitry can be separated into four sections: the rectifier, filter, regulator, and protective or control circuitry associated with the proper functioning of the supply. We will examine typical circuitry of the four sections and look at applications from the schematics of a major manufacturer in this chapter.

12–2 Rectifier Circuits

The nonlinear property of the semiconductor diode is used to convert ac voltages to unipolar or single polarity voltages. Either single-phase or three-phase ac voltages are used depending on the amount of power demanded by the load. Parameters that we must be concerned about in designing or applying dc power supplies are dc output voltage, dc output current, diode PIV, diode current, output ripple, transformer secondary voltage ripple frequency, and transformer efficiency. Table 12–1 shows the values of these parameters for the common rectifier configurations. We will look at each configuration in more detail.

TABLE 12–1
Summary of Rectifier Circuits and Parameters

Parameters	Single-Phase Half-Wave	Center-Tapped Full Wave	Single-Phase Bridge Full Wave	Three-Phase Half-Wave	Three-Phase Bridge	Six-Phase Star	Comments
DC voltage, V_{dc}	V_{dc}	V_{dc}	V_{dc}	V_{dc}	V_{dc}	V_{dc}	Average value of rectified output voltage
DC current, I_{dc}	$^*I_{dc}$	I_{dc}	I_{dc}	I_{dc}	I_{dc}	I_{dc}	Average value of rectified current for highly inductive load
Number of diodes	1	2	4	3	6	6	
Diode PIV	$3.14 V_{dc}$	$3.14 V_{dc}$	$1.57 V_{dc}$	$2.09 V_{dc}$	$1.05 V_{dc}$	$2.09 V_{dc}$	
Average current per diode I_D	I_{dc}	$I_{dc}/2$	$I_{dc}/2$	$I_{dc}/3$	$I_{dc}/3$	$I_{dc}/6$	Assuming highly inductive load
Ripple frequency	f	$2f$	$2f$	$3f$	$6f$	$6f$	f is frequency of ac source
Percent ripple	121%	48%	48%	18.3%	4.2%	4.2%	V_{rms} of ripple/V_{dc}
Transformer secondary voltage per leg	$2.22 V_{dc}$	$1.11 V_{dc}$	$1.11 V_{dc}$	$0.855 V_{dc}$	$0.428 V_{dc}$	$0.740 V_{dc}$	rms value
Transformer efficiency, η_T	0.286	0.636	0.90	0.675	0.636	0.951	dc watts output/ac input

*Resistive load.

12–2–1 Single-Phase Half-Wave Rectifier

The simplest rectifier is the single-phase half-wave configuration of Fig. 12–1(a). If the input voltage is sinusoidal, the load voltage will be a half sinusoid as in Fig. 12–1(b). Let's calculate the parameters of Fig. 12–1 for this rectifier. The sine wave input voltage is:

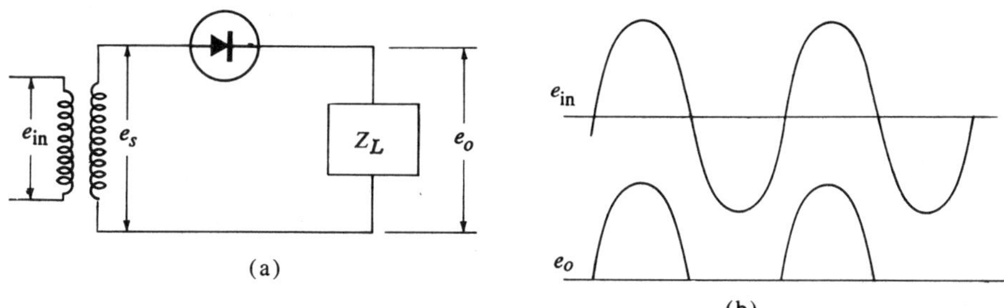

Fig. 12–1 (a) Single-phase half-wave rectifier, (b) rectified load voltage.

$$e_{in} = E_{max} \sin \omega t \qquad \textbf{(12.1)}$$

where E_{max} is the maximum value of the voltage, ω is the frequency in rad/s. The average value of the rectified load voltage is given by the equation

$$V_{av} = \frac{1}{2\pi} \int_0^{\pi} E_{max} \sin \omega t \, d\omega t = \frac{E_{max}}{\pi} = V_{dc}$$

The diode PIV must be at least as great as E_{max} so

$$\text{PIV} = E_{max} = \pi V_{dc} = 3.14 \text{ V}_{dc}$$

To calculate percent ripple, we must look at the rms value of the ac ripple waveform. Figure 12–2 shows the ripple waveform for the single-phase half-wave rectifier.

$$\text{Percent ripple} = \frac{V_{rms} \text{ of ripple}}{V_{dc}} \times 100 = \frac{V_{rip}}{V_{dc}} \times 100 \qquad \textbf{(12.2)}$$

The rms value of the half-wave output voltage is given by the equation

$$V_{rms} = \sqrt{\frac{1}{2\pi} \int_0^{\pi} E_{max} \sin^2 \omega t \, d\omega t} = \frac{E_{max}}{2} = \frac{\pi}{2} \text{ V}_{dc} \qquad \textbf{(12.3)}$$

Then

$$V_{rip} = \sqrt{V_{rms}^2 - V_{dc}^2} \tag{12.4}$$

$$= \sqrt{\frac{\pi^2}{4} V_{dc}^2 - V_{dc}^2} = V_{dc}\sqrt{\frac{\pi^2}{4} - 1}$$

$$V_{rip} = V_{dc}\sqrt{1.42} = 1.21\,\mathrm{V_{dc}} \tag{12.5}$$

From Eq. (12.2),

$$\text{Percent ripple} = \frac{1.21\,\mathrm{V_{dc}}}{V_{dc}} \times 100 = 121\%$$

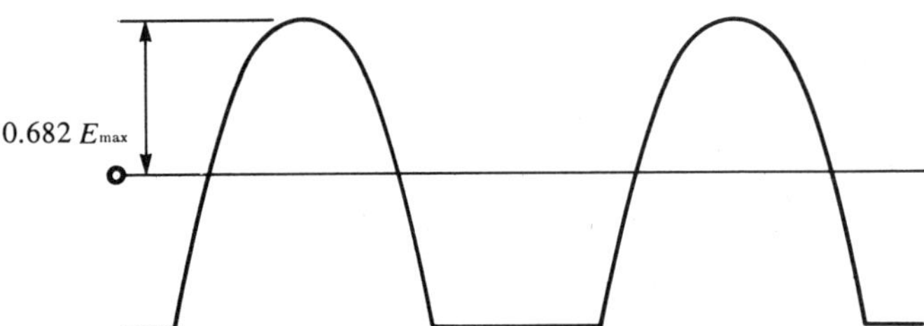

Fig. 12–2 AC ripple waveform of half-wave rectifier of Fig. 12–1(a).

The rms value of the transformer secondary voltage is determined from the equation

$$V_s = \sqrt{\frac{1}{\pi}\int_0^{\pi} E_{max}^2 \sin^2 \omega t \, d\omega t} = 0.707 E_{max} \tag{12.6}$$

Since $E_{max} = \pi V_{dc}$

$$V_s = 0.707\pi V_{dc} = 2.22\,\mathrm{V_{dc}} \tag{12.7}$$

The transformer efficiency of rectification is defined as the ratio of dc power output to ac volts-amperes input, or,

$$\eta_T = \frac{V_{dc} I_{dc}}{V_s I_s} \tag{12.8}$$

For the half-wave rectified current waveform

$$I_s = I_{rms} = \sqrt{\frac{1}{2\pi}\int_0^{\pi} \frac{E_{max}}{R^2} \sin^2 \omega t \, d\omega t} = 0.5\,\frac{E_{max}}{R}$$

$$= 0.5\pi\,\frac{V_{dc}}{R}$$

or

$$I_s = 0.5\pi I_{dc} \qquad \textbf{(12.9)}$$

Substituting from Eqs. (12.7) and (12.9)

$$\eta_T = \frac{V_{dc}I_{dc}}{2.22V_{dc}\pi\, 0.5I_{dc}} = \frac{1}{3.49} = 0.286$$

We have gone over these simple calculations to get an idea of how values in Table 12–1 are determined. Similar calculations are used to determine values of the parameters for other rectifier configurations.

The single-phase half-wave rectifier is used only in circuits where high ripple, poor regulation, and inefficiency can be tolerated. If the load current requirements are not too high, a filter capacitor across the output terminals can provide sufficient smoothing of the ripple to make this simple power supply acceptable.

12–2–2 Center-Tapped Full-Wave Rectifier

Another diode and transformer secondary can be added to the half-wave rectifier circuit of Fig. 12–1(a) to form the center-tapped full-wave rectifier of Fig. 12–3.

The center-tapped transformer provides the voltage, A-N, 180 degrees out of phase with the voltage B-N. When point A is positive with respect to point N, diode D_A is forward-biased and conducts; at the same time point B is

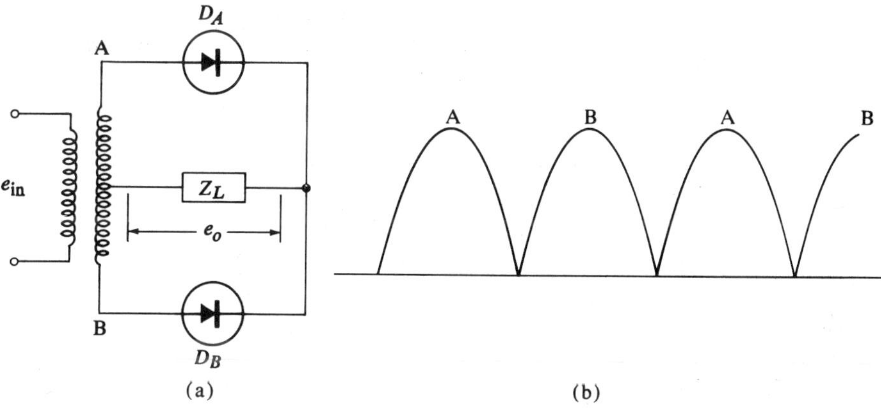

Fig. 12–3 (a) Center-tapped full-wave rectifier, (b) load voltage waveform.

negative with respect to point N so diode D_B is reverse-biased and maintains an open circuit. One half-cycle later, point A will be negative with respect to point N, diode D_B will be forward-biased and conducting. Load voltage will be alternately provided by conduction paths through D_A and D_B as shown in Fig. 12–3(b).

The center-tapped full-wave rectifier is a considerable improvement over the single-phase half-wave rectifier. Ripple is down to 48% while transformer efficiency is up to 0.636. The added cost of an extra diode and a center-tapped transformer must be considered. The PIV rating of the diode must be equal to the peak-to-peak value of the ac voltage from A to N or B to N.

12–2–3 The Single-Phase Bridge Rectifier

The single-phase bridge rectifier has the advantage of the same ripple as the center-tapped full-wave rectifier and increased efficiency without the expensive center-tapped transformer. Four diodes are required but the PIV is reduced by a factor of 2 over the center-tapped full-wave circuit.

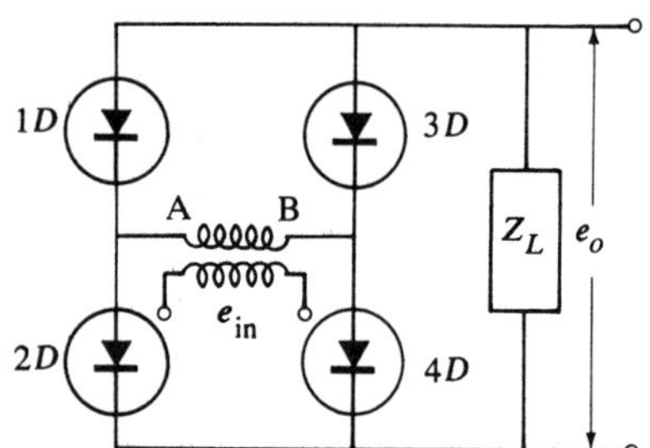

Fig. 12–4 Single-phase bridge rectifier.

Figure 12–4 shows the single-phase bridge rectifier circuit. When the voltage at point A is positive with respect to point B, diodes $2D$ and $3D$ are forward-biased and conducting. Diodes $1D$ and $4D$ are reverse-biased and effectively open-circuited. Figure 12–5(a) shows the conduction path for this condition. When the voltage at point B is positive with respect to the voltage at point A, diodes $1D$ and $4D$ are forward-biased and conducting. Diodes $2D$ and $3D$ are now reverse-biased and effectively open-circircuited. Figure 12–5(a) shows the conduction path for this condition.

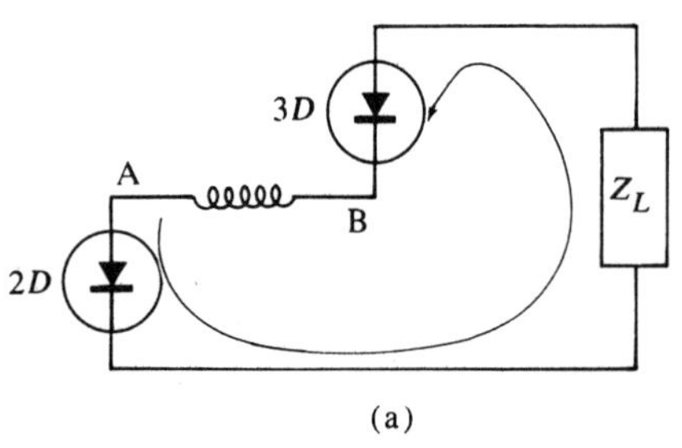

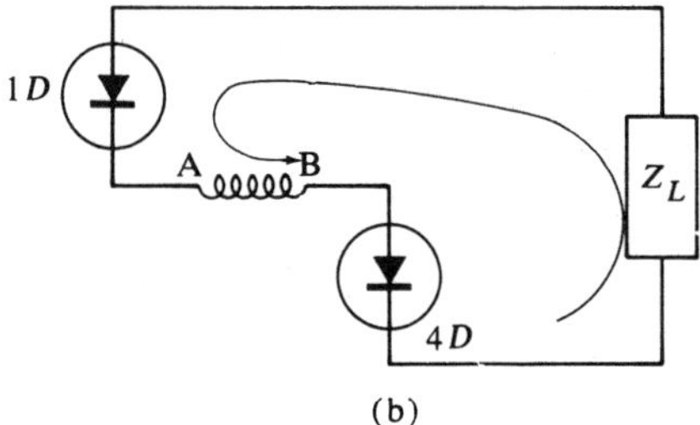

Fig. 12–5 (a) Circuit of Fig. 12–4 when V_{AB} is positive, (b) circuit of Fig. 12–4 when V_{BA} is positive.

The bridge is the more common single-phase full-wave rectifier configuration. The diode PIV rating is half that of the center-tapped full-wave rectifier for the same output voltage.

12–2–4 The Three-Phase Half-Wave Rectifier

As load requirements increase, the use of single-phase power is no longer practical and economical. Three-phase line voltages of 230 or 460 volts become the primary source for rectification. The simplest configuration is the 3-phase half-wave rectifier of Fig. 12–6(a). A diode is connected to each leg of the Y-secondary of the power transformer. Figure 12–6(b) shows the phase relationship between the leg-to-neutral voltages of the transformer secondary. The

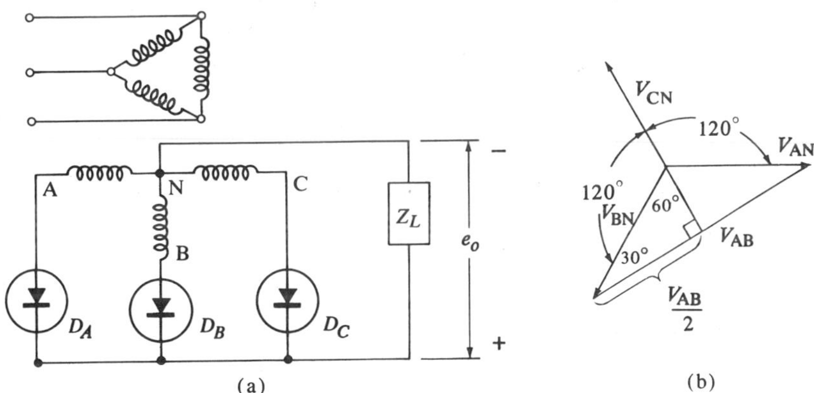

Fig. 12–6 (a) Three-phase half-wave rectifier, (b) three-phase line voltage.

voltages are of equal magnitude and 120 degrees apart. The leg-to-leg voltages V_{AB}, V_{BC}, and V_{CA} are larger than the leg-to-neutral voltages. The ratio can be determined by the triangle from *A-N-B* of Fig. 12–7(b). A perpendicular bisector is drawn from *N* to V_{AB}. Then:

$$\cos 30^\circ = \frac{V_{AB}}{2} \Big/ V_{BN} \qquad \textbf{(12.10)}$$

or

$$V_{AB} = 2 \cos 30^\circ \; V_{BN} \qquad \textbf{(12.11)}$$

but

$$\cos 30° = \frac{\sqrt{3}}{2}$$

so

$$V_{AB} = 2\,\frac{\sqrt{3}}{2}\,V_{BN} = \sqrt{3}\,V_{BN} \tag{12.12}$$

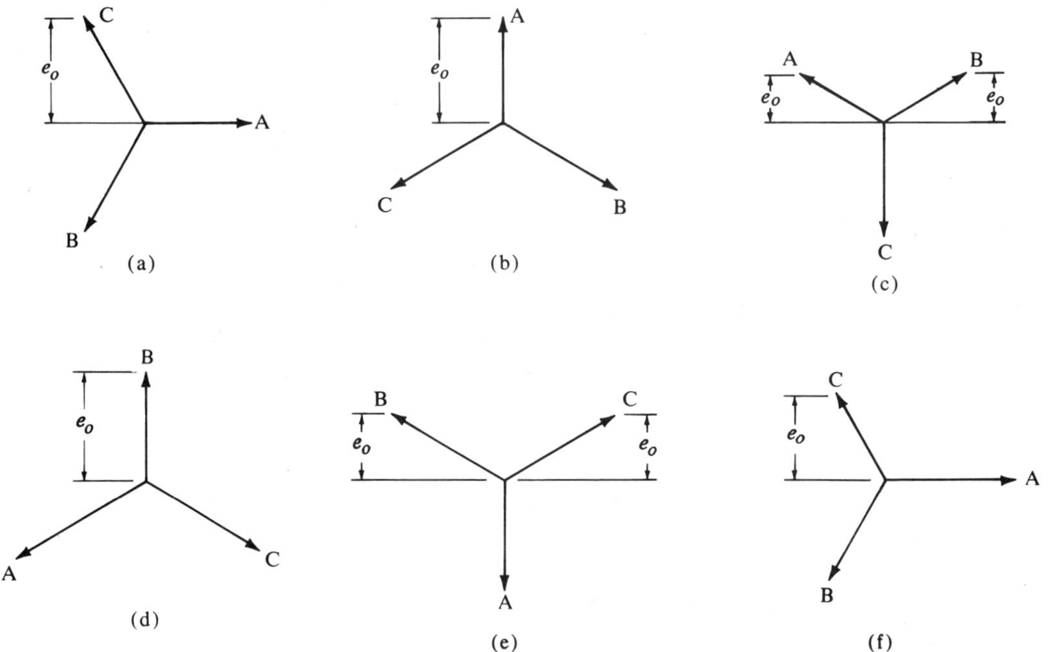

Fig. 12–7 Output of three-phase half-wave rectifier.

The leg-to-leg voltage is 1.732 times larger than the leg-to-neutral voltage of a 3-phase Y-secondary transformer.

The diode that has the largest positive bias voltage at any given time is the one that conducts and only one diode conducts at a time. Figure 12–7 shows the leg voltages at different times during one complete rotation of the phasors. In Fig. 12–7(a), diode D_C will be most positive and conducting. The output voltage will be the vertical component of the V_{CN} phasor. Ninety degrees later at Fig. 12–7(b) diode D_A will be the most positive and hence conducting. At this time, e_o is equal to V_{AN}. In Fig. 12–6(c), 60 degrees later diodes D_A and D_B are equally forward-biased. This is the point of "commutation" when conduction is switching from D_A to D_B. Sixty degrees later, in Fig. 12–6(d), diode D_B is the most positively biased and conducting. The output voltage is equal to V_{BN}. Sixty degrees later, diodes V_B and V_C are equally biased. This is the point of "commutation" where conduction is switching from D_B to D_C. At 90 degrees later D_C is the most positively biased and back at the same point as Fig. 12–6(a).

Figure 12–8 shows the continuous output voltage with the points of Fig. 12–7 labeled. Notice that the ripple has completed three cycles during one rotation of the phasor diagram. The ripple frequency is 3 times the line frequency.

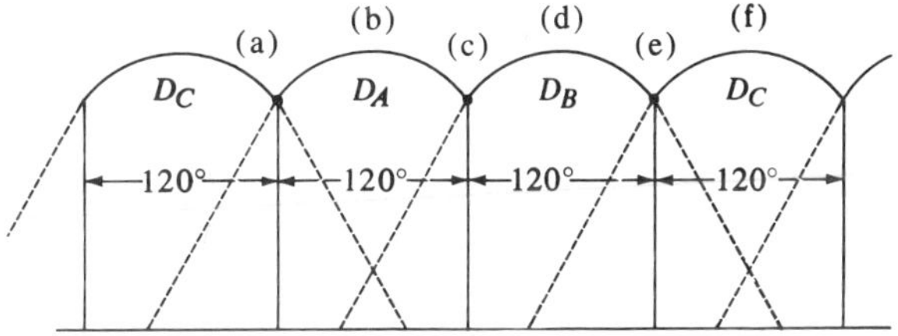

Fig. 12–8 Output voltage waveform as plotted from Fig. 12–7.

12–2–5 The Three-Phase Bridge Rectifier

Just as the bridge rectifier is used in many cases of single-phase rectification, a bridge rectifier is used quite often in three-phase rectifiers (Fig. 12–9). Two diodes must be conducting at all times to provide the complete path for current flow through the load. The bar graph of Fig. 12–11 shows which diodes are conducting at the instants the phasor diagrams of Fig. 12–10(a)–(f) were drawn. The two diodes that will be conducting at any given time are determined by the following rule:

The diode whose anode voltage is most positive will conduct. The diode whose cathode voltage is most negative will conduct.

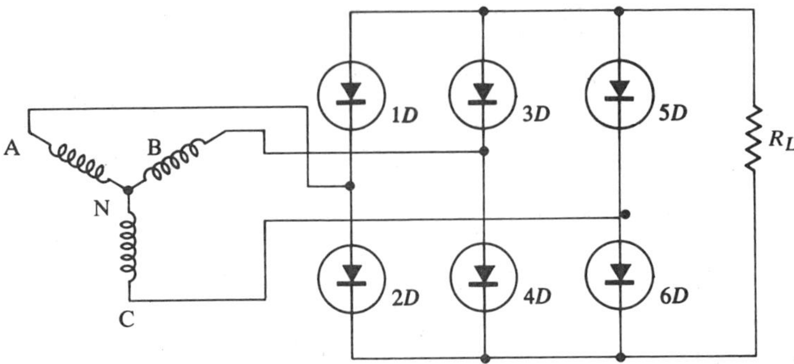

Fig. 12–9 Three-phase bridge rectifier.

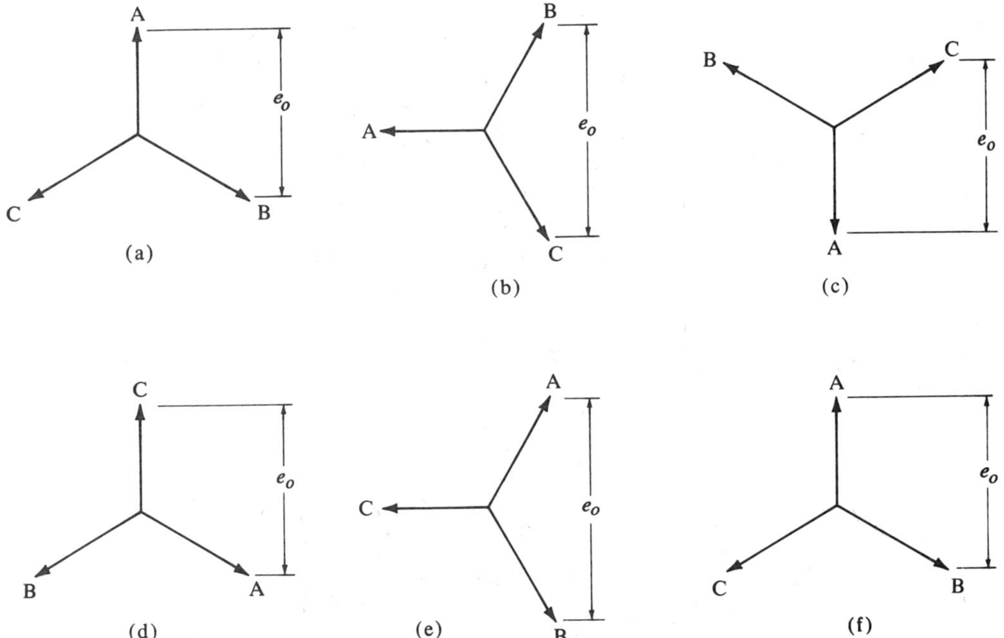

Fig. 12–10 Instantaneous phasor diagrams of bridge rectifier voltage.

In Fig. 12–10(a), leg voltage A is most positive. Since this voltage is applied at the anode of $2D$, $2D$ will be conducting. Leg voltages B and C are equally negative so $3D$ and $5D$ will both try to conduct. This is the point of "commutation." Conduction is switching from $3D$ to $5D$;

this is the point *a* in Fig. 12–11. In Fig. 12–10(b), leg voltage *B* is the most positive. Since voltage *B* is applied at the anode of 4*D*, 4*D* is conducting. Leg voltage *C* is the most negative so 5*D* will also be conducting. Continue the analysis through Fig. 12–10(f) and verify the graph of Fig. 12–11. Notice that commutation occurs six times during a complete cycle of line voltage. Each diode conducts for 120 degrees of each cycle. The output voltage of the bridge rectifier is shown in Fig. 12–12.

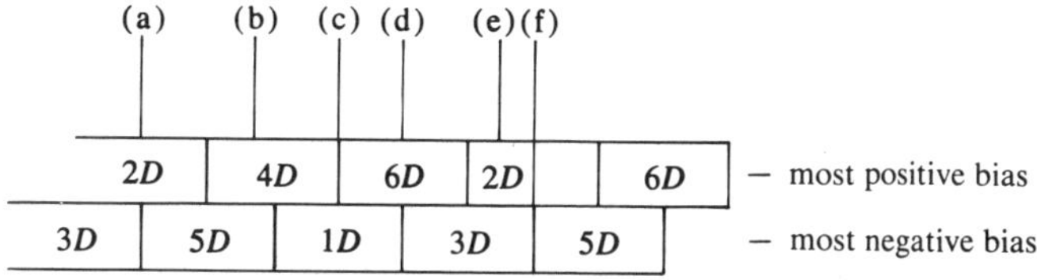

Fig. 12–11 Conducting diodes at the instant of phasor diagrams in Fig. 12–10.

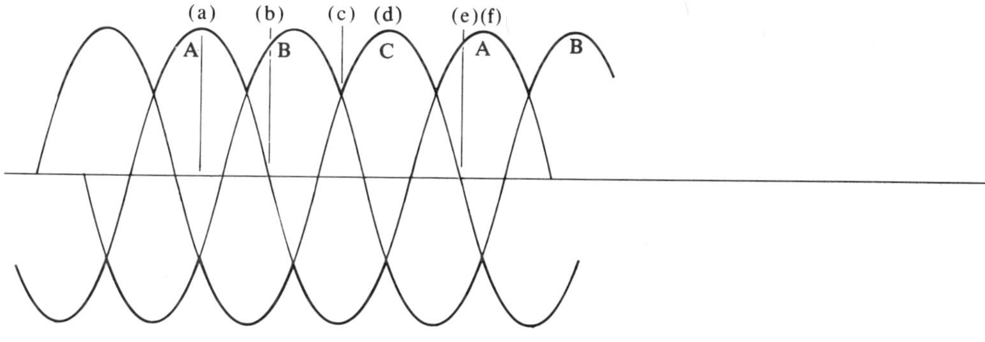

Fig. 12–12 Output voltage of three-phase bridge rectifier.

12–2–6 The Six-Phase Star Rectifier

Another rectifier circuit often used in industrial power supplies is the six-phase star type. If the three secondary windings of a three-phase transformer are center-tapped and all three center taps are connected together, six sinusoidal output voltages are obtained. As shown in Fig. 12–13,

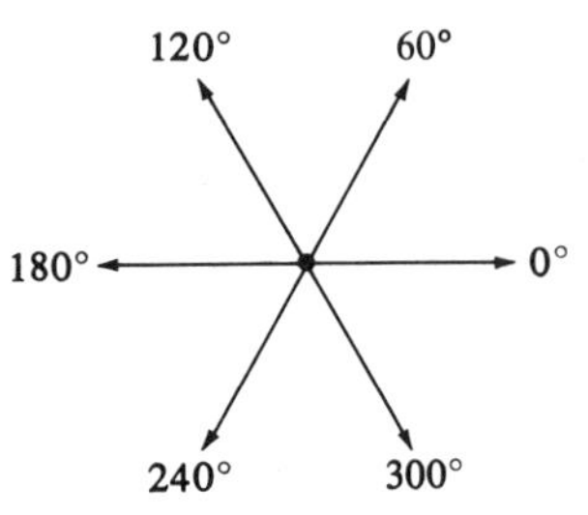

Fig. 12–13 Output voltages of six-phase transformer.

each voltage output is 60 degrees out of phase with the preceding output. A diode is placed in each leg of the transformer secondary. The result is effectively six half-wave rectifiers. The diode with the most positive anode voltage will conduct load current. Commutation occurs every 60 degrees as is shown in Fig. 12–14(b). Each diode conducts for only 60 degrees during each cycle of line voltage. The ripple frequency is six times the line frequency. The percent ripple of the star rectifier is reduced to 4.2%. The star rectifier is often drawn in the circular form of Fig. 12–14(a).

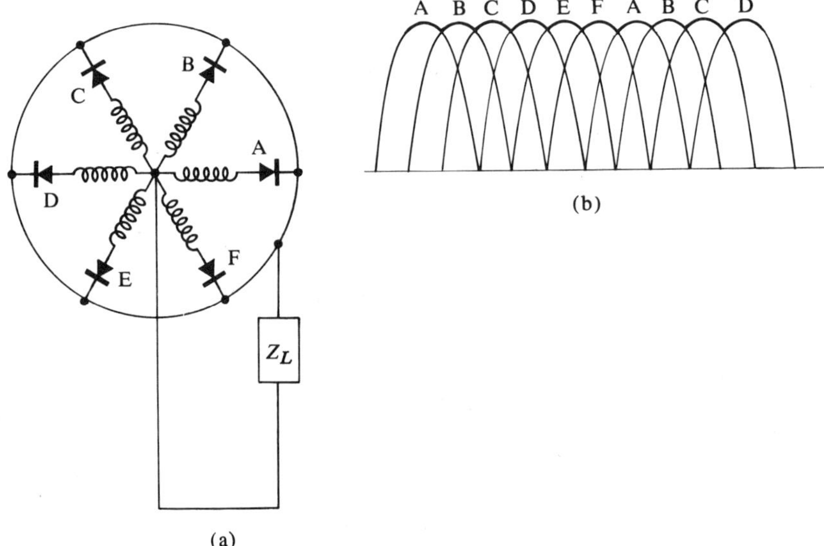

Fig. 12–14 (a) Six-phase star rectifier, (b) output voltage of six-phase star rectifier.

12–2–7 Design of Rectifier Circuits

Choice of the proper rectifier depends on the specifications of the user. If low ripple is required, the 3-phase bridge or 6-phase star rectifier should be used. If high values of dc voltage are required, the 3-phase bridge rectifier allows for a smaller transformer. If high output current is required, the 6-phase star minimizes the current demand of each diode. By reviewing the data of Table 12–1, you can point out other advantages and disadvantages of the various rectifiers based on user specifications. The designer must choose the most economical circuit that satisfies user specifications. An example will illustrate design of a rectified dc power supply using the data of Table 12–1.

Example 12–1. An industrial load of 0.3-Ω power requires 120 V dc. The line power system is 440 V, 3 phase, 60 Hz. Specify the components for a bridge rectifier, a 6-phase star rectifier, and indicate your choice.

If we look first at the bridge rectifier, using the data of Table 12–1:

$$\text{Diode PIV} = 1.05 \times 120 = 126 \text{ V}$$

$$\text{Diode average current} = \frac{120}{0.3 \times 3} = 134 \text{ A}$$

$$\text{Number of diodes} = 6$$

$$\text{Transformer secondary voltage} = 0.428 \times 120 = 51.5 \text{ V}$$

$$\text{Transformer turns ratio, } \frac{N_P}{N_S} = \frac{440}{\sqrt{3} \times 51.5} = 4.8{:}1$$

$$\text{Transformer efficiency} = 63.6\%$$

$$\text{Ripple} = 4.2\%$$

For the 6-phase star rectifier:

$$\text{Diode PIV} = 2.09 \times 120 = 252 \text{ V}$$

$$\text{Diode average current} = \frac{120}{0.3 \times 6} = 67 \text{ A}$$

$$\text{Number of diodes} = 6$$

$$\text{Transformer secondary voltage} = 0.740 V_{dc} = 88.6 \text{ V}$$

$$\text{Transformer turns ratio, } \frac{N_P}{N_S} = \frac{440}{\sqrt{3} \times 88.6} = 2.9{:}1$$

$$\text{Transformer efficiency} = 95.1\%$$

$$\text{Ripple} = 4.2\%$$

The better choice of rectifier for this application is the 6-phase star. Our decision is based on the smaller transformer required.

12–3 Filtering and Regulation

The ideal dc power supply would provide a constant output regardless of the amount of current drawn by the load. Of course this ideal device does not exist. The output of the rectifiers we discussed in the previous section was not constant. The small amount of voltage variation due to the ripple is often enough to disturb the operation of industrial devices or systems. The power supply can be improved by filtering the ac signals produced by the ripple from the rectified output. Capacitor and inductor networks are used to perform the filtering.

The output of the rectifier circuit depends on the value of the ac source voltage. Variations in the source voltage will be reflected in the output of the rectifier. Source voltage variations may be due to other heavy loads switching onto the power line or losses in the transformer and diodes due to loading of the power supply itself. In either case, the customer specifications might allow only a limited amount of variation in the power supply output.

Circuitry must be added to the rectifier filter to regulate the output voltage. We will look at the techniques for filtering and regulation in the sections that follow.

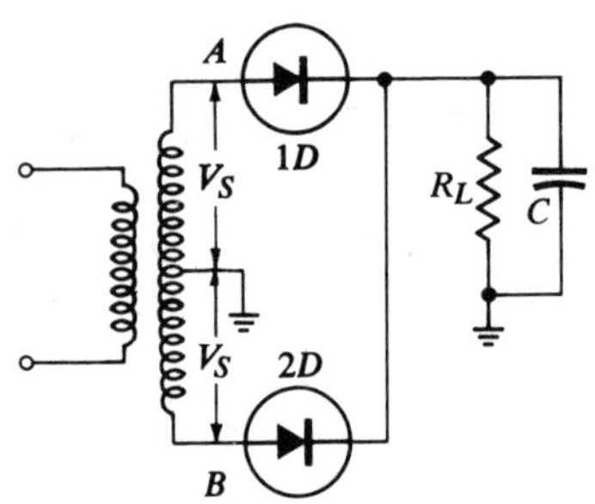

Fig. 12–15 Capacitor filtering of rectifier voltage.

12–3–1 Filtering

The most commonly used filter is a simple capacitor placed across the load. Figure 12–15 shows the center-tapped full-wave rectifier with a filter capacitor. The capacitor acts as an energy storage device. When the source voltage is first applied to the rectifier circuit, the capacitor charges up to the peak value of the transformer secondary voltage. Current to charge the capacitor flows through the diode and transformer secondary. This is a very low resistance path. The RC time constant (see chapter 4) is very small so the capacitor charges quickly. Once the capacitor is fully charged, the diodes are reverse-biased and current flow from the transformer is blocked. This action is displayed in Fig. 12–16. Assuming that power is applied when

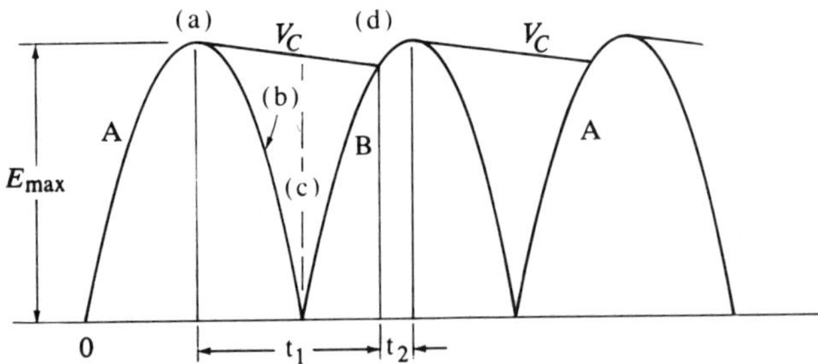

Fig. 12–16 Filtering of full-wave rectified voltage.

the source voltage is zero and going positive on the "A" half-cycle, during the first 90 degrees, the capacitor charges to E_{max}. If we looked at the diode voltage at this time, the anode is at E_{max} and the cathode is at E_{max}, or the voltage across the ideal diode would be zero. No diode current would be flowing. At 135°, point (*b*) of Fig. 12–15, $V_S = 0.707\ E_{max}$ while V_c is still near E_{max}. The diode voltage, from Fig. 12–17, will be $V_s - V_C$. The diode is reverse-biased and cut off. The capacitor will start to discharge through the load at the time constant R_LC. If this time constant is made long compared to the period of the source voltage, then very little of the stored energy of the capacitor will be dissipated. At the point (*c*) of Fig. 12–16, the rectifier voltage reaches zero and starts to go positive on the "*B*" half cycle. The value of V_S is still less than V_C and no diode current flow is allowed. At point (*d*) of Fig. 12–16, the rectifier voltage equals the capacitor voltage; V_S then becomes larger than V_C and the diode $2D$ is forward-biased. The capacitor is then recharged to E_{max} again and the cycle repeats.

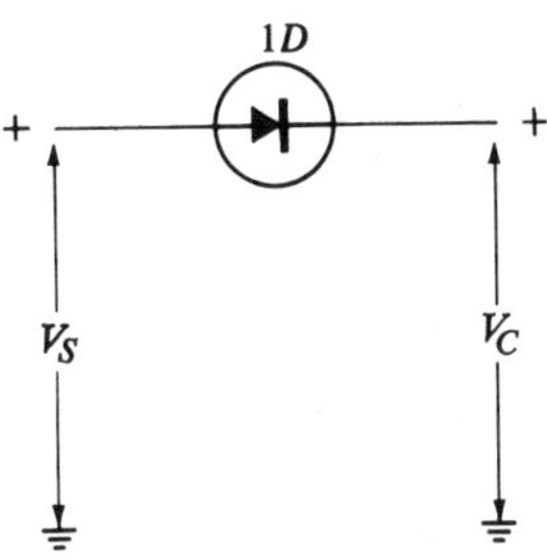

Fig. 12–17 Voltage at diode 1*D* in Fig. 12–15.

The amount of ripple in the load voltage is reduced by the filtering action of the capacitor. A quantitative measure of the improvement requires calculation of the new ripple factor or percentage ripple.

$$\text{Percent ripple} = \frac{100}{\sqrt{3}(4fCR_L - 1)} \qquad \textbf{(12.13)}$$

Equation (12.13)* can be used for the center-tapped full-wave filtered output. The larger the R_LC time constant, the lower the percent ripple. Similar equations for the calculation of ripple for other rectifiers with capacitive filters can be found in engineering handbooks.

The capacitive filter is usually adequate for low- and medium-current applications. At high-current levels, the initial surge of current to charge the capacitor poses a problem. The diodes and capacitors required to withstand the surge current are sometimes large and expensive. An inductor is placed in the output lead of the rectifier to smooth the current (see chapter 4). Figure 12–18 shows the *LC* filter in a center-tapped full-wave rectified power supply.

---

*From Lloyd P. Hunter, *Handbook of Semiconductor Electronics* (New York, McGraw-Hill Book Company, 1970), pp. 10, 17.

$$\text{Percent ripple} = \frac{100\sqrt{2}}{12\omega^2 LC} \qquad \textbf{(12.14)}$$

The percent ripple can be determined from Eq. (12.14).

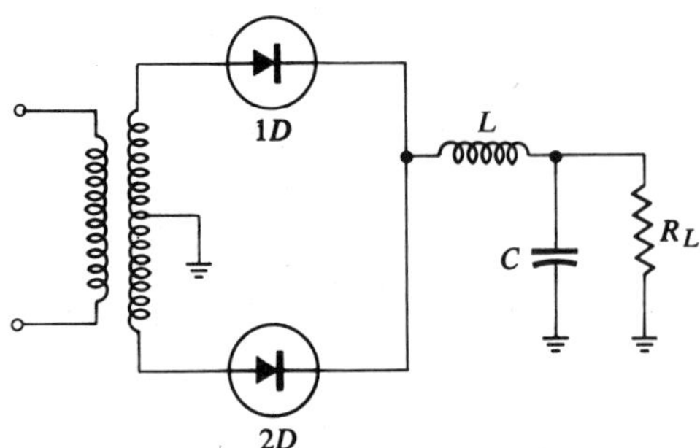

Fig. 12–18 Inductor-capacitor filtering of rectified voltage.

If sufficient filtering is not achieved with a single *LC* filter, other sections may be added in cascade until the percent ripple is below the specified level. Care must be taken to avoid resonance (see chapter 4) at harmonics of the ripple frequency.

12–3–2 Regulation of Power Supplies

The rectified and filtered output of the dc power supply is expected to remain constant over the range of output current specified by the customer. Special circuitry must be designed to make up for variations in the line voltage and losses in the power supply components if the constant output is to be achieved. Even then, minor variations in the output will occur. The quality criterion for specifying the variation in output voltage is "percent regulation."

Percent regulation

$$= \frac{(\text{No load voltage} - \text{Full load voltage}) \times 100}{\text{Full load voltage}} \qquad \textbf{(12.15)}$$

The percent regulation varies from 0.10% for highly regulated supplies to 10% for lesser regulated supplies.

The elements of a regulated power supply as shown in Fig. 12–19 are: a source, regulator, reference, error amplifier, and load. A portion of the output voltage or current is sampled, compared with a reference signal, amplified, and

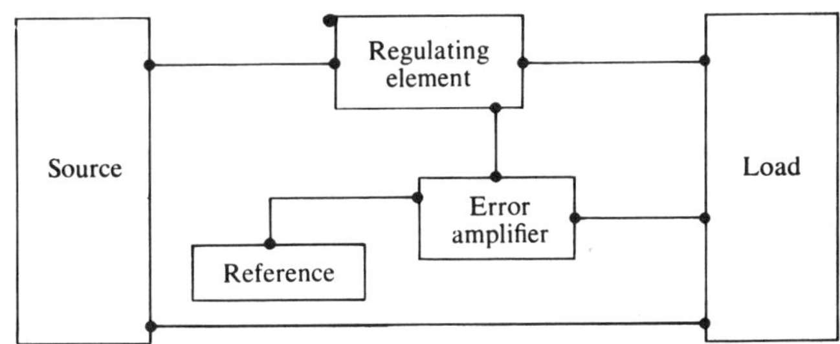

Fig. 12–19 Block diagram of a series-regulated power supply.

fed back to the regulating element. The regulating element is adjusted to make up for any difference between the desired and actual output. The block diagram of Fig. 12–19 shows the regulating element in series with the load. The regulating element could also be shunted across the load. The zener diode regulator of Fig. 3–4 in an example of shunt regulation. Series regulation is usually used when variable output voltages are required. Let's look at a simplified series voltage-regulated power supply.

Figure 12–20 shows the simplified schematic of a series voltage-regulated power supply. The reference voltage, e_{ref}, is provided by zener diode $1D$. A portion of the output voltage, e_f, is fed back to the base of $2Q$ by potentiometer $2R$. The base-to-emitter voltage at $2Q$ is $V_f - V_{ref}$. Transistor $2Q$ acts as the amplifier of the error voltage. Transistor $1Q$ is the regulating element. The regulating action can be seen by considering what happens when E_{out} exceeds

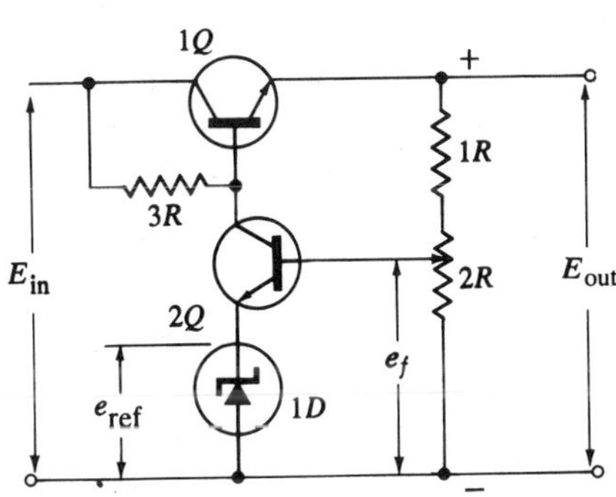

Fig. 12–20 Simplified series voltage-regulated power supply.

the desired value. Voltage e_f is increased by applying additional forward bias to 2Q. The collector current of 2Q is increased by drawing current through resistor 3R. The base-to-emitter voltage of 1Q is reduced by the increased voltage drop across 3R. The resulting decrease in emitter current of 1Q causes an increase in its collector-to-emitter voltage, thus reducing the load voltage to the desired value. We will look at the design of some commercial power supplies in the next section.

12–4 High-Current Regulated DC Power Supply

Figure 12–21 is a photograph of the MOD LH-119 high-current regulated power supply made by Lambda Electronics Corporation. The manufacturer's specifications for this supply are:

Voltage range—0–10 V dc
Current range—0–9.0 A at 30°C
Voltage regulation—0.015%
Current regulation—less than 15 mA with voltage cross-over
—0.05% precision regulated

Figure 12–22 is a block diagram of the power supply. The circuit schematic is shown in Fig. 12–23.

Courtesy of Lambda Electronics Corporation

Fig. 12–21 Commercial DC power supply.

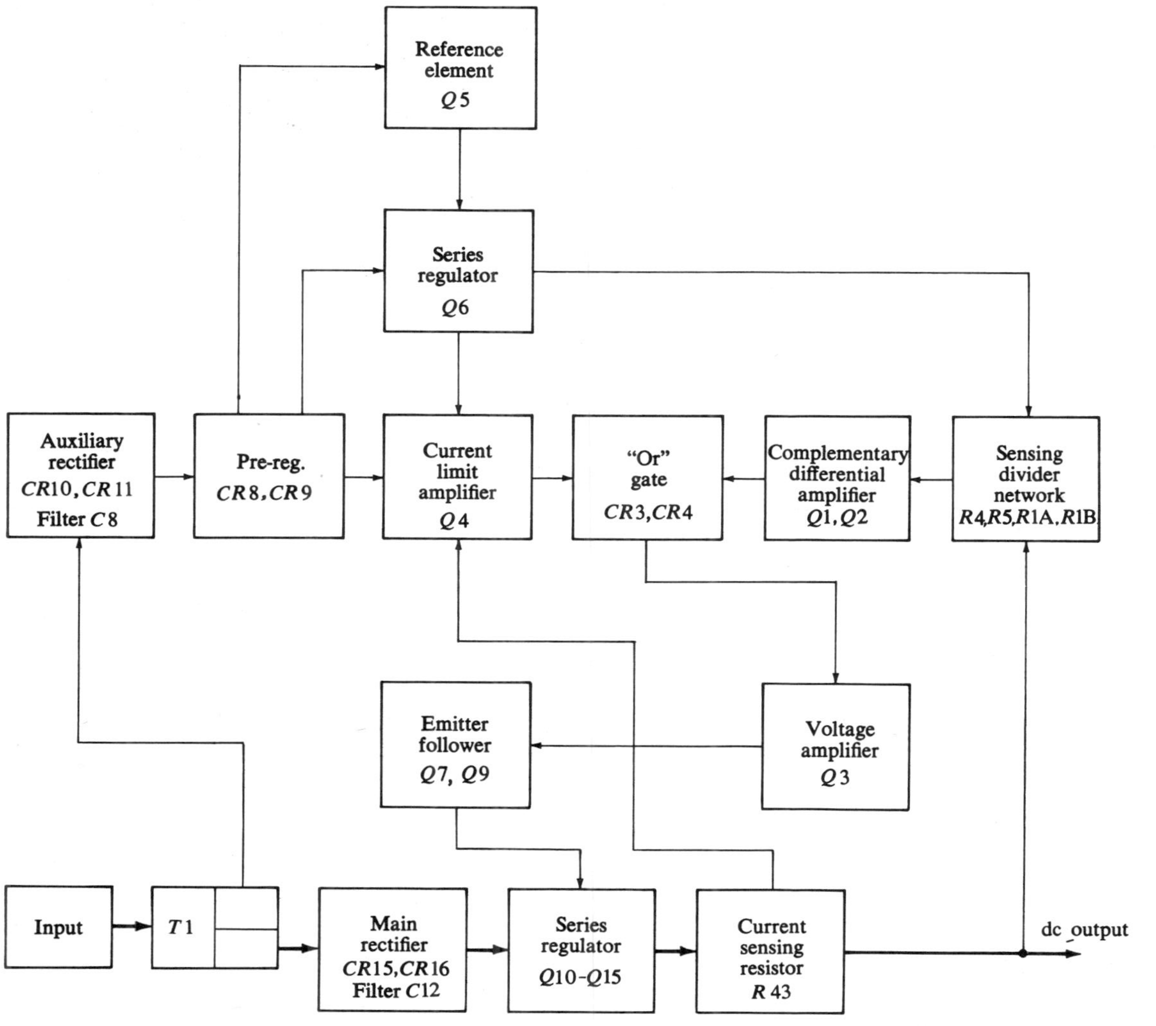

Courtesy of Lambda Electronics Corporation

Fig. 12–22 Block diagram of regulated power supply.

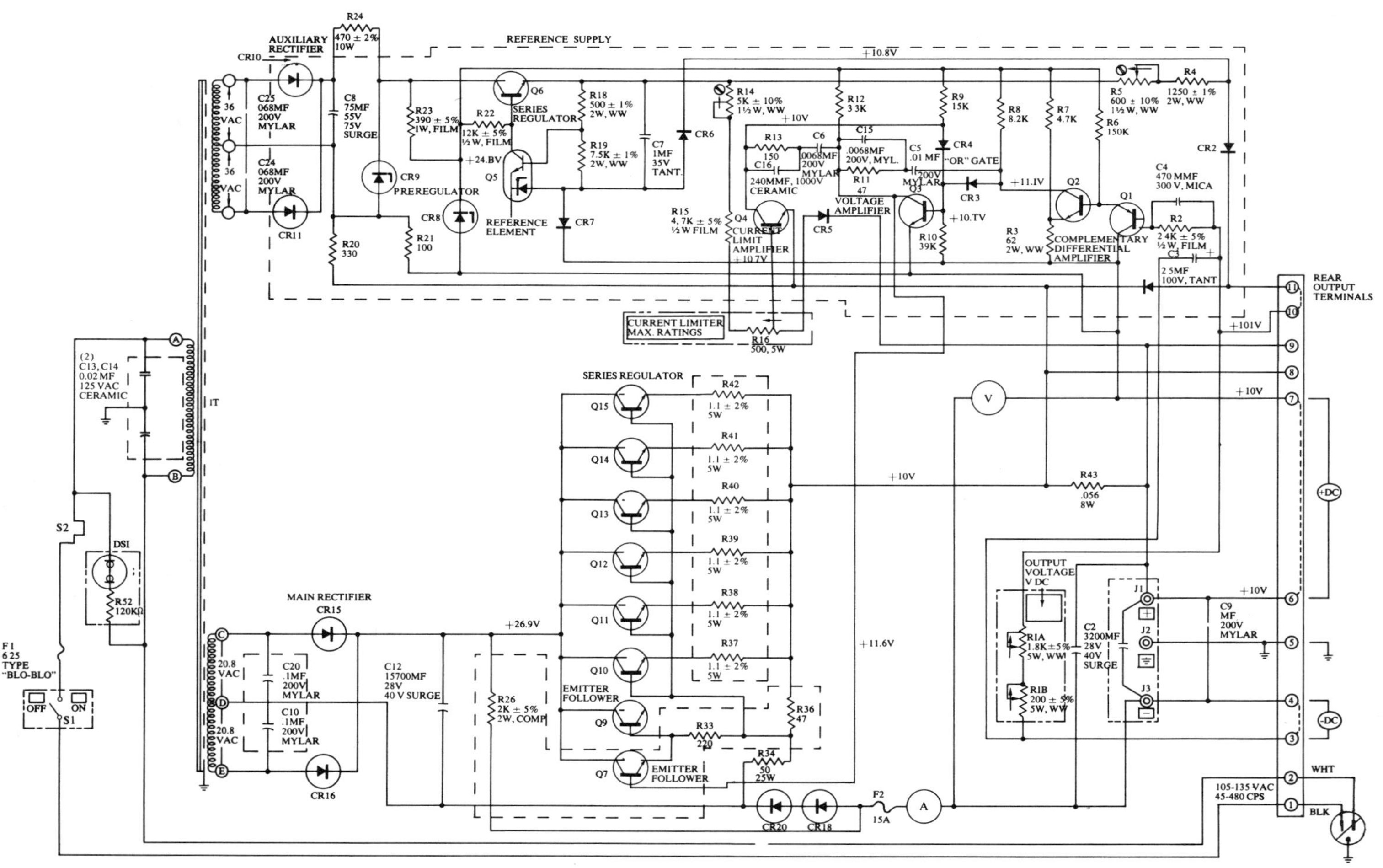

Courtesy of Lambda Electronics Corporation

Fig. 12–23 Regulated power supply schematic.

The circuit can be separated into two sections: the power output section in the bottom half of the schematic and the reference and control. Each section has its own rectifier and filter circuit.

The single-phase 120-V ac voltage is applied at the primary of transformer 1*T*. Diodes 15*CR* and 16*CR* form a bi-phase half-wave rectifier for the power section. Diodes 10*CR* and 11*CR* form the bi-phase half-wave rectifier for the control section. Capacitors 12*C* and 8*C* filter the rectified outputs. A zener diode regulator is used in the control reference section to "preregulate" the filtered output for more precise control. A cascaded zener regulator using 8*CR*, 9*CR*, 23*R*, and 24*R* reduces the ripple of the filtered output voltage. This is a common technique. Figure 12–24 shows how the ripple is reduced at the output of each zener regulator. The preregulated voltage provides the bias supply for all transistors in the control section.

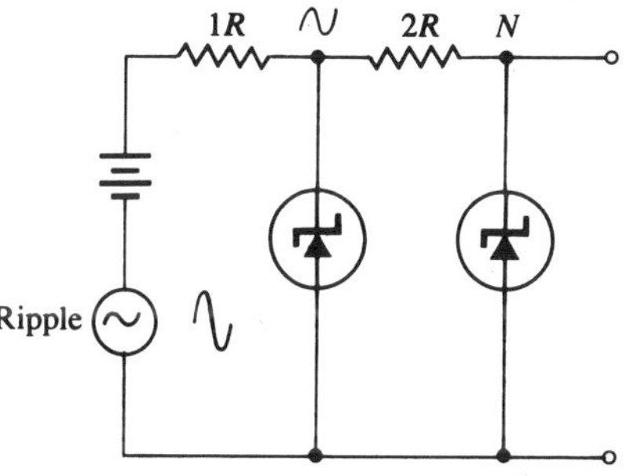

Fig. 12–24 Cascaded zener regulators to reduce ripple.

Reference element 5*Q* contains a zener diode and transistor in one package, which helps to reduce the effects of temperature change and provides a stable reference voltage for the series regulating element, transistor 6*Q*.

A change in the load voltage is sensed by the current divider containing 4*R*, 5*R*, 2*CR*, *RIA*, and *RIB*. Variation in the load voltage causes a change at the input of 1*Q*. Transistors 1*Q* and 2*Q* form a complementary differential amplifier. The output of 2*Q* is proportional to the difference between the reference voltage and load voltage. The output of 2*Q* is fed into the voltage amplifier at the base of 3*Q*. The output signal from 3*Q* goes to the power section, where it is amplified by emitter followers 7*Q* and 8*Q*. The amplified signal from the emitter followers controls the bias current of transistors 10*Q*-15*Q*. Six transistors are paralleled to handle the high-current output to the load. Variation of the bias current of the series regulating transistors 10*Q*-15*Q* changes the voltage drop across the transistors to adjust the load voltage.

This power supply is designed to operate at constant voltage up to a present current level and then "cross over" to constant current limiting. The crossover is achieved by using a diode "OR" gate at the input to the voltage amplifier 3*Q*. Figure 12–25 shows how the OR gate controls the regulation. Resistor 43*R* is in series with the load and regulating transistors so it senses the level of output current. Resistor 43*R* is in the emitter circuit of the amplifier of 4*Q*

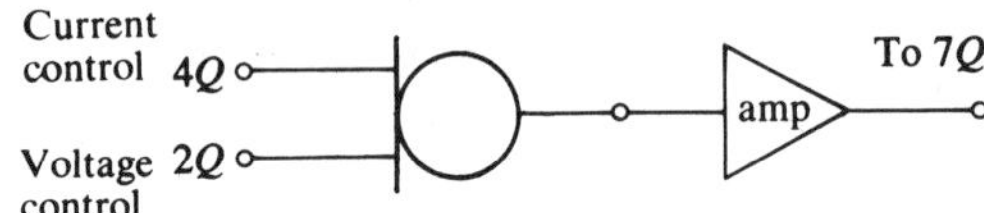

Fig. 12–25 Logic circuit for crossover in commercial power supply.

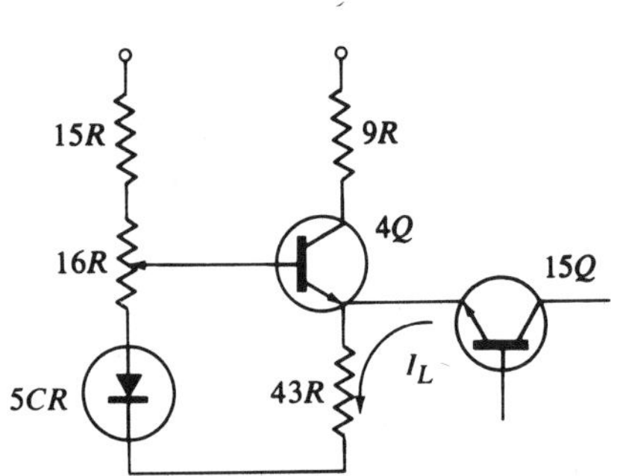

Fig. 12–26 Current limiting circuit of commercial power supply.

as shown in Fig. 12–26. Potentiometer 16R is adjusted so that 4Q is saturated as long as the load current, I_L, is less than a predetermined value. The saturation of 4Q reverse-biases 4CR and the OR gate is controlled by 3CR. The output of the OR gate provides the input signal to the voltage amplifier 3Q to control the regulating transistors 10Q-15Q. When the load current reaches the preset level, 4Q is driven into cutoff. Diode 4CR conducts and 3CR is reverse-biased. The OR gate is now controlled by 4CR and independent of the output of 2Q. The signal going from the voltage amplifier 3Q to control the regulating transistors maintains the load current at the fixed maximum level.

You can relate the functions of this high-quality commercial power supply to the simplified circuit of Fig. 12–20. The complexity and cost can be directly related to the high-output current and good regulation of the power supply.

12–5 Digital Control of DC Power Supplies

Computers and digital programmers, such as the Modicon Model 084, are rapidly replacing "conventional" control systems in many automated industrial operations. New interfacing equipment and techniques have been developed to match the digital output of the control system to the machine or power control device. The result has been that machines and power controls must be responsive to electronic commands, especially remote commands. Remote control is commonly called programmable control. Programmability is becoming a prerequisite to the successful marketing of industrial electronic devices and systems. Power supplies are no exception. Most quality dc power supplies have the built-in option of remote control of the output or programmability. We indicated the need of such a programmable voltage source in the digital phase control system of Fig. 7–14. Let's look at the operational approach to analyzing regulated dc voltage sources and a digitally controlled power supply system.

12-5-1 The Operational Power Supply

We identified the essential elements of a regulated dc power supply in sec. 12-3-3 as an amplifier, reference, and feedback mechanism. These are the same features of computational operational amplifier circuits. The language of op-amps can be used to explain the control of dc power supplies. Figure 12-27 shows the operational representation of a regulated dc power supply. E_i is the reference

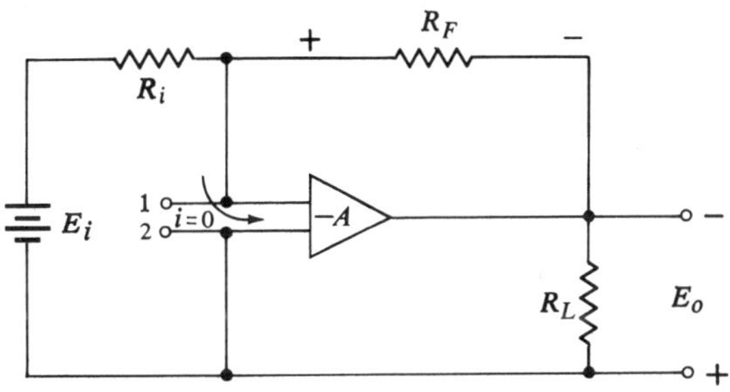

Fig. 12-27 Operational representation of regulated DC voltage supply.

voltage, R_f is the feedback resistor, and the operational amplifier provides the gain. The operational amplifier has been used for many years in analog computer circuits to perform integration, summing, multiplication, and other mathematical operations. The development of integrated circuit op-amps has made this versatile device adaptable to use in many other electronic applications. The gain of the amplifier is so large that we can assume that the input voltage at terminals 1 and 2 is virtually zero so that both input terminals are at ground potential. This "virtual" ground assumption has been used successfully with very little error to analyze op-amp circuits. The other assumed characteristic of op-amps is infinite input impedance so that no current flows into the ungrounded terminal 1. Using the assumptions, with reference to Fig. 12-27,

$$\frac{E_i}{R_1} = -\frac{E_f}{R_f} = -\frac{E_o}{R_f} \tag{12.16}$$

or

$$E_o = -E_i \frac{R_f}{R_i} \tag{12.17}$$

The output voltage can be controlled by varying E_i, R_f, or R_i. The configuration of Fig. 12–28 feeds back a signal proportional to the output voltage so that the load voltage is regulated. If the feedback voltage is taken from a resistor in series with the load, as in Fig. 12–28, the signal fed back is proportional to the output current so that load current is regulated.

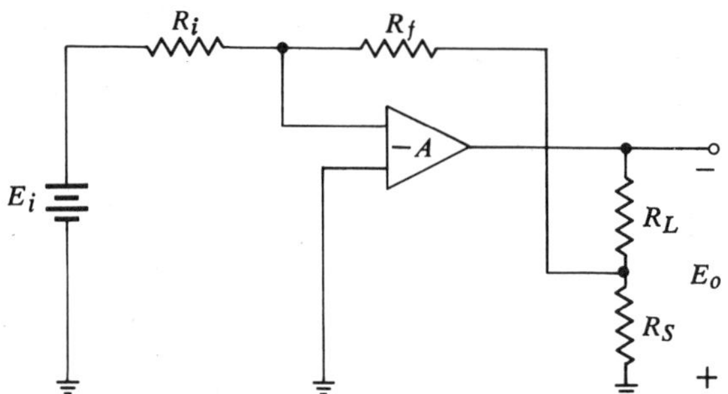

Fig. 12–28 Operational representation of regulated DC current supply.

The output of the voltage-regulated supply of Fig. 12–28 can be digitally controlled by switching fixed resistors into the feedback and input loop. This technique is used by Kepco in the DPD (Digital Programming Decade) of Fig. 12–29. The 8–4–2–1 binary code is used to select the proper

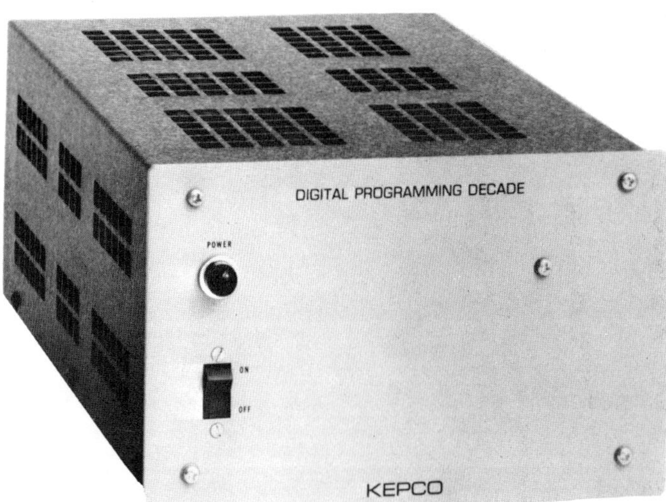

Courtesy of Kepco, Inc.

Fig. 12–29 Kepco DPD digital programmer.

resistors. Figure 12–30 shows the multiple decade 8–4–2–1 resistors in the feedback loop with current selector resistors in the input circuit. The switches are controlled by relays actuated upon command from a programmer. Let's look at how the entire system works.

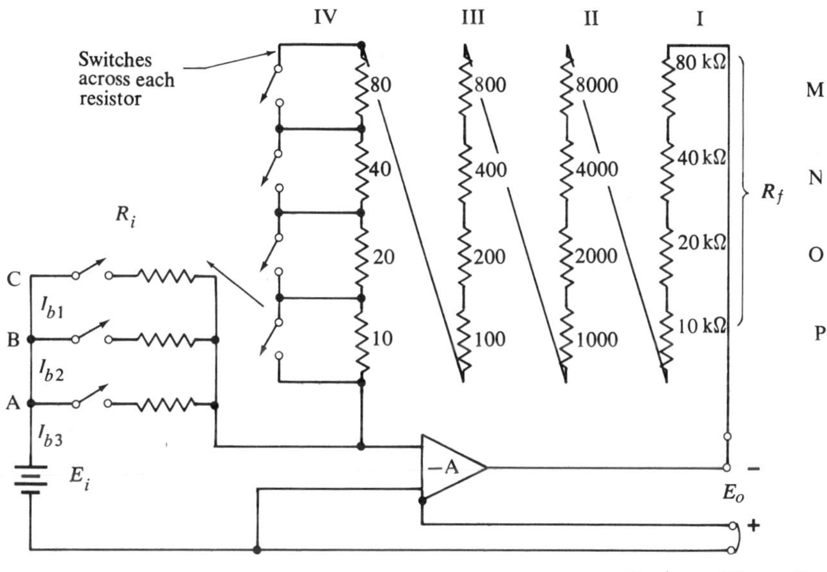

Courtesy of Kepco, Inc.

Fig. 12–30 Basic schematic of Kepco DPD digital programming decade.

12–5–2 A Digitally Controlled Power Supply System

A good example of a digitally controlled dc power supply is the Kepco Digital Programming System. The system is composed of the four parts shown in the block diagram of Fig. 12–31. The desired output voltage is entered manually through a keyboard or directly from magnetic tape or direct computer control. The serial digital commands are entered into a register where they are stored, converted to parallel form, displayed, and transmitted to the programming decade. The digital programming decade converts the parallel coded digital command to an analog output, which is used to control the programmable regulated power supply. The desired voltage is available at the output of the power supply. A closer look at each part of the system will help us understand the techniques used to perform each

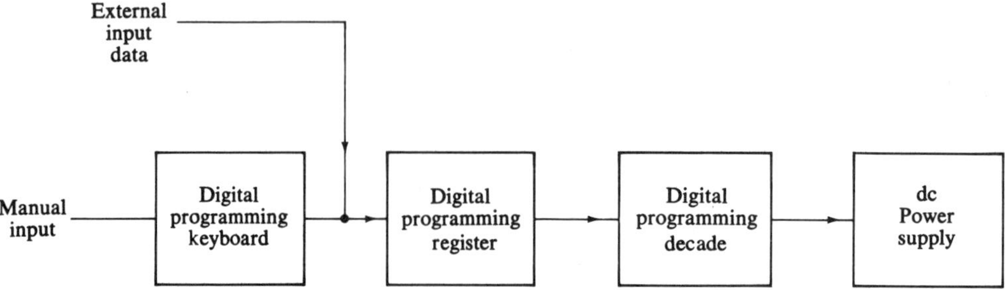

Fig. 12–31 Block diagram of Kepco digital programming system.

function. The simple manual keyboard is shown in Fig. 12–32. The desired output voltage is entered in decimal form.

The Digital Programming Register (DPR) of Fig. 12–33 is used for the storage and display of digital command data.

Courtesy of Kepco, Inc.

Fig. 12–32 Kepco digital programming keyboard.

Courtesy of Kepco, Inc.

Fig. 12–33 Kepco digital programming register.

The data enters serially, is converted for parallel output, displayed visually, and transmitted to the digital programming decade. Integrated circuit *RTL* logic is used to perform all the necessary signal processing. Figure 12–34 shows the basic decimal to 8–4–2–1 logic decoder. The Motorola RTL dual and quad input gates use MIL-STD 806B logic symbols.

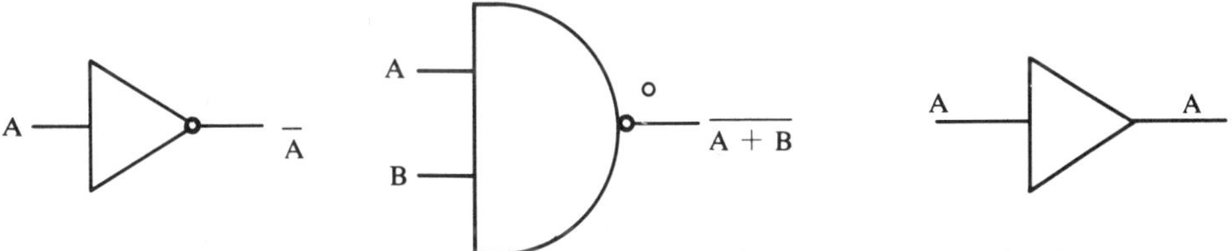

The 8–4–2–1 outputs are combined with the decade signals, *X*1, *X*10, or *X*100 to control the reed relay switches in the resistor circuits of Fig. 12–30.

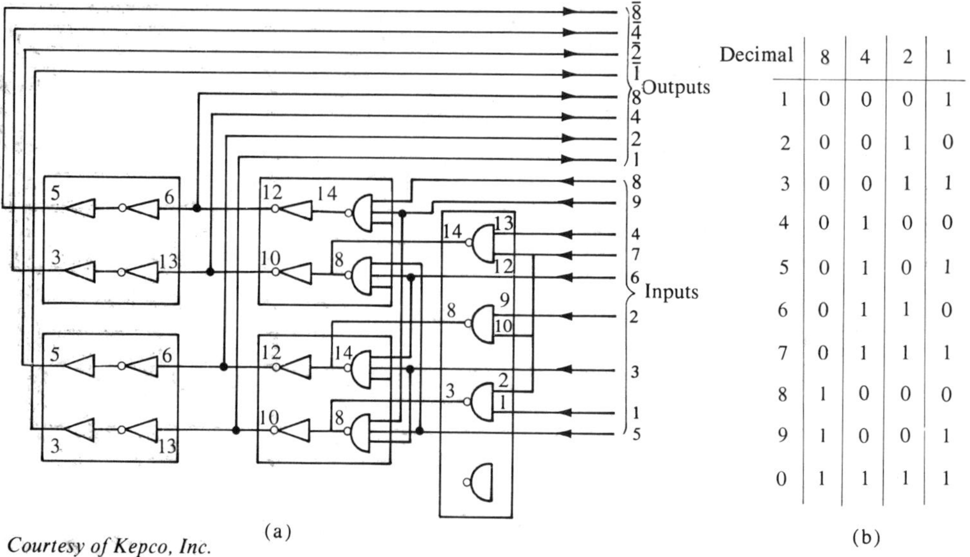

Decimal	8	4	2	1
1	0	0	0	1
2	0	0	1	0
3	0	0	1	1
4	0	1	0	0
5	0	1	0	1
6	0	1	1	0
7	0	1	1	1
8	1	0	0	0
9	1	0	0	1
0	1	1	1	1

Courtesy of Kepco, Inc.

Fig. 12–34 (a) Basic decimal to 8-4-2-1 decoder used in DPR, (b) truth table.

The output voltage of the Digital Programming Decade is determined by closing the proper input resistor switches and opening the proper switches in the feedback loops. From Eq. (12.16),

$$E_o = -\frac{E_i}{R_i} R_f \tag{12.17}$$

Since E_i is fixed, E_i/R_i can be viewed as a current generator $I_b = -E_i/R_i$. The output voltage can be programmed by

$$E_o = +I_b R_f \tag{12.18}$$

The three values of I_b available are: 0.1 mA, 1.0 mA, and 10.0 mA.

Switches A, B, and C of Fig. 12–30 are normally open. Switches M, N, O, and P in each of the four decades of R_f are normally closed. Let's look at the switch action for the following decimal voltage values entered at the keyboard.

Example 12–2. E_o desired is equal to 1.234 V.

From the truth table of Fig. 12–35(b), open R_f switches will be:

$$\begin{array}{ll} \text{I}-P & \rightarrow 10\ \text{k}\Omega \\ \text{II}-O & \rightarrow 2\ \text{k}\Omega \\ \text{III}-O, P & \rightarrow 300\ \Omega \\ \text{IV}-N & \rightarrow 40\ \Omega \\ \hline R_f = & 12{,}340\ \Omega \end{array}$$

Switch C will be closed, so $I_b = 0.1$ mA. Then $E_o = \underset{I_b}{(1 \times 10^{-4})}\ \underset{R_f}{(12{,}340)} = 1.234$ V

Example 12–3. E_o desired is equal to 27.89 V. Open R_f switches will be:

$$\begin{array}{ll} \text{I}-O & \rightarrow 20\ \text{k}\Omega \\ \text{II}-N, O, P & \rightarrow 7\ \text{k}\Omega \\ \text{III}-M & \rightarrow 800\ \Omega \\ \text{IV}-M, P & \rightarrow 90\ \Omega \\ \hline R_f = & 27{,}890\ \Omega \end{array}$$

Switch B will be closed, so $I_b = 1.0$ mA. Then $E_o = \underset{I_b}{(1.0 \times 10^{-3})}\ \underset{R_f}{(27{,}890)} = 27.89$ V

The DPD is connected to a programmable dc power supply by disconnecting the normal feedback element and difference amplifier and replacing them with the DPD as shown in Fig. 12–35. The DPD then regulates the output voltage.

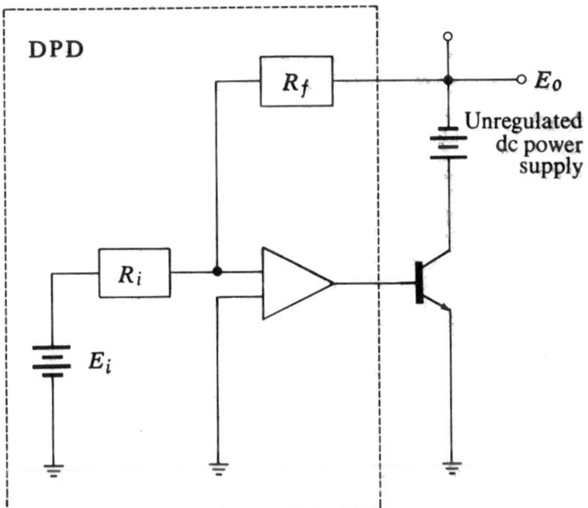

Fig. 12–35 Simplified diagram of DPD controlling unregulated dc power supply.

12–6 The Electronic Crowbar

The current and voltage limiting built into a power supply such as the LH-119 of section 12–4 does not protect the supply from transients that sometimes occur in the line or load circuit. Overvoltage protection must be provided by external circuitry. A very popular, high-speed switching circuit for the protection of dc power supplies is the *electronic crowbar.* Figure 12–36 shows the schematic of an SCR crowbar. The UJT relaxation oscillator consisting of $1Q$, $2R$, $3R$, $4R$, $5R$, and $2C$ is designed so the voltage at

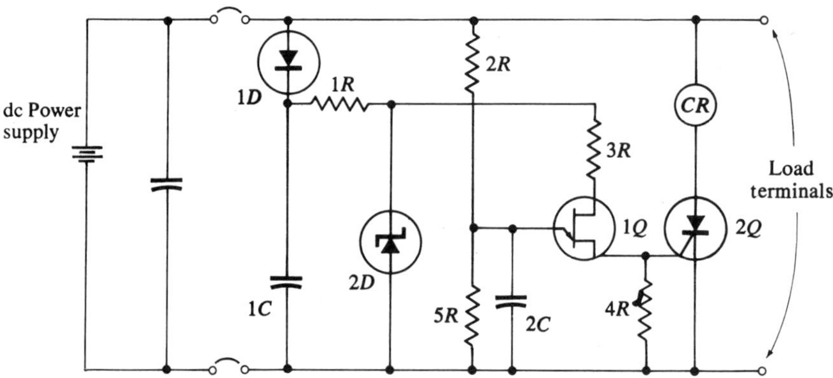

Fig. 12–36 SCR crowbar circuit.

the emitter of the UJT is slightly less than the peak point voltage. The difference is controlled by adjusting $2R$. Any transient voltages that cause the emitter voltage to exceed the peak point voltage will trigger the UJT, delivering a pulse to the gate of the SCR. The SCR switches on within a few microseconds, applying the relay coil across the power supply lines. The SCR and coil act to dissipate the energy in the transient until the contact can be opened, disconnecting the power supply. Resistor $1R$ and zener diode $2D$ provide a stable bias supply for the relaxation oscillator.

12–7 Summary

In this chapter, we have looked at the circuitry associated with dc power supplies. Rectification, filtering, regulation, and limiting or protective circuits were discussed. Both analog and digital-controlled supplies were presented. The commercial supplies that we analyzed embodied many of the design techniques and specifications that you are likely to encounter in your work with dc power supplies.

EXERCISES

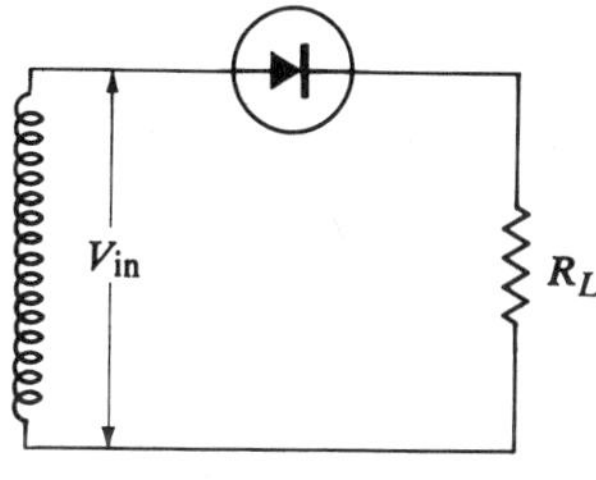

Fig. 12–37

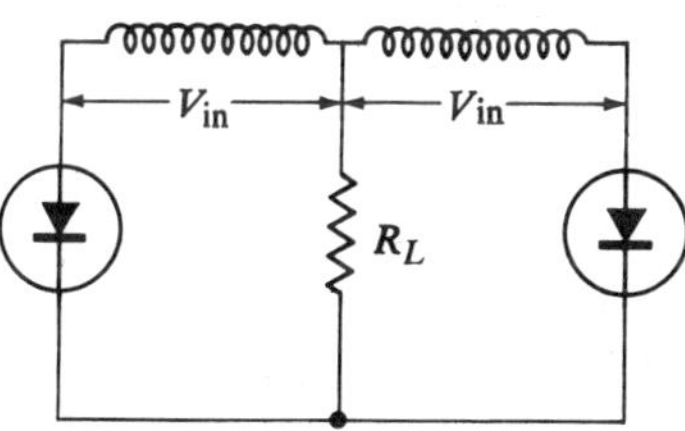

Fig. 12–38

1. What are the four main sections of a dc power supply?

2. The resistive load of Fig. 12–37 requires an average voltage of 140 V. Assuming zero-volt drop across the diode during conduction and $R_L = 2.0\ \Omega$, determine:
   (a) I_{dc}
   (b) V_{in}
   (c) Diode PIV
   (d) ac input power

3. The center-tapped full-wave rectifier of Fig. 12-38 is used to provide 200 V dc to the 5.0-Ω load.
   (a) Specify the diode current and voltage rating.
   (b) Determine V_{in}.
   (c) The frequency of V_{in} is 400 Hz. What is the ripple frequency?
   (d) What is the percent ripple?
   (e) What is the transformer efficiency?

4. A highly inductive load requires 150 V dc at 50A dc. Three-phase, 440 V is available. Design the rectifier circuit using each configuration of Table 12–1 and determine the following for each configuration: (assume zero-volt drop across the diodes during conduction):
   (a) Number of diodes
   (b) Diode PIV
   (c) Average current per diode
   (d) Ripple frequency
   (e) Percent ripple
   (f) Transformer efficiency
   (g) Transformer turns ratio
   (h) Transformer secondary leg voltage

5. Choose the best rectifier design of problem 4 and explain your choice.

6. An inductive load requires a rectifier with minimum ripple content. Which rectifier configuration would you recommend?

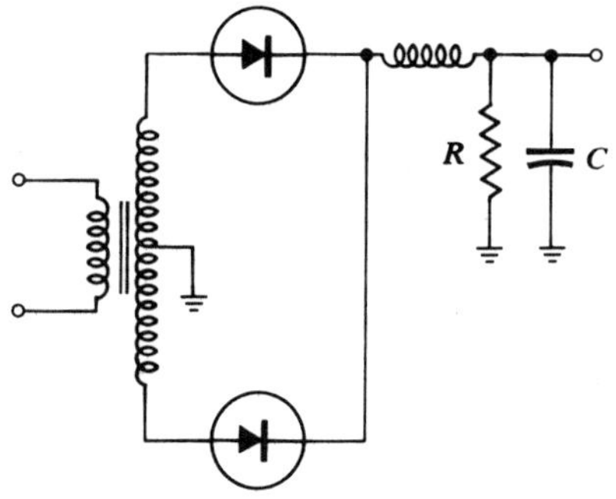

7. A 50-kW inductive load requires a rectified power supply. How much ac power must be provided at the transformer for:
   (a) A 3-phase half-wave rectifier?
   (b) A 3-phase bridge rectifier?
   (c) A 6-phase star rectifier?
   Assume zero volts dropped across the diodes during conduction.

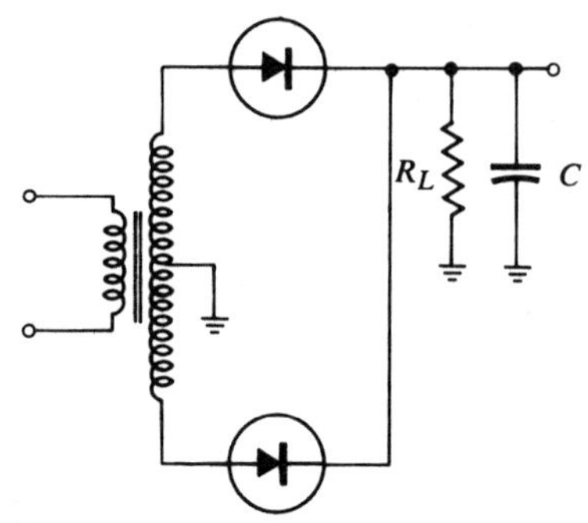

Fig. 12-39

8. The single-phase full-wave rectifier of Fig. 12-39 requires a filter to reduce the percent ripple in the load voltage. The input voltage is at 60 Hz, R_L is 40 Ω. Specify the value of C for 10% ripple.

9. The single-phase full-wave rectifier of Fig. 12-40 requires a filter to reduce the percent ripple in the load voltage. The input voltage is at 60 Hz, R_L is 40 Ω, L is 0.2 H. Specify C for 10% ripple.

Fig. 12-40

10. A regulated dc power supply puts out 14 V unloaded. At maximum load, the output is 8.5 V. What is the percent regulation?

11. A customer requests a 40-V dc supply. Under maximum load, the voltage is allowed to drop 10%. What is the percent regulation?

12. A purchased dc po er supply is rated at 0.6% regulation up to maximum current output. The no-load voltage is specified at 120 V. What is the minimum load voltage if the maximum current is not exceeded?

13. Specify the switch settings in the programmer of Fig. 12-30 if the desired output voltage is 9.312 V.

14. Specify the switch settings in the programmer of Fig. 12-30 if the desired output voltage is 14.21 V.

15. If SCR $2Q$ in the crowbar circuit of Fig. 12-36 is shorted, what will be the effect on the load voltage?

16. If resistor $2R$ of Fig. 12-36 is increased, the level of transient protection will be decreased or increased. Which one and why?

Appendix A

Basic Semiconductor Physics

Throughout the text we have used the terminal characteristics of semiconductor devices to design industrial electronic circuits. A more comprehensive understanding of semiconductor devices requires a study of the physical properties of the devices. Most readers will have been introduced to the physics of semiconductors in earlier electronics courses, physics courses, or metallurgy. In this appendix, we will offer a brief review of the physical properties of semiconductor devices. The reader encountering this subject for the first time should refer to a textbook.*

A–1 Structure of Matter

A review of the basic structure of matter will provide a good background for our later discussion of the electrical properties of semiconductor materials.

A–1–1 The Atomic Model

Our understanding of the structure of matter is based on the Bohr atom model. This model depicts electrons in orbit about a nucleus much like the earth orbits about the sun. The nucleus contains protons and neutrons. The charge of an electron is -1.6×10^{-19} coulombs. The charge of a proton is $+1.6 \times 10^{-19}$ coulombs. The neutron is uncharged but accounts for half the mass of the atom. Figure A–1 shows the atomic model of hydrogen, the lightest of all known elements. Each atom is electrically neutral so the number of electrons in orbit about the nucleus is exactly equal to the number of protons in the nucleus. We classify elements according to the number of protons in the nucleus. The electrical properties of elements are determined by the number and distribution of electrons in orbit about its nucleus. The mass of a proton or neutron is 1.67×10^{-27}

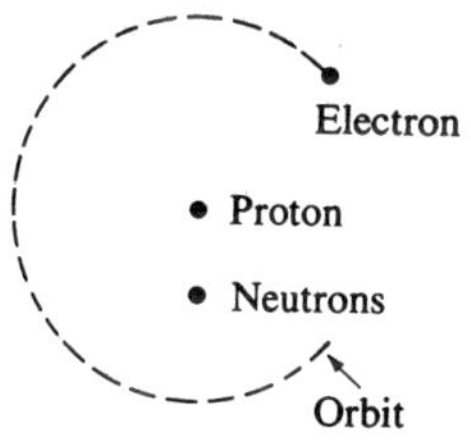

Electron
Proton
Neutrons
Orbit

Fig. A–1 Atomic model of hydrogen.

*See, for example, Louis H. Lenert, *Semiconductor Physics, Devices, and Circuits* (Columbus, Ohio, Charles E. Merrill Publishing Company, 1968).

kilograms. The mass of the electron is 9.11×10^{-31} kilograms. Since protons and neutrons are more than a thousand times heavier than electrons, the mass of an element is determined by the weight of its nucleus. There are two properties of elements that we use to classify them: the number of protons in the nucleus is called the *atomic number* and the total number of protons and neutrons in the nucleus is called the *atomic mass*. The periodic table of elements (found in most chemistry texts) can be used to determine the atomic number and mass of any element.

A–1–2 Energy Bands

We showed the atomic model for hydrogen in Fig. A–1. Since there is only one electron in orbit we had no problem sketching the model. Elements with greater numbers of electrons in orbit require more careful consideration. Electrons are distributed in orbits of increasing radius with a maximum number in each orbit. The orbits are grouped into shells, each shell containing a specified number of orbits. Figure A–2 shows the arrangement of electron shells in the atomic model. The shells are labeled in increasing

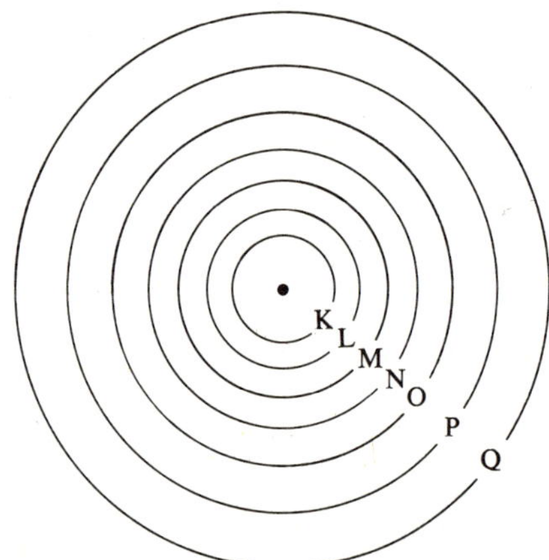

Fig. A–2 Electron shells in atomic models of elements.

radius as K, L, M, N, O, P, and Q. Within each shell there are subshells. Table A–1 shows the number of subshells and maximum number of electrons in each electron shell.

TABLE A–1
Electron Distribution in Atomic Models

Shell	Number of Subshells	Maximum Number of Electrons
K	1	2
L	2	2,6
M	3	2,6,10
N	4	2,6,10,14
O	4	2,6,10,14
P	3	2,6,10
Q	2	2,6

The electrons in orbit about the nucleus experience a force tending to pull them into the nucleus due to the attraction of particles of opposite charge. In order to counteract this force, the electrons move at a velocity such that the centrifugal force of the circular motion is equal to the force of attraction. The electron in orbit has potential energy equal to the work that would be performed if it fell to the nucleus. The electron also has kinetic energy due to its motion in orbit. The kinetic energy is positive. The potential energy is negative, indicating the internal forces involved. Since electrons further from the origin move at slower speeds to offset the force of attraction of the nucleus, as we get farther and farther from the nucleus the potential energy, though negative, decreases in magnitude. Decreasing the magnitude of a negative number represents an increase in the number. The result is that the total energy associated with an electron in orbit increases as the distance of the electron from the nucleus increases. We can then associate energy levels with each electron subshell. Figure A–3 shows the energy level diagram of cadmium. The maximum energy is zero at infinite radius. The vertical position of the horizontal lines represents the levels of energy of the atom corresponding to the electronic shells. The length of the horizontal lines represents the diameter of the shell. The electrons fill the orbital subshells at the lower energy levels first so that the higher level orbits are often empty or partially filled. The highest energy subshell to be filled or partially filled contains the valence electrons. As we shall see later, these valence electrons play an important role in the electrical behavior of elements. The diagram of Fig. A–3 is an adequate description of a single atom of cadmium. As other atoms are grouped together to form

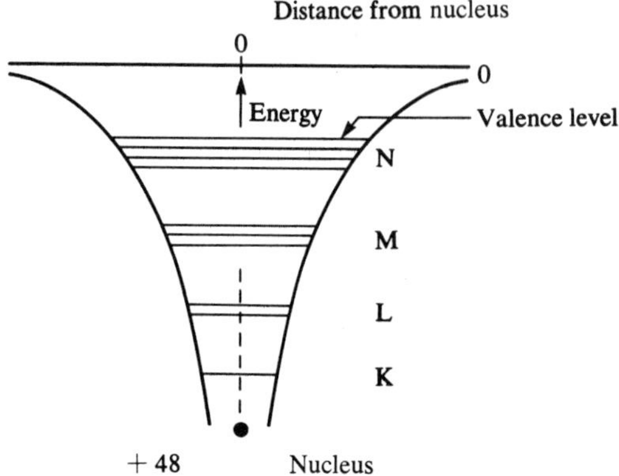

Fig. A–3 Energy level diagram of Cd atom.

crystals, the energy levels are not so neatly defined. The energy levels become energy bands with each band containing as many energy levels as the number of atoms in the crystal. Figure A–4 shows how the energy level diagram is modified to an energy band diagram when the effect of several atoms is considered.

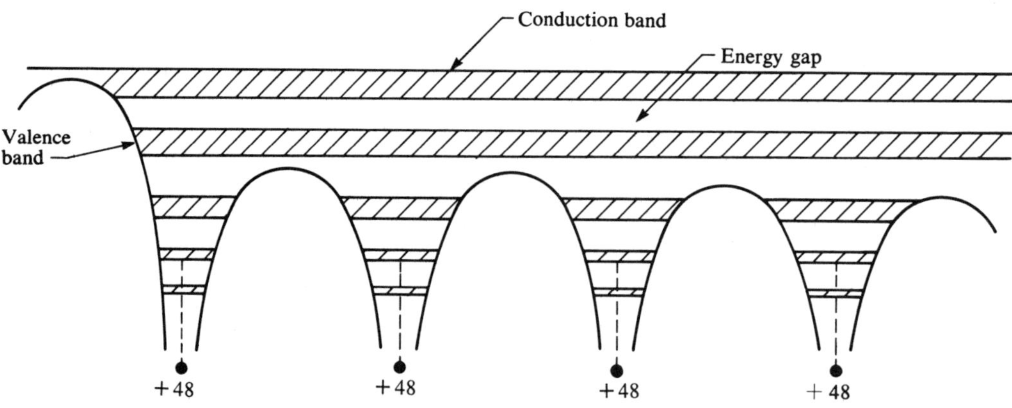

Fig. A–4 Energy band diagram of Cd atoms.

A–2 Electrical Properties of Materials

Our understanding of the basic structure of matter can be used to explain differences in the electrical properties of materials. We will then concentrate on the electrical properties of those semiconductor materials used in the manufacture of electronic devices.

A–2–1 Conductors, Insulators, and Semiconductors

Electrons are not fixed in the energy bands of Fig. A–4, but jump around from one band to another. The probability of an electron being in a given energy band has been formulated in terms of the "quantum" behavior of atomic particles. This higher study of atomic physics is called *quantum mechanics*. We can distinguish between conductors, insulators, and semiconductors by examining the behavior of electrons in the valence energy band. Electrons in the valance band are in constant agitation due to thermal energy. Some valence electrons acquire sufficient energy to escape the attraction of the nucleus and become free or conduction electrons. Electrons in this state are in the conduction band. It has been shown that there is a definite gap between the upper valence band and the lower conduction band. This gap is characteristic of the element involved and the temperature. Even though liquids, solids, and gasses all possess the same properties, we will concern ourselves only with solids. Table A–2 shows the energy gap of three solids. The energy units used are electron-volts:

$$1 \text{ electron-volt} = 1.602 \times 10^{-19} \text{ joules}$$

TABLE A–2
Energy Gap of Three Solids at Room Temperature

Element	Symbol	Atomic Number	Atomic Mass	Energy Gap (eV)
Carbon	C	6	12	6.5
Silicon	Si	14	28	1.11
Copper	Cu	29	63	—

Six and one-half electron-volts of energy must be supplied to an electron in the valence band of a carbon atom before it can escape from the atom and become free for conduction. This added energy may be in the form of light, heat, or electric potential.

An electron in the valence band of a silicon atom requires a smaller increase in energy to become free for conduction. The energy gap for copper is undefined. Electrons in the valence band and conduction band are almost the same as far as conduction capability is concerned. This represents an overlap in the valence and conduction energy bands. Figure A–5 shows the diagrams of these three solids.

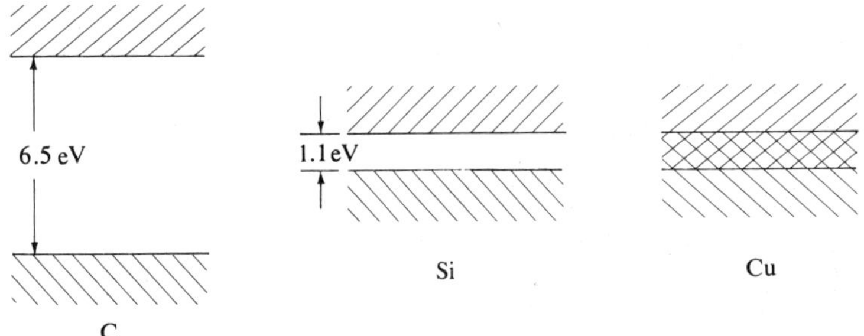

Fig. A–5 Energy gap diagrams of three solids at room temperature.

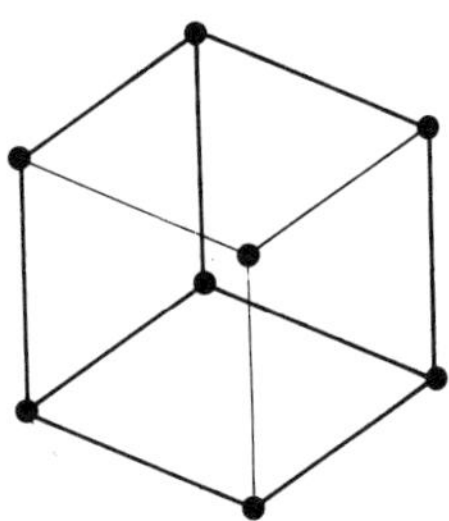

Fig. A–6 Cubic cell structure of silicon.

Solids with large energy gaps such as carbon are classified as insulators. There are very few electrons in the conduction band due to the large amount of energy that must be acquired. Solids with overlapping valence and conduction energy bands are called conductors due to the excess of free electrons available for conduction. Solids with energy gaps in the neighborhood of 1 eV are called semiconductors. The semiconductor that we will be most concerned with is silicon.

A–2–2 Intrinsic Silicon

The silicon atom has 14 electrons: 2 in the K shell, 8 in the L shell, and 4 in the M shell. The 4 electrons in the outer shell determine the crystal structure of pure silicon material. The basic cell structure of silicon is the symmetrical cube of Fig. A–6. Each silicon atom shares an electron in its outer shell with 4 of its closest neighbors so that the outer subshell is completely filled with 8 electrons. The result is a strong, symmetrical *covalent bonding* of the silicon atoms. Figure A–7 shows the electron sharing between atoms. It is from this structure that an electron must acquire sufficient energy to become a free electron. Each time an electron breaks a bond and becomes free for conduction, it leaves behind a vacancy called a *hole*. Figure A–8 demonstrates this action.

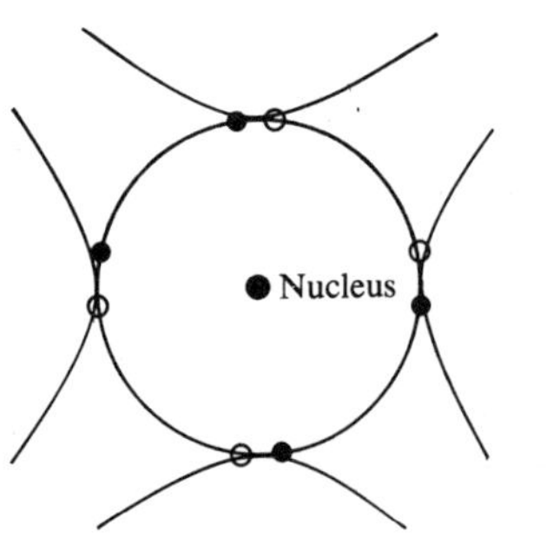

Fig. A–7 Covalent bonding of silicon atoms.

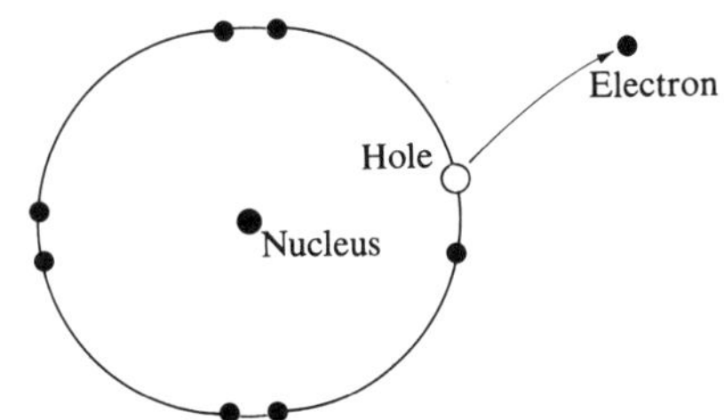

Fig. A–8 Electron-hole pair generation.

This concept of a hole has caused much confusion in semiconductor physics. A hole is just the vacancy caused by an electron escaping the covalent bond crystal structure and becoming a free electron. Since the ideal silicon material would contain no free electrons due to the symmetrical cubical bonding, each free electron generates a hole. We often refer to "electron-hole" pairs in pure or "intrinsic" silicon. The electrical properties of intrinsic silicon are determined by the concentration of electron-hole pairs. At room temperature, 300°K, the concentration of electron-hole pairs is 1.5×10^{10} per cm^3. The equation is:

$$N_i = n_i = p_i = 1.5 \times 10^{+10} \frac{\text{pairs}}{\text{cm}^3} \qquad \textbf{(A.1)}$$

where N_i = intrinsic concentration in electron-pairs per cm^3

n_i = number of free electrons per cm^3

p_i = number of holes per cm^3

The intrinsic concentration is a function of temperature and energy gap.

The conductivity of intrinsic silicon can be determined by the equation:

$$\sigma = qn_i\mu_n + qp_i\mu_p \qquad \textbf{(A.2)}$$

where q = charge of an electron

μ_n = mobility of electrons, $\frac{\text{number}}{\text{cm}^3}$

μ_p = mobility of holes, $\frac{\text{number}}{\text{cm}^3}$

σ = conductivity, $\frac{\text{siemens}}{\text{cm}}$

The mobility constants μ_n and μ_p are proportional to the average velocity of electrons and holes in the silicon material when an external potential is applied.

$$\text{At } 300^\circ \text{ K}, \mu_n = 1500 \frac{\text{cm}^2}{\text{V-s}}$$

$$\mu_p = 500 \frac{\text{cm}^2}{\text{V-s}}$$

The intrinsic conductivity of silicon at room temperature is 4.8×10^{-6} S/cm.

A–2–3 Doping of Silicon

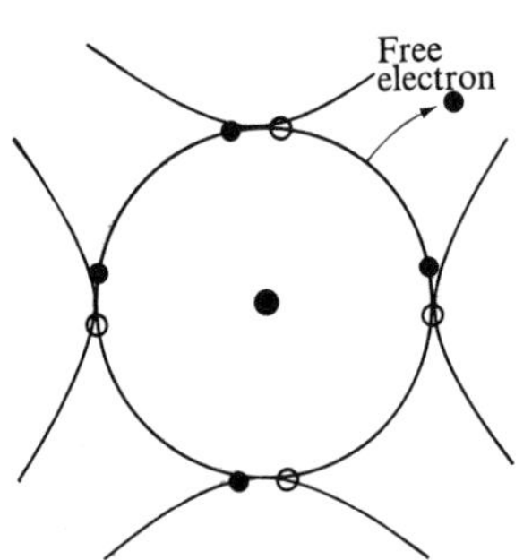

Fig. A–9 An Sb atom in the silicon covalent bonding structure.

From Eq. (A.2) we can vary the conductivity of silicon by changing n or p. This requires changing the temperature on energy gap of the material. We choose to change the conductivity by introducing impurities into the intrinsic silicon material. Let's look at type-V elements in the periodic table of elements. These elements, phosphorus, arsenic, and antimony, all have five electrons in their outer subshells. If a type-V element replaces a silicon atom in the covalent bond of Fig. A–7, one free electron is generated. Figure A–9 shows that if an antimony atom is added to the covalent bond structure of silicon, the antimony atom shares an electron with four of its closest neighbors but the fifth electron in the outer subshell is free for conduction. Each antimony atom added to the silicon material donates one free electron. Type-V elements are called *donor impurities* because they donate electrons. The silicon is called n-type material after the type-V element is added. The process of adding impurities to the intrinsic semiconductor material is called doping. The total number of electrons in the silicon material after doping is given by the equation:

$$N_n = N_D + n_n \qquad \textbf{(A.3)}$$

where

$$N_D = \frac{\text{number of atoms of donor}}{\text{cm}^3}$$

$$N_n = \frac{\text{total number of electrons}}{\text{cm}^3}$$

$$n_n = \frac{\text{number of electron-hole pairs}}{\text{cm}^3}$$

The total number of holes must still be the same as the number of free electrons due to the silicon electron-hole pairs, or

$$p_n = n_n \qquad \textbf{(A.4)}$$

where p_n is the concentration of holes in the negative-doped silicon. Even after doping, the product of the concentration of holes and the concentration of free electrons remains constant. The equation is:

$$N_i^2 = N_n p_n \qquad \textbf{(A.5)}$$

This says that the concentration of electron-hole pairs has decreased due to doping. Equation (A.2) is modified to

$$\sigma_n = qN_n\mu_n + qp_n\mu_p \qquad \textbf{(A.6)}$$

after doping. In most cases, N_n is so much greater than p_n that we use

$$\sigma_n = qN_n\mu_n \qquad \textbf{(A.7)}$$

to approximate the conductivity of n-type silicon. Type-III elements such as aluminum, gallium, and boron are also used for doping. The effect is analogous to that of type-V elements. Type-III elements have three electrons in their outer subshell. When a Type-III element such as gallium replaces a silicon atom in the covalent bond structure of Fig. A–7, the gallium atom will share an electron with each of its four closest neighbors but a hole will be generated due to the electron vacancy as shown by Fig. A–10. Each gallium atom adds one hole to the silicon material. Type-III elements are called acceptors. The silicon is called p-type after doping with an acceptor-type material. The total number of holes in the semiconductor after doping is given by the equation:

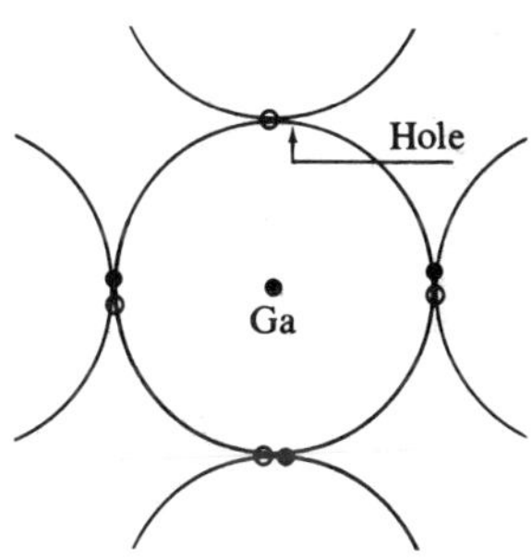

Fig. A–10 Hole generated by the addition of gallium to intrinsic silicon.

$$N_p = N_A + p_i \qquad \textbf{(A.8)}$$

where $N_A = \dfrac{\text{number of atoms of acceptor}}{\text{cm}^3}$

$N_p = \dfrac{\text{total number of holes}}{\text{cm}^3}$

and

$$n_p = p_p \qquad \textbf{(A.9)}$$

where n_p = concentration of free electrons in P-type material. Equation (A.2) is modified to

$$\sigma_p + qN_p\mu_p + qn_p\mu_n \qquad \textbf{(A.10)}$$

after doping. In most cases, N_p is much greater than n_p so the conductivity is closely approximated by

$$\sigma_p = qN_p\mu_p \qquad \textbf{(A.11)}$$

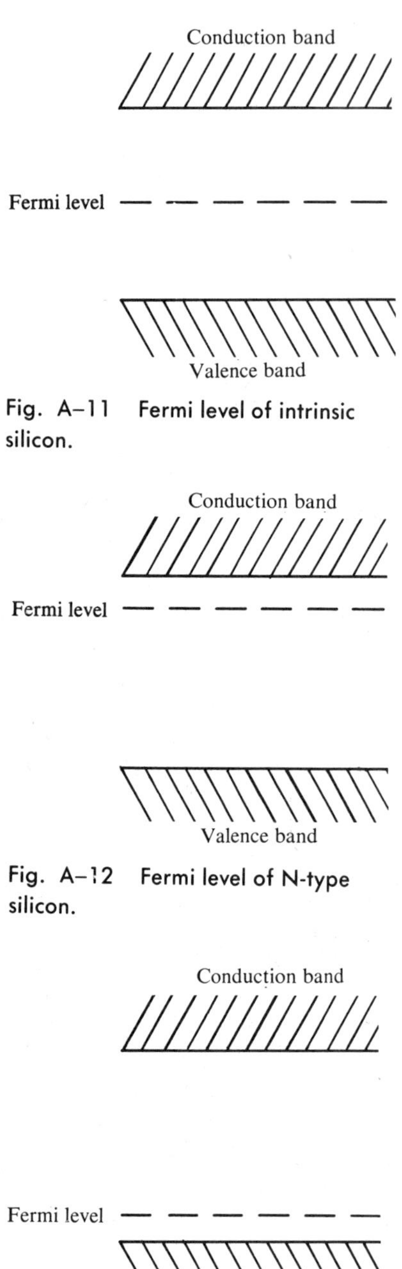

Fig. A–11 Fermi level of intrinsic silicon.

Fig. A–12 Fermi level of N-type silicon.

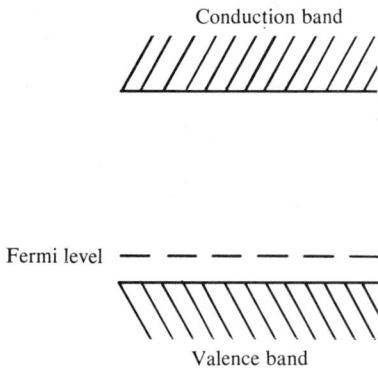

Fig. A–13 Fermi level of P-type silicon.

A–3 PN Junction Devices

Doped semiconductor materials are used to manufacture the electronic devices whose terminal characteristics we have used in the design of industrial electronics circuits. The basic process is the formation of the PN junction. We will look at the PN junction and its application to electronic devices.

A–3–1 The Diode Junction

Doping of intrinsic silicon affects the energy band diagrams we discussed earlier. The Fermi level is changed by the addition of impurities. The Fermi energy level is the level at which the probability of finding an electron is 50%, as shown in Fig. A–11. When donors are added to the intrinsic material, the probability of electrons being in the conduction band is increased so the Fermi level moves toward the conduction band as shown in Fig. A–12. When acceptors are added to the intrinsic material, some of the valence electrons combine with acceptor atoms so that the number of free electrons is decreased. The probability of finding electrons in the conduction band is decreased and the Fermi level moves toward the valence band as shown in Fig. A–13. A PN junction is formed at the interface of a P-type and N-type bar of silicon. The Fermi level must remain constant across the junction. The energy level diagram must be shifted to align their Fermi levels at the junction as shown in Fig. A–14. A potential difference occurs at the junction due to the energy level shift. Electrons near the junction roll down the *potential hill* to combine with holes in the P-type material. Holes are pushed up the potential hill to combine with electrons in the N-type material. The junction reaches a stable condition with an excess of positive ions on the N side of the junction and negative ions on the P side. These ions are not free for conduction so the region near the junction is "deplete" of conduction electrons or holes. This is known as the depletion region. The potential difference at the junction is called the barrier potential. For silicon, the barrier potential is approximately 0.6 to 0.7 volts.

The PN junction is stabilized with no external voltage applied. If an external voltage is applied to forward-bias the junction as in Fig. A–15(a), the barrier potential is reduced as in Fig. A–16(a). More electrons roll down the potential hill, more holes are pushed up the potential hill,

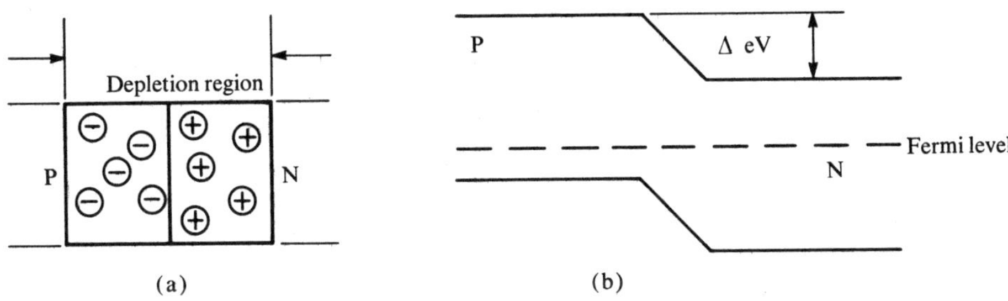

Fig. A–14 (a) PN semiconductor junction; (b) energy level diagram.

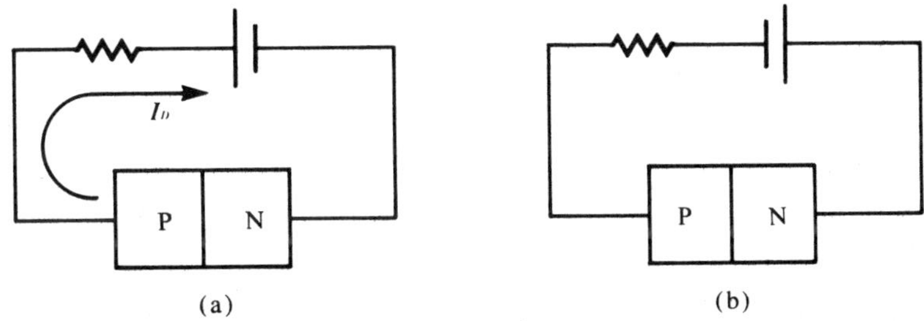

Fig. A–15 (a) Forward-biased PN junction; (b) reverse-biased PN junction.

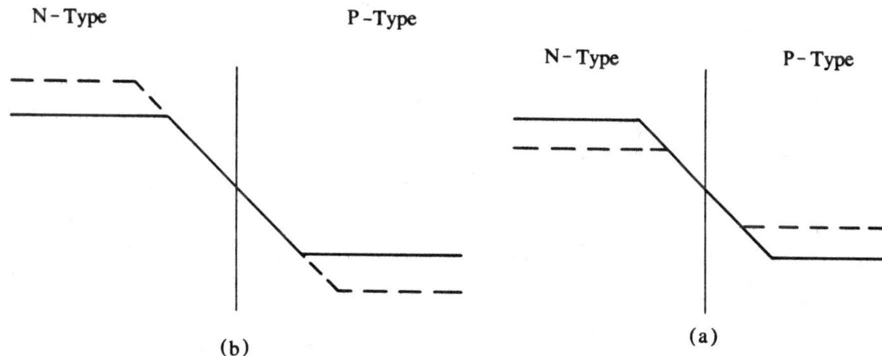

Fig. A–16 Barrier potential variation due to (a) forward biasing, (b) reverse biasing.

and current flows through the external circuit. If an external voltage is applied to reverse-bias the junction, as in Fig. A–15(b), the barrier potential is increased. Very little current flows in the external circuit. This accounts for the nonlinear diode characteristics of Fig. 2–2.

A–3–2 The Bipolar Transistor

The bipolar transistor is made of two PN junctions. The emitter is very heavily doped and acts as the source of carriers. The base is very lightly doped and acts to modulate the flow of carriers. The collector is moderately doped and receives the carriers from the emitter. Figure A–17 shows the PNP bipolar transistor and its barrier potentials. The NPN configuration is also used and analogous diagrams could be drawn. External voltage sources are connected to the transistor junctions in order to obtain amplification.

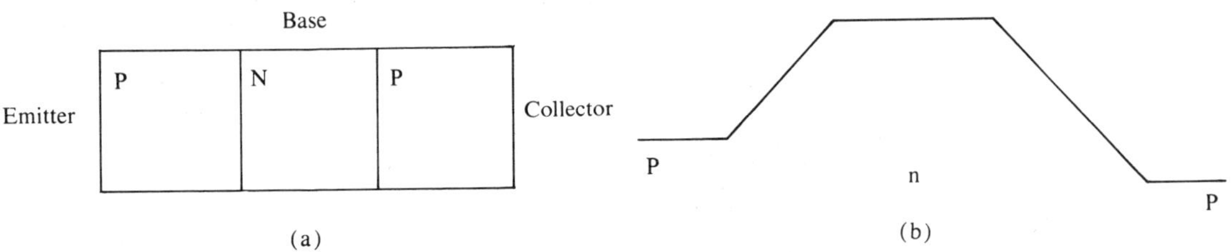

Fig. A–17 (a) Bipolar transistor junction model, (b) barrier potential.

Figure A–18 shows that the emitter-base junction is forward-biased, the collector-base junction is reverse-biased.

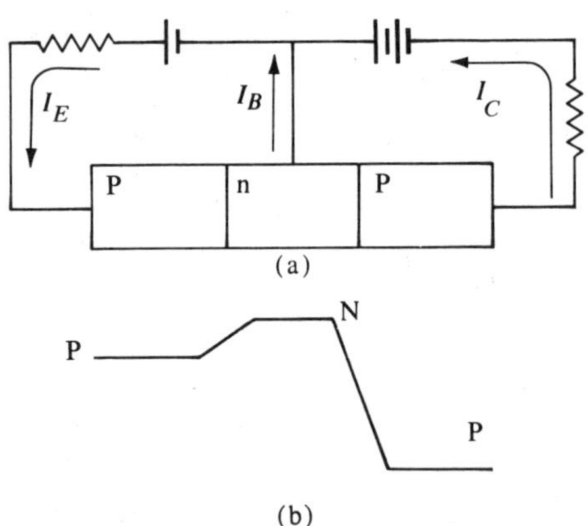

Fig. A–18 (a) Bipolar transistor biased for amplification; (b) barrier potential with biasing.

At the forward-biased emitter-base junction the barrier potential is reduced, making it easier for holes to climb the potential hill. The emitter region is heavily doped so that a large number of holes make the trip up the hill and into the base region. The base region is very lightly doped so that the number of electrons that roll down the hill from base to emitter is small. They represent a small leakage current. The large number of holes entering the base region will follow two paths: some will combine with electrons in the n-type silicon, others will drift through the base region and reach the potential hill at the collector-base junction. Bipolar transistors are designed so that most of the holes make it through the base region to the collector junction. Once the holes reach the steep potential hill at the collector junction, they are carried into the collector region by the barrier potential. These holes drift through the collector material to the external terminal constituting current flow in the circuit. Notice that the current flow from base to collector is made up of holes. Since the base is an N-type silicon, holes are the *minority carriers*. There is also a small current flow from collector to base due to holes in the P-type collector making it up the potential hill to the base region. Figure A–19 shows the current flow in the transistor of Fig. A–18. Another representation of these currents is Fig. A–20. The equations describing these currents are:

$$I_E = I_B + I_C \qquad \textbf{(A.12)}$$

$$I_C = \alpha I_E + I_{CBO} \qquad \textbf{(A.13)}$$

$$I_B = (1 - \alpha)I_E - I_{CBO} \qquad \textbf{(A.14)}$$

Alpha, α, is the portion of holes that make it through the base region without recombination and reach the collector

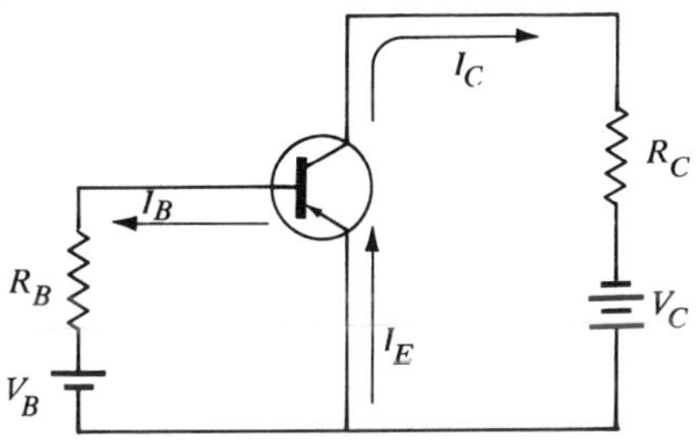

Fig. A–19 PNP transistor as an amplifier.

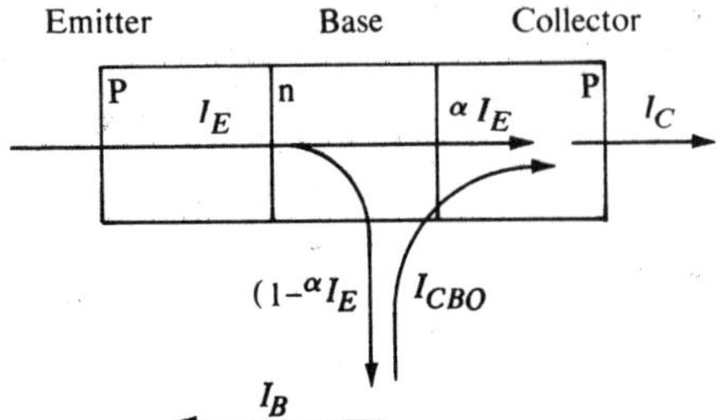

Fig. A–20 Current flow at bipolar transistor junction.

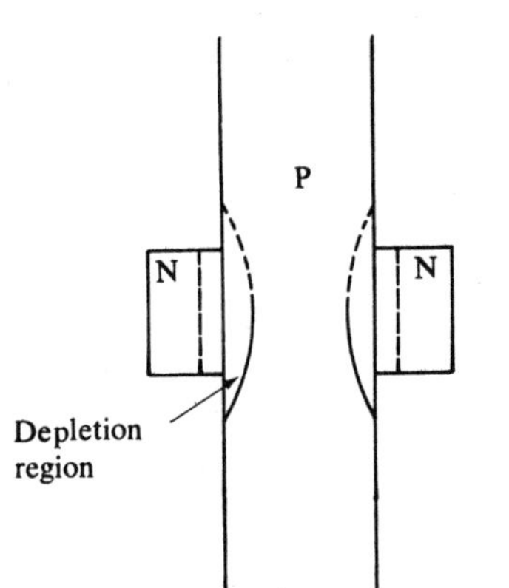

Fig. A–21 Depletion region at JFET junction.

junction. The value of α is usually very close to 1 in order to get current gain during amplification. I_{CBO} is the leakage current due to electrons in the collector making it up the potential hill to the base region. This is the current that plays an important role in the thermal stability of transistors.

A-3-3 Field-Effect Transistors

There are two types of field-effect transistors based on different semiconductor effects. We will look at them separately.

The junction field-effect transistor, JFET, is based on conductance modulation of a semiconductor channel. Figure A–21 shows the cross section of a JFET model. Heavily doped N-type silicon and lightly doped P-type silicon materials are used to form PN junctions. Without any external bias, depletion regions will be formed on either side of the junctions by the action described in Sec. A–3. The width of the depletion region will be less on the heavily doped N side of the junction and greater on the P side of the junction. If an external voltage is applied across the P-type silicon bar, current will flow through the channel formed by the depletion regions inside the bar. The amount of current flow will be determined by the cross-sectional area of the channel. The cross-sectional area of the channel will be determined by the width of the depletion region formed at the PN junction. Figure A–22 shows the JFET model with external bias. The terminals at the ends of the silicon P-type bar are called the drain and source terminals. The flow of conventional current is from source to drain in this P-channel JFET. If the gate terminals are connected

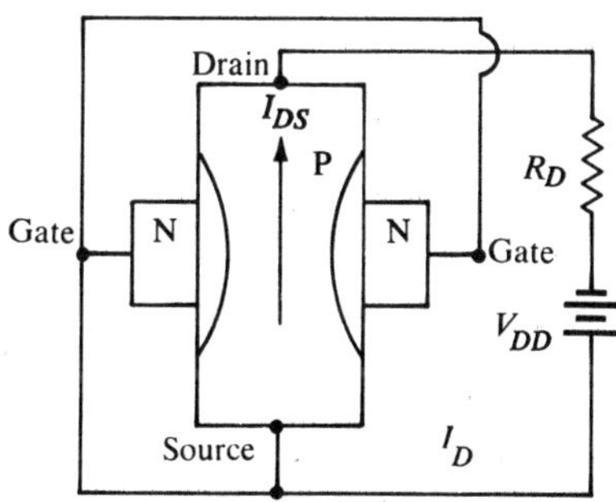

Fig. A–22 JFET model without gate bias.

together and a reverse bias is applied at the gate, the depletion region within the silicon bar will be widened, as shown in Fig. A–23. The cross-sectional area through which the drain-to-source current flows is reduced. Reducing the cross-sectional area of the channel has the effect of increasing the channel resistance or decreasing the channel conductance. Figure A–24 shows how the drain-to-source current varies as the external voltage source in the gate circuit of Fig. A–23 is varied. The slope of this characteristic curve is the ac conductance of the JFET. Values of 2,000 to 10,000 μS are common. The point at which the drain current goes to zero in Fig. A–24 is called the *turn-off voltage.* This represents the point at which the depletion regions from either side of the silicon bar meet in the center, shutting off all current flow through the bar.

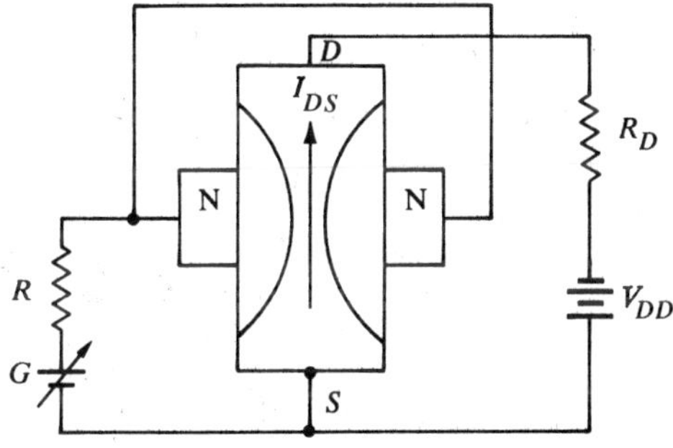

Fig. A–23 JFET model with gate bias.

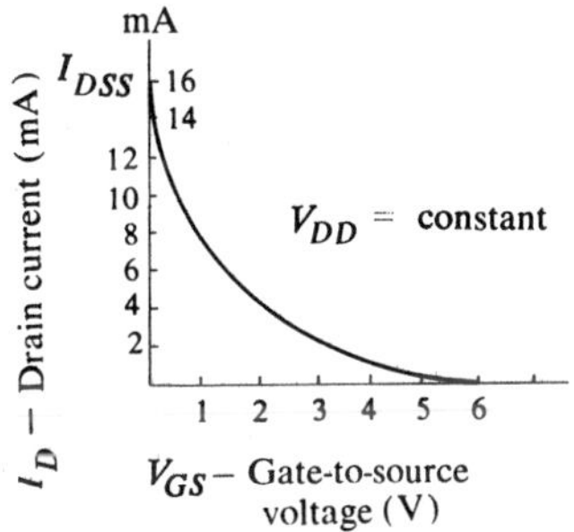

Fig. A–24 Conductance characteristics of JFET.

Surface-type field-effect transistors such as the insulated gate field-effect transistor (IGFET) and the metal oxide semiconductor field-effect transistor (MOSFET) do not use the same gate structure to modulate the channel conductance. Figure A–25 shows the cross section of a typical MOSFET. The drain and source terminals are connected to heavily doped N-type silicon regions. A moderately

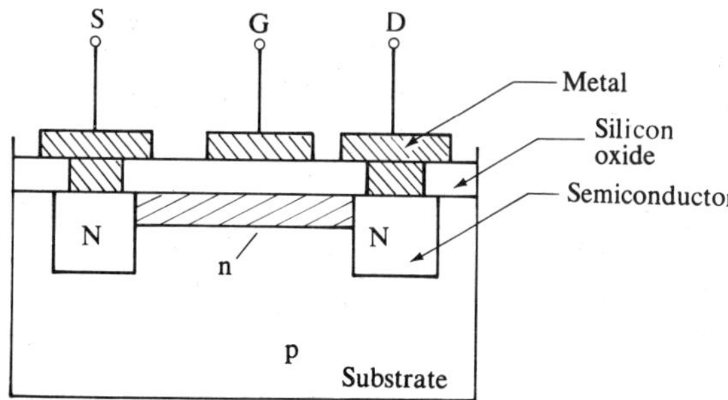

Fig. A–25 MOSFET structure.

doped N channel is formed between the two regions. A thin layer of silicon oxide insulating material separates the gate terminal from the N channel. Unlike the JFET, there is no PN junction at the gate so that either positive or negative external gate voltages may be applied. If the gate terminal is connected to the source and a voltage is applied at the drain-to-source terminals, as in Fig. A–27, a fixed amount of drain current will flow. This is I_{DSS} in Fig. A–26. If a negative gate-to-source voltage is applied, negative charges will accumulate on the metal surface of the gate. Just as in a capacitor, a corresponding number of positive charges will be induced on the other side of the insulator. The positive charges will combine with the electrons in the N-type channel. The channel will become depleted and the resistivity will increase. The increased resistivity will reduce the amount of current flow in the channel just as in the JFET. The conductance characteristics of the MOSFET operating in the depletion mode are essentially the same as the JFET characteristics of Fig. A–24. If a positive gate-to-source voltage is applied to the MOSFET of Fig. A–27, positive charges will accumulate on the surface of the metal at the gate. Corresponding negative charges will be induced on the other side of the insulating material. The addition of elec-

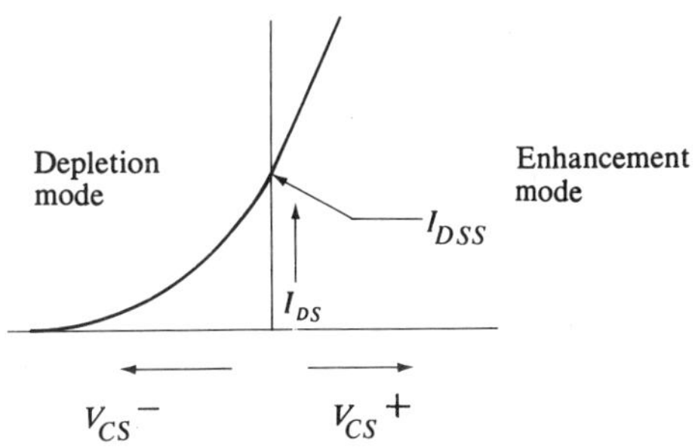

Fig. A–26 Conductance characteristics of the MOSFET.

trons to the N-type channel will decrease the resistance of the channel. Decreasing the resistance of the channel allows more current to flow between the drain and source. The electron density of the channel is *enhanced.* This is the enhancement mode of operation. Figure A–25 shows the conductance characteristics of the MOSFET in both the *enhancement* and *depletion* modes.

A–3–4 PNPN Devices

The PNPN devices that we discussed in the text all operate as switches. They are either in the high-resistance "off" condition or the low-resistance "on" condition. The SCR, PUT, SCS, and SUS are typical of the family of PNPN devices. These devices are best described by the two-transistor model. We will analyze the SCR using this model. Other PNPN switches can be analyzed in a similar manner.

Figure A–28 shows the two-transistor model of the SCR with the terminals identified. From Eq. (A.13),

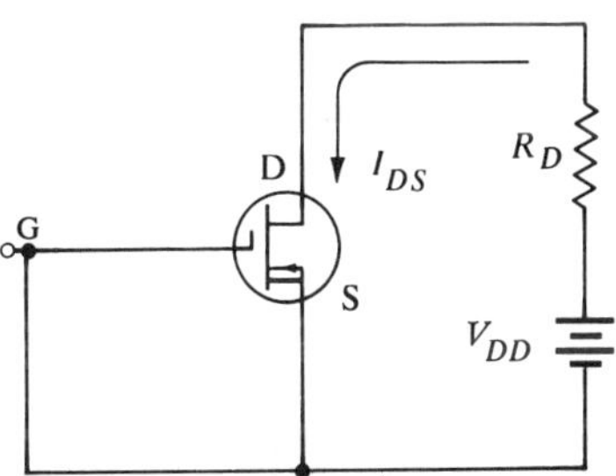

Fig. A–27 MOSFET without gate bias.

$$I_{C_1} = \alpha I_{E_1} + I_{CBO_1} \quad \textbf{(A.15)}$$

$$I_{C_2} = \alpha I_{E_2} + I_{CBO_2} \quad \textbf{(A.16)}$$

But $\quad I_{E_1} = I_{E_2} = I_A$ and $I_{C_1} = I_{B_2} = I_A - I_{C_2}$

so Eqs. (A.15) and (A.16) can be rearranged to

$$I_{CBO_1} = -I_{C_2} + (1 - \alpha_1) I_A \quad \textbf{(A.17)}$$

$$I_{CBO_1} = I_{C_2} - \alpha_2 I_A \quad \textbf{(A.18)}$$

In matrix form

$$\begin{bmatrix} I_{CBO_1} \\ I_{CBO_2} \end{bmatrix} = \begin{bmatrix} -1 & 1 - \alpha_1 \\ 1 & -\alpha_2 \end{bmatrix} \begin{bmatrix} I_{C_2} \\ I_A \end{bmatrix}$$

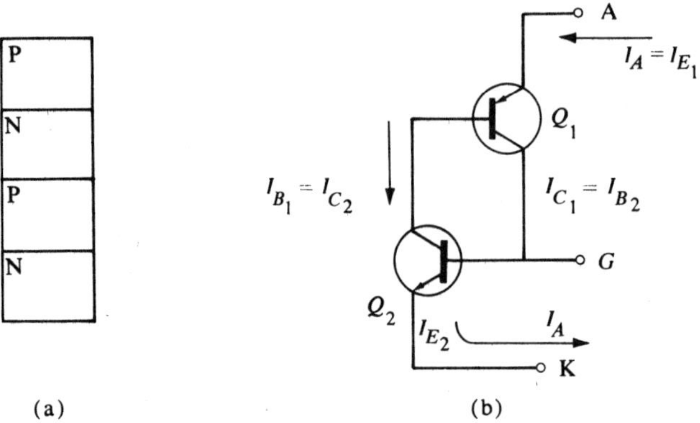

Fig. A-28 SCR PNPN junction model; (b) two-transistor model of SCR.

Using the method of determinants to solve for I_A,

$$I_A = \frac{\begin{vmatrix} -1 & I_{CBO_1} \\ 1 & I_{CBO_2} \end{vmatrix}}{\begin{vmatrix} -1 & 1 - \alpha_1 \\ 1 & -\alpha_2 \end{vmatrix}} \tag{A.20}$$

or

$$I_A = \frac{I_{CBO_2} + I_{CBO_1}}{1 - (\alpha_1 + \alpha_2)} \tag{A.21}$$

We recognize the unstable condition of the SCR by examining the denomination of Eq. (A.21). When $\alpha_1 + \alpha_2 = 1$, the anode current goes to infinity. This represents the "on" condition. The gain of the NPN transistor is controlled by its emitter current as shown in Fig. A-29. We can

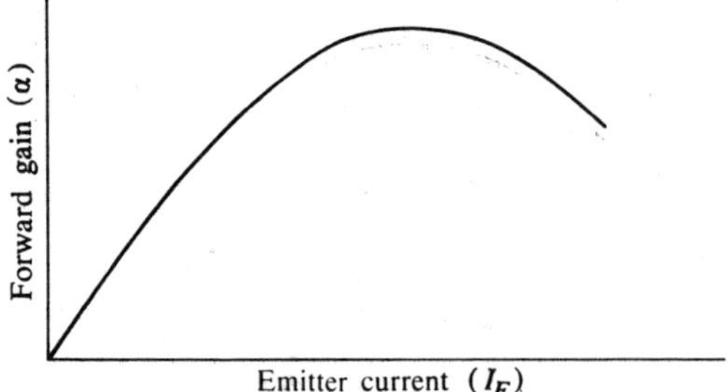

Fig. A-29 Variation of α with I_E.

cause the SCR to switch on by increasing the base current of $2Q$ (the gate current of the SCR) or by increasing the collector current of $1Q$ (exceeding the forward blocking voltage).

Appendix B

1N5411 DIAC

Si all-diffused three-layer trigger diode type used for triac phase-control circuits for lamp dimming, universal-motor speed, and heat controls. JEDEC DO-26, **Outline** No. 66.

MAXIMUM RATINGS

Peak Pulse Current, Forward or Reverse (t_p = 30 μs, df = 0.004)		2	A
Device Dissipation (Tc up to 75°C)		0.5	W
Temperature Range:			
Operating (Junction)	T_j(opr)	−40 to 100	°C
Storage	T_{STG}	−40 to 150	°C
Lead-Soldering Temperature (10 s max)	T_L	255	°C

CHARACTERISTICS (At case temperature = 25°C)

Breakover Voltage, Forward or Reverse	$V_{(BO)}$	29 to 35	V
Breakover-Voltage Symmetry	$\lvert +V_{(BO)}\rvert - \lvert -V_{(BO)}\rvert$	±3	V
Breakback Voltage Change, Forward or Reverse (I_{BO} (forward or reverse) = 10 mA)	ΔV	5 min	V
Peak Breakover Current	$I_{(BO)}$	50 max	μA

DIAC 40583

Si all-diffused three-layer trigger diode type used for triac phase-control circuits for lamp dimming, universal-motor speed, and heat controls. JEDEC DO-26, **Outline** No. 66. This type is identical with type 1N5411 except for the following item:

CHARACTERISTICS (At case temperature = 25°C)

Breakover Voltage, Forward or Reverse	$V_{(BO)}$	27 to 37 V

Courtesy of RCA Solid State Division

Fig. B–1 DIAC characteristics.

Type	Peak Forward Blocking Voltage, V_{DRM} R_{GK} = 1000 Ohms T_J = −40°C to +110°C	Working and Repetitive Peak Reverse Voltage, V_{RRM} T_J = −40°C to +110°C
C106Y1, C106Y2, C106Y3, C106Y4	30 volts	30 volts
C106F1, C106F2, C106F3, C106F4	50 volts	50 volts
C106A1, C106A2, C106A3, C106A4	100 volts	100 volts
C106G1, C106G2, C106G3, C106G4	150 volts	150 volts
C106B1, C106B2, C106B3, C106B4	200 volts	200 volts

MAXIMUM ALLOWABLE RATINGS AND CHARACTERISTICS

RMS Forward Current, On-state I_F . . . 2.0 Amperes
Average Forward Current, On-state $I_{F(AV)}$ . . . Depends on conduction angle (see chart)
Peak One Cycle Surge Forward Current (Non-repetitive), I_{FM} (surge) . . . 15 Amperes
I^2t (for fusing) . . . 0.5 Ampere² seconds (for times ≥ 1.5 milliseconds)
Peak Reverse Gate Voltage, V_{GRM} . . . 6 Volts
Operating Temperature T_J . . . −40°C to +110°C
Forward and Reverse Blocking Current* . . . $I_{FX}I_{RX}$ Typ. 10.0/Max. 100 μAdc
Holding Current‡ . . . I_{HX} Typ. 0.3/Max. 3 mAdc
Turn-off Time* . . . t_{off} Typ. 40/Max. 100 μsec
dv/dt* . . . Typ. 20V/μsec
di/dt . . . Max. 50 A/μsec

‡T_L = +25°C, R_{GK} = 1000 ohms
*T_L = +110°C, R_{GK} 1000 ohms

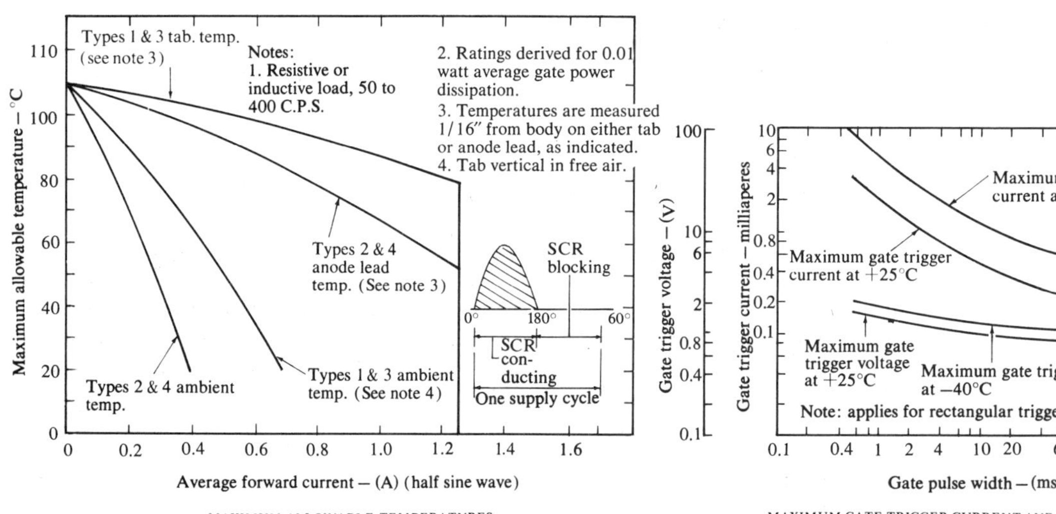

MAXIMUM ALLOWABLE TEMPERATURES

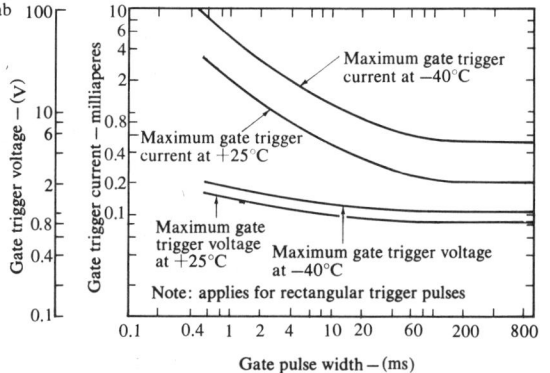

MAXIMUM GATE TRIGGER CURRENT AND VOLTAGE VARIATION WITH TRIGGER PULSE WIDTH

Courtesy of General Electric Semiconductor Dept., Syracuse, N. Y.

Fig. B–2 C106 SCR characteristics.

Programmable Unijunction Transistor

2N6027
2N6028

ABSOLUTE MAXIMUM RATINGS: (25°C)

Voltage	
*Gate-Cathode Forward Voltage	+40 V
*Gate-Cathode Reverse Voltage	−5 V
*Gate-Anode Reverse Voltage	+40 V
*Anode-Cathode Voltage	±40 V
Current	
*DC Anode Current†	150 mA
Peak Anode, Recurrent Forward	
(100 μsec pulse width, 1% duty cycle)	1 A
*(20 μsec pulse width, 1% duty cycle)	2 A
Peak Anode, Non-recurrent Forward	
(10 μsec)	5 A
*Gate Current	±20 mA
Capacitive Discharge Energy††	250 μJ
Power	
*Total Average Power†	300 mW
Temperature	
*Operating Ambient†	
Temperature Range	−50°C to + 100°C

†Derate currents and powers 1%/°C above 25°C
††E = 1/2 CV^2 capacitor discharge energy with no current limiting.

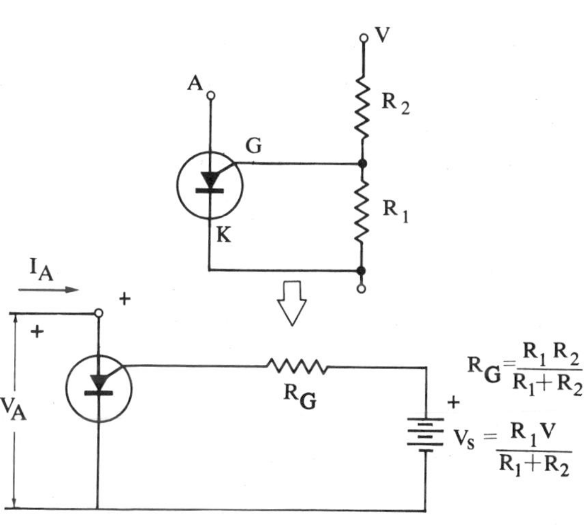

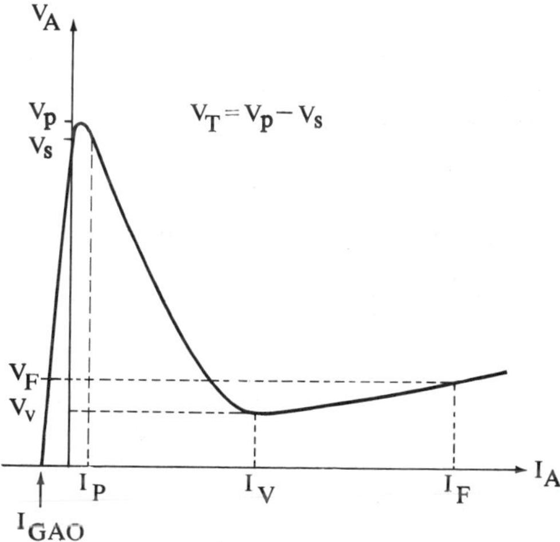

Courtesy of General Electric Semiconductor Dept., Syracuse, N. Y.

Fig. B–3 PUT characteristics.

ELECTRICAL CHARACTERISTICS: (25°C) (unless otherwise specified)

			2N6027 (D13T1)		2N6028 (D13T2)	
		Fig. No.	Min.	Max.	Min.	Max.
***Peak Current** (V_s = 10 Volts)	I_P	3				
(R_G = 1 Meg)				2		.15 μA
(R_G = 10 k)				5		1.0 μA
***Offset Voltage** (V_s = 10 Volts)	V_T	3				
(R_G = 1 Meg)			.2	1.6	.2	.6 Volts
(R_G = 10 k)			.2	.6	.2	.6 Volts
***Valley Current** (V_s = 10 Volts)	I_V	3				
(R_G = 1 Meg)				50		25 μA
(R_G = 10 k)			70		25	μA
(R_G = 200 Ω)			1.5		1.0	mA
Anode Gate-Anode Leakage Current						
*(V_s = 40 Volts, T = 25°C)	I_{GAO}	4		10		10 nA
(T = 75°C)				100		100 nA
Gate to Cathode Leakage Current						
(V_s = 40 Volts, Anode-cathode short)	I_{GKS}	5		100		100 nA
***Forward Voltage** (I_F = 50 mA)	V_F			1.5		1.5 Volts
***Pulse Output Voltage**	V_O	6	6		6	Volts
Pulse Voltage Rate of Rise	t_r	6		80		80 nsecs.

*JEDEC registered data

Courtesy of General Electric Semiconductors Dept., Syracuse, N.Y.

Fig. B–3 PUT Characteristics (continued).

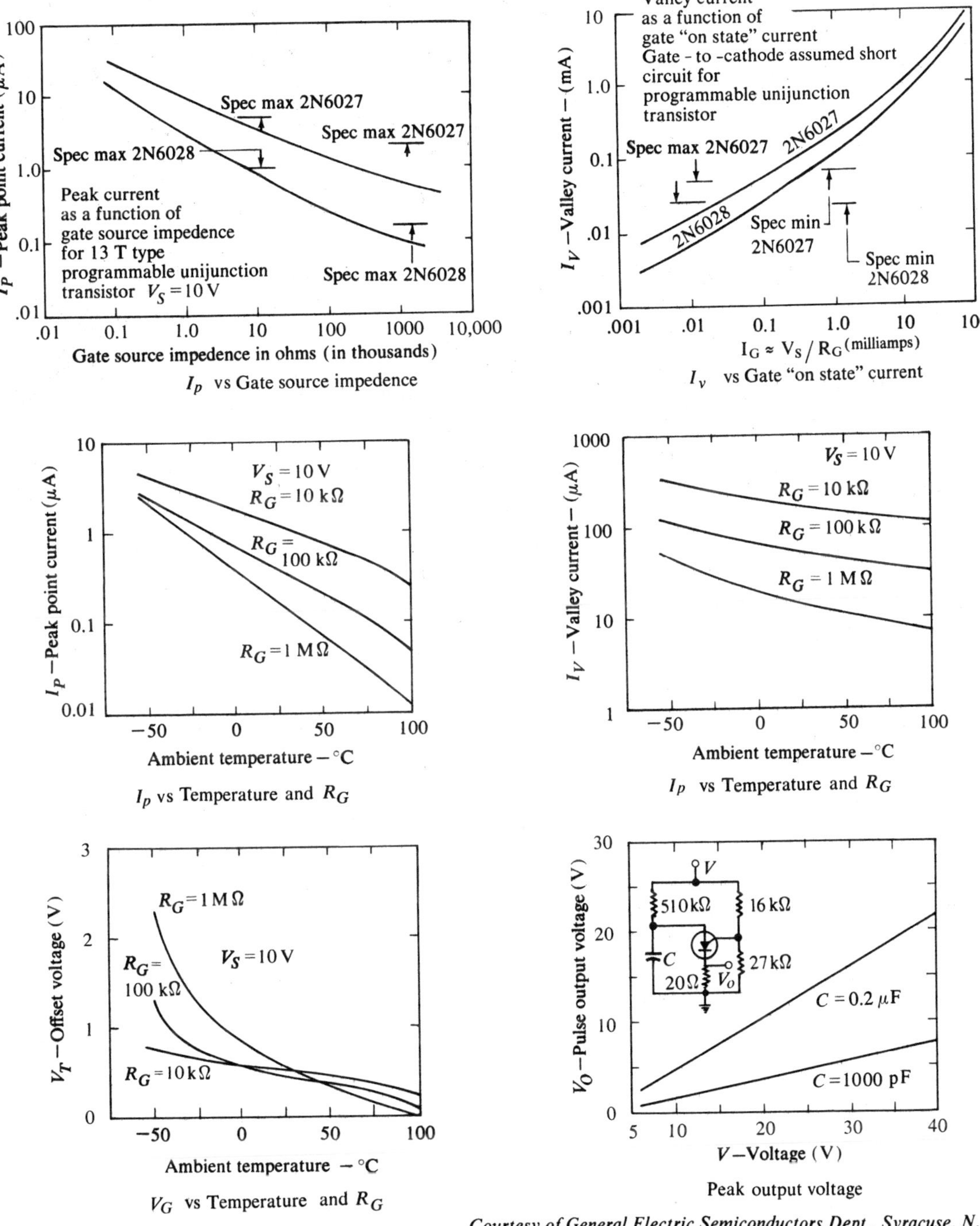

Courtesy of General Electric Semiconductors Dept., Syracuse, N.Y.

Fig. B–3 PUT characteristics (continued).

55A RMS SCR up to 1200V

- Guaranteed di/dt rating of 50 amps/μsec
- Guaranteed dv/dt rating of 200 volts/μsec
- Reduced Package Size, Mounted on a 5/16 Inch Stud

Type	Peak Forward Blocking Voltage, V_{FOM} $T_C = -40°C$ to + 120°C	Peak Forward Voltage, PFV $T_C = -40°C$ to + 120°C	Repetitive Peak Reverse Voltage, V_{RRM} $T_C = -40°C$ to + 120°C	Non-Repetitive Peak Reverse Voltage, < 5 Millisec V_{RMS}* $T_C = -40°C$ to + 120°C
C145F	50 volts	75 volts	50 volts	75 volts
C145A	100 volts	150 volts	100 volts	150 volts
C145B	200 volts	300 volts	200 volts	300 volts
C145C	300 volts	400 volts	300 volts	400 volts
C145D	400 volts	500 volts	400 volts	500 volts
C145E	500 volts	600 volts	500 volts	600 volts
C145M	600 volts	720 volts	600 volts	720 volts
C145S	700 volts	840 volts	700 volts	840 volts
C145N	800 volts	960 volts	800 volts	960 volts
C145T	900 volts	1080 volts	900 volts	1080 volts
C145P	1000 volts	1200 volts	1000 volts	1200 volts
C145PA	1100 volts	1320 volts	1100 volts	1320 volts
C145PB	1200 volts	1440 volts	1200 volts	1440 volts

*Values apply for zero or negative gate voltage only. Maximum case to ambient thermal resistance for which maximum V_{FOM} and V_{ROM} ratings apply equals 1.3°C/watt.

MAXIMUM ALLOWABLE RATINGS AND CHARACTERISTICS

RMS Forward Current, On-state I_F . . . . 55 Amperes
Average Forward Current, On-state $I_{F(AV)}$ . . . . Depends on conduction angle (see chart)
Peak One Cycle Surge Forward Current (Non-repetitive), I_{FM} (surge) . . . . 700 Amperes
I^2t (for fusing) . . . . 2000 Ampere^2 seconds (for times $\geq$ 1.5 milliseconds)
Operating Temperature T_J . . . . $-40°C$ to $+125°C$
Turn-off Time . . . . not specified
dv/dt ($T_J = +120°C$) . . . . Min. 200V/μsec

Courtesy of General Electric Semiconductor Dept., Syracuse, N. Y.

Fig. B–4 C-145 SCR characteristics.

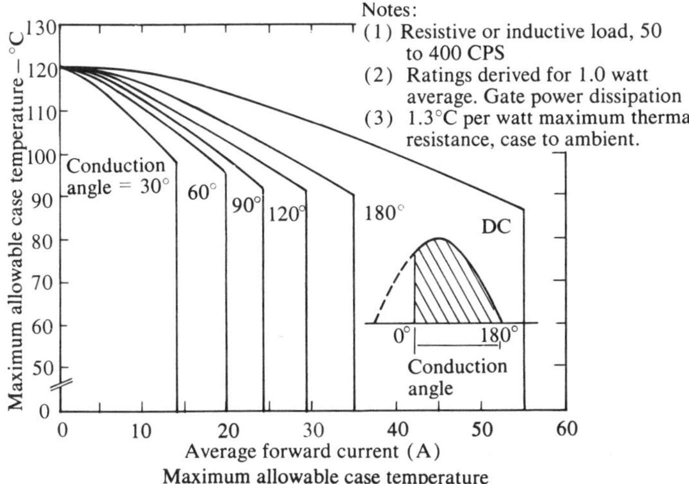

Maximum allowable case temperature

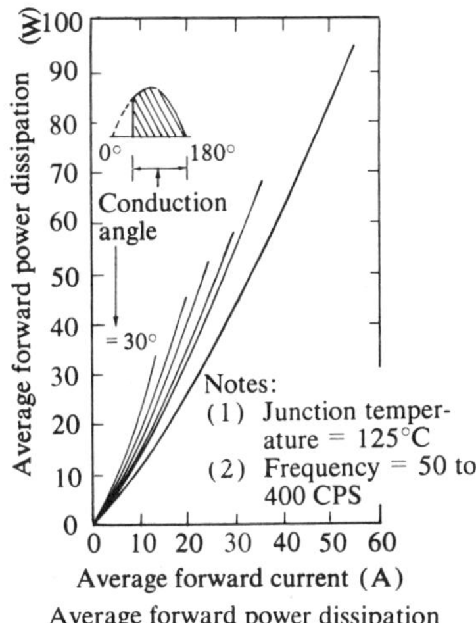

Average forward power dissipation

Courtesy of General Electric Semiconductors Dept., Syracuse, N.Y.

Fig. B–4 C–145 SCR characteristics (continued).

235A RMS SCR Up to 1300V

- Range of Turn-off Times
- Guaranteed dv/dt (200V/μsec)
- Immunity to Forward Voltage Destruction

Type	Peak Forward Blocking Voltage, V_{FOM}* $T_C = -40°C$ to $+120°C$	Repetitive Peak Reverse Voltage, V_{RRM}* $T_C = -40°C$ to $+120°C$	Non-Repetitive Peak Reverse Voltage, <5.0 Millisec. V_{RMS} $T_C = -40°C$ to $+120°C$
C180, C181, C185A	100 volts	100 volts	200 volts
C180, C181, C185B	200 volts	200 volts	300 volts
C180, C181, C185C	300 volts	300 volts	400 volts
C180, C181, C185D	400 volts	400 volts	500 volts
C180, C181, C185E	500 volts	500 volts	600 volts
C180, C181M	600 volts	600 volts	720 volts
C180, C181S	700 volts	700 volts	840 volts
C180, C181N	800 volts	800 volts	950 volts
C180, C181T	900 volts	900 volts	1075 volts
C180, C181P	1000 volts	1000 volts	1200 volts
C180, C181PA	1100 volts	1100 volts	1325 volts
C180, C181PB	1200 volts	1200 volts	1450 volts
C180RC	1300 volts	1300 volts	1550 volts

*Ratings apply for zero or negative gate voltage. Maximum heatsink thermal resistance for which maximum PRV ratings apply equal 1.5°/watt.

Courtesy of General Electric Semiconductors Dept., Syracuse, N.Y.

Fig. B–5 C180, C181, C185 SCR characteristics.

MAXIMUM ALLOWABLE RATINGS AND CHARACTERISTICS

RMS Forward Current, On-state I_F . . . . 235 Amperes
Average Forward Current, On-state $I_{F(AV)}$ . . . . Depends on conduction angle (see chart)
Peak One Cycle Surge Forward Current (Non-repetitive), I_{FM} (surge) . . . . C180, 185 3500 Amperes
C181 2500 Amperes
I^2t (for fusing) . . . . C180, 185 32,000 Ampere^2 seconds (for times $\geq$ 1.5 milliseconds)
C181 25,000 Ampere^2 seconds (for times $\geq$ 1.5 milliseconds)
Operating Temperature T_J . . . . −40°C to + 125°C
dv/dt (T_J = +120°C) . . . . Min. 200V/μsec
di/dt . . . . Switching from 500V or less, Max. 100A/μsec
Switching from 1000V to 500V, Max. 75A/μsec
Switching from 1300V to 1000V, Max. 50A/μsec

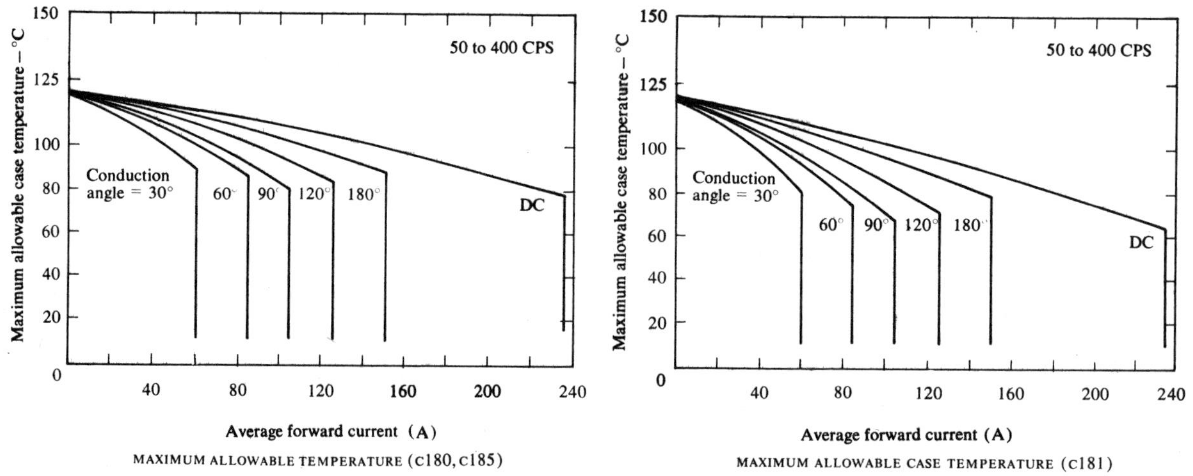

MAXIMUM ALLOWABLE TEMPERATURE (C180, C185)

MAXIMUM ALLOWABLE CASE TEMPERATURE (C181)

Courtesy of General Electric Semiconductors Dept., Syracuse, N.Y.

Fig. B–5 C180, C181, C185 SCR characteristics (continued).

2N3650–2N3653

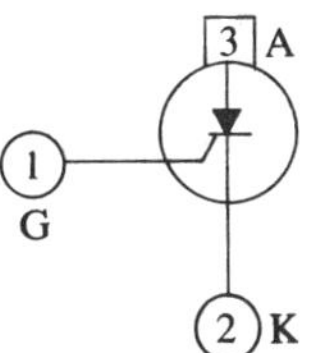

Si all-diffused types used for high-speed switching applications such as power inverters, switching regulators, and high-current pulse applications. JEDEC TO-48, **Outline** No. 20. See **Mounting Hardware** for desired mounting arrangement.

MAXIMUM RATINGS

	2N3650	2N3651	2N3652	2N3653	
V_{ROSM}	150	300	400	500	V
V_{DSOM}	150	300	400	500	V
V_{RROM}	100	200	300	400	V
V_{DROM}	100	200	300	400	V
I_{TSM} (1 cycle of principal voltage at 60 Hz)		180			A
$I_{T\,(AV)}$ (conduction angle = 180°)		25			A
$I_{T\,(RMS)}$		35			A
di/dt (V_{DM} = $V_{(BO)O}$, I_{GT} = 200 mA, tr = 0.1 μs		400			A/μs
P_{GM} (10 μs max)		40			W
$P_{G\,(AV)}$ (10 ms max)		1			W
T_{STG}		−65 to 150			°C
$T_{C\,(opr)}$		−65 to 120			°C
T_S (10 s max)		225			°C

CHARACTERISTICS (At maximum electrical ratings at T_C = 25°C)

	2N3650	2N3651	2N3652	2N3653	
$V_{(BO)O}$ (T_C = 120°C)	100 min	200 min	300 min	400 min	V
I_{DOM} (V_{DO} = V_{DROM}, T_C = 120°C)	6 max	6 max	5.5 max	4 max	mA
I_{RROM} (V_{RO} = V_{DRROM}, T_C = 120°C)	6 max	6 max	5.5 max	4 max	mA
v_T (i_T = 25 A)		2.05 max			V
I_{GT}:					
V_D = 6 Vdc, R_L = 4 Ω		80 typ; 180 max			mA
V_D = 6 Vdc, R_L = 2 Ω, T_C = −65°C		150 typ; 500 max			mA
V_{GT}:					
V_D = 6 Vdc, R_L = 4 Ω		1.5 typ; 3 max			V
V_D = V_{DROM}, R_L = 200 Ω, T_C = 120°C		0.25 min			V
V_D = 6 Vdc, R_L = 2 Ω, T_C = −65°C		2 typ; 4.5 max			V
i_{HO}:					
T_C = 25°C		75 typ; 150 max			mA
T_C = −65°C		150 typ; 350 max			mA
Critical dv/dt (V_{DO} = V_{DROM}, exponential rise, T_C = 120°C)		200 min			V/μs
tq, Rectangular Pulse (V_{DX} = V_{DROM}, i_T = 10 A, tp = 50 μs, I_{GT} = 200 mA at turn-on, − di/dt = 5 A/μs, dv/dt = 200 V/μs, V_{RX} = 15 min, V_{GK} = 0 at turn-off, T_C = 120°C)		11 typ; 15 max			μs
tq, Half-Sinosoidal Waveform (V_{DX} = V_{DROM}, i_T = 100 A, tp = 1.5 μs, I_{GT} = 200 mA, dv/dt = 200 V/μs, V_{RX} = 30 V_{min}, V_{GK} = 0 at turn-off, T_C = 115°C)		12 typ; 15 max			μs
$\Theta_{J\text{-}C}$		1.7 max			°C/W

Courtesy of RCA Solid State Division Fig. B–6 SCR characteristics.

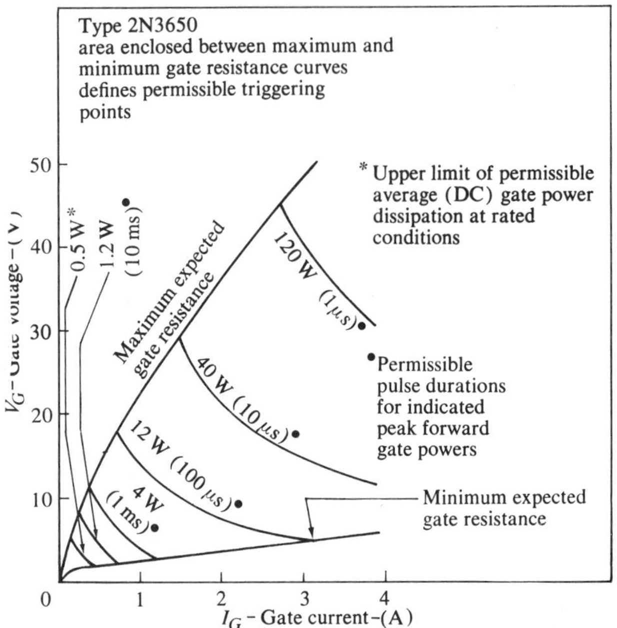

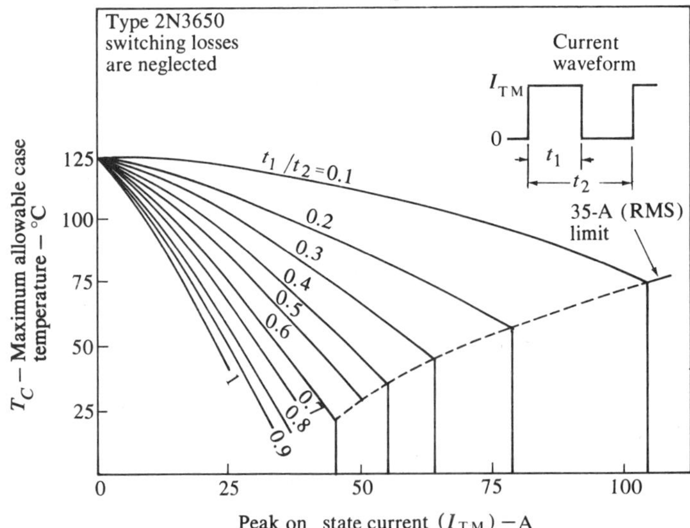

Courtesy of RCA Solid State Division

Fig. B–6 SCR characteristics (continued)

40773 2.5A, 200V
40774 2.5A, 400V

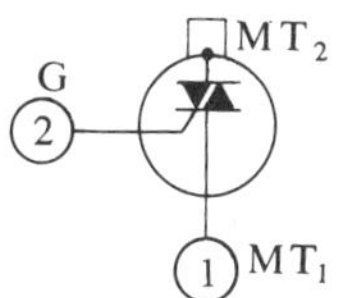

Si gate-controlled full-wave types used for control-systems application in airborne and ground-support type equipment. JEDEC TO-5 (modified), **Outline** No. 7. See **Mounting Hardware** for desired mounting arrangement.

MAXIMUM RATINGS (For sinusoidal supply voltage up to 400 Hz with resistive or inductive load)

	40773	40774	
V_{DROM}* (T_J = −50 to 100°C)	200	400	V
$I_{T(RMS)}$ (conduction angle = 360°):			
T_C = 90°C	2.5		A
T_A = 25°C without heat sink	0.5		A
I_{TSM} (1 cycle sinusoidal principal voltage):			
400 Hz	200		A
60 Hz	100		A
di/dt (V_{DM} = V_{DROM}, I_{GT} = 80 mA, t_r = 0.1 μs)	150		A/μs
I_{GTM}‡ (1 μs max)	4		A
P_{GM} (1 μs max, I_{GTM} < 4 A)	16		W
$P_{G(AV)}$	0.2		W
T_{STG}	−50 to 150		°C
T_C	−50 to 100		°C
T_L (10 s max)	225		°C

CHARACTERISTICS (At maximum electrical ratings at T_C = 25°C)

I_{DROM}* (T_J = 100°C, V_{DROM} = max rated value)	0.2 typ; 4 max	mA
V_{TM}* (i_T = 30 A peak)	1.6 typ; 2.25 max	V
I_{HO}* (initial principal current = 150 mAdc, v_D = 12 V	15 typ; 30 max	mA
Commutating dv/dt* (v_D = V_{DROM}, $I_{T(RMS)}$ = 2.5 A, commutating di/dt = 8.9 A/ms, gate unenergized, T_C = 90°C	3 min; 10 typ	V/μs
Critical dv/dt* (v_D = V_{DROM}, exponential voltage rise, T_C = 100°C)	30 min; 150 typ	V/μs
I_{GT}*‡ (v_D = 12 Vdc, R_L = 30 Ω):		
I^+ mode, V_{MT2} positive, V_G positive	15 typ; 25 max	mA
I^- mode, V_{MT2} positive, V_G negative	25 typ; 40 max	mA
III^+ mode, V_{MT2} negative, V_G positive	25 typ; 40 max	mA
III^- mode, V_{MT2} negative, V_G negative	15 typ; 25 max	mA
V_{GT}*‡ (v_D = 12 Vdc, R_L = 30 Ω)	1 typ; 2.2 max	V
V_{GT}*‡ (v_D = V_{DROM}, R_L = 125 Ω, T_C = 100°C)	0.2 min	V
t_{gt} (v_D = V_{DROM}, I_{GT} = 80 mA, t_r = 0.1 μs, i_T = 10 A peak)	1.8 typ; 2.5 max	μs
Θ_{J-C} (steady-state)	4 max	°C/W

*For either polarity of main terminal 2 voltage (V_{MT2}) with reference to main terminal 1.
‡For either polarity of gate voltage (V_G) with reference to main terminal 1.

Courtesy of RCA Solid State Division

Fig. B–7 TRIAC characteristics.

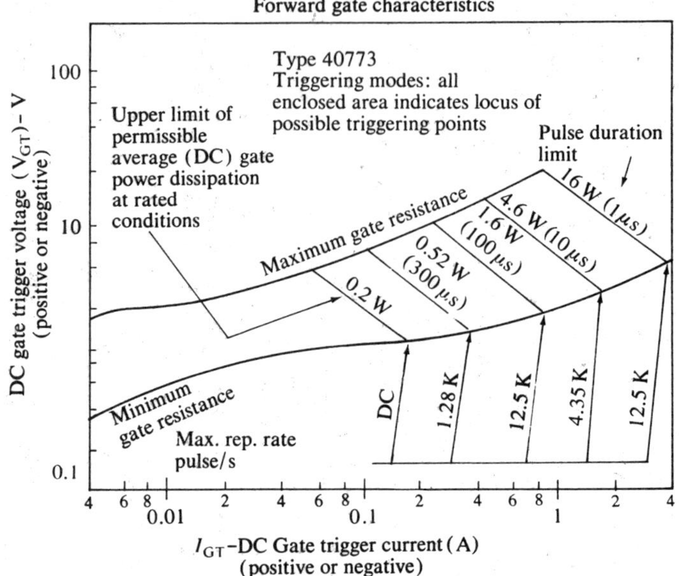

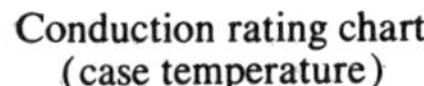

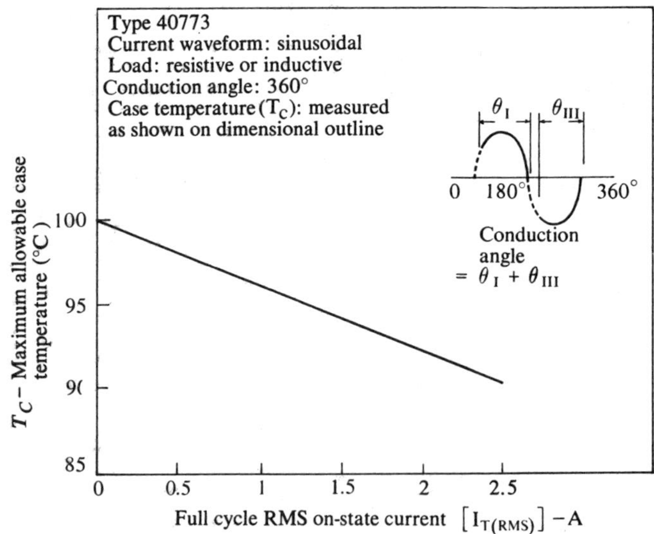

Courtesy of RCA Solid State Division

Fig. B–7 TRIAC characteristics (continued).

2N5567 10A, 200V
2N5568 10A, 400V

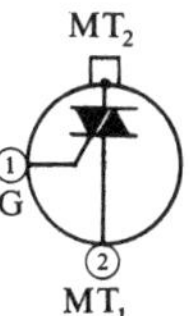

Si gate-controlled full-wave types used for control of ac loads in applications such as heating controls, motor controls, arc-welding equipment, light dimmers, and power switching systems. See **Mounting Hardware** for desired mounting arrangement. **Outline** No. 36.

<table>
<tr><th colspan="4">MAXIMUM RATINGS (For sinusoidal supply voltage at 50/60 Hz with resistive or inductive load)</th></tr>
<tr><td></td><td>2N5567</td><td>2N5568</td><td></td></tr>
<tr><td>V_{DROM}* (T_J = −65°C to 100°C)</td><td>200</td><td>400</td><td>V</td></tr>
<tr><td>$I_{T(RMS)}$ (T_C = 85°C, conduction angle = 360°)</td><td colspan="2">10</td><td>A</td></tr>
<tr><td>I_{TSM}:</td><td colspan="2"></td><td></td></tr>
<tr><td>1 cycle of principal voltage at 60 Hz</td><td colspan="2">100</td><td>A</td></tr>
<tr><td>1 cycle of principal voltage at 50 Hz</td><td colspan="2">85</td><td>A</td></tr>
<tr><td>I_{GTM}‡ (1 μs max)</td><td colspan="2">4</td><td>A</td></tr>
<tr><td>di/dt (V_{DM} = V_{DROM}, I_{GT} = 160 mA, t_r = 0.1 μs)</td><td colspan="2">150</td><td>A/μs</td></tr>
<tr><td>P_{GM}‡ (1 μs max, I_{GTM} ≤ 4 A peak)</td><td colspan="2">16</td><td>W</td></tr>
<tr><td>$P_{G(AV)}$</td><td colspan="2">0.5</td><td>W</td></tr>
<tr><td>T_{STG}</td><td colspan="2">−65 to 150</td><td>°C</td></tr>
<tr><td>T_C (opr)</td><td colspan="2">−65 to 100</td><td>°C</td></tr>
<tr><td>T_T (10 s max)</td><td colspan="2">225</td><td>°C</td></tr>
<tr><th colspan="4">CHARACTERISTICS</th></tr>
<tr><td>I_{DROM}* (T_J = 100°C, V_{DROM} = max rated value)</td><td colspan="2">0.1 typ; 2 max</td><td>mA</td></tr>
<tr><td>V_{TM}* (i_T = 14 A peak, T_C = 25°C)</td><td colspan="2">1.35 typ; 1.65 max</td><td>V</td></tr>
<tr><td>I_{HO}* (initial principal current = 500 mAdc):</td><td colspan="2"></td><td></td></tr>
<tr><td>T_C = 25°C</td><td colspan="2">15 typ; 30 max</td><td>mA</td></tr>
<tr><td>T_C = −65°C</td><td colspan="2">75 typ; 200 max</td><td>mA</td></tr>
<tr><td>Commutating dv/dt* (v_D = V_{DROM}, $I_{T(RMS)}$ = 10 A, commutating di/dt = 5.4 A/ms, gate unenergized at T_C = 85°C)</td><td colspan="2">2 min; 5 typ</td><td>V/μs</td></tr>
<tr><td>Critical dv/dt* (v_D = V_{DROM}, exponential voltage rise, T_C = 100°C)</td><td>30 min; 150 typ</td><td>20 min; 100 typ</td><td>V/μs</td></tr>
<tr><td>I_{GT}*‡ (v_D = 12 Vdc, R_L = 30 Ω, T_C = 25°C):</td><td colspan="2"></td><td></td></tr>
<tr><td>I^+ mode, V_{MT2} positive, V_G positive</td><td colspan="2">10 typ; 25 max</td><td>mA</td></tr>
<tr><td>I^- mode, V_{MT2} positive, V_G negative</td><td colspan="2">20 typ; 40 max</td><td>mA</td></tr>
<tr><td>III^+ mode, V_{MT2} negative, V_G positive</td><td colspan="2">20 typ; 40 max</td><td>mA</td></tr>
<tr><td>III^- mode, V_{MT2} negative, V_G negative</td><td colspan="2">10 typ; 25 max</td><td>mA</td></tr>
<tr><td>I_{GT}*‡ (v_D = 12 Vdc, R_L = 30 Ω, T_C = −65°C):</td><td colspan="2"></td><td></td></tr>
<tr><td>I^+ mode, V_{MT2} positive, V_G positive</td><td colspan="2">45 typ; 100 max</td><td>mA</td></tr>
<tr><td>I^- mode, V_{MT2} positive, V_G negative</td><td colspan="2">80 typ; 150 max</td><td>mA</td></tr>
<tr><td>III^+ mode, V_{MT2} negative, V_G positive</td><td colspan="2">80 typ; 150 max</td><td>mA</td></tr>
<tr><td>III^- mode, V_{MT2} negative, V_G negative</td><td colspan="2">45 typ; 100 max</td><td>mA</td></tr>
<tr><td>V_{GT}*‡ (v_D = 12 Vdc, R_L = 30 Ω):</td><td colspan="2"></td><td></td></tr>
<tr><td>T_C = 25°C</td><td colspan="2">1 typ; 2.5 max</td><td>V</td></tr>
<tr><td>T_C = −65°C</td><td colspan="2">2 typ; 4 max</td><td>V</td></tr>
<tr><td>V_{GT}*‡ (v_D = V_{DROM}, R_L = 125 Ω, T_C = 100°C)</td><td colspan="2">0.2 min</td><td>V</td></tr>
<tr><td>t_{gt} (v_D = V_{DROM}, I_{GT} = 160 mA, 0.1 μs, tr, i_T = 15 A peak, T_C = 25°C)</td><td colspan="2">1.6 typ; 2.5 max</td><td>μs</td></tr>
<tr><td>$\theta_{J\text{-}C}$ (steady-state)</td><td colspan="2">1 max</td><td>°C/W</td></tr>
</table>

*For either polarity of main terminal 2 voltage (V_{MT2}) with reference to main terminal 1.
‡For either polarity of gate voltage (V_G) with reference to main terminal 1.

Courtesy of RCA Solid State Division Fig. B–8 TRIAC characteristics.

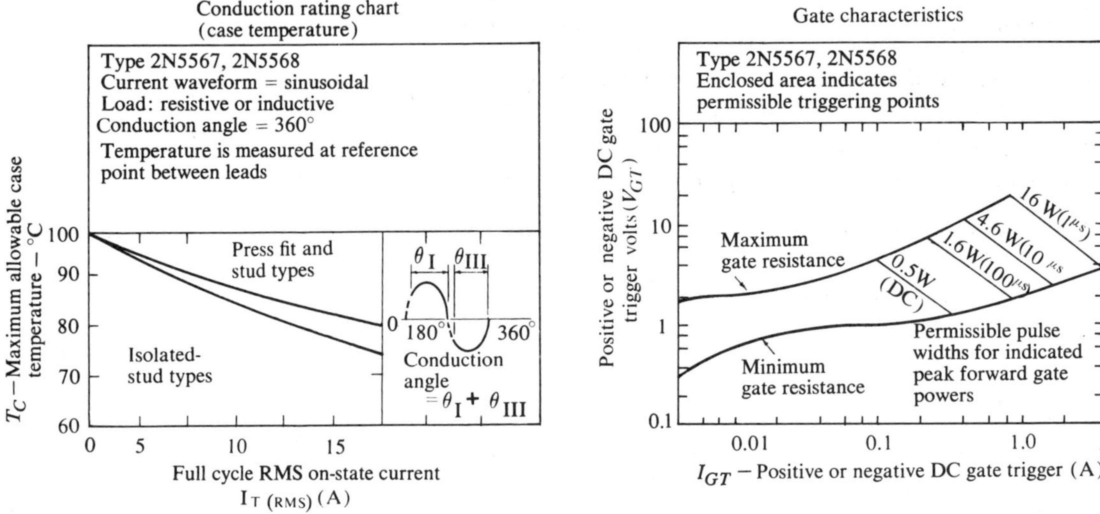
Conduction rating chart
(case temperature)
Type 2N5567, 2N5568
Current waveform = sinusoidal
Load: resistive or inductive
Conduction angle = 360°
Temperature is measured at reference point between leads
Press fit and stud types
Isolated-stud types
θI θIII
0 180° 360°
Conduction angle = θI + θIII
TC — Maximum allowable case temperature — °C
100
90
80
70
60
0 5 10 15
Full cycle RMS on-state current
IT (RMS) (A)
Gate characteristics
Type 2N5567, 2N5568
Enclosed area indicates permissible triggering points
Positive or negative DC gate trigger volts (VGT)
100
10
1
0.1
Maximum gate resistance
Minimum gate resistance
16 W(1 μs)
4.6 W(10 μs)
1.6 W(100 μs)
0.5W (DC)
Permissible pulse widths for indicated peak forward gate powers
0.01 0.1 1.0
IGT — Positive or negative DC gate trigger (A)

Courtesy of RCA Solid State Division

Fig. B–8 TRIAC characteristics (continued).

2N2646 (SILICON)
2N2647

V_{BB} = 35 V
I_e = 50 mA
P_D = 300 mW

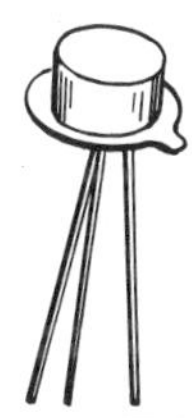

Silicon annular PN unijunction transistors designed for use in pulse and timing circuits, sensing circuits and thyristor trigger circuits.

CASE 22A (Lead 3 connected to case)

MAXIMUM RATINGS (T_A = 25°C unless otherwise noted)

Rating	Symbol	Value	Unit
RMS Power Dissipation*	P_D	300*	mW
RMS Emitter Current	I_e	50	mA
Peak Pulse Emitter Current**	i_e	2**	Amp
Emitter Reverse Voltage	V_{B2E}	30	Volts
Interbase Voltage	V_{B2B1}	35	Volts
Operating Junction Temperature Range	T_J	−65 to +125	°C
Storage Temperature Range	T_{stg}	−65 to +150	°C

*Derate 3.0 mW/°C increase in ambient temperature. The total power dissipation (available power to Emitter and Base-Two) must be limited by the external circuitry.

**Capacitor discharge – 10 μF or less, 30 volts or less.

Courtesy of Motorola Inc.

Fig. B–9 Unijunction transistor characteristics.

2N2646, 2N2647 (*continued*)

ELECTRICAL CHARACTERISTICS (T_A = 25°C unless otherwise noted)

Characteristics		Symbol	Min	Typ	Max	Unit
Intrinsic Standoff Ratio (V_{B2B1} = 10 V) (Note 1)	2N2646	η	0.56	—	0.75	—
	2N2647		0.68	—	0.82	
Interbase Resistance (V_{B2B1} = 3 V, I_E = 0)		R_{BB}	4.7	7.0	9.1	K ohms
Interbase Resistance Temperature Coefficient (V_{B2B1} = 3 V, I_E = 0, T_A = −55°C to +125°C)		αR_{BB}	0.1	—	0.9	%/°C
Emitter Saturation Voltage (V_{B2B1} = 10 V, I_E = 50 mA) (Note 2)		$V_{EB1(sat)}$	—	3.5	—	Volts
Modulated Interbase Current (V_{B2B1} = 10 V, I_E = 50 mA)		$I_{B2(mod)}$	—	15	—	mA
Emitter Reverse Current (V_{B2E} = 30 V, I_{B1} = 0)	2N2646	I_{EO}	—	0.005	12	μA
	2N2647		—	0.005	0.2	
Peak Point Emitter Current (V_{B2B1} = 25 V)	2N2646	I_P	—	1.0	5.0	μA
	2N2647		—	1.0	2.0	
Valley Point Current (V_{B2B1} = 20 V, R_{B2} = 100 ohms) (Note 2)	2N2646	I_V	4.0	6.0	—	mA
	2N2647		8.0	10	18	
Base-One Peak Pulse Voltage (Note 3, Figure 3)	2N2646	V_{OB1}	3.0	5.0	—	Volts
	2N2647		6.0	7.0	—	

NOTES

1. Intrinsic standoff ratio, η is defined by equation:

$$\eta = \frac{V_P - V_{(EB1)}}{V_{B2B1}}$$

Where V_P = Peak Point Emitter Voltage
V_{B2B1} = Interbase Voltage
$V_{(EB1)}$ = Emitter to Base-One Junction Diode Drop (≈0.5 V @ 10 μA)

2. Use pulse techniques: PW ≈ 300 μs duty cycle ≤2% to avoid internal heating due to interbase modulation which may result in erroneous readings.

3. Base-One Peak Pulse Voltage is measured in circuit of Fig. 3. This specification is used to ensure minimum pulse amplitude for applications in SCR firing circuits and other types of pulse circuits.

Courtesy of Motorola Inc.

Fig. B–9 Unijunction transistor characteristics (continued).

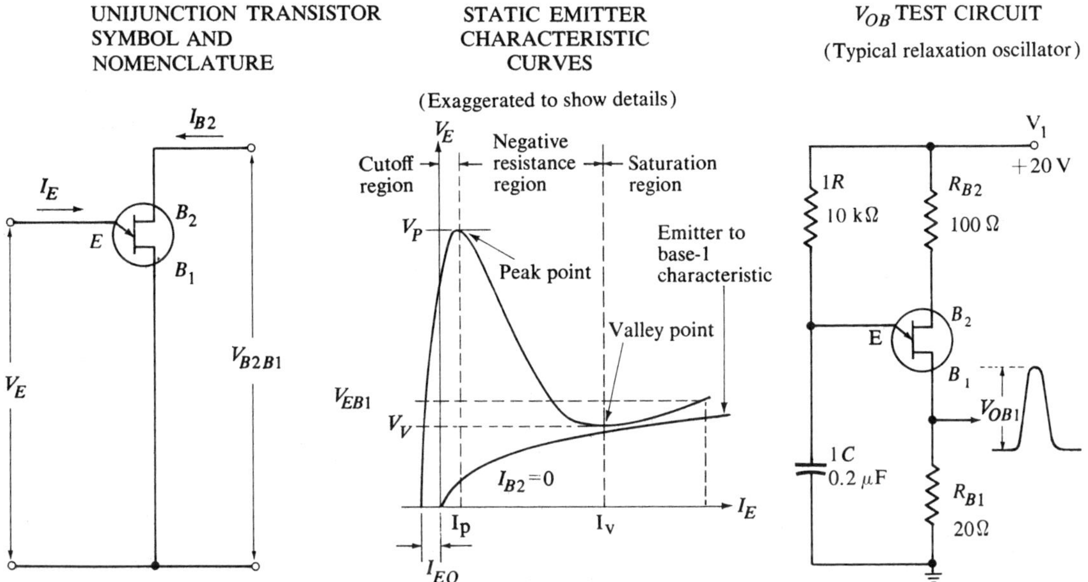

Courtesy of Motorola Inc.

Fig. B–9 Unijunction transistor characteristics (continued).

MPS6516 thru MPS6519, MPS6522, MPS6523

ELECTRICAL CHARACTERISTICS (T_A = 25°C unless otherwise noted)

Characteristic		Symbol	Min	Typ	Max	Unit
Collector-Emitter Breakdown Voltage (I_C = −0.5 mAdc, I_B = 0)	MPS6516 thru MPS6518	BV_{CEO}	−40	—	—	Vdc
	MPS6519, MPS6522, MPS6523		−25	—	—	
Emitter-Base Breakdown Voltage (I_B = −10 μAdc, I_C = 0)		BV_{EBO}	−4	—	—	Vdc
Collector Cutoff Current		I_{CBO}				μAdc
(V_{CB} = −30 Vdc, I_E = 0)	MPS6516 thru MPS6518		—	—	−0.05	
(V_{CB} = −20 Vdc, I_E = 0)	MPS6519, MPS6522, MPS6523		—	—	−0.05	
(V_{CB} = −30 Vdc, I_E = 0, T_A = 60°C)	MPS6516 thru MPS6518		—	—	−1.0	
(V_{CB} = −20 Vdc, I_E = 0, T_A = 60°C)	MPS6519, MPS6522, MPS6523		—	—	−1.0	
DC Current Gain		h_{FE}				—
(I_C = −100 μAdc, V_{CE} = −10 Vdc)	MPS6522		100	—	—	
	MPS6523		150	—	—	
(I_C = −2 mAdc, V_{CE} = −10 Vdc)	MPS6516		50	—	100	
	MPS6517		90	—	180	
	MPS6518		150	—	300	
	MPS6519		250	—	500	
	MPS6522		200	—	400	
	MPS6523		300	—	600	
(I_C = −100 mAdc, V_{CE} = −10 Vdc)	MPS6516		30	—	—	
	MPS6517		60	—	—	
	MPS6518		90	—	—	
	MPS6519		150	—	—	
Collector-Emitter Saturation Voltage (I_C = −50 mAdc, I_B = −5 mAdc)		$V_{CE(sat)}$	—	—	−0.5	Vdc

Courtesy of Motorola Inc.

Fig. B–10 Bipolar transistor characteristics.

ELECTRICAL CHARACTERISTICS (T_A = 25°C unless otherwise noted) (continued)

Characteristic		Symbol	Min	Typ	Max	Unit
Current Gain – Bandwidth Product		f_T				MHz
(I_C = −2 mAdc, V_{CE} = −10 Vdc)	MPS6516, MPS6517		—	200	—	
	MPS6518, MPS6519		—	340	—	
	MPS6522, MPS6523		—	340	—	
(I_C = −10 mAdc, V_{CE} = −10 Vdc)	MPS6516, MPS6517		—	270	—	
	MPS6518, MPS6519		—	420	—	
	MPS6522, MPS6523		—	420	—	
Output Capacitance		C_{ob}				pF
(V_{CB} = −10 Vdc, I_E = 0, f = 100 kHz)	MPS6516 thru MPS6519		—	—	4.0	
	MPS6522, MPS6523		—	—	4.0	
Wideband Noise Figure (V_{CE} = −5 Vdc, I_C = −10 μAdc, R_S = 10 kohms, Power Bandwidth = 15.7 kHz, 3 dB points @ 10 Hz and 10 kHz)	MPS6522, MPS6523	NF	—	1.8	3.0	dB

*Pulse Test: Pulse Width < 30 μs, duty cycle ≤ 2%

WIDEBAND NOISE FIGURE VS SOURCE IMPEDANCE

Figure 1 — MPS6516 thru MPS6519

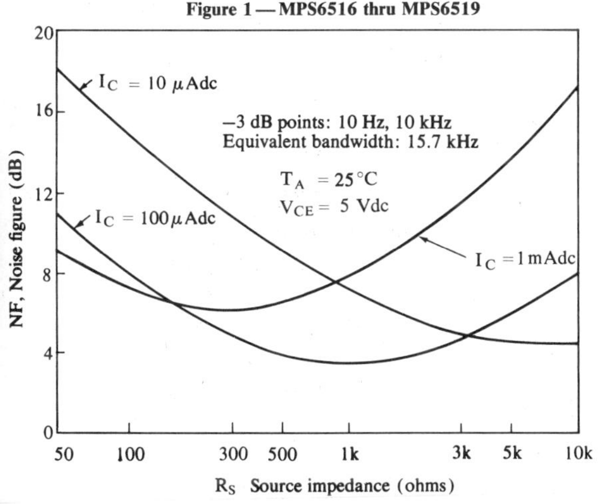

Figure 2 — MPS6522, MPS6523

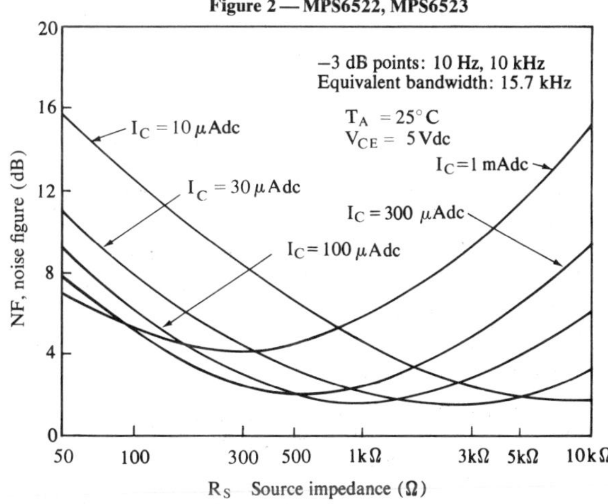

Courtesy of Motorola Inc.

Fig. B–10 Bipolar transistor characteristics (continued).

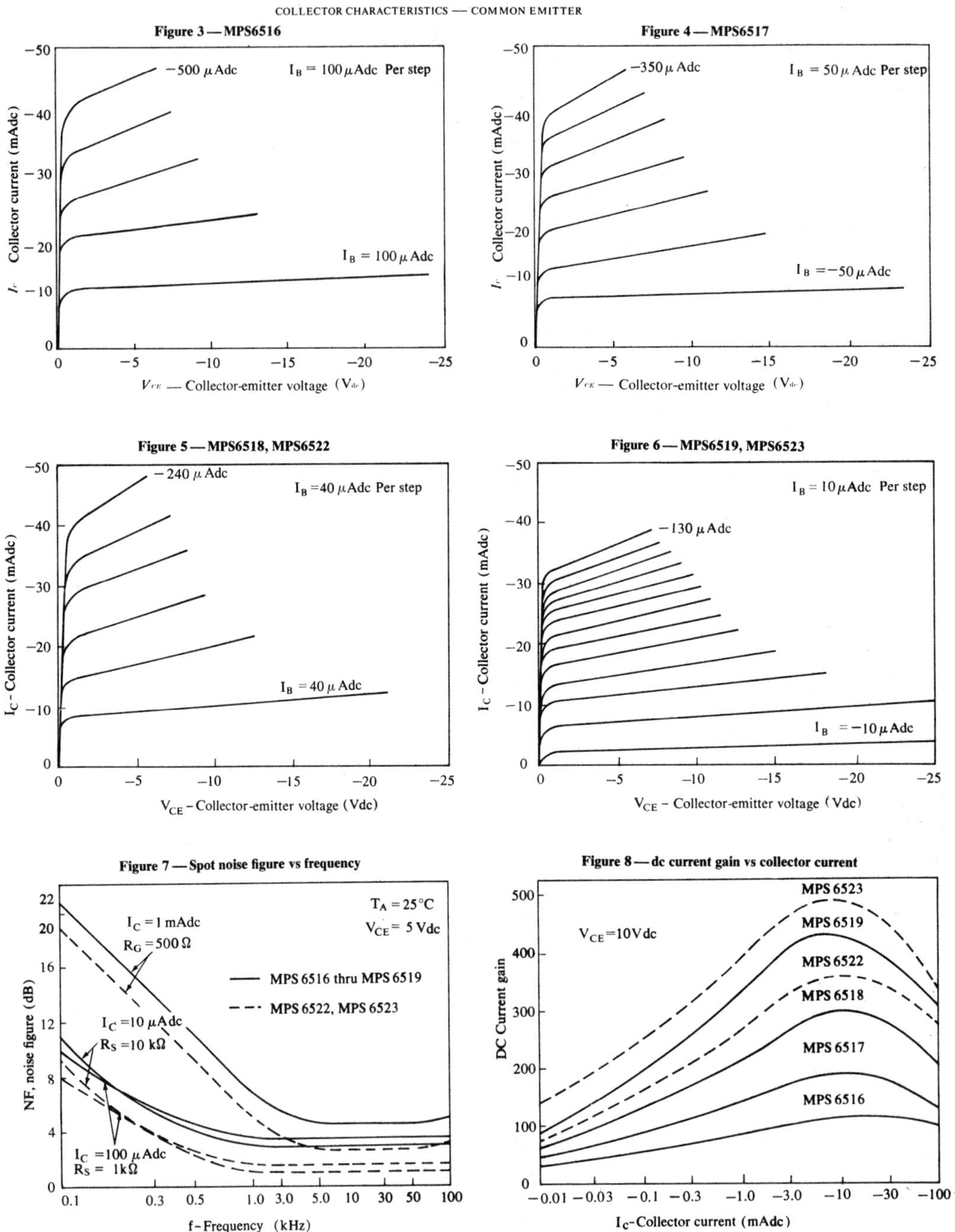

Courtesy of Motorola Inc.

Fig. B–10 Bipolar transistor characteristics (continued).

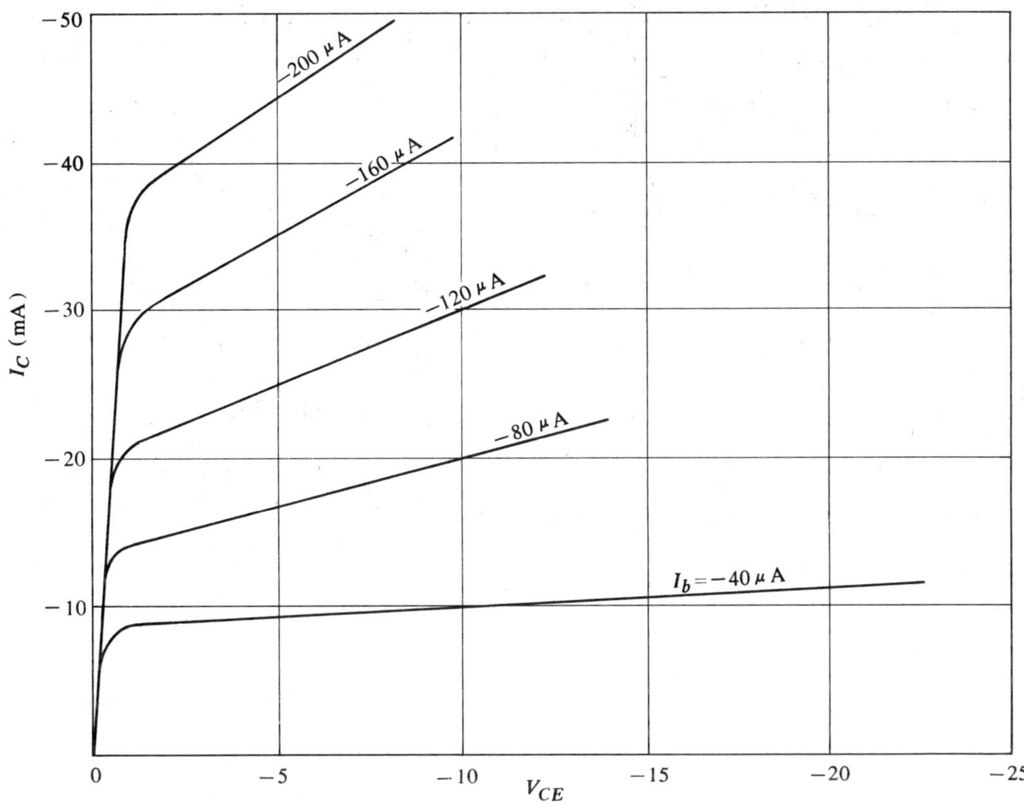

Fig. B–11 Bipolar transistor worksheet.

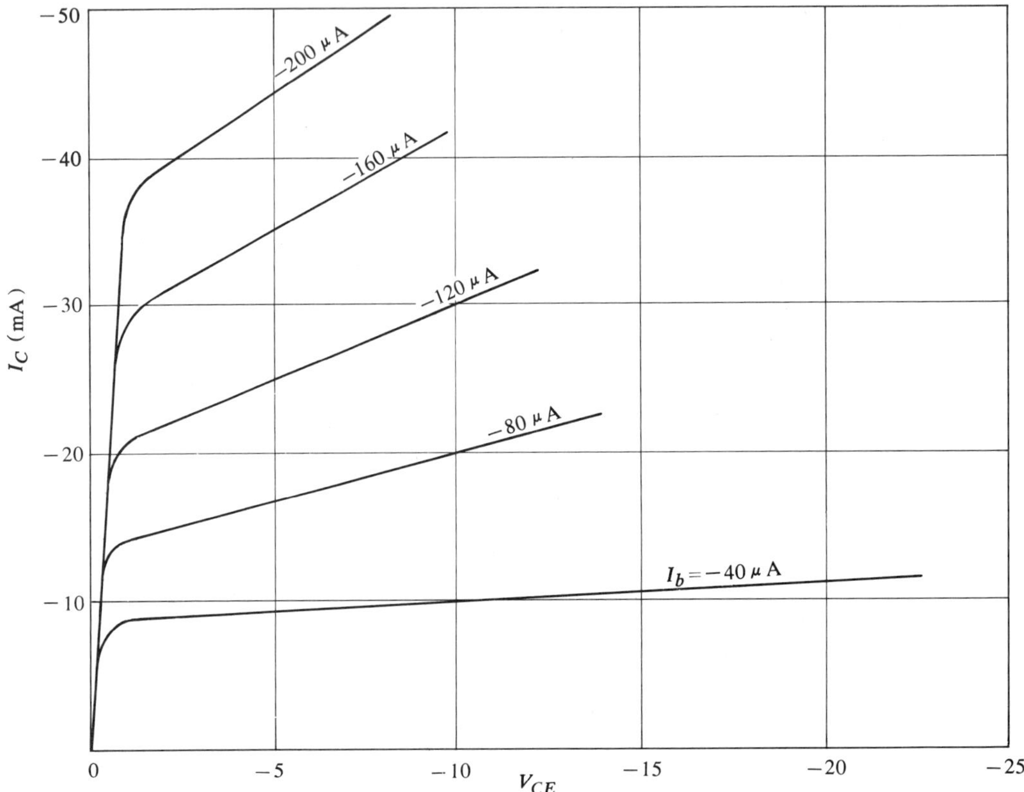

Fig. B–11 Bipolar transistor worksheet.

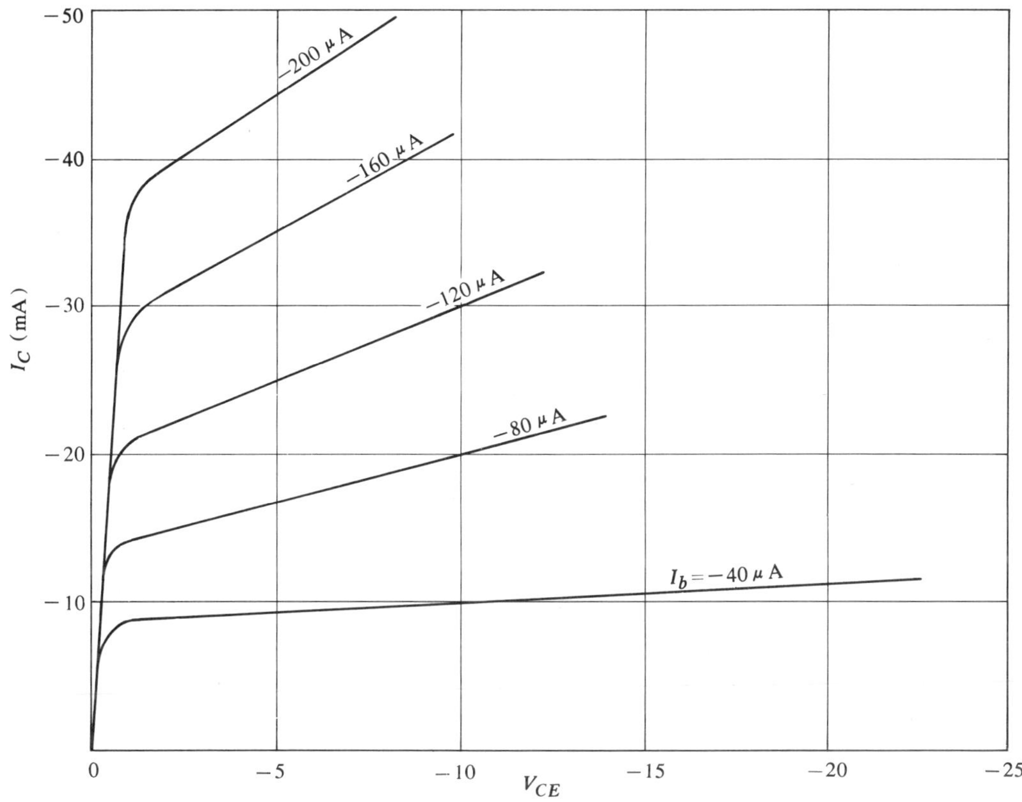

Fig. B–11 Bipolar transistor worksheet.

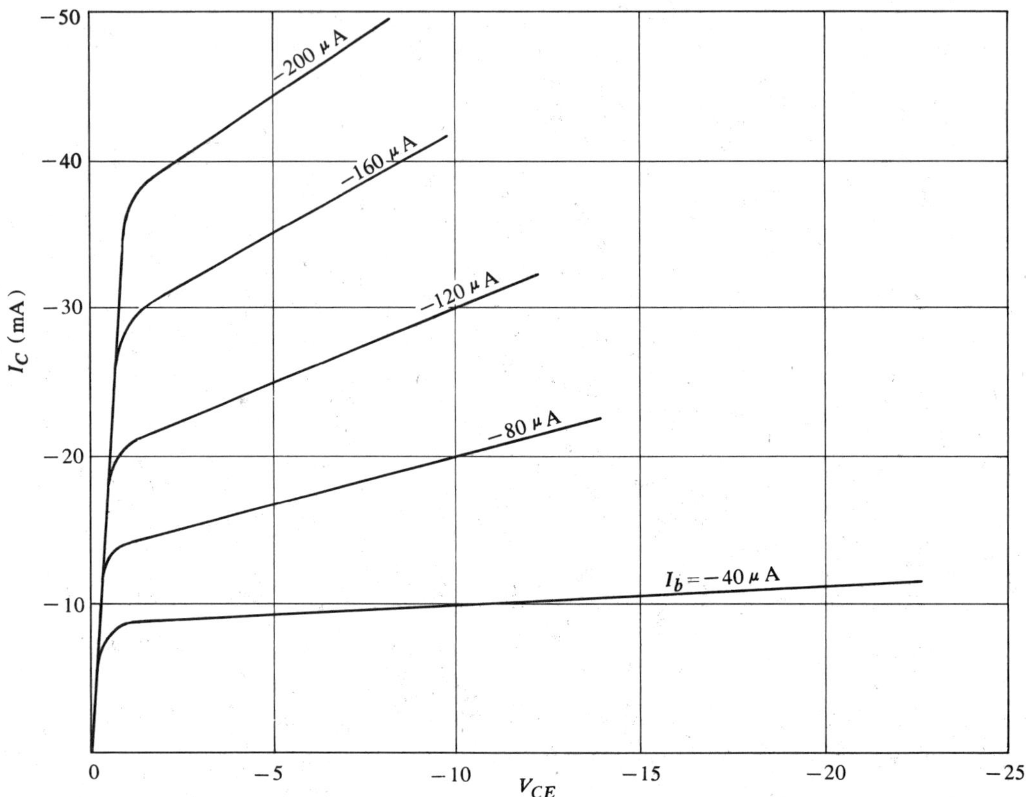

Fig. B–11 Bipolar transistor worksheet.

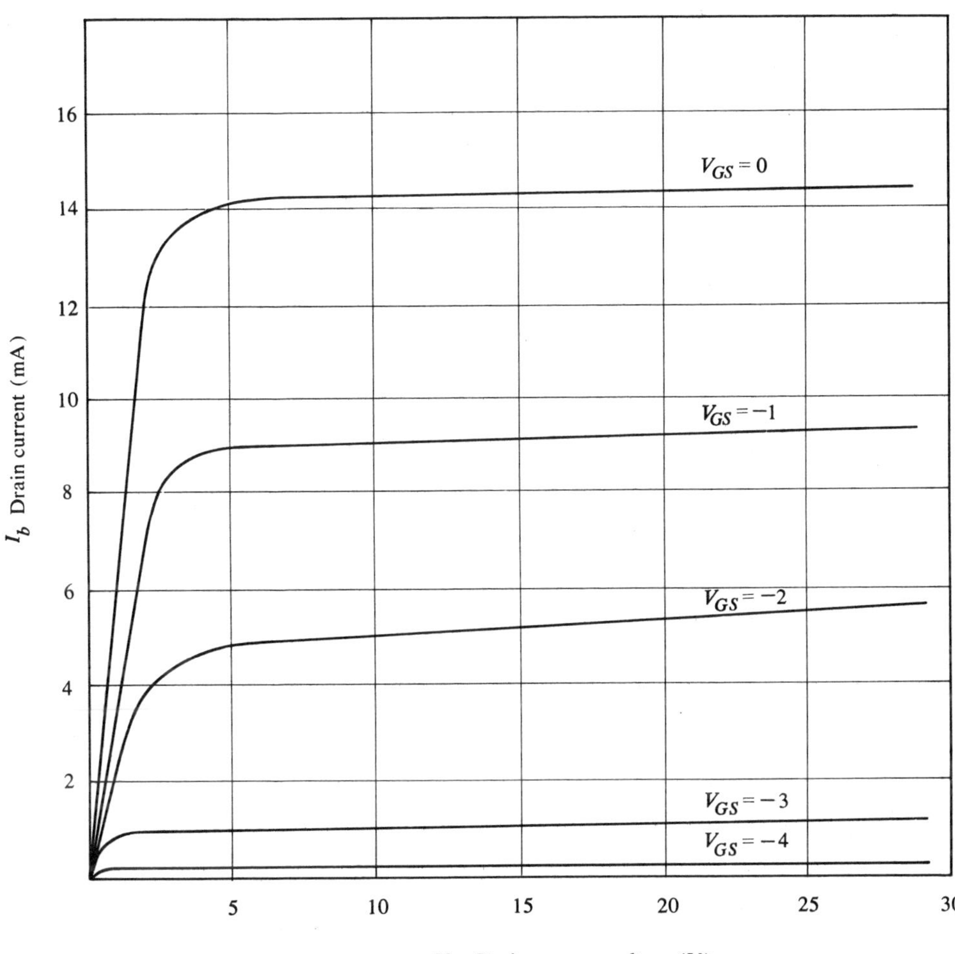

Fig. B–12 JFET worksheet.

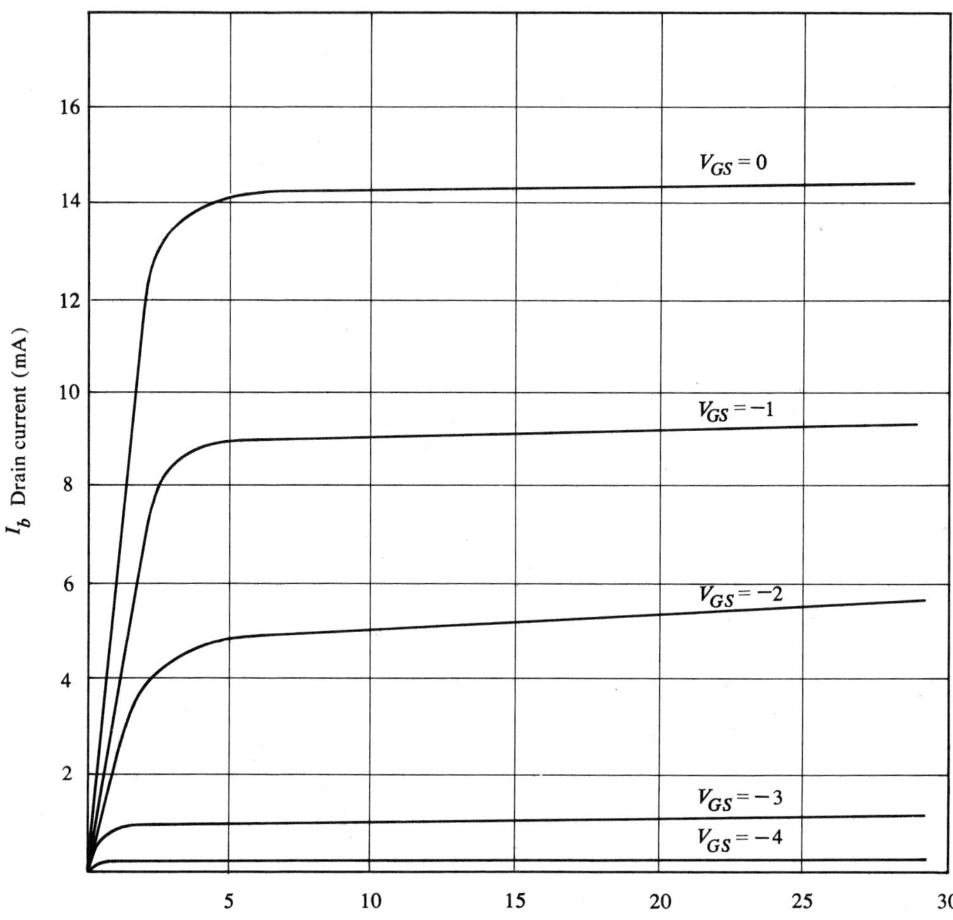

Fig. B–12 JFET worksheet.

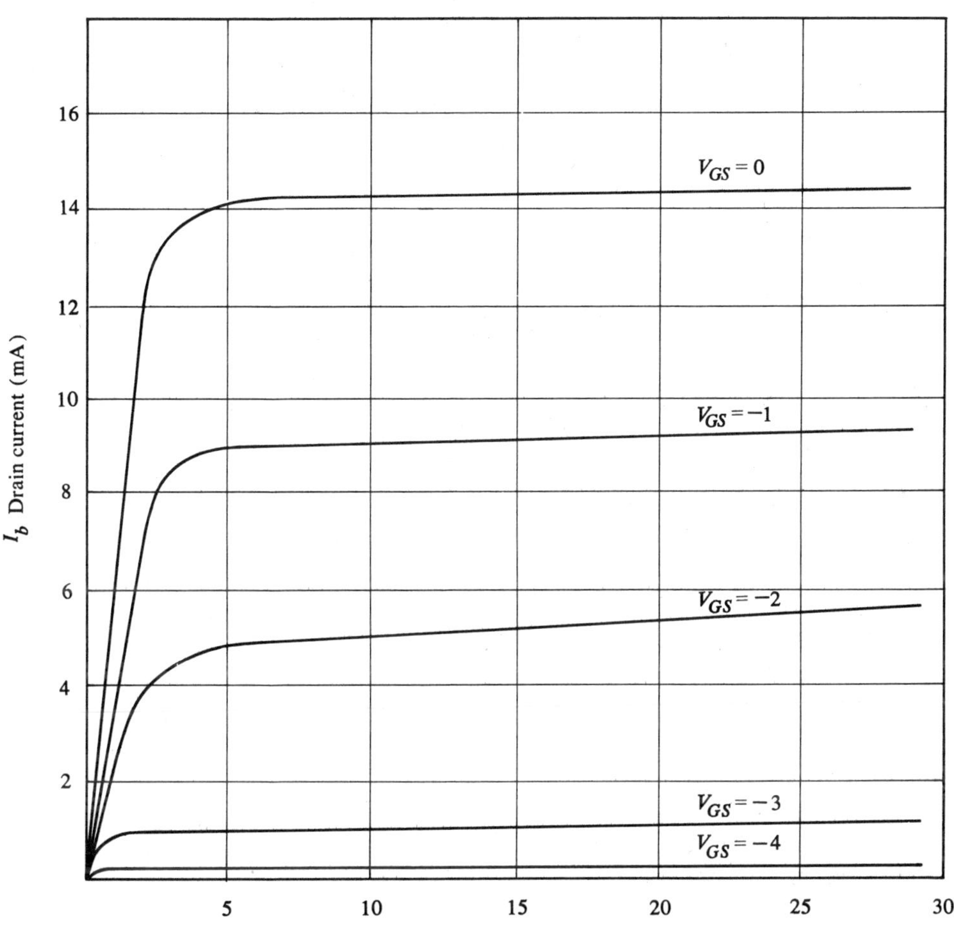

Fig. B–12 JFET worksheet.

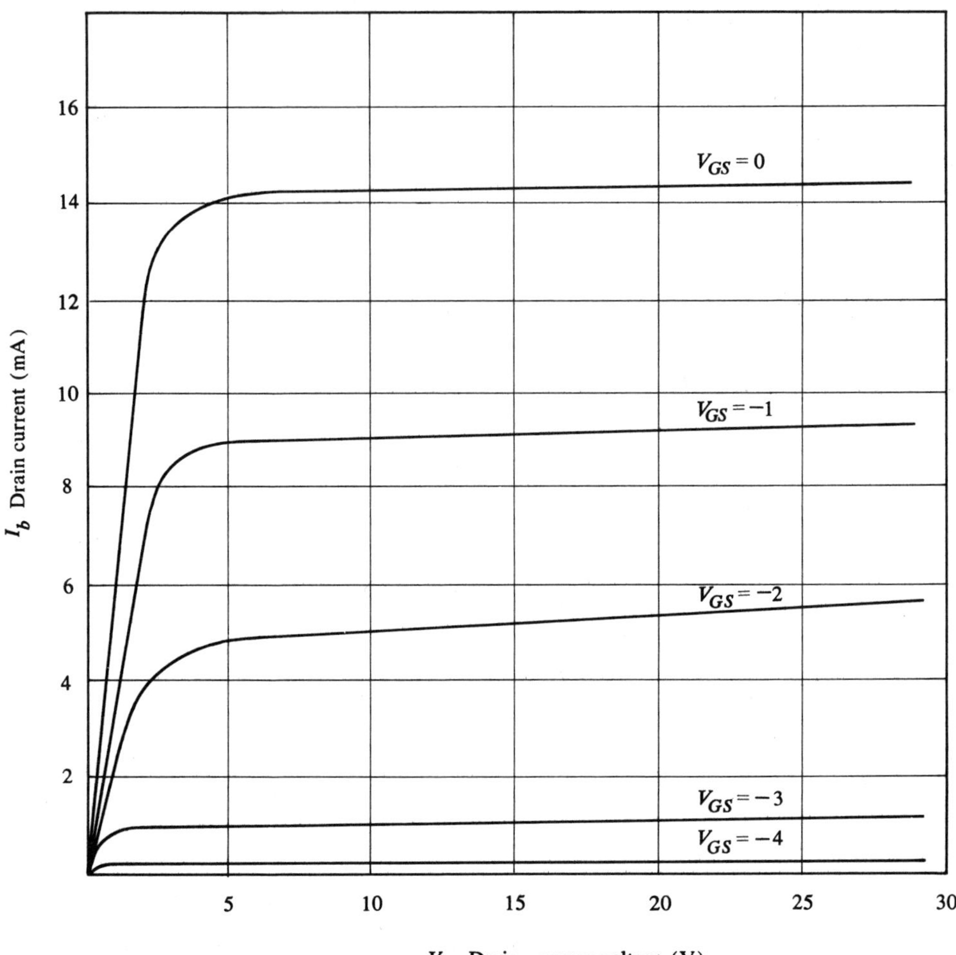

Fig. B–12 JFET worksheet.

Appendix C

TABLE C–1
Values of e^{-x}

x	e^{-x}	x	e^{-x}	x	e^{-x}	x	e^{-x}
0.00	1.000	0.35	0.705	0.70	0.497	1.50	0.223
0.01	0.990	0.36	0.698	0.71	0.492	1.60	0.202
0.02	0.980	0.37	0.691	0.72	0.487	1.70	0.183
0.03	0.970	0.38	0.684	0.73	0.482	1.80	0.165
0.04	0.961	0.39	0.677	0.74	0.477	1.90	0.150
0.05	0.951	0.40	0.670	0.75	0.472	2.00	0.135
0.06	0.942	0.41	0.664	0.76	0.468	2.10	0.122
0.07	0.932	0.42	0.657	0.77	0.463	2.20	0.111
0.08	0.923	0.43	0.651	0.78	0.458	2.30	0.100
0.09	0.914	0.44	0.644	0.79	0.454	2.40	0.091
0.10	0.905	0.45	0.638	0.80	0.449	2.50	0.082
0.11	0.896	0.46	0.631	0.81	0.445	2.60	0.074
0.12	0.887	0.47	0.625	0.82	0.440	2.70	0.067
0.13	0.878	0.48	0.619	0.83	0.436	2.80	0.061
0.14	0.869	0.49	0.613	0.84	0.432	2.90	0.055
0.15	0.861	0.50	0.607	0.85	0.427	3.00	0.050
0.16	0.852	0.51	0.600	0.86	0.423	3.10	0.045
0.17	0.844	0.52	0.595	0.87	0.419	3.20	0.041
0.18	0.835	0.53	0.589	0.88	0.415	3.30	0.037
0.19	0.827	0.54	0.583	0.89	0.411	3.40	0.033
0.20	0.819	0.55	0.577	0.90	0.407	3.50	0.030
0.21	0.811	0.56	0.571	0.91	0.403	3.60	0.027
0.22	0.803	0.57	0.566	0.92	0.399	3.70	0.025
0.23	0.795	0.58	0.560	0.93	0.395	3.80	0.022
0.24	0.787	0.59	0.554	0.94	0.391	3.90	0.020
0.25	0.779	0.60	0.549	0.95	0.387	4.00	0.018
0.26	0.771	0.61	0.543	0.96	0.383	4.10	0.017
0.27	0.763	0.62	0.538	0.97	0.379	4.20	0.015
0.28	0.756	0.63	0.533	0.98	0.375	4.30	0.014
0.29	0.748	0.64	0.527	0.99	0.372	4.40	0.012
0.30	0.741	0.65	0.522	1.00	0.368	4.50	0.011
0.31	0.733	0.66	0.517	1.10	0.333	5.0	0.007
0.32	0.726	0.67	0.512	1.20	0.301	6.0	0.002
0.33	0.719	0.68	0.507	1.30	0.273	7.0	0.001
0.34	0.712	0.69	0.502	1.40	0.247		

TABLE C–2

Integrals

(1) $\int x^n \, dx = \dfrac{x^{n+1}}{n+1}, \quad n \neq -1$

(2) $\int \dfrac{dx}{x} = \ln x$

(3) $\int \dfrac{dx}{ax+b} = \dfrac{1}{a} \ln (ax+b)$

(4) $\int (ax+b)^n \, dx = \dfrac{(ax+b)^{n+1}}{a(n+1)}, \quad n \neq -1$

(5) $\int \dfrac{dx}{x^2+a^2} = \dfrac{1}{a} \tan^{-1} \dfrac{x}{a}$

(6) $\int e^x \, dx = e^x$

(7) $\int x e^x \, dx = e^x (x-1)$

(8) $\int \sin x \, dx = -\cos x$

(9) $\int \sin^2 x \, dx = \frac{1}{2}(x - \sin x \cos x)$

(10) $\int \cos x \, dx = \sin x$

(11) $\int \cos^2 x \, dx = \frac{1}{2}(x + \sin x \cos x)$

(12) $\int \cos^m x \sin x \, dx = -\dfrac{\cos^{m+1} x}{m+1}$

(13) $\int \tan x \, dx = -\ln \cos x$

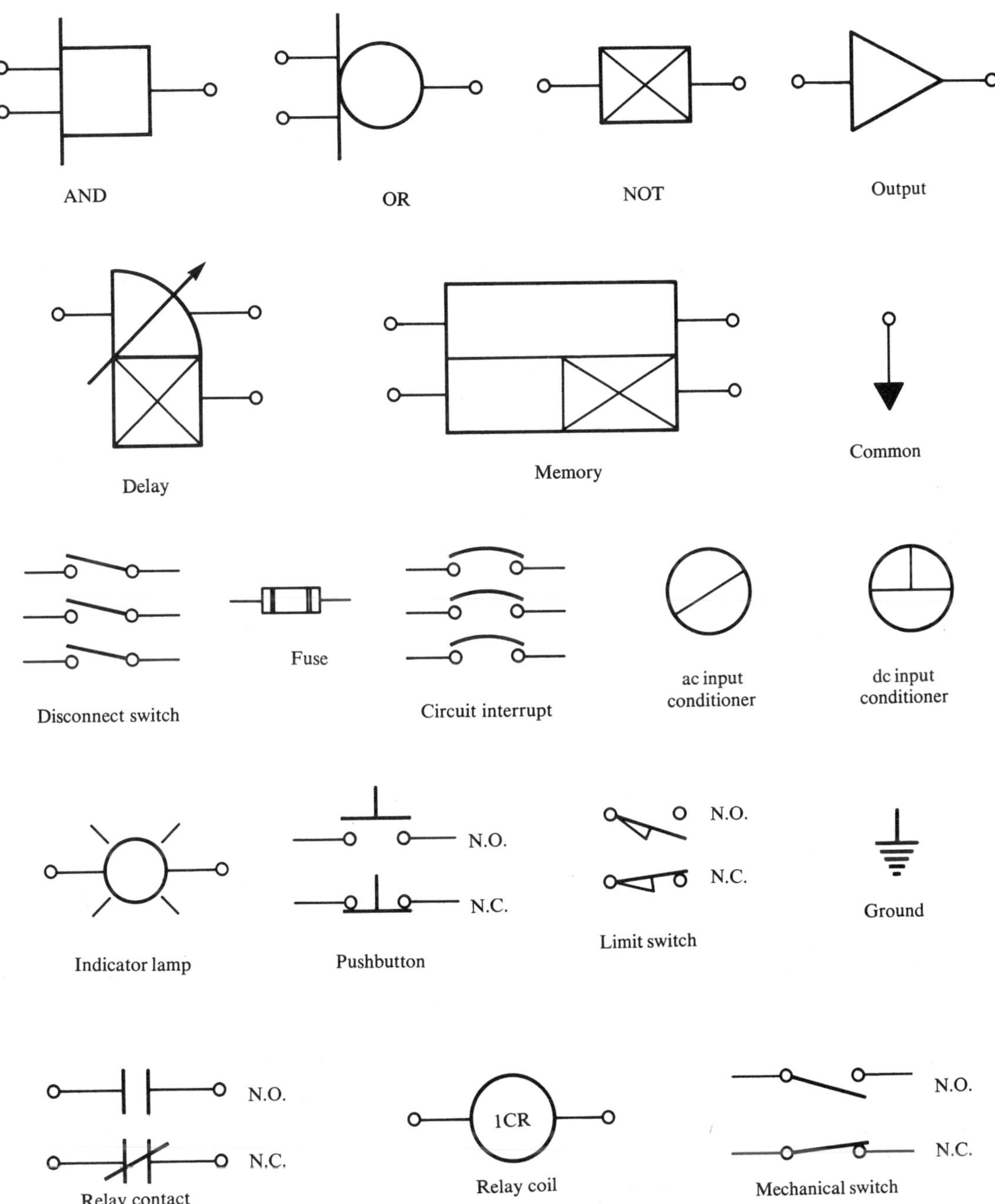

Fig. C-1 NEMA–JIC graphic symbols.

Appendix D

DEVICE	CIRCUIT SYMBOL	DEVICE	CIRCUIT SYMBOL
Rectifier diode	 Anode, Cathode	P - channel MOSFET	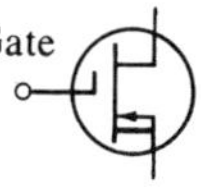 Gate, Drain, Source
Zener diode	 Anode, Cathode	Unijunction transistor	 Gate, Base 2, Base 1
Thyrector diode		Programmable unijunction transistor	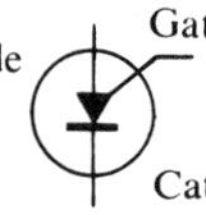 Anode, Gate, Cathode
Shockley diode	 Anode, Cathode	Silicon controlled rectifier	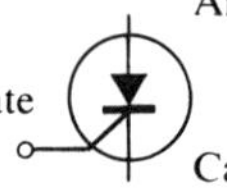 Anode, Gate, Cathode
PNP transistor	 Base, Collector, Emitter	Complementary silicon controlled rectifier	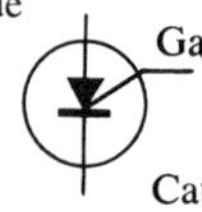 Anode, Gate, Cathode
NPN transistor	 Base, Collector, Emitter	Light - activated SCR	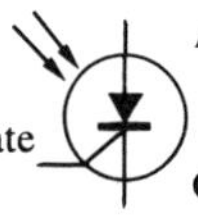 Anode, Gate, Cathode
N - channel JFET	 Gate, Drain, Source	TRIAC	 Anode 2, Gate, Anode 1
P - channel JFET	 Gate, Drain, Source	Silicon controlled switch	 Anode, Anode gate, Cathode gate, Cathode
N - channel MOSFET	 Gate, Drain, Source	Silicon unilateral switch	 Anode, Gate, Cathode
DIAC		Silicon bilateral switch	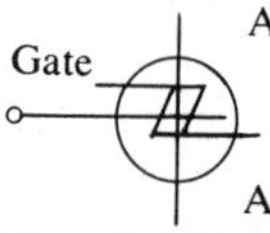 Gate, Anode 2, Anode 1

Fig. D–1 Semiconductor devices and symbols.

Answers to Selected Problems

Answers to Selected Problems

Chapter 1

1–2 The thyrector diode has no forward conducting characteristics. The thyrector diode acts like a reverse-biased rectifier regardless of the polarity of the voltage applied.

1–5 The shape of the output characteristic curves is the same for a similar NPN or PNP bipolar transistor. The polarity of the collector current, base current, and collector-to-emitter voltage is reversed due to the semiconductor current flow.

1–6 The JFET gate-to-source voltage curves are always in the depletion mode of biasing, while the MOSFET gate-to-source voltage curves are in both the depletion and enhancement modes of biasing.

1–7 The input impedance of the MOSFET is usually higher than the input impedance of the bipolar transistor.

1–8 Thyristors are usually more efficient and economical.

1–10 The emitter current.

1–13 The terminal characteristics are plots of the currents and voltages at the device terminals made from laboratory measurements.

Chapter 2

2–1 When the anode potential is higher than the cathode potential.

2–5 $I_D = 83$ mA, $i_d = 0.005 \sin \omega t$, $I_d = 0.083 + 0.005 \sin \omega t$

2–7 $I_{max} = 30$ mA, $I_{av} = 19.1$ mA, PIV = 60 V

2–8 $\theta_{CA} = 2.28°C/W$

2–9 $\theta_{SA} = 1.78\,°C/W$

2–11 $P_J = 58$ W $T_J = 167.5°C$

2–14 $I_{L_{max}} = 0.058A$

2–17 $R_1 = 270\ \Omega$, $V_i = 6.32$ V

2–19 $R_{E_{max}} = 650\ \Omega$

2–22 $R_B = 105.5\ k\Omega$, peak-to-peak base current = 0.71 mA, peak-to-peak collector current = 14.2 mA

2–24 The amplifier is saturated. The maximum sinusoidal output voltage is zero.

2–27 $R_1 = 400\ \Omega, R_2 = 1100\ \Omega$

Chapter 3

3–1 To boost the electrical output to a power level sufficient to handle load requirements.

3–2 Higher current capacity at low cost; isolation of control and power circuits.

3–4 The relay would be actuated.

3–5 Transistor $3Q$ would be saturated forcing $4Q$ into cutoff.

3–11 To suppress the transients that would be generated due to inductive kick.

Chapter 4

4–1 5×10^{-4} joules

4–4 $V = 222$ V

4–6 $V_C(t) = 5.2$ V at $t = 3$ ms, $V_{max} = 5.2$ V
$V_C(t) = 4.25$ V, $t = 5$ ms

4–11 $V_C(t) = 15.73$ V at $t = 2$ s
$V_C(t) = 60.7$ V at $t = 5$ s

4–17 1.125×10^{-7} joules

Chapter 5

5–1 Power delivered to the load can be controlled by varying the firing angle of the SCR.

5–3 $L = 3.25 \times 10^{-6}$ H

5–5 No

5–6 $I_{rms} = 51.3$ A, $R_L = 3.14\ \Omega$ $P_L = 14.3$ kW

5–12 $T_{C_{max}} = 96°C$, $\theta_{CA} = 31.6°C/W$

5–14 Exchange the position of R_5 and R_P.

5–18 Because the stand-off ratio can be controlled by the selection of external resistors.

Chapter 6

6–2 $R_Z = 470\ \Omega$, $1\ R = 9 \times 10^5\ \Omega$, $1\ C = 0.004$ F

6–9 $C_1 = 0.454 \times 10^{-6}$ F

6–13 The relay would be energized at all times.

6–14 Provides faster switching at higher current level.

Chapter 7

7–2 25.9, 93.5, 79.5, 72.6, 25.9, 100.0

7–6 55.0 A at all angles.

7–8 Alpha(min) = 96°, alpha(max) = 110°, V_{FOM} = 325 V, I_{rms} = 23.0 A

7–12 The *A/D* converter digitizes the analog sensor output. The *D/A* converter changes the digital control signal to an analog signal for the phase shift controller.

7–14 Increased firing angle.

Chapter 8

8–1 1010110

8–2 11100011

8–3 309

8–4 634

8–5 If the most positive of the two outputs is assigned logic 1, it is a positive logic assignment.

8–10 $D = \bar{A}\bar{B}C + A\bar{B}C + AB\bar{C} + ABC$

$E = \bar{A}\bar{B}\bar{C} + \bar{A}BC + AB\bar{C}$

8–12 256

Chapter 9

9–4 L = 1, lamp on; A = 1, switch closed.

A	B	L
0	0	0
0	1	1
1	1	0
1	0	1

$L = \bar{A}B + A\bar{B}$

9–7 L = 1, lamp on: A = 1, contact closed.

A	B	C	L
0	0	0	0
0	0	1	1
0	1	1	0
0	1	0	1
1	1	0	0
1	0	0	1
1	0	1	0
1	1	1	1

$L = \bar{A}\bar{B}C + \bar{A}\bar{C}B + A\bar{B}\bar{C} + ABC$

9–12 The programmable controller usually makes logical decisions and provides output control, it has no arithmetic capability, has simple programming language oriented toward the industrial designer with relay logic experience.

Chapter 10

10–1 HP = 0.0458

10–3 Universal motor.

10–5 Alpha(max) = 80.5°, alpha(min) = 20.5°, maximum firing angle is 180°, minimum firing angle is 34.4°.

10–9 The speed will increase.

10–10 A thyrector diode to suppress voltage transients.

10–14 Reverses the direction of rotation of the motor shaft.

Chapter 11

11–1 The amplitude of the ac voltage can be controlled as well as the phase delay.

11–3 To convert the sinusoidal line voltage waveform into a square wave output.

11–6 Alpha(max) = 138°, alpha(min) = 6.9°, V_{max} = V_{min} = 34 V

Chapter 12

12–2 I_{dc} = 80.0 A, V_{in} = 310 V_{rms}; PIV = 440 V, P_{in} = 38.9 kW

12–5 6-phase star rectifier; high transformer efficiency, low ripple, low diode current.

12–6 Either 3-phase bridge or 3-phase star.

12–8 $C = 12.25 \times 10^{-3}$ F

12–9 $C = 41.5 - 10^{-6}$ F

12–11 11.1% regulation.

12–15 DC supply will be disconnected by activation of the breaker, load voltage will go to zero.

12–16 Level will be increased due to lower voltage division ratio of $R_5/(R_5 + R_2)$.

Index

Index

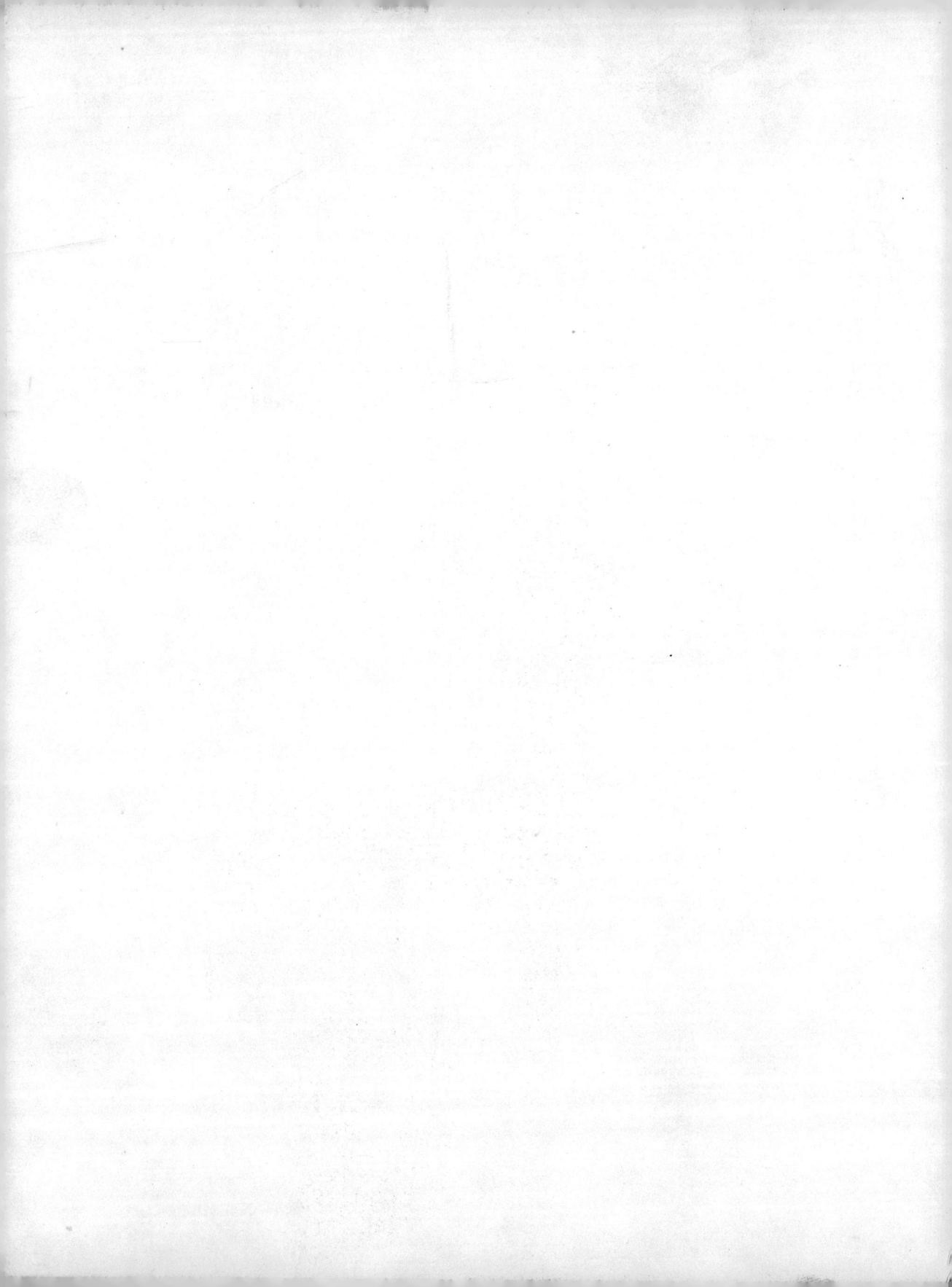